Author Index

physica status solidi (a)

AKADEMIE-VERLAG · BERLIN

THEORIA CUM PRAXI

Classification Scheme

1. Structure of Crystalline Solids
   1.1 Perfectly Periodic Structure
   1.2 Solid-State Phase Transformations
   1.3 Alloys. Metallurgy
   1.4 Microstructure (Magnetic Domains See 18; Ferroelectric Domains See 14.4.1)
   1.5 Films
   1.6 Surfaces
2. Non-Crystalline State
3. Crystal Growth
4. Bonding Properties
5. Mössbauer Spectroscopy
6. Lattice Dynamics. Phonons
7. Acoustic Properties
8. Thermal Properties
9. Diffusion
10. Defect Properties (Irradiation Defects See 11)
    10.1 Metals
    10.2 Non-Metals
11. Irradiation Effects (X-Ray Diffraction Investigations See 1 and 10)
12. Mechanical Properties (Plastic Deformations See 10)
    12.1 Metals
    12.2 Non-Metals
13. Electron States
    13.1 Band Structure
    13.2 Fermi Surfaces
    13.3 Surface and Interface States
    13.4 Impurity and Defect States
    13.5 Elementary Excitations (Phonons See 6)
        13.5.1 Excitons
        13.5.2 Plasmons
        13.5.3 Polarons
        13.5.4 Magnons
14. Electrical Properties. Transport Phenomena
    14.1 Metals. Semi-Metals
    14.2 Superconductivity. Superconducting Materials and Devices
    14.3 Semiconductors
        14.3.1 Films
        14.3.2 Surfaces and Interfaces
        14.3.3 Devices. Junctions (Contact Problems See 14.3.4)
        14.3.4 High-Field Phenomena, Space-Charge Effects, Inhomogeneities, Injected Carriers (Electroluminescence See 20.3; Junctions See 14.3.3)
    14.4 Dielectrics
        14.4.1 Ferroelectrics
15. Thermoelectric and Thermomagnetic Properties
16. Photoconductivity. Photovoltaic Effects
17. Emission of Electrons and Ions
    17.1 Field Emission Microscope Investigations
18. Magnetic Properties
    18.1 Paramagnetic Properties
    18.2 Ferromagnetic Properties
        18.2.1 Ferromagnetic Films
    18.3 Ferrimagnetic Properties
    18.4 Antiferromagnetic Properties

(Continued on cover three)

physica status solidi (a)

applied research

Volume 68 · Number 1 · Pages 1 to 328, K1 to K104, and A1 to A8

November 16, 1981

PSSA 68(1) 1—328, K1—K104, A1—A8 (1981)

ISSN 0031-8965

AKADEMIE-VERLAG · BERLIN

Subscriptions and orders for single copies should be directed

in the GDR:

to the Postzeitungsvertrieb, or
to the Akademie-Verlag, DDR-1086 Berlin, Leipziger Straße 3 – 4;

in the other socialist countries:

to a book-shop for foreign language literature or
to the competent news-distributing agency;

in the FRG and BERLIN (WEST):

to a book-shop or
to the wholesale distributing agency Kunst und Wissen, Erich Bieber OHG, D-7000 Stuttgart 1, Wilhelmstr. 4 – 6;

in the other Western European countries:

to Kunst und Wissen, Erich Bieber GmbH, CH-8008 Zürich, Dufourstr. 51;

in USA and CANADA:

to Verlag Chemie International, Inc., Plaza Centre, Suite E, 1020 N. W. 6th Street, Deerfield Beach, FL 33441, USA;

in other countries:

to the international book- and journal-selling trade,
to Buchexport, Volkseigener Außenhandelsbetrieb der Deutschen Demokratischen Republik, DDR-7010 Leipzig, Postfach 160, or
to the Akademie-Verlag, DDR-1086 Berlin, Leipziger Straße 3 – 4.

Editorial Note:

"physica status solidi (a)" undertakes that an original paper *accepted for publication* before the 23rd of any month will be published within 50 days of this date unless the author requests a postponement. In special cases there may be some delay between *receipt* and *acceptance* of a paper due to the review and, if necessary, revision of the paper.

Schriftleiter und verantwortlich für den Inhalt:
Professor Dr. Dr. h.c. P. Görlich, DDR-1020 Berlin, Neue Schönhauser Str. 20 bzw. DDR-6900 Jena, Schillbachstr. 24.

Verlag:
Akademie-Verlag, DDR-1086 Berlin, Leipziger Str. 3–4: Fernruf 2236221 und 2236229; Telex-Nr.: 114420; Bank: Staatsbank der DDR, Berlin, Kto.-Nr.: 6836-26-20712.

Chefredakteur:
Dr. H.-J. Hänsch.

Redaktionskollegium:
Prof. Dr. E. Gutsche, Dr. H.-J. Hänsch, Dr. H. Lange, Dr. S. Oberländer

Anschrift der Redaktion:
DDR-1020 Berlin, Neue Schönhauser Str. 20; Fernruf: 2823380.

Veröffentlicht unter der Lizenznummer 1620 des Presseamtes beim Vorsitzenden des Ministerrates der Deutschen Demokratischen Republik.

Gesamtherstellung:
VEB Druckerei „Thomas Müntzer", DDR-5820 Bad Langensalza.

Erscheinungsweise:
Die Zeitschrift „physica status solidi (a)" erscheint jeweils am 16. eines jeden Monats. Jährlich erscheinen 6 Bände zu je 2 Heften.

Bezugspreis:
Je Band 165,– M zuzüglich Versandspesen (Preis für die DDR: 130,– M).

Bestellnummer dieses Bandes: 1085/68.

AN (EDV) 20735

Contents

Review-Article

A Review Article for Volume 68 will appear in No. 2 of this Volume.

Original Papers

Short Notes

Pre-Printed Titles

Systematic List

Subject classification:	Corresponding papers begin on the following pages (pages given in italics refer to the principle subject classification):
1.1	179, *239*
1.2	*29*, *173*, *193*, 233, *245*
1.3	*269*, *K43*
1.4	*109*, *123*, *261*, *281*
1.5	135, *K11*, *K55*, *K87*
2.	221, *K5*, *K29*
3.	K87
4.	109, *321*
8.	*11*, *301*
9.	*209*, *289*, *K37*
10.	*123*, *K15*, *K37*
10.1	*19*, *91*, *109*, *159*, 269, K21
10.2	*45*, *53*, *77*, *129*, *143*, *253*, *261*, *309*
11.	*45*, *69*, *179*, *187*
12.1	*K21*
12.2	245
13.1	K49, K75
13.4	*103*, *199*, 261, *K71*
14.1	*135*, K67
14.2	301, *K33*, *K45*, *K59*
14.3	103, *143*, *K75*, *K93*
14.3.1	*227*, *K63*, *K101*
14.3.2	*83*
14.3.3	*39*, 83, 253, *K49*, *K97*
14.3.4	*39*, *61*, *83*, *K97*
14.4	*11*, *117*, *129*
14.4.1	*173*, *245*, *K81*
18.	*233*
18.1	*K67*
18.2	*159*, *221*, 289, K67, *K93*
18.3	*97*, *K1*
18.4	*11*, K81
19.	*K25*, *K29*, *K71*
20.	*K5*, K87
20.1	77, 129, *153*, *K11*
20.2	*K63*
20.3	77, *199*, *309*

Contents of Volume 68 Continued on Page 331

Original Papers

phys. stat. sol. (a) 68, 11 (1981)

Subject classification: 18.4; 22.8

Laboratoire de Chimie Minérale Structurale, associé au C.N.R.S. n° 200, Université René Descartes, Paris[1]) *(a)*
et Laboratoire de Chimie Minérale C, Chimie des Matériaux ERA 314, Université des Sciences et Techniques du Languedoc, Montpellier[2]) *(b)*

Structure magnétique de $Fe_2Te_3O_9$ à 4,2 K

Par

M. WINTENBERGER (a) et J. C. JUMAS (b)

Le composé $Fe_2Te_3O_9$, de groupe spatial Pnma ($Z = 2$), où les ions Fe^{3+} sont en position générale, s'ordonne magnétiquement avec un vecteur de propagation $\boldsymbol{q} = 0{,}125_6\boldsymbol{b}^*$. A partir des intensités magnétiques mesurées on a évalué approximativement les angles entre les moments. On a d'autre part recherché les configurations magnétiques possibles dans cette structure en construisant une matrice d'interaction qui fait intervenir huit intégrales d'échange isotropes qui correspondent aux liaisons Fe–O–Fe ou Fe–O–Te–O–Fe. Dans cette hypothèse $\boldsymbol{q}$ peut prendre soit la forme q_x, 0, q_z soit la forme 0, q_y, 0, et l'on retrouve dans ce dernier cas les angles entre moments déduits de l'expérience.

The compound $Fe_2Te_3O_9$, with space group Pnma ($Z = 2$), where Fe^{3+} ions are on a general position, orders magnetically with a propagation vector $\boldsymbol{q} = 0.125_6\boldsymbol{b}^*$. From intensity measurements it is possible to deduce approximate values of the angles between magnetic moments. The magnetic configurations which can occur in this structure can be predicted by building an interaction matrix which involves eight isotropic exchange integrals, corresponding to Fe–O–Fe or Fe–O–Te–O–Fe bonds. In this model $\boldsymbol{q}$ is either of type q_x, 0, q_z or of type 0, q_y, 0. In this latter case one gets the same angles between moments as those which have been found experimentally.

1. Introduction

Trois composés définis ont été mis en évidence dans le système Fe_2O_3–TeO_2:

1. $Fe_2Te_4O_{11}$, dont la structure [1] comporte des groupements Fe_2O_{10} formés de deux octaèdres FeO_6 partageant une arête. Ces groupements sont reliés entre eux par des motifs Te_2O_6 et des bandes infinies Te_6O_{16}.
2. $Fe_2Te_3O_9$ qui comporte des groupements Fe_2O_9 constitués de deux octaèdres FeO_6 partageant une face. Ces groupements sont reliés par des motifs TeO_3 [2].
3. Fe_2TeO_5 dont la structure contient des feuillets formés d'octaèdres FeO_6 partageant des arêtes et des sommets. Ces feuillets sont reliés par des motifs TeO_3 [2].

Nous avons décrit antérieurement les propriétés magnétiques de ces composés [3, 4], les spectres Mössbauer de $Fe_2Te_3O_9$ et $Fe_2Te_4O_{11}$ [4], ainsi que les structures magnétiques de Fe_2TeO_5 [3] et $Fe_2Te_4O_{11}$ [4]. Dans ce dernier composé les moments des deux atomes de fer d'un même groupement Fe_2O_{10} sont antiparallèles. Dans Fe_2TeO_5 on peut rendre compte de la structure antiferromagnétique en faisant intervenir seulement trois intégrales d'échange négatives entre premiers voisins, associées aux liaisons Fe–O–Fe telles que l'angle Fe–O–Fe soit supérieur à 125°, les intégrales d'échange qui font intervenir des angles Fe–O–Fe voisins de 90° étant probablement beaucoup plus faibles en valeur absolue par suite d'une compensation entre l'échange direct Fe–Fe, antiferromagnétique, et l'échange Fe–O–Fe à 90°, ferromagnétique.

---

[1]) 4 Avenue de l'Observatoire, 75270 Paris Cedex 06, France.
[2]) Place Eugène Bataillon, 34060 Montpellier Cédex, France.

Nous complétons ces études par la détermination de la structure magnétique de $Fe_2Te_3O_9$. Pour ce composé la spectroscopie Mössbauer indique $T_N = 34$ K; les spectres pour $T < T_N$ ne peuvent s'interpréter avec un seul sextuplet [4].

2. Détermination de la structure magnétique par diffraction neutronique

$Fe_2Te_3O_9$ a pour groupe spatial Pnma avec $Z = 2$. Les huit atomes de fer sont en position 8d,

$$x = 0{,}225\,, \qquad y = 0{,}055\,, \qquad z = 0{,}059$$

(nous avons déplacé l'origine de $\frac{1}{2}$ suivant l'axe b par rapport à [2]).

Les diagrammes de poudre ont été enregistrés sur le dispositif à multidétecteur du réacteur Siloé au C.E.N. de Grenoble, à 4,2 et 77 K, avec la longueur d'onde $\lambda = 2{,}475$ Å (Fig. 1).

Sur ces diagrammes on décèle la présence d'une certaine quantité de $Fe_2Te_4O_{11}$ dans l'échantillon. Il y a de plus quelques raies très faibles d'une impureté non identifiée. Ceci, joint au fait que les raies nucléaires de $Fe_2Te_3O_9$ sont nombreuses, rend difficile une mesure précise des intensités magnétiques. Nous n'avons exploité quantitativement que la région des diagrammes correspondant à $\theta < 23{,}5°$, pour laquelle la contribution de $Fe_2Te_4O_{11}$ est connue [4].

Le Tableau 1 permet de comparer les intensités nucléaires calculées avec les intensités observées à 77 K, pour les raies qui ne sont pas affectées par la présence de $Fe_2Te_4O_{11}$.

Les raies magnétiques observées à 4,2 K s'indexent avec un vecteur de propagation $\boldsymbol{q}$ de composantes 0, q_y, 0 avec $q_y = 0{,}125_6 b^*$, compte tenu des paramètres de maille

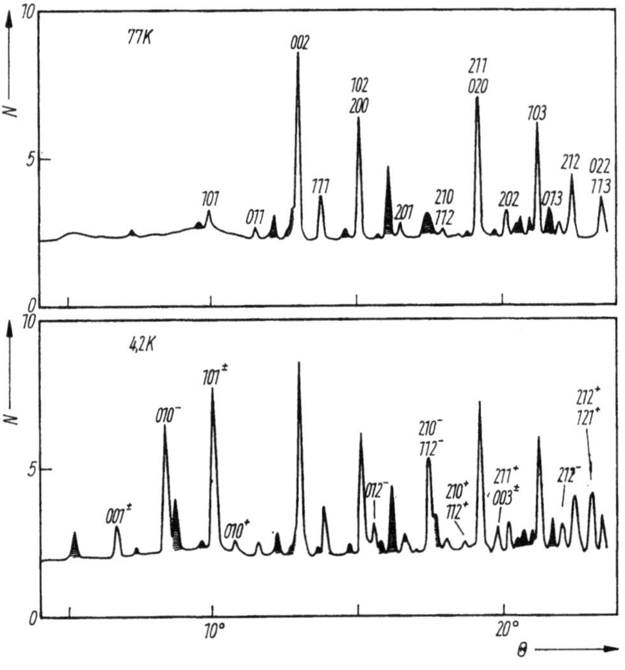

Fig. 1. Diagrammes de diffraction neutronique de $Fe_2Te_3O_9$. Les raies d'impuretés sont hachurées

Tableau 1

Intensités nucléaires observées (après correction de Lorentz et normalisation) et calculées

hkl	I_{calc}	I_{obs}
1 0 1	9,75	12
0 1 1	2,1	8,4
0 0 2	165	172
1 1 1	34,7	41
1 0 2/2 0 0	139,5	123
1 1 2/2 1 0	9	11,5
2 1 1/0 2 0	285,7	240
2 0 2	52	51
1 0 3	216,8	241
2 1 2	116	186
1 1 3/0 2 2	137,5	139

mesurés à cette température

$$a = 9{,}47\,, \qquad b = 7{,}48\,, \qquad c = 10{,}98\ \text{Å}\,.$$

Ceci conduit à envisager une structure du type spirale où le moment du site i ($i = 1$ à 8, les sites étant numérotés dans l'ordre des Tables Internationales) de la maille origine est donné par

$$\boldsymbol{S}_i = \boldsymbol{S}(\boldsymbol{q})_i + \boldsymbol{S}(-\boldsymbol{q})_i$$

avec

$$\boldsymbol{S}(\boldsymbol{q})_i = \frac{\mu}{2}\,(\boldsymbol{u} - i\boldsymbol{v})\exp i\varphi_i \quad \text{et} \quad \boldsymbol{S}(-\boldsymbol{q})_i = \boldsymbol{S}^*(\boldsymbol{q})_i$$

d'où

$$\boldsymbol{S}_i = \boldsymbol{u}\cos\varphi_i + \boldsymbol{v}\sin\varphi_i\,,$$

μ étant la longueur du moment magnétique et $\boldsymbol{u}$ et $\boldsymbol{v}$ deux vecteurs unitaires orthogonaux. Tous les moments sont contenus dans le plan $(\boldsymbol{u}, \boldsymbol{v})$.

Le facteur de structure des satellites magnétiques $\pm\boldsymbol{q}$ d'un noeud $\boldsymbol{H}_N$ du réseau nucléaire peut alors s'écrire

$$\boldsymbol{F}(\boldsymbol{H}_N \pm \boldsymbol{q}) = \mu f(\boldsymbol{H} \pm \boldsymbol{q})\,\frac{\boldsymbol{u} \mp i\boldsymbol{v}}{2}\exp(2\pi i\boldsymbol{H}_N\cdot\boldsymbol{r}_i)\exp[i(\pm 2\pi\boldsymbol{q}\cdot\boldsymbol{r}_i \mp \varphi_i)]$$

ou f est le facteur de forme magnétique et où $\boldsymbol{r}_i$ repère le site i.

Si l'on pose

$$\alpha_i = -2\pi\boldsymbol{q}\cdot\boldsymbol{r}_i + \varphi_i\,,$$

$\boldsymbol{F}(\boldsymbol{H}_N \pm \boldsymbol{q})$ est proportionnel à $\sum_i \exp(2\pi i\boldsymbol{H}_N\cdot\boldsymbol{r}_i)\exp \mp\alpha_i$, $\exp(2\pi i\boldsymbol{H}_N\cdot\boldsymbol{r}_i)$ est le terme qui intervient dans le facteur de structure cristallographique. Il peut donc y avoir des extinctions systématiques si $\alpha_i - \alpha_j = 0$ ou π pour des sites i et j qui se correspondent par un élément de symétrie avec translation.

Par ailleurs l'angle entre les moments et le vecteur de diffusion intervient sous la forme d'un facteur $(1 + \cos^2\omega)/4$ qui multiplie $|\boldsymbol{F}|^2$, ω étant l'angle de la normale $\boldsymbol{w}$ au plan $(\boldsymbol{u}, \boldsymbol{v})$ avec le vecteur de diffusion. Quelque soit ω ce facteur n'annule jamais une intensité magnétique.

Tableau 2

Angles de Bragg et intensités observés et calculés pour les raies magnétiques

$h\,k\,l\pm$	θ_{Bragg} (calc.)	θ_{Bragg} (obs.)	I_{calc}	I_{obs}
0 0 1 ±	6,58°	6,62°	9	8,5 ± 1,5
1 0 0 ±	7,60°	—	0	—
0 1 0 ⁻	8,32°	8,32°	43	51 ± 6
1 0 1 ±	10,01°	10,02°	81	85 ± 10
0 1 1 ⁻	10,56°	—	0	—
0 1 0 ⁺	10,73°	10,75°	13	7,4 ± 4
1 1 0 ⁻	11,24°	—	0	—
0 1 1 ⁺	12,57°	—	0	—
1 1 1 ⁻	13,0°	—	4	—
0 0 2 ±	13,10°	—	0	—
1 1 0 ⁺	13,15°	—	0	—
1 1 1 ⁺	14,70°	—	1,2	—
1 0 2 ±	15,15°	—	1,1	—
2 0 0 ±	15,20°	—	2,9	—
0 1 2 ⁻	15,53°	15,55°	68	41 ± 20
2 0 1 ±	16,60°	—	0,8	—
0 1 2 ⁺	17,0°	17,0°	19,5	non mesurable
1 1 2 ⁻ 2 1 0 ⁻	17,35° 17,40°	17,42°	161	140 ± 50
0 2 0 ⁻	18,05°	—	0	—
2 1 1 ⁻	18,60°	—	2	—
2 1 0 ⁺ 1 1 2 ⁺	18,70° 18,70°	18,65°	48	23 ±15
0 2 1 ⁻	19,25°	—	≈0	—
1 2 0 ⁻	19,66°	—	0	—
0 0 3 ± 2 1 1 ⁺	19,80° 19,90°	19,80°	44	42 ± 12
2 0 2 ±	20,2°	—	3,8	—
0 2 1 ⁺	20,67°	—	10	—
1 2 1 ⁻	20,78°	—	≈0	—
1 0 3 ±	21,3°	—	19,6	—
0 1 3 ⁻	21,57°	—	0	—
2 1 2 ⁻	21,97°	22,0°	85	45 ± 20
1 2 0 ⁺	22,03°	—	0	—
0 2 2 ⁻	22,53°	—	0	—
0 1 3 ⁺	22,70°	—	0	—
1 1 3 ⁻	22,97°	—	≈0	—
1 2 1 ⁺ 2 1 2 ⁺	23,06° 23,1°	23,05°	143	153 ± 30
3 0 0 ±	23,11°	—	0	—
3 0 1 ±	23,18°	—	6,6	—

Expérimentalement on voit que 010^- est forte, alors que $100^\pm$, 110^- et 110^+ sont absentes. Ceci suggère que $hk0^\pm$ n'existe que pour h pair, s'est-à-dire $\alpha_i = \alpha_j$ (soit $\varphi_i = \varphi_j$) pour des sites i et j qui se correspondent par le miroir a. De même le fait que 011^-, 011^+ et $002^\pm$ sont absentes suggère que $\alpha_i = \alpha_j + \pi$ pour des sites i et j qui se correspondent par le miroir n. Les angles α vérifient alors les relations

$$\alpha_1 = \alpha_8 = \alpha_3 + \pi = \alpha_6 + \pi\,,$$

$$\alpha_2 = \alpha_7 = \alpha_5 + \pi = \alpha_4 + \pi\,.$$

Et il reste à déterminer $\alpha_2 - \alpha_1$ et la direction $\boldsymbol{w}$. Le facteur de forme utilisé pour Fe^{3+} est celui de Watson et Freeman [5].

On pourrait en principe déterminer $\alpha_2 - \alpha_1$ à partir du rapport des intensités de 010^+ et 010^- car ω a la même valeur pour ces deux raies, mais 010^+ est faible et sa mesure est peu précise. On peut seulement dire que

$$255° < \alpha_2 - \alpha_1 < 285° .$$

Pour trouver la direction de $\boldsymbol{w}$ nous avons tout d'abord examiné les possibilités que $\boldsymbol{w}$ soit parallèle à $\boldsymbol{a}$, $\boldsymbol{b}$ ou $\boldsymbol{c}$. La direction $\boldsymbol{a}$ ne convient pas. La direction $\boldsymbol{b}$ associée à la valeur $\alpha_2 - \alpha_1 = 285°$ fournit un assez bon accord entre le calcul et l'observation. Si l'on évalue le moment μ à partir des raies $001^{\pm}$, 010^-, $101^{\pm}$ et $121^+/212^+$, pour lesquelles les mesures d'intensité sont les plus précises, on trouve $\mu = 4{,}1\mu_B$. Si l'on place $\boldsymbol{w}$ sur l'axe c on obtient le meilleur accord entre l'observation et le calcul pour $\alpha_2 - \alpha_1 = 255°$ et $\mu = 4{,}6\mu_B$ (Tableau 2). Si l'on déplace $\boldsymbol{w}$ dans le plan $(\boldsymbol{b}, \boldsymbol{c})$ on constate que si $\boldsymbol{w}$ va de c vers $\boldsymbol{b}$ l'angle $\alpha_2 - \alpha_1$ évolue de façon continue de 255° vers 285° et que μ diminue progressivement.[3])

Compte tenu de la valeur du champ magnétique hyperfin $H = 52{,}5$ T déduite des spectres Mössbauer [4] une valeur de $\mu = 4{,}1\mu_B$ paraît peu probable et $\boldsymbol{w}$ doit se trouver sur l'axe c ou à son voisinage; $\alpha_2 - \alpha_1$ doit être voisin de 255° mais n'est pas déterminé avec une grande précision.

Les phases φ_i ont alors les valeurs suivantes:

$$\varphi_1 = \varphi_8 ,$$

$$\varphi_2 = \varphi_7 = \varphi_1 + \alpha_2 - \alpha_1 + 2\pi q_y \left(\tfrac{1}{2} - 2y\right) \approx \varphi_1 + 272{,}5° ,$$

$$\varphi_3 = \varphi_6 = \varphi_1 + \pi + \pi q_y = \varphi_1 + 202{,}5° ,$$

$$\varphi_4 = \varphi_5 = \varphi_2 + \pi - \pi q_y \approx \varphi_1 + 70° .$$

3. Discussion

Les configurations magnétiques possibles dans un composé peuvent être prévues si l'on définit les interactions magnétiques qui entrent en jeu. On peut alors, tout au moins dans les cas simples où l'on fait intervenir un petit nombre d'intégrales d'échange isotrope J_{ij} et où les réseaux de Bravais distincts sont peu nombreux, trouver des inégalités entre les J_{ij} en écrivant que la structure effectivement observée correspond au minimum d'énergie. Dans le cas de $Fe_2Te_3O_9$ la structure n'a pas été résolue avec une grande précision; d'autre part elle comporte huit réseaux de Bravais et doit faire intervenir un nombre élevé de J_{ij}. On ne peut donc pas espérer obtenir de ce type d'étude des informations précises sur les valeurs relatives des J_{ij}. Il a cependant paru intéressant de l'entreprendre, surtout parce qu'au vu de la structure cristalline on pouvait s'attendre à ce qu'il y aît compétition entre les intégrales d'échange qui propagent la structure aussi bien dans les directions $\boldsymbol{x}$ et $\boldsymbol{z}$ que dans la direction $\boldsymbol{y}$, et à ce que le vecteur $\boldsymbol{q}$ puisse avoir une direction quelconque.

Nous avons fait l'hypothèse qu'il n'existe que des intégrales d'échange isotrope. Nous avons traité le problème par la méthode de comparaison de phase, décrite en détail par Bertaut [6]. Nous en résumons le principe.

---

[3]) Pour une collection d'intensités magnétiques donnée le moment devrait rester approximativement constant quel que soit le modèle choisi si les raies prises en compte étaient assez nombreuses. Ici lorsque μ diminue la somme des intensités calculées pour les raies qui ne sont pas utilisées dans l'évaluation de μ décroît, ces intensités restant compatibles avec les observations.

Si $\boldsymbol{r}_i$ et $\boldsymbol{r}_j$ repèrent des atomes magnétiques appartenant aux réseaux de Bravais i et j, entre lesquels existe une intégrale d'échange isotrope J_{ij} ($J_{ij} < 0$ si le couplage est antiferromagnétique) et si $\boldsymbol{r}_{i0}$ et $\boldsymbol{r}_{j0}$ fixent l'origine sur chacun des réseaux de Bravais, on construit une matrice d'interaction dont les éléments sont

$$\eta_{ij}(\boldsymbol{q}) = \sum_{\boldsymbol{r}_j} J_{ij} \exp\left[2\pi i \boldsymbol{q}(\boldsymbol{r}_{j0} - \boldsymbol{r}_j)\right].$$

Cette matrice est hermitique. Ses valeurs propres λ_{p} sont telles que l'énergie magnétique $W = -2\lambda$. On recherche des vecteurs propres qui soient des vecteurs colonne $\boldsymbol{Q}_{\mathrm{p}}(\boldsymbol{q})$ dont les composantes $\boldsymbol{Q}_{\mathrm{p}j}(q)$ aient la forme

$$\boldsymbol{Q}_j(q) = \frac{\mu}{2}(\boldsymbol{u} + i\boldsymbol{v})\, Q_j \quad \text{avec} \quad Q_j = \exp -i\varphi_j .$$

Le moment $\boldsymbol{S}$ en $\boldsymbol{r}_{j0}$ vaut

$$\boldsymbol{S} = \boldsymbol{Q}_j(\boldsymbol{q}) + \boldsymbol{Q}_j(-\boldsymbol{q}) = \boldsymbol{Q}_j(\boldsymbol{q}) + \boldsymbol{Q}_j^*(\boldsymbol{q}) = \boldsymbol{u}\cos\varphi_j + \boldsymbol{v}\sin\varphi_j .$$

(On note la correspondance $\boldsymbol{Q}_j(\boldsymbol{q}) = \boldsymbol{S}_j^*(\boldsymbol{q})$.) On doit vérifier le système

$$(\eta - \lambda_{\mathrm{p}})\, \boldsymbol{Q}_{\mathrm{p}} = 0 .$$

Si on multiplie ces équations respectivement par $\boldsymbol{Q}_{\mathrm{p}1}^{-1}, \ldots, \boldsymbol{Q}_{\mathrm{p}j}^{-1}$ on obtient le système

$$-\lambda_{\mathrm{p}} + \sum_j \eta_{ij} \boldsymbol{Q}_{\mathrm{p}j} \cdot \boldsymbol{Q}_{\mathrm{p}i}^{-1} = 0 .$$

Symboliquement ces équations doivent avoir tous leurs termes identiques ou conjugués, ce que l'on écrira $(i) = (j)$ ou $(i) = (j^*)$. Ceci permet de trouver des relations entre les phases φ_i.

Outre l'intégrale d'échange J_{17} entre deux atomes de fer dont les octaèdres de coordination partagent une face, nous avons admis que nous prendrions en compte les intégrales d'échange entre deux ions Fe^{3+} que l'on peut relier par un parcours Fe–O–Te–O–Fe tel que les distances Te–O soient des distances de liaison chimique. Il n'y a alors pas de termes J_{12} ou J_{18}, qui feraient intervenir des distances Te–O $> 2{,}6$ Å.

Les intégrales d'échange J_{1j} et les éléments de matrice $\eta_{1j}(\boldsymbol{q})$ sont définis dans le Tableau 3. L'application des opérations de symétrie du groupe Pnma permet de déduire des η_{1j} les autres éléments de la matrice η.

Il n'est pas possible d'écrire un système de relations $(i) = (j)$ ou $(i) = (j^*)$ compatibles entre elle si $\boldsymbol{q}$ est quelconque. Cela n'est possible que dans deux cas:

Cas 1: $q_x = q_z = 0$. On peut alors obtenir

$$(1) = (2^*) = (3) = (4^*) = (5^*) = (6) = (7^*) = (8)$$

avec les phases

$$\begin{aligned} &\varphi_1, && \varphi_5 = \varphi_4 + \pi n, \\ &\varphi_2, && \varphi_6 = \varphi_3 + \pi n, \\ &\varphi_3 = \varphi_1 + \pi q_y + \pi m, && \varphi_7 = \varphi_2 + \pi n, \\ &\varphi_4 = \varphi_2 - \pi q_y + \pi m, && \varphi_8 = \varphi_1 + \pi n, \end{aligned}$$

où m et n sont des entiers.

Si l'on pose arbitrairement $\varphi_1 = 0$ il vient

$$\begin{aligned} \lambda = (-1)^m \{&2\cos\pi q_y [J_{13} + J'_{13} + (-1)^n J_{16}] + \\ &+ \cos(\pi q_y - \varphi_2)\,[2J_{14} + (-1)^n (J_{15} + J'_{15})]\} + \\ &+ (-1)^n [J_{17}\cos\varphi_2 + J'_{17}\cos(2\pi q_y - \varphi_2)] . \end{aligned}$$

Tableau 3
Environnement de l'atome 1 et éléments η_{1j} de la matrice d'interaction

réseau j	coordonnées	J_{1j}	η_{1j}
3	$\bar{x}, \frac{1}{2} + y, \bar{z}$	J_{13}	$[1 + \exp(2\pi i q_y)] \times$
	$\bar{x}, -\frac{1}{2} + y, \bar{z}$	J_{13}	$\times [J_{13} + J'_{13} \exp(-2\pi i q_x)]$
	$1 - x, \frac{1}{2} + y, \bar{z}$	J'_{13}	
	$1 - x, -\frac{1}{2} + y, \bar{z}$	J'_{13}	
4	$\frac{1}{2} - x, \bar{y}, \frac{1}{2} + z$	J_{14}	$J_{14}[1 + \exp(2\pi i q_z)]$
	$\frac{1}{2} - x, \bar{y}, -\frac{1}{2} + z$	J_{14}	
5	$\bar{x}, \bar{y}, \bar{z}$	J_{15}	$J_{15} + J'_{15} \exp(-2\pi i q_x)$
	$1 - \bar{x}, \bar{y}, \bar{z}$	J'_{15}	
6	$\frac{1}{2} - x, \frac{1}{2} + y, \frac{1}{3} + z$	J_{16}	$J_{16}\{1 + \exp[2\pi i(q_y + q_z)]\}$
	$\frac{1}{2} - x, -\frac{1}{2} + y, -\frac{1}{2} + z$	J_{16}	
7	$x, \frac{1}{2} - y, z$	J_{17}	$J_{17} + J'_{17} \exp(2\pi i q_y)$
	$x, -\frac{1}{2} - y, z$	J'_{17}	

La phase φ_2 est définie par la relation

$$(-1)^m \sin(\pi q_y - \varphi_2)\,[2J_{14} + (-1)^n (J_{15} + J'_{15})] + \\ + (-1)^n [J'_{17} \sin(2\pi q_y - \varphi_2) - J_{17} \sin\varphi_2] = 0$$

qui est nécessaire pour que λ soit réel, et exprime en même temps que $\partial\lambda/\partial\varphi_2 = 0$.

Cette relation définit tg φ_2 donc fournit, pour chacun des quatre cas dépendant des parités de m et n, deux solutions soit φ_2 et $\varphi'_2 = \varphi_2 + \pi$.

Cas 2: $-q_y = 0$. On obtient

$$(1) = (2) = (3^*) = (4^*) = (5^*) = (6^*) = (7) = (8)$$

ave les phases

$$\varphi_1\,, \qquad \varphi_5 = \varphi_3 + \pi b\,,$$
$$\varphi_2 = \pi(q_x + q_z) + \pi a\,, \qquad \varphi_6 = \varphi_4 + \pi b\,,$$
$$\varphi_3\,, \qquad \varphi_7 = \varphi_1 + \pi b\,,$$
$$\varphi_4 = \varphi_3 + \pi(q_x + q_z) + \pi a\,, \qquad \varphi_8 = \varphi_2 + \pi b\,,$$

où a et b sont des entiers.

En faisant $\varphi_1 = 0$ on trouve

$$\lambda = (-1)^b (J_{17} + J'_{17}) + \cos\varphi_3[2J_{13} + (-1)^b J_{15}] + \\ + 2\cos(\varphi_3 + \pi q_x)\cos\pi q_z (-1)^a [J_{14} + (-1)^b J_{16}] + \\ + \cos(\varphi_3 + 2\pi q_x)\,[2J'_{13} + (-1)^b J'_{15}]$$

avec la condition

$$\sin\varphi_3[2J_{13} + (-1)^b J_{15}] + \\ + (-1)^a 2\sin(\varphi_3 + \pi q_x)\cos\pi q_z[J_{14} + (-1)^b J_{16}] + \\ + \sin(\varphi_3 + 2\pi q_x)\,[2J'_{13} + (-1)^b J'_{15}] = 0$$

qui fournit deux valeurs φ_3 et $\varphi_3' = \varphi_3 + \pi$.

L'analyse du modèle d'interactions isotropes que nous avons défini montre qu'effectivement le vecteur de propagation $\boldsymbol{q}$ ne peut être quelconque et que si $q_y \neq 0$ alors $q_x = q_z = 0$. De plus elle permet de déterminer toutes les phases φ_i, et de constater que la structure observée expérimentalement correspond bien à une des structures prévues dans le cas $q_y \neq 0$ (n pair et m impair).

Si l'on écrit que la structure observèe ($\varphi_2 - \varphi_1 \approx 272{,}5°$) est plus stable que la solution correspondant à $\varphi_2 - \varphi_1 = 272{,}5° + \pi$ on trouve, en tenant compte de l'égalité liant $\varphi_2 - \varphi_1$ èt q_y

$$J_{17} > J'_{17}\,.$$

On peut vérifier que cette inégalité reste valable lorsque $\varphi_2 - \varphi_1$ reste compris entre 270 et 300°. Elle signifie que l'intégrale d'échange entre deux ions Fe^{3+} dont les octaèdres de coordination partagent une arête est moins fortement négative que celle qui fait intervenir des groupements TeO_3 entre deux ions Fe^{3+} situés le long d'une même rangée de direction [010].

Remerciements

Les auteurs remercient les personnes du laboratoire de diffraction neutronique du C.E.N.G. qui ont contribué à la réalisation des expériences et tout particulièrement S. Quezel.

Bibliographie

[1] F. PERTLIK, Tschermaks mineralog. petrogr. Mitt. **18**, 39 (1972).
[2] R. ASTIER, E. PHILIPPOT, J. MORET et M. MAURIN, Rev. Chim. Minérale **13**, 359 (1975).
[3] J. C. JUMAS, M. WINTENBERGER et E. PHILIPPOT, Mat. Res. Bull. **12**, 1063 (1977).
[4] J. C. JUMAS, L. FOURNES, M. WINTENBERGER et E. PHILIPPOT, J. Solid State Chem., **39**, 39 (1981).
[5] R. E. WATSON et A. J. FREEMAN, Acta cryst. **14**, 27 (1961).
[6] E. F. BERTAUT, J. Physique **35**, 659 (1974).

(Received July 23, 1981)

phys. stat. sol. (a) 68, 19 (1981)

Subject classification: 10.1; 21.1

Institut für Festkörperphysik der Universität Wien[1])

Analysis of the Stacking Faults of Dissociated Glide Dislocations in F.C.C. Metals[2])

By

A. KORNER and H. P. KARNTHALER

In f.c.c. crystals the dislocations gliding on {111} planes are split into partial dislocations bounding a stacking fault ribbon. In principle the nature of these faults could be either intrinsic or extrinsic. The weak-beam method of electron microscopy is used in combination with bright-field images to study both, the partials and the stacking faults. The different contrast intensities of the partials observed in the weak-beam images are a good means to determine the sign of their Burgers vector. Several hundred dissociated glide dislocations are studied in deformed single crystals of pure Ag and Cu–Al alloys. All the analysed faults are intrinsic and not a single exception is encountered. This result can be interpreted by differences in the nucleation mechanism of faults formed by shear.

In k. f. z.-Kristallen sind die Versetzungen, die auf {111}-Ebenen gleiten, in Partialversetzungen aufgespalten, die ein Stapelfehlerband einschließen. Im Prinzip könnten diese Stapelfehler entweder von intrinsischer oder extrinsischer Natur sein. Elektronenmikroskopische Methoden, und zwar Dunkelfeldabbildungen mit Hilfe schwach angeregter Strahlen ("weak-beam") in Kombination mit Hellfeldaufnahmen, werden zur Untersuchung sowohl der Partialversetzungen als auch der Stapelfehler herangezogen. Der relative Unterschied der Kontrastintensitäten der Partialversetzungen erweist sich bei der Bestimmung des Vorzeichens ihres Burgersvektors als nützlich. Mehrere hundert aufgespaltene Gleitversetzungen werden in verformten Einkristallen aus reinem Ag und aus Cu–Al-Legierungen untersucht. Dabei zeigt es sich, daß alle analysierten Stapelfehler ausnahmslos von intrinsischer Art sind. Dieses Ergebnis kann aufgrund von Unterschieden im Keimbildungsmechanismus der Fehler, die bei Gleitung entstehen, interpretiert werden.

1. Introduction

In f.c.c. crystals stable stacking faults can occur on {111} planes and there are two possible types of them: First, intrinsic faults corresponding to the removal of a {111} plane and second, extrinsic faults corresponding to the insertion of an extra {111} plane. The difference in the displacement associated with these two faults is $\pm 2a/3$ ⟨111⟩ and, since this is not a lattice vector, the two different faults are not crystallographically equivalent.

When the stacking faults are extended, like it is frequently the case in alloys of low stacking-fault energy, the nature of the faults can be determined by electron-microscopy methods. In this case it is necessary that the faults intersect the surface and are inclined to it. To analyse the type of these faults their characteristic fringe contrast is studied in both bright-field and ordinary dark-field images [1 to 3]; for a recent review see [4]. High voltage electron-microscopy methods allow to analyse the nature of faults that do not intersect the foil surface by obtaining a systematic set of reflection conditions [5]. Dislocation loops which are formed by clustering of point defects can contain a stacking fault and depending on the nature of the fault

[1]) Boltzmanngasse 5, A-1090 Wien, Austria.

[2]) Dedicated to Prof. Dr. DDr. h. c. Erich Schmid on the occasion of his 85th birthday.

they are referred to as intrinsic (vacancy) type and extrinsic (interstitial) type, respectively. Methods which are similar to the ones mentioned above have been described in the literature to distinguish the two different types of loops (e.g. [6, 7]).

In f.c.c. metals the dislocations gliding on {111} planes are split into partial dislocations bounding a stacking fault ribbon. In principle the nature of these faults could be either intrinsic or extrinsic. When bright-field images are used the investigations are restricted to large separations of the partials ($\gtrsim 20$ nm); therefore mainly the faults occurring at extended dislocation nodes have been studied in the past. Dislocations showing extrinsic faults are rare and only observed in the vicinity of a characteristic threefold node configuration. A comparison of the stacking-fault energies γ leads to the result that γ for the two different types of faults is about the same [8, 9].

It is the purpose of this paper to determine the nature of the stacking faults of dissociated glide dislocations. The weak-beam method of electron microscopy [10] is used in combination with bright-field images to analyse both the partials and the faults unambiguously. Simple rules are worked out which are generally applicable; the dislocations can be either parallel or inclined to the foil, and the splitting width rather narrow ($\gtrsim 3$ nm). As an example the method is applied to glide dislocations which are observed in plastically deformed pure Ag.

2. Theoretical Aspects

The sketch of Fig. 1 illustrates the arrangement of the partial dislocations bounding an intrinsic and an extrinsic fault, respectively. The atoms of the primary glide plane (111) are shown as hard spherical balls and the corners of the Thompson tetrahedron are marked with ABC. The Burgers vector $\boldsymbol{b}$ of a total dislocation is

$$\boldsymbol{b} = 1/2\,\mathrm{BC} \equiv a/2\,[10\bar{1}]\,.$$

(i) When $\boldsymbol{b}$ dissociates into $\boldsymbol{b} = \boldsymbol{b}_1 + \boldsymbol{b}_2$ where

$$\boldsymbol{b}_1 = 1/2\,\mathrm{B}\delta \equiv a/6\,[2\bar{1}\bar{1}] \quad \text{and}$$

$$\boldsymbol{b}_2 = 1/2\,\delta\mathrm{C} \equiv a/6\,[11\bar{2}]$$

the fault between the partials is intrinsic, since the sequence of the layers in the f.c.c. lattice is changed from *abcabcabc* to *abcacabca*.

(ii) When $\boldsymbol{b}$ dissociates into $\boldsymbol{b} = \boldsymbol{b}_2 + \boldsymbol{b}_1$ (reversed sequence to (i)) the enclosed fault is extrinsic. As illustrated in Fig. 1 this dissociation would lead to a high energy fault (*abcaabcab*) which can be sheared to form an extrinsic fault (*abcbabcab*). It should be mentioned that an extrinsic fault can be formed by shear avoiding the high energy arrangement, since two consecutive intrinsic shears of neighbouring planes

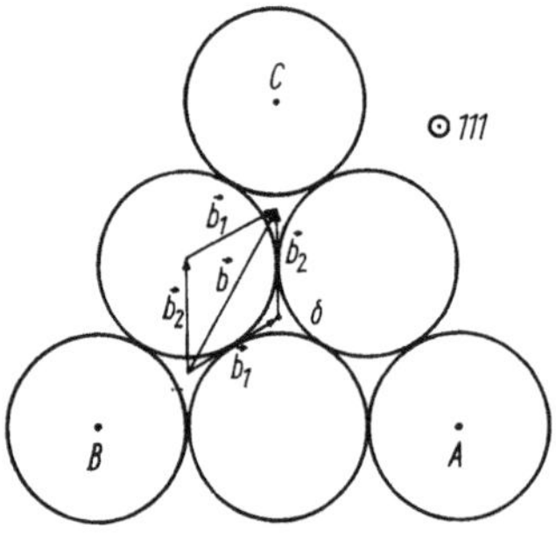

Fig. 1. Arrangement of the partial dislocations bounding an intrinsic ($\boldsymbol{b} = \boldsymbol{b}_1 + \boldsymbol{b}_2$) and an extrinsic ($\boldsymbol{b} = \boldsymbol{b}_2 + \boldsymbol{b}_1$) fault, respectively. The atoms of the primary glide plane (111) are shown as hard spherical balls; the centre and corners of the Thompson tetrahedron are marked with δ and ABC, respectively

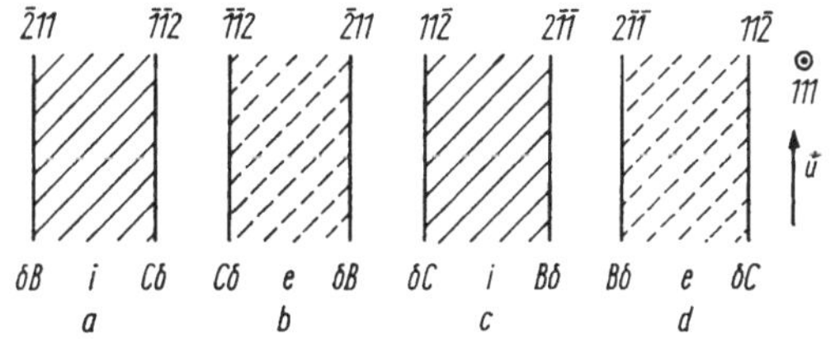

Fig. 2. In the case of dissociated glide dislocations with total $\boldsymbol{b} = \pm a/2\,[10\bar{1}]$ the partials and the stacking faults can be arranged in four different ways. The intrinsic (i) and extrinsic (e) faults are indicated by hatched areas (e with broken lines). a) and b) $\boldsymbol{b} = a/2\,[\bar{1}01]$; c) and d) $\boldsymbol{b} = a/2\,[10\bar{1}]$

would also lead to an extrinsic fault (cf. the book by Hirth and Lothe [11] for a more detailed discussion).

In the case of dissociated glide dislocations the distinction between the two types of faults can be made by using the so-called "Roman-Greek axiom" [11]: "Viewing the perfect dislocation (e.g. 1/2 BC) from outside the tetrahedron and along the positive sense of its line the intrinsic stacking fault arrangement is achieved by placing the Roman-Greek partial (1/2 Bδ) on the right and the Greek-Roman partial (1/2 δC) on the left." (For an extrinsic fault the sequence of the partials is reversed.)

From this it follows that in the case of a dissociated glide dislocation with a total $\boldsymbol{b} = \pm a/2\,[10\bar{1}]$ there are four different ways in which the partials and the faults can be arranged. This is illustrated in the sketches of Fig. 2; the faults are indicated by hatched areas and the positive sense of the dislocation line is marked with $\boldsymbol{u}$. The arrangements of Fig. 2a and b have the same sign of the total Burgers vector ($\boldsymbol{b} = +1/2$ CB) but show different types of stacking faults (intrinsic and extrinsic faults are indicated by full and broken lines, respectively). Fig. 2c and d show the intrinsic and extrinsic arrangements, when $\boldsymbol{b} = -1/2$ CB $\equiv +1/2$ BC. The nature of the faults of dissociated dislocations can therefore be analysed by a method which allows to resolve the partials and to determine their Burgers vector $\boldsymbol{b}_p$ (including the sign).

3. Experimental Observations

3.1 Experimental procedure

Single crystals of silver (purity 5N) were grown in graphite moulds under vacuum using a modified Bridgman technique. The crystals, 8 mm in diameter and approximately 100 mm long, were deformed in tension at room temperature in an Instron tensile tester into stage II of the work-hardening curve. They were then sectioned either parallel to the primary glide plane (111) or parallel to a plane 30° inclined to [111] using a spark cutter. Thin foils were prepared by standard techniques and examined at 100 kV in a Philips EM 300 fitted with a goniometer stage. The primary glide dislocations were imaged in different reflections $\boldsymbol{g}$ to analyse $\boldsymbol{b}$ and the stacking faults. To resolve the partial dislocations the $\boldsymbol{g}$ ($3\boldsymbol{g}$) weak-beam diffraction conditions were used for the reflections $\boldsymbol{g} = \langle 220 \rangle$; thus reducing the contrast width of the dislocations to about 2 nm (deviation parameter $s = 0.18\ \mathrm{nm}^{-1}$).

For the printing the convention is used to print all micrographs and diffraction patterns with the emulsion side of the photographic film upwards. Therefore, taking into account the relative rotation and the inversion between the diffraction pattern and the micrographs, they can be readily compared with each other. Fig. 3 shows a stereographic projection of a schematic Kikuchi map centered on [001]. This Kikuchi map corresponds to the "Roman-Greek axiom" given in Section 2. It should be mentioned that the indexing of the map of Fig. 3 agrees with the one used by Loretto and Smallman [12], but it is different from the indexing shown in the book by Hirsch et al. [13].

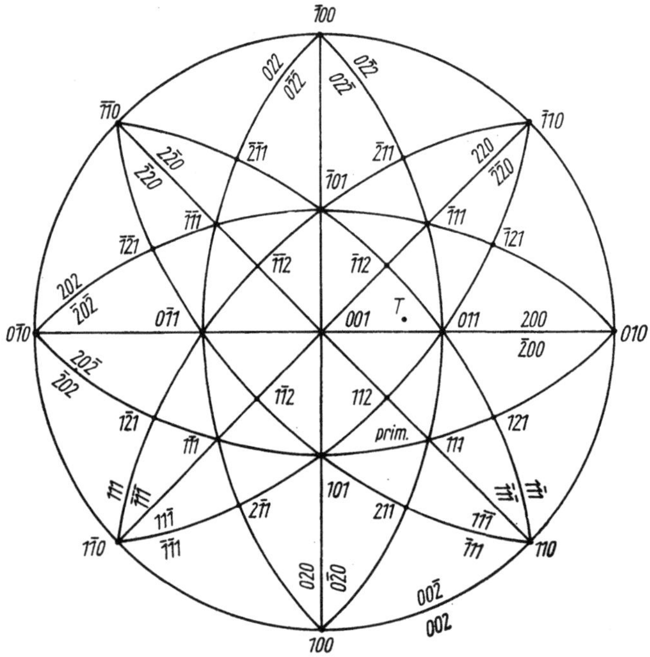

Fig. 3. Stereographic projection of a schematic Kikuchi map of the positive hemisphere centered on [001]. The primary glide plane (111) and the tensile axis T are indicated

3.2 Experimental results

The micrographs of Fig. 4 show the same area in different weak-beam and bright-field reflections. The foil is parallel to the primary glide plane (foil normal FN $\approx$ beam direction BD $\approx$ [111]). Most of the dislocations are primary glide dislocations ($\boldsymbol{b} = \pm a/2\,[10\bar{1}]$), many of them are forming dipoles and faulted dipoles. In the weak-beam images of Fig. 4a and b $\boldsymbol{g}$ is parallel to $\boldsymbol{b}$, therefore both partials of the dissociated glide dislocations are in contrast; there is no contrast of the stacking fault bounded by the partials, since $\boldsymbol{g} \perp [111]$. Some of the primary dislocations are marked in Fig. 4a (cf. A, B, and C); dislocation A shows several constrictions along its length indicating jogs; dislocations B and C have opposite sign and form a widely separated dipole (the distance between the dislocations B and C is about 70 nm).

In Fig. 4a and b the partials show a characteristic contrast behaviour. Although $\boldsymbol{g} \cdot \boldsymbol{b}_\mathrm{p} = 1$ for both partials of the glide dislocations, the contrast of the partials is not the same (e.g. in Fig. 4a, the left hand partial of dislocation A has a stronger contrast than the partial on the right hand side). This relative difference in contrast depends on both the sign of $\boldsymbol{b}$ and $\boldsymbol{g}$. The contrast is reversed when either the sign of the total $\boldsymbol{b}$ is reversed (cf. the dislocations B and C) or the sign of $\boldsymbol{g}$ is reversed (cf. the contrast of dislocations A, B, and C in Fig. 4a and b).

Fig. 4c shows a bright-field image taken in the same reflection $\boldsymbol{g} = [\bar{2}02]$ as Fig. 4a. It should be pointed out that in this case the bright-field contrast of the glide dislocations, which is asymmetric in respect to the dislocation line, corresponds to the contrast of the weak-beam images (e.g. the strong contrast of dislocation A appears on the left hand side in both Fig. 4a and c; for the dislocations B and C the result is analogous). From the asymmetry of the bright-field contrast the sign of the total $\boldsymbol{b}$

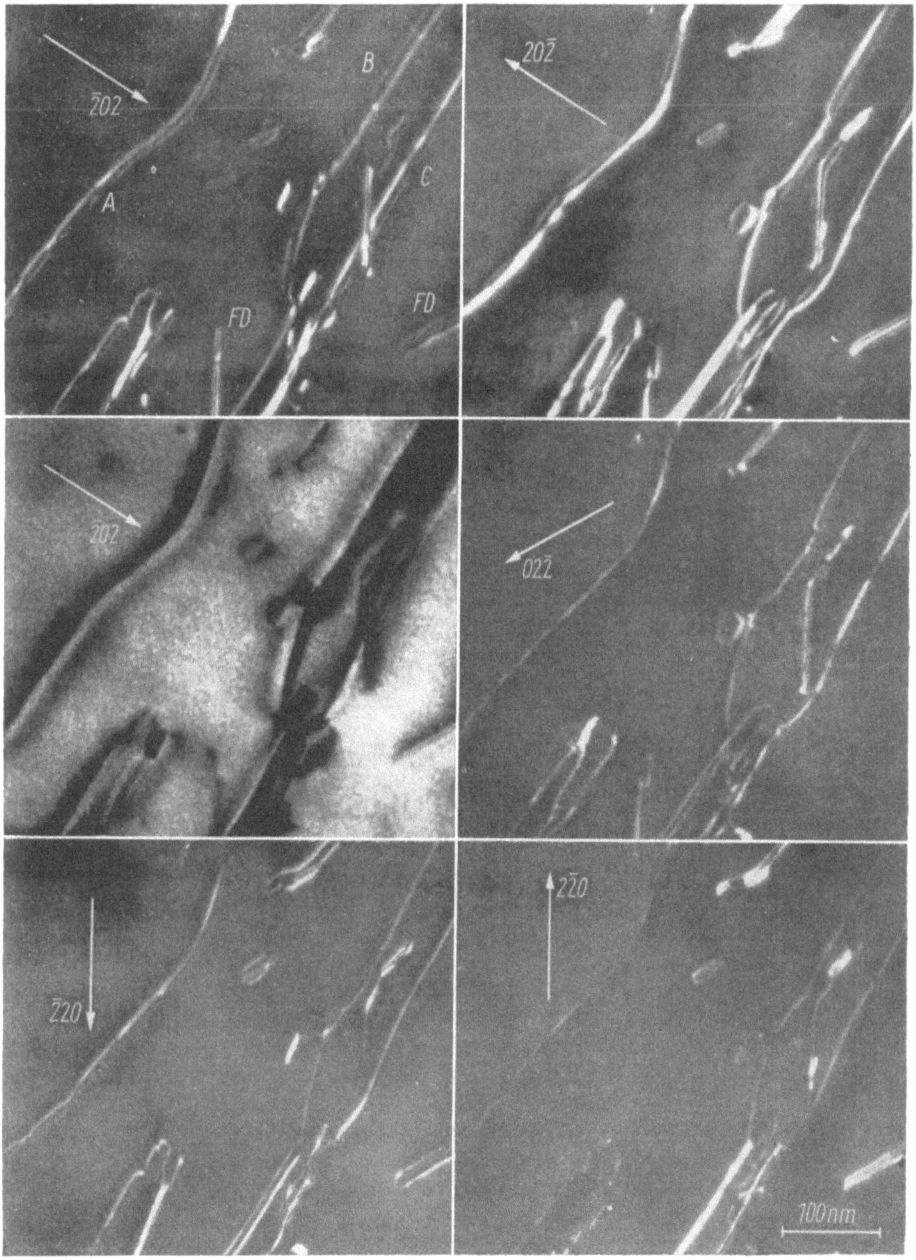

Fig. 4. Ag, (111) foil. Micrographs showing the same area in different reflections $\boldsymbol{g}$. Some of the glide dislocations (cf. A, B, and C) and some of the faulted dipoles (FD) are marked. a) and b) Weak-beam images with $\boldsymbol{g} = \pm [\bar{2}02]$. The relative difference of the contrast of the partials, which depends on the sign of $\boldsymbol{g}$, is clearly visible. c) Bright-field image corresponding to a), $\boldsymbol{g} = [\bar{2}02]$. d) Weak-beam image with $\boldsymbol{g} = [02\bar{2}]$. e) and f) Weak-beam images with $\boldsymbol{g} = \pm [\bar{2}20]$

can be determined, since FN and $\boldsymbol{g}$ are known and the positive sense of the dislocation line $\boldsymbol{u}$ is assumed to run from bottom to top of the micrograph [12]. This leads to $\boldsymbol{b} = a/2\,[\bar{1}01]$ for dislocation A and C, whereas $\boldsymbol{b} = a/2\,[10\bar{1}]$ for dislocation B. This result can be checked by a comparison with a computer simulation of the bright-field images [14].

In Fig. 4d $\boldsymbol{g} = [02\bar{2}]$ yielding $\boldsymbol{g} \cdot \boldsymbol{b}_1 = 0$ and $\boldsymbol{g} \cdot \boldsymbol{b}_2 = \pm 1$. Therefore $\boldsymbol{b}_p$ of the partials can be determined with the exception of its sign (e.g. $\boldsymbol{b}_2 = a/6\,[\bar{1}\bar{1}2]$ for the right hand partial of dislocation A and C and for the left hand partial of B). In the different reflections of Fig. 4 the dipoles and faulted dipoles FD (some are marked in Fig. 4a) show a characteristic contrast behaviour which was described previously (e.g. [15, 16]). As expected the faulted dipoles are aligned along $\pm[\bar{1}10]$ or $\pm[01\bar{1}]$ and are completely out of contrast when $\boldsymbol{g}$ is parallel to their length.

In Fig. 4e and f $\boldsymbol{g} \cdot \boldsymbol{b}_1 = \pm 1$ and $\boldsymbol{g} \cdot \boldsymbol{b}_2 = 0$, since $\boldsymbol{g} = \pm[\bar{2}20]$. From this it follows that $\boldsymbol{b}_1 = a/6[\bar{2}11]$. A comparison of the contrast intensities of the images of Fig. 4e and f taken in $+\boldsymbol{g}$ and $-\boldsymbol{g}$, respectively, shows a change in contrast which is analogous to the one of Fig. 4a and b. The relatively strong contrast of the partials in Fig. 4e corresponds to their contrast behaviour in Fig. 4a, since the angle between their $\boldsymbol{g}$-vectors ($[\bar{2}20]$ and $[\bar{2}02]$) is acute.

Since the micrographs of Fig. 4 allow to resolve the partials and to determine $\boldsymbol{b}_p$ (including the sign) the nature of the stacking faults can be worked out. This yields the result that all glide dislocations of Fig. 4 have intrinsic stacking faults.

Fig. 5 shows a weak-beam image of a dissociated glide dislocation; the glide plane (111) is 30° inclined to the foil (FN $\approx$ [153]). The character of the dislocation is near edge orientation ($\theta \approx 95°$) and both partials are in contrast, since $\boldsymbol{g} = [\bar{2}02]$ is

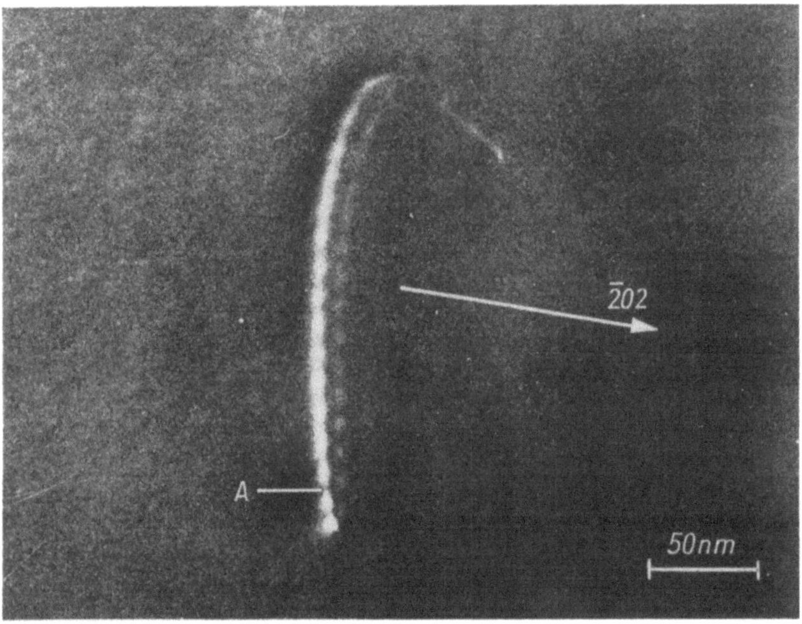

Fig. 5. Ag, (153) foil, BD $\approx$ [111]. Weak-beam image of a dissociated glide dislocation (near edge orientation) which is inclined to the foil (about 30°); $\boldsymbol{g} = [\bar{2}02]$. Although $\boldsymbol{g} \cdot \boldsymbol{b}_p = 1$ for both partials the left hand partial has a stronger contrast than the partial on the right hand side. When $\boldsymbol{b}_p$ is known this can be used to analyse the nature of the fault

parallel to $\boldsymbol{b}$ and they are imaged undistorted, since BD $\approx$ [111]. (It is interesting to note that near the top end the dislocation lies on the cross-slip plane). Since the contrast oscillates as a function of depth, there are certain depths where the partial on the left hand side has little or no contrast (e.g. near A where the dislocation approaches the surface). At these certain depths the relative contrast intensities of the partials are reversed. But it is unlikely that in the case of a dislocation being parallel to the foil this rather localized effect will disturb the general result of the relative intensities. Firstly, since the effective extinction distance is rather small ($\xi_{g\text{eff}} = 5.6$ nm) the depth of the dislocation in the foil varies in most cases by a few $\xi_{g\text{eff}}$, and secondly, the contrast oscillations are smeared out near the centre of the foil. The analysis of the dislocation imaged in Fig. 5 leads to the result that the stacking fault is intrinsic.

4. Discussion

From the results of Fig. 4 and 5 the contrast behaviour of the four different arrangements of Fig. 2 can be deduced. The sketches of Fig. 6a to f illustrate schematically the contrast of the partials which is observed in weak-beam images using the six $\langle 220 \rangle$ reflections. The stacking faults are indicated by hatched regions (intrinsic and extrinsic faults by full and broken lines, respectively); but it should be noted that in the corresponding weak-beam images the faults have no contrast, since $\boldsymbol{g} \cdot \boldsymbol{R} = 0$ (shear of the faults $\boldsymbol{R} = \pm a/3$ [111]). Fig. 6g shows computer simulations of bright-field images (FN $\approx$ [131], BD = [111]), they correspond to Fig. 6a as they have the same $\boldsymbol{g}$. For the computations [17] the dislocations are assumed to be dissociated edge dislocations (splitting width 8.4 nm, foil thickness $t = 80$ nm). The different intensities of the partials observed in the weak-beam micrographs are also indicated (e.g. a bold line indicates a "strong" contrast of the partial).

In Fig. 6a to f the sketches of the arrangements having extrinsic faults are drawn analogously to the intrinsic configurations; which means that the different intensities of the partials agree with the asymmetric contrast of the corresponding bright-field images. This assumption is in agreement with recent results of computer simulated weak-beam images [23].

It follows from Fig. 6 that the relative contrast intensities are reversed when either the sign of $\boldsymbol{g}$ is reversed or the sign of $\boldsymbol{b}$ is reversed or the nature of the stacking fault is changed. It should be mentioned that the contrast behaviour of the partials showing different intensities in the weak-beam images is similar to the one occurring in the case of superdislocations which was discussed in detail by Wilkens and Hornbogen [18].

Therefore several methods can be deduced from Fig. 6 to analyse the nature of the faults by determining $\boldsymbol{b}_p$ and their sign. Each method requires a minimum of two micrographs and two of the possible combinations are given in the following:

(i) Two weak-beam images, one where $\boldsymbol{g}$ is parallel to $\boldsymbol{b}$, and another one where $\boldsymbol{g}$ is not parallel to $\boldsymbol{b}$ (to determine the sign from the different intensities and $\boldsymbol{b}_p$, respectively).

(ii) Two weak-beam images with $\pm\boldsymbol{g}$ where $\boldsymbol{g}$ is not parallel to $\boldsymbol{b}$. (In principle even two bright-field images could be used in this case, but the difference of their contrast intensities is much less pronounced.)

By using one of these methods or a combination of them, the type of the stacking fault of a dissociated glide dislocation can be identified unambiguously and rather quickly. Numerous glide dislocations have been studied by weak-beam methods in pure Ag and low stacking fault energy alloys (Cu–Al alloys within the solid solution range: Cu–5, 10, and 15 at% Al). The analysis of several hundred glide dislocations

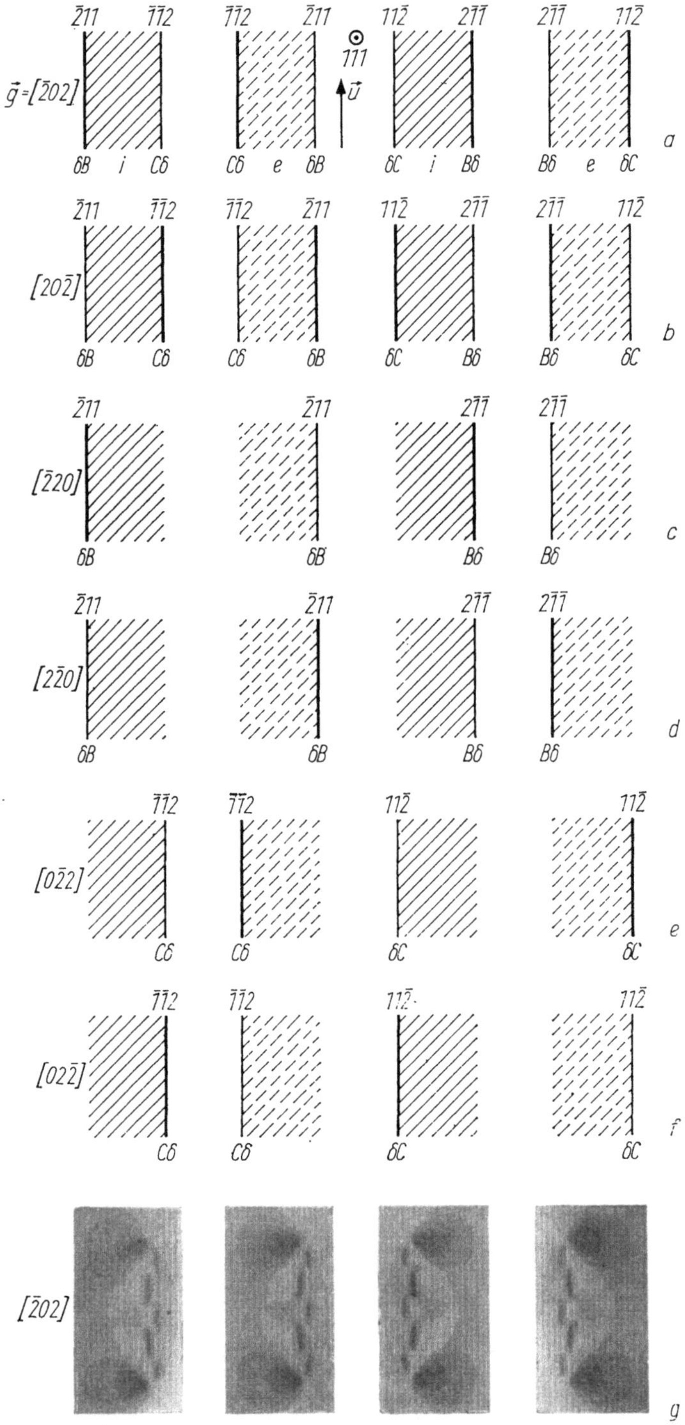
$\bar{2}11$
$\bar{1}\bar{1}2$
$11\bar{2}$
$2\bar{1}\bar{1}$
$\bar{1}11$
$\vec{u}$
$\vec{g}=[\bar{2}02]$
$[20\bar{2}]$
$[\bar{2}20]$
$[2\bar{2}0]$
$[0\bar{2}2]$
$[02\bar{2}]$
$[\bar{2}02]$
δB
$C\delta$
δC
$B\delta$
i
e
a
b
c
d
e
f
g

Fig. 6

Fig. 6. The contrast behaviour of the four different arrangements of a dissociated glide dislocation ($\boldsymbol{b} = \pm a/2$ [$\bar{1}$01] (111)) corresponding to Fig. 4. a) to f) Schematic representations of the weak-beam contrast using different $\boldsymbol{g}$-vectors. The relative contrast intensities of the partials are indicated. The areas of the faults are hatched (N.B. No contrast of the faults is observed in the corresponding images, since $\boldsymbol{g} \cdot \boldsymbol{R} = 0$). g) Simulated bright-field images of dissociated edge dislocations which are inclined to the foil, length of the dislocation line $l \approx 160$ nm, FN $\approx$ [131]; $\boldsymbol{g} = [\bar{2}02]$

leads to the general result that only intrinsic stacking faults are encountered. It is interesting to note that at least in our observations this result has never been violated. This seems to be in contradiction to the reports in the literature showing the occurrence of extrinsic stacking faults [9, 19, 20]. But one has to be aware of the fact that the dissociated dislocations having extrinsic stacking faults are always observed at a characteristic three fold node configuration and are not individual glide dislocations. This is in agreement with observations showing that nodes can act as nucleation centres; since it has been studied by weak-beam methods that for instance cross-slip and the transformation of a Lomer-Cottrell lock into a dislocation glissile on (001) can nucleate at a constricted node (e.g. [21, 22]). In this context it should be mentioned that in a specimen deformed into stage II the glide dislocations are frequently jogged (cf. dislocation A in Fig. 4) and some of the jogs are constricted (at least to the resolution of the weak-beam image). In spite of this the careful investigation of the dislocations in the neighbourhood of the jogs did not show the occurrence of extrinsic faults.

Finally one can discuss why the dissociated glide dislocations do not show extrinsic stacking faults. Calculations [11] based on a model of hard spherical balls bound by central forces yield the same fault energies for both configurations ($\gamma_i = \gamma_e$), and this result agrees with the experimental observations [8, 9]. Contrary to this the analysis of the dissociated dislocations shows a large difference in the frequency of the occurrence of the two fault types, which could arise from their different nucleation mechanism. As described in Section 2 an intrinsic fault can be formed by a single shear operation whereas an extrinsic fault requires two consecutive intrinsic shears of neighbouring planes. This indicates that it is more difficult to form an extrinsic fault than an intrinsic fault by shear. Therefore the experimental results which show that all the analysed glide dislocations have intrinsic faults are in agreement with these considerations.

5. Conclusions

1. To study the nature of the stacking faults occurring in dissociated glide dislocations different methods are worked out by which the partials and the faults can be analysed unambiguously and rather quickly.

2. When a standardized procedure of indexing the micrographs is used the different contrast intensities of the partials allow to determine the sign of their $\boldsymbol{b}_p$. The contrast behaviour of images taken in different reflections is similar when the angle between their $\boldsymbol{g}$-vectors is acute.

3. Several hundred dissociated glide dislocations have been studied in pure Ag and Cu–Al alloys (Cu–5, 10, and 15 at% Al) and all the analysed faults are intrinsic.

4. Dislocations showing extrinsic faults, which were described in the literature, are rare and only observed at a characteristic three fold node configuration. The result that individual glide dislocations have always intrinsic faults, can be interpreted by differences in the nucleation mechanism of intrinsic and extrinsic faults which are formed by shear.

Acknowledgements

The authors would like to thank Prof. K. Lintner for the provision of laboratory facilities and Prof. P. Skalicky for his permission to use the computer programme (described in [17]) which was used to simulate the bright-field images of Fig. 6g.

References

[1] A. Art, R. Gevers, and S. Amelinckx, phys. stat. sol. **3**, 697 (1963).
[2] R. Gevers, A. Art, and S. Amelinckx, phys. stat. sol. **3**, 1563 (1963).
[3] J. van Landuyt, R. Gevers, and S. Amelinckx, phys. stat. sol. **18**, 167 (1966).
[4] S. Amelinckx, Dislocations in Solids, Vol. 2, Ed. F. R. N. Nabarro, North-Holland Publ. Co., Amsterdam/New York/Oxford 1979.
[5] L. J. Chen and G. Thomas, phys. stat. sol. (a) **25**, 193 (1974).
[6] R. E. Smallman, K. H. Westmacott, and J. H. Coiley, J. Inst. Metals **88**, 127 (1959).
[7] D. J. Mazey and R. S. Barnes, Phil. Mag. **17**, 387 (1968).
[8] P. C. J. Gallagher, phys. stat. sol. **16**, 95 (1966).
[9] P. C. J. Gallagher and J. Washburn, Phil. Mag. **14**, 971 (1966).
[10] D. J. H. Cockayne, Z. Naturf. **27a**, 452 (1972).
[11] J. P. Hirth and J. Lothe, Theory of Dislocations, McGraw-Hill Publ. Co., New York 1968.
[12] M. H. Loretto and R. E. Smallman, Defect Analysis in Electron Microscopy, Chapman & Hall, Ltd., London 1975.
[13] P. B. Hirsch, A. Howie, R. B. Nicholson, D. W. Pashley, and M. J. Whelan, Electron Microscopy of Thin Crystals, Butterworths, London 1965.
[14] H. P. Karnthaler, P. M. Hazzledine, and M. S. Spring, Acta metall. **20**, 459 (1972).
[15] E. Wintner and H. P. Karnthaler, Phil. Mag. **36**, 1317 (1977).
[16] E. Wintner and H. P. Karnthaler, Acta metall. **26**, 941 (1978).
[17] P. Skalicky, H. P. Karnthaler, and H. Priemer, phys. stat. sol. (a) **22**, 167 (1974).
[18] M. Wilkens and E. Hornbogen, phys. stat. sol. **4**, 557 (1964).
[19] M. H. Loretto, Phil. Mag. **10**, 467 (1964).
[20] P. C. J. Gallagher and J. Washburn, Phil. Mag. **15**, 969 (1967).
[21] H. P. Karnthaler, Phil. Mag. **A38**, 141 (1978).
[22] H. P. Karnthaler and A. Korner, Strength of Metals and Alloys, Ed. P. Haasen, V. Gerold, and G. Kostorz, Pergamon Press, Oxford 1979 (p. 35).
[23] P. Pongratz, private communication.

(Received July 28, 1981)

phys. stat. sol. (a) 68, 29 (1981)

Subject classification: 1.2; 22.8

Departamento de Física, Universidad del País Vasco, Bilbao[1])

Structural Phase Transitions in $(CH_3NH_3)_2ZnCl_4$ Crystals[2])

By

J. M. Pérez-Mato, J. L. Mañes, J. Fernández, J. Zúñiga, M. J. Tello, C. Socías, and M. A. Arriandiaga

Two structural phase transitions in $(CH_3NH_3)_2ZnCl_4$ at 426 and 484 K are discovered and investigated by X-ray diffraction, calorimetric, dielectric, thermal expansion, and optical measurements. The crystals are shown to be ferroelastic below 426 K. This behaviour is explained by a symmetry change Pnma–$P2_1/a$, which is consistent with the compound structure observed at room temperature. The possible soft mode associated to the displacive part of the transition at 426 K is discussed, and a phenomenological approach following Landau theory is carried out.

Es werden zwei strukturelle Phasenübergänge in $(CH_3NH_3)_2ZnCl_4$ bei 426 und 484 K durch Röntgenstrahl-Interferenz, kalorimetrische, dielektrische und optische Messungen, sowie durch Wärmeausdehnung festgestellt und untersucht. Die Kristalle sind unterhalb 426 K ferroelastisch. Dieses Verhalten erklärt sich aus einer Symmetrieänderung Pnma–$P2_1/a$, was mit der beobachteten Verbundstruktur bei Raumtemperatur übereinstimmt. Die mögliche kritische Mode des Übergangs bei 426 K wird diskutiert und eine phänomenologische Untersuchung nach der Landau-Theorie durchgeführt.

1. Introduction

Bis-alkyl-ammonium zinc chlorides (C_nZn) belong to a large family of compounds with the general formula $(C_nH_{2n+1}NH_3)_2MX_4$ which have been extensively investigated in the last few years because of the large variety of structural and magnetic phases they exhibit (see e.g., [1] and references cited therein). Recent structural studies show that the short chain compounds with X = Cl and M = Mn, Cd, Fe, Cu exhibit perovskite structures which consist of infinite layers of corner-sharing chlorine octahedra with one layer of alkyl-ammonium groups attached on each side [2, 3]. Depending on the length of the organic chains different structures and chain packings can be realized [1, 4]. Most of the observed phase transitions can be described by mechanisms based on an interplay of deformations of the anion octahedra, changes in the hydrogen bonding scheme, thermal motion of the whole alkyl-ammonium groups, and conformational changes of the alkyl chains [5 to 8]. An interesting feature is that the sequence and nature of these structural phase transitions are functions of n, M, and X [9 to 11]. Thus for short chains the structural phase transitions are mainly characterized by a dynamic disorder where the short chains move on conical surfacse among two or more potential wells of the octahedral cavity [1, 12]. In this case, together with the more or less hindered dynamic order–disorder mechanisms, melting of the alkyl-ammonium chains occurs (conformational changes).

Another interesting point of view deals with the possibility of condensing lattice modes in short chain compounds [8]. From a systematic study of several structural changes in this family Geick and Strobel [6] found that some of these phase transitions are partly of the displacive type and partly of the order–disorder type and that the condensing modes which involve rotational motions of the alkyl chains are not necessarily the driving mechanism for the continuous phase transitions.

[1]) Apdo. 644, Bilbao, Spain.

[2]) Work supported in part by Spain-Comisión Asesora de Investigación Científica y Técnica.

Compounds with M = Co and Zn show two important differences if compared with other members of this family. On one hand, there are no infinite corner-sharing layers of chlorine octahedra, but instead unassociated Cl_4 tetrahedra are present [13]. On the other hand, intercalation of the chains for long chain compounds [4] occurs. This paper is the first attempt of a study about the structural phase transitions in the bis-methyl-ammonium zinc chloride (MAZnCl). At room temperature MAZnCl belongs to the monoclinic $P2_1/a$ (c, unique axis) space group and shows a strong orthorhombic pseudostructure [13]. The structure consists of not disordered $CH_3NH_3^+$ and $ZnCl_4^{2-}$ ions with a slight distortion for the tetrahedral $ZnCl_4^{2-}$ ions.

It is our purpose to show the experimental evidence of two structural phase transitions above room temperature in this compound. From the structural and symmetry analysis we found that the compound can be related with the A_2BX_4(β-K_2SO_4) family. The experimental results relative to thermal, dielectric, and optical measurements as well as the ferroelastic character of the room temperature phase are shown. We also present a study of the condensing mode for the ferro–paraelastic phase transition together with a free-energy expansion that accounts for the symmetry changes in the crystal. Finally the thermal, elastic, and dielectric properties are briefly discussed.

2. X-Ray Structure Analysis

The room temperature structure was analysed by single crystal diffraction with a four circle automatic diffractometer (Syntex $P2_1$) and refined to final residual reliability index values of $R = 0.034$ and $R_w = 0.033$ $(w = 1/\sigma^2(F))$ for 2361 intensities greater than $3\sigma(I)$. The calculated structure is in agreement with that of Mororin and Emerson [13]. The unit cell parameters were found to be $a = 10.865$ Å, $b = 7.627$ Å, $c = 12.661$ Å, $\gamma = 96.58°$. Fig. 1 shows a schematic projection of the structure along the c-axis. The structure mainly consists of $ZnCl_4^{2-}$ tetrahedra and methyl-ammonium chains which are linked to the Cl^- ions by H-bonds. The Cl^- tetrahedra show a small distortion which is probably due to the presence of H-bonds. It is interesting to notice that each methyl-ammonium group has four Cl^- ions in the same range length. This fact opens the possibility of H-bonds to more than three Cl^- ions which would account for the disorder of the organic groups in the high temperature phase. However, it must be noticed that one of the two independent methyl-ammonium groups has a N–H ... Cl bond length which is much shorter than the other one. Fig. 2 shows the stereoscopic projection of the hydrogen bond scheme revealed by the X-ray analysis, which does not allow the layer-type structure to occur as in other members of this family with M = Cd, Mn, or Cu.

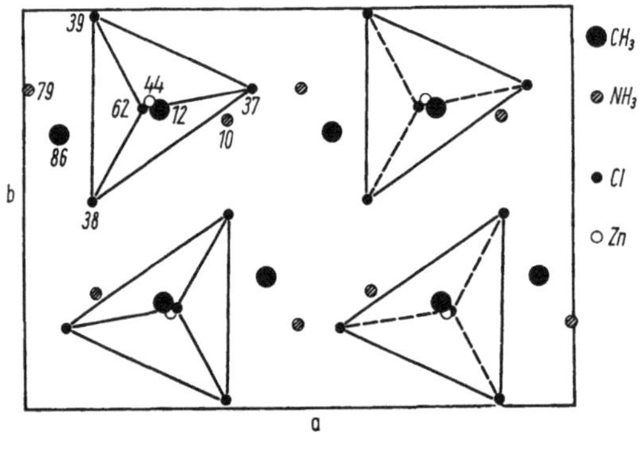

Fig. 1. Schematic c-projection of room temperature phase in (C_1Zn)

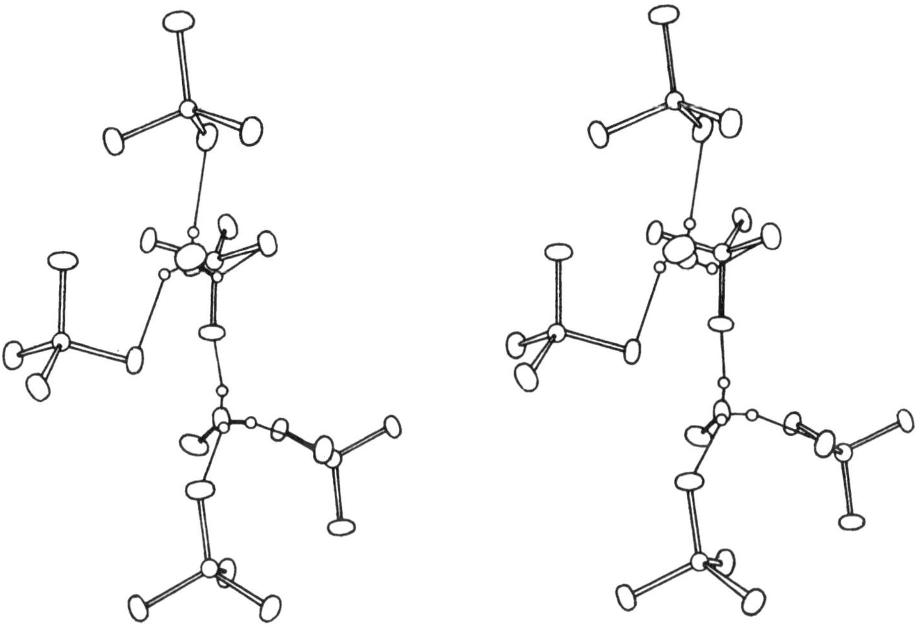

Fig. 2. Stereographic projection of room temperature phase along the *c*-axis showing the H-bonding scheme

3. Experimental Results

Experimental details about crystal growth and techniques as well as the measurement accuracy were previously described [14]. Conventional X-ray, chemical analysis, microscopic observations, and IR spectroscopy in the 400 to 4000 cm^{-1} region were used as characterization techniques. Crystal growth yielded two types of crystal samples: some grew as thin platelets with the monoclinic axis along their plane and others presented prismatic forms with the monoclinic axis perpendicular to the larger face of the prism. Under conoscopic observation the crystal shows a moderate inclined dispersion and therefore both the acute and obtuse bisectrices lie in the crystallographic symmetry planes. Two methods, compensation or relative retardation, were used for the optical measuremnets in function of the appearance of twinning in the crystal samples. Problems related with the path difference (Γ) measurements were explained by Fousek et al. [15] for a mmm-2/m point symmetry change.

As the samples were hygroscopic the experiments were carried out in dry argon or nitrogen atmosphere. Fig. 3 shows the specific heat curves revealing the existence of three phases above room temperature: phase I above 483 K, phase II between

Table 1

Thermodynamic functions and thermal hysteresis for the two phase transitions present in $(CH_3NH_3)_2ZnCl_4$

T (K)	ΔT (K)	ΔH (kJ/mol)	ΔS (J/mol K)
426	19	0.29 ± 0.03	0.68 ± 0.06
483	11	8.7 ± 0.1	18.1 ± 0.2

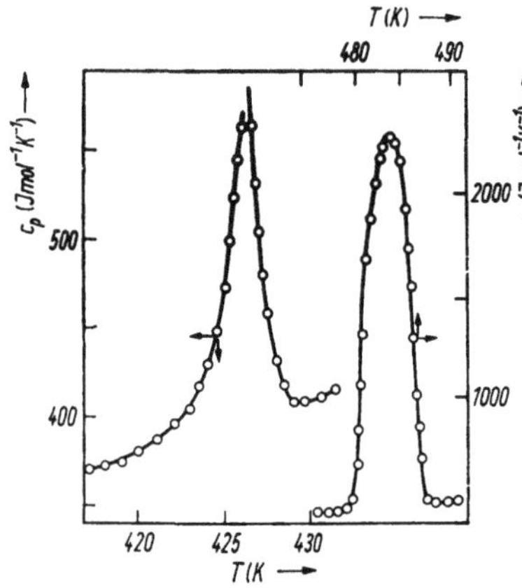

Fig. 3. Specific heat curve showing the two-phase transitions above room temperature

483 and 426 K, and phase III below 426 K. The observed transition entropy and enthalpy values for these transitions, as well as their thermal hysteresis (ΔT) are shown in Table 1. It must be noted that crystals melt at 563 K and decomposition immediately occurs. This fact was checked by means of direct optical observations of the melting point. Both phase transitions show a strong thermal hysteresis of 22 K (I → II) and 11 K (II → III) in a dynamical regime at a rate of 8 K/min. Slight variations in ΔT were observed as a function of the heating rate, similar to those found for methyl-ammonium cadmium and manganese compounds [10]. It was also noticed that when heating above 484 K the phase transition at 426 K is not reproducible. In order to recover this phase transition cooling must start before reaching 484 K and the crystal must be kept in the intermediate phase for five or more minutes. Finally when crystals were observed in vacuum, a substantial decrease of the phase transition temperatures was detected. Thus, at 0.5 Torr the II → III phase transition occurs at 408 K and the melting point is lowered down to 473 K.

Measurements of the dielectric constant and loss tangent for single crystals in the c-axis direction were made. It must be noted that cycling through the phase transitions affects the value of the dielectric constant and both phase transitions show significant thermal hysteresis in agreement with the calorimetric measurements. The dielectric constant shows a step at 426 K and increases when going to the high temperature phase with an inflection point at 484 K. Therefore, tan δ shows also a step discontinuity at 426 K but a peak appears where the inflection point of the dielectric constant is located.

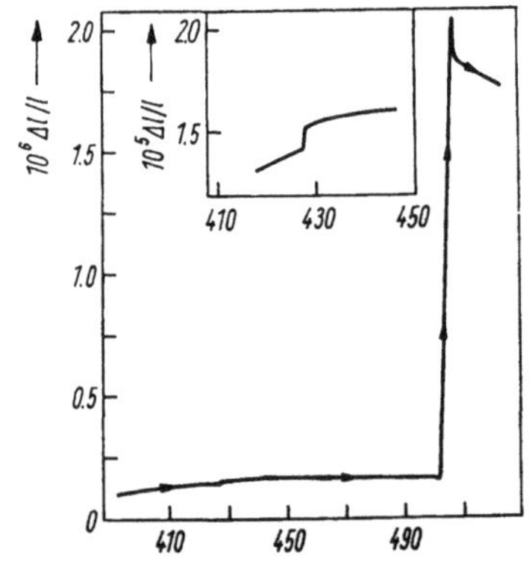

Fig. 4. Thermal expansion along the c-axis

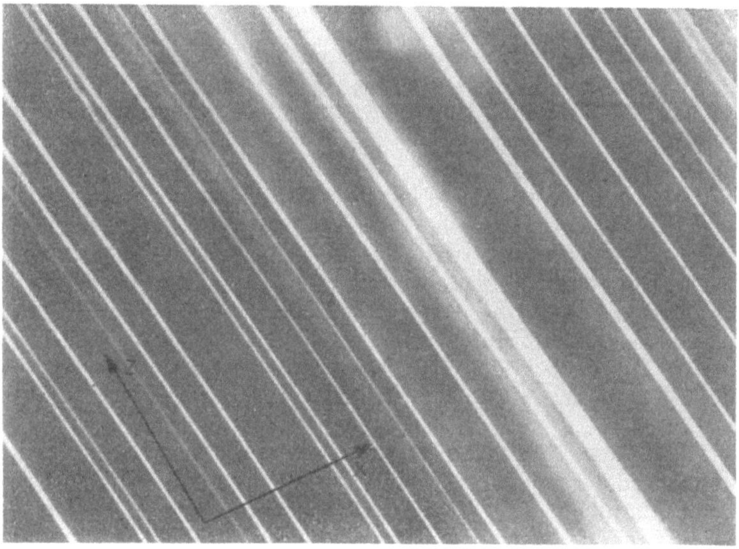

Fig. 5. Ferroelectric domains in the monoclinic phase

Thermal expansion measurements along the c-axis were made on single domain platelets of uniform thickness. Samples were replaced by new ones each time a single measuring run was over in order to avoid the influence of domain formation and non-reconstructive internal cracks when heating above 484 K. The results are shown in Fig. 4. One observes discontinuous changes in thermal expansion at both transition temperatures. This fact together with the relatively large thermal hysteresis observed, suggests a first-order character for both phase transitions.

Optical observations on plates along the c-axis have shown that a domain structure (see Fig. 5) can be induced and moved across the crystal by means of a mechanical stress. In this case the applied stress must be greater than for copper or manganese compounds. In Fig. 6 the temperature dependence of the spontaneous path difference $\delta_s\Gamma(T)$ for the II → III phase transition is shown. It was obtained from the actual value in the monoclinic phase by subtraction of the Γ-value extrapolated from the II-phase [15]. One observes a sudden increase in Γ which is related with a similar effect in the thermal expansion and dielectric constant. Above 484 K no measurement was possible because the crystals become opaque.

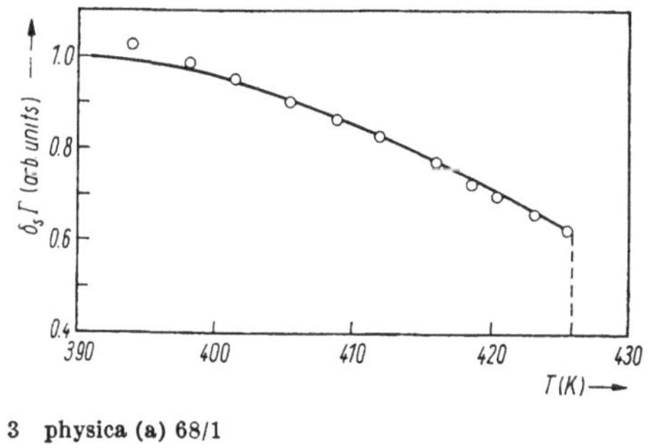

Fig. 6. Spontaneous path difference for the II–III phase transition. The continuous line shows the theoretical behaviour given by (6) in the text

4. Symmetry Coordinates and Soft Mode

The experimental results described above prove the existence of two phase transitions above room temperature. The structural data available up to now permit a more extensive discussion of the II → III phase transition at 426 K. From the monoclinic room temperature structure and making use of the small displacement theory for the ferroparaelastic phase transitions [16] the Pnma space group can be assigned to the II-phase. This point is supported by the recent X-ray determination of the high temperature phase in $(C_3H_7NH_3)_2ZnCl_4$ crystals [17]. Also, it must be noted that this is a common symmetry at room temperature in A_2BX_4 compounds with tetrahedral coordination [18]. In Fig. 7 we present a schematic projection along the c-axis of the assumed structure for this phase. One can see that only a small shift of the y-coordinate for some heavy ions distinguishes this arrangement from the one presented in Fig. 1 for the III-phase. This transition can be taken to be ferrodistortive and the order parameter will transform according to the B_{1g} irreducible representation of the Pnma space group.

Since the Pnma → $P2_1/a$ phase transition was not observed up to now in the $(C_nH_{2n+1}NH_3)_2MX_4$ compounds it does not appear in the previously published theoretical picture of this family [6]. In order to get a microscopic description of the order parameter the symmetry coordinates must be determined. The structure has four chemical formula units per unit cell and therefore there exist 156 branches for the dispersion relations (only librations for CH_3^- and NH_3^+ groups are considered). From Fig. 7 two groups of ions can be considered. On one hand we have Zn^+, $Cl^-(1)$, $Cl^-(2)$, $CH_3^-(1)$, $CH_3^-(2)$, $NH_3^+(1)$, and $NH_3^+(2)$ ions (α-type) with $y = 0.25$ or 0.75 and four equivalent positions. On the other hand, we have the $Cl^-(3)$ and $Cl^-(4)$ ions and the other six Cl^- ions related by symmetry operations which we call β-type. The polar and axial vector representations for the mmm point group lead to the invariance of the corresponding subspaces for the displacements and librations along or around the three orthogonal directions. Therefore for $\boldsymbol{k} = \boldsymbol{0}$ the 156-dimensional U-space of displacements and librations can be described in terms of invariant subspaces,

$$U = \sum_{\alpha=1}^{7} (U_{\alpha x} + U_{\alpha y} + U_{\alpha z}) + \sum_{R\alpha=1}^{4} (U_{R\alpha x} + U_{R\alpha y} + U_{R\alpha z}) + \\ + U_{\beta x} + U_{\beta y} + U_{\beta z} \,, \qquad (1)$$

where $U_{\alpha i}$ correspond to the atomic displacements of the four α-ions related by symmetry operations. The associated subspaces are therefore four-dimensional and α goes from 1 to 7 since there are seven α-type independent ions. $U_{R\alpha i}$ are associated with the librations of those ionic groups with librational freedom. $U_{\beta i}$ are the eight-dimensional subspaces for the β-type ions.

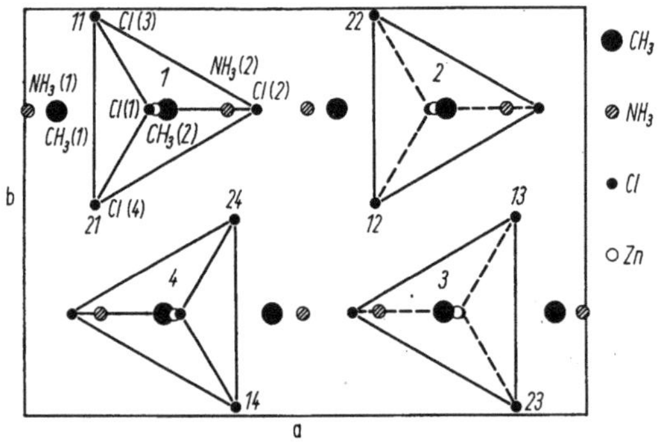

Fig. 7. Schematic c-projection of the idealized phase II

With the above general considerations in mind and making use of the usual group techniques we can find the subspaces which transform according to the irreducible representations of mmm whose basis are the symmetry adopted modes we are looking for. The soft mode for the Pnma → $P2_1/a$ phase transition is given by a linear combination of the symmetry modes e_i that transform according to B_{1g}.

For the α-ions $e_{1\alpha} = \frac{1}{2}(u_1^y + u_2^y - u_3^y - u_4^y)$.

For the rigid ionic groups

$$e_{2\alpha} = \frac{1}{2}(R_1^x - R_2^x + R_3^x - R_4^x),$$
$$e_{3\alpha} = \frac{1}{2}(R_1^z + R_2^z + R_3^z + R_4^z).$$

For the β-ions

$$\left.\begin{aligned} e_4 &= \tfrac{1}{8}(u_{11}^x - u_{12}^x + u_{13}^x - u_{14}^x - u_{21}^x + u_{22}^x - u_{23}^x + u_{24}^x),\\ e_5 &= \tfrac{1}{8}(u_{11}^y + u_{12}^y - u_{13}^y - u_{14}^y + u_{21}^y + u_{22}^y - u_{23}^y - u_{24}^y),\\ e_6 &= \tfrac{1}{8}(u_{11}^z + u_{12}^z + u_{13}^z + u_{14}^z + u_{21}^z + u_{22}^z + u_{23}^z + u_{24}^z),\end{aligned}\right\} \quad (2)$$

where u_j^i stands for the displacement from the equilibrium position along the i-axis of the j-ion. R_j^i has a similar meaning in the case of librations (see Fig. 7 for the atom labels).

Thus the soft mode would involve translational motions u^y for the α-ions, librational motions R^x and R^z for the ionic groups, and translational motions u^x, u^y, and u^z for the Cl^- ions of the β-type. Though a detailed structure determination of the intermediate phase has not yet been made in order to choose which of these symmetry coordinates are relevant for the phase transition, obvious geometrical considerations permit to point out that the displacements of the $CH_3^-(1)$ group in the y-direction are probably the most important ones for the displacive part of this phase transition. It must also be noted that the displacements of the $Cl^-(2)$ ions account for the small distortion of the Cl^--tetrahedra at room tempearture.

5. Phenomenological Approach

As it can be seen from the decomposition of the symmetrized square of the vector representation for the mmm point group, the u_6 component of the strain tensor transforms according to the B_{1g} representation and can be therefore identified with the thermodynamic order parameter for the transition. From a geometrical point of view it represents the difference between the monoclinic angle and $\pi/2$. This fact is in agreement with the ferroelastic character for the crystal in the distorted phase.

Since the active representation is one-dimensional the free-energy expansion as a function of the order parameter, polarization P_i, and strain u_i can be readily written as

$$\begin{aligned} F = {} & F_0 + \tfrac{1}{2}a(T - T_0)\,u_6^2 + \tfrac{1}{4}\beta u_6^4 + \tfrac{1}{6}Wu_6^6 + \\ & + \tfrac{1}{2}C_{11}^0u_1^2 + \tfrac{1}{2}C_{22}^0u_2^2 + \tfrac{1}{2}C_{33}^0u_3^2 + \tfrac{1}{2}C_{44}^0u_4^2 + \tfrac{1}{2}C_{55}^0u_5^2 + \\ & + C_{12}^0u_1u_2 + C_{13}^0u_1u_3 + C_{23}^0u_2u_3 + \\ & + \tfrac{1}{2}(\chi_{11}^0)^{-1}P_x^2 + \tfrac{1}{2}(\chi_{22}^0)^{-1}P_y^2 + \tfrac{1}{2}(\chi_{33}^0)^{-1}P_z^2 + \\ & + \delta_1u_4u_5u_6 + [\eta_1u_1 + \eta_2u_2 + \eta_3u_3]\,u_6^2 + \\ & + [\tfrac{1}{2}n_{11}u_1^2 + \tfrac{1}{2}n_{22}u_2^2 + \tfrac{1}{2}n_{33}u_3^2 + \tfrac{1}{2}n_{44}u_4^2 + \tfrac{1}{2}n_{55}u_5^2 + \\ & + n_{12}u_1u_2 + n_{13}u_1u_3 + n_{23}u_2u_3]\,u_6^2 + \\ & + \gamma_1P_xP_yu_6 + [t_1P_x^2 + t_2P_y^2 + t_3P_z^2]\,u_6^2, \end{aligned} \quad (3)$$

where $a > 0$. Sixth-order terms must be included due to the first-order character of the transition. Minimization of the free-energy expansion with respect to u_i and P_i leads, besides to the trivial solution corresponding to the orthorhombic phase, to

$$\begin{aligned} u_6 &= -\frac{\beta'}{2W'}\left[1 + \left(1 - \frac{4W'a(T - T_0)}{\beta'^2}\right)^{1/2}\right], \\ u_i &= A_i u_6^2 \qquad (i = 1, 2, 3)\,, \\ P_i &= u_4 = u_5 = 0 \qquad (i = 1, 2, 3)\,, \end{aligned} \tag{4}$$

where A_i, β', and W' are directly related with the coefficients in expansion (3) and $\beta' < 0$, $W' > 0$.

The phase transition temperature T_c and the finite jump of the order parameter at T_c can be determined from the equilibrium conditions together with the condition $F(T_c) = F_0$. Thus

$$T_c = T_0 + \frac{3}{16}\frac{\beta'^2}{aW'}\,, \qquad u_6^2(T) = -\frac{3}{4}\frac{\beta'}{W'}\,. \tag{5}$$

Using (4) and (5) we obtain for the temperature dependence of the order parameter

$$u_6^2(T) = \frac{2}{3}\,u_6^2(T_c)\left[1 + \left\{1 - \frac{3}{4}\left(\frac{T - T_0}{T_c - T_0}\right)\right\}^{1/2}\right], \tag{6}$$

where as usual $(T_c - T_0)$ must be of the order of the transition thermal hysteresis.

The clamped dielectric susceptibilities and elastic constant polarization in the distorted phase are obtained from the corresponding second derivatives of the free-energy expansion,

$$\left.\begin{aligned} &\text{a)} \quad C_{ij}^{P} = C_{ij}^{0} + n_{ij}u_6^2 \qquad (ij = 11, 12, 13, 44, 55, 22, 33, 23)\,, \\ &\text{b)} \quad C_{ij}^{P} = 2\eta_i u_6 + 2R_i u_6^3 \qquad (ij = 16, 26, 36)\,, \\ &\text{c)} \quad C_{45}^{P} = \delta_1 u_6\,, \\ &\text{d)} \quad (\chi_{11}^{U})^{-1} = (\chi_{11}^{0})^{-1} + t_1 u_6^2 - \frac{\gamma_1^2}{t_2 + (\chi_{22}^{0})^{-1}\,u_6^{-2}}\,, \\ &\text{e)} \quad (\chi_{22}^{U})^{-1} = (\chi_{22}^{0})^{-1} + t_2 u_6^2 - \frac{\gamma_1^2}{t_1 + (\chi_{11}^{0})^{-1}\,u_6^{-2}}\,, \\ &\text{f)} \quad (\chi_{12}^{U})^{-1} = -D_1 u_6^{-1} + D_2 u_6 - D_3 u_6^3\,, \\ &\text{g)} \quad (\chi_{33}^{0})^{-1} = (\chi_{33}^{0})^{-1} + t u_6^2\,, \end{aligned}\right\} \tag{7}$$

where

$$\begin{aligned} R_i &= \sum_{ij} n_{ij} A_j\,, \\ D_1 &= \frac{1}{\gamma_1}\,(\chi_{11}^{0})^{-1}\,(\chi_{22}^{0})^{-1}\,, \\ D_2 &= [\gamma_1 - (\chi_{11}^{0})^{-1}\,t_2 - (\chi_{22}^{0})^{-1}\,t_1]\,, \\ D_3 &= t_1 t_2\,. \end{aligned}$$

The expressions can be directly compared with experimental results since the crystal is not piezoelectric.

Finally the C^P_{66} coefficient (the inverse of the susceptibility associated with the order parameter) shows a clear difference from the other macroscopic constants as it follows a typical Curie law,

$$\left.\begin{aligned} C^P_{66} &= a(T - T_0)\,; & T > T_c\,, \\ C^P_{66} &= 2\beta u_6^2 + 4W u_6^4\,; & T < T_c\,. \end{aligned}\right\} \tag{8}$$

6. Discussion

The structural and symmetry analyses confirm the difference between MAZnCl and other members of the $(C_nH_{2n+1}NH_3)_2MX_4$ family with M = Mn, Cd, Cu, etc. This difference is so significant that in some aspects, MAZnCl can be considered to belong to the group of tetrahedrally coordinated A_2BX_4 (β-K_2SO_4) compounds rather than to the one previously mentioned, account taken of the dissimilitudes arising from the presence of an organic group.

The occurrence of spontaneous strain (shear strain component in the (001) plane) is a primary effect and therefore u_6 can be considered to be a macroscopic order parameter.

In this case the optical polarization coefficients (B_i) show the same type of behaviour as the strain components (4) and the indicatrix rotation angle in the lowest approximation is proportional to the order parameter [15]. Within the same approximation the change in birefringence (measured in the rotating system) is given by

$$\delta_s\,\Delta n'_{21} = \alpha_1 B_1 + \alpha_2 B_2 + \alpha \tan^2 z\varphi\,.$$

Since all the terms are proportional to u_6^2 we can expect $\delta_s\,\Delta n'_{12} \sim u_6^2$. As u_3 shows the same behaviour, it can be assumed that $\delta_s\Gamma(T)$ obeys the temperature behaviour given by (6). This prediction is compared with the experimental results in Fig. 6.

The temperature dependence for the strain components is summarized in (4). These expressions predict a positive or negative step discontinuity as a function of the positive or negative values for A_i, which is consistent with the experimental results for the u_3 component in Fig. 4 taking $A_3 < 0$.

Different types of behaviour are predicted in (7) for the tensor magnitudes of the crystal, from the linear dependence on the order parameter for C_{45} to the more complex behaviour of χ_{11} and χ_{22}. For χ_{33} a step discontinuity is observed at T_c, but no definite comparison with the predicted behaviour in (7) is possible because the samples show a large increase of their conductivity above T_c. Also it should be taken into account that, as a consequence of the I → II transition, the various magnitudes of the crystal have a significant temperature dependence above T_c, against the assumptions of the phenomenological formalism we are making use of.

In order to describe the I → II phase transition and its influence on the one studied here (with the introduction of the possible coupling of the order parameters for both phase transitions) structural studies for the I-phase must be carried out. It would be also important to have an experimental confirmation of the Curie law (8) obtained for the C^P_{66} coefficient.

Acknowledgements

One of us (J.Z.) gratefully acknowledges the facilities found at the Institut de Cristallographie de l'Université de Lausanne for X-ray measurements and helpful discussions with Dr. G. Chapuis.

References

[1] R. KIND, S. PLESKO, H. AREND, R. BLINC, B. ZECKS, J. SELINGER, B. LOZAR, J. SLAK, A. LEVSTIK, C. FILIPIC, V. ZAGAR, G. LAHAJNAR, F. MILIA, and G. CHAPUIS, J. chem. Phys. **71**, 2118 (1979).
[2] R. D. WILLET, O. L. LILES JR., and C. MICHELSON, Inorg. Chem. **6**, 1885 (1967).
[3] E. R. PETERSON and R. D. WILLET, J. chem. Phys. **56**, 1879 (1972).
[4] M. R. CIAJOLO, P. CORRADINI, and V. PAVONE, Acta cryst. B **33**, 553 (1977).
[5] R. KIND and J. ROOS, Phys. Rev. B **13**, 45 (1976).
[6] R. GEICK and K. STROBEL, J. Phys. C **10**, 4221 (1977).
[7] R. BLINC, B. ZEKS, and R. KIND, Phys. Rev. B**17**, 3409 (1978).
[8] J. PETZELT, J. Phys. Chem. Solids **36**, 1005 (1975).
[9] M. A. ARRIANDIAGA, M. J. TELLO, J. FERNÁNDEZ, H. AREND, and J. ROOS, phys. stat. sol. (a) **48**, 53 (1978).
[10] M. J. TELLO, M. A. ARRIANDIAGA, and J. FERNANDEZ, Solid State Commun. **24**, 299 (1977).
[11] E. H. BOCANEGRA, M. J. TELLO, M. A. ARRIANDIAGA, and H. AREND, Solid State Commun. **17**, 1221 (1975).
[12] C. SOCÍAS, J. HERREROS, M. A. ARRIANDIAGA, J. FERNÁNDEZ, and M. J. TELLO, Ferroelectrics **25**, 453 (1980).
[13] B. MOROBIN and K. EMERSON, Acta cryst. B**32**, 294 (1976).
[14] C. SOCÍAS, M. A. ARRIANDIAGA, M. J. TELLO, J. FERNÁNDEZ, and P. GILI, phys. stat. sol. (a) **57**, 405 (1980).
[15] J. FOUSEK, C. KONAK, and G. ERRANDONEA, J. Phys. C **12**, 3197 (1979).
[16] H. P. WEBER, B. C. TOFIELD, and P. F. LIAO, Phys. Rev. B **11**, 1152 (1975).
[17] J. ZÚÑIGA, Doctoral Thesis, Bilbao 1980.
[18] H. AREND, P. MURALT, S. PLESKO, and D. ALTERMATT, Ferroelectrics **24**, 297 (1980).

(Received November 6, 1980)

phys. stat. sol. (a) 68, 39 (1981)

Subject classification: 14.3.3 and 14.3.4; 22.1.2

Institute of Semiconductors, Academy of Sciences of the Ukrainian SSR, Kiev[1])

Negative Resistance and Current Filament in Avalanche Diodes

By

V. I. FINEBERG, R. V. KONAKOVA, L. V. SHCHERBINA, and N. S. BOLTOVETS

The static and impulse current–voltage characteristics of silicon reverse biased p^+–n–n^+ diodes, are investigated. In the region of the avalanche breakdown a negative differential resistance, which leads to a current filament, is observed in these diodes. The influence is studied of the impurity concentration in the base and the base width on the value of the critical current, which determines the appearence of a negative resistance section at the current–voltage (I–U) characteristics. It is shown, that the process of the formation of a high-current density filament in the static regime develops in the microplasmas region. Under the switching on of diodes in the working scheme the generation of relaxation oscillations is observed. It may lead to catastrophic failure of diodes in schemes with a high parasite capacity of mounting.

Die statische und Impuls-Strom–Spannungscharakteristik von in Sperrichtung gepolten p^+–n–n^+-Dioden werden untersucht. Im Bereich des Lawinendurchbruchs wird in diesen Dioden ein negativer differentieller Widerstand beobachtet, der sich zu einem Stromkanal ausbildet. Der Einfluß der Störstellenkonzentration der Basis und der Basisweite auf den kritischen Stromwert, der das Auftreten eines Abschnittes mit negativem Widerstand in der Strom–Spannungs(I–U)-Charakteristik bestimmt, wird untersucht. Es wird gezeigt, daß der Prozeß der Ausbildung eines Kanals mit hoher Stromdichte sich im statischen Regime im Bereich der Mikroplasmen entwickelt. Beim Einschalten der Dioden in der Betriebsschaltung wird die Erzeugung von Relaxationsschwingungen beobachtet. Dies kann zu katastrophalen Fehlern der Dioden in Schaltungen mit hohen parasitären Schaltkapazitäten führen.

1. Introduction

One of the interesting phenomena in modern microelectronics is the negative differential resistance (NDR), as well as the current filament, both appearing under avalanche break-down conditions [1 to 6, 11 to 20]. Whereas NDR is promising for the fabrication of microwave generators of new types and other devices, the current filament phenomenon in p–n-junctions under avalanche breakdown essentially influences the reliability of avalanche transit time diodes (IMPATT) and p–i–n diodes [1].

The NDR phenomenon under avalanche breakdown is characterized by the critical density j_{cr}, which determines, according to [15, 18], the appearanche of the NDR section on the I–U characteristic. The values of j_{cr} obtained in some works [2 to 6], are essentially smaller than the calculated ones. Such a mismatch may be explained by the long duration of the pulses used by the authors to measure I–U characteristics. To obtain on isothermic I–U characteristic of the diode the pulse duration must be considerably less than the thermal relaxation time which is 10^{-6} s for silicon back mesa-structures with p–n junction diameter of the order of 60 μm [7], and the microplasmas switching time which is in the range of 10^{-6} to 10^{-7} s [8]. Note that although in [5] the pulse duration is less than 10^{-7} s, the values of j_{cr} are not given there.

[1]) Prospekt Nauki 117, 252028 Kiev, USSR.

The aim of the present paper is to study the mechanism of NDR forming in silicon p^+–n–n^+ diodes, and also to investigate static and pulse I–U characteristics and to find out the relation between the parameters characterising these I–U characteristics and the reason of catastrophic failures of the devices under voltages close to the breakdown voltage.

The results of the experimental study of the pulse I–U characteristics of diodes at pulse duration of 15 ns and generation of relaxation oscillations on the samples under investigation are given in this paper.

2. Description of the Experimental Techniques and Samples

The experiment was performed on silicon diodes of p^+–n–n^+ type with back mesa-structure with a diameter of (50 to 70) μm. The base length (epitaxial n-silicon film) for one of the samples under investigation was 1.8 μm with the impurity concentration 2.5×10^{14} cm^{-3} in the n-region, and for all others it changed in the limits from 4 to 5.2 μm with impurity concentrations from 6×10^{13} to 2.5×10^{14} cm^{-3} in the n-region.

Pulse I–U characteristics were measured by means of the method described in [9]. The use of the pulse accumulator on the cable delay lines in this method gives an opportunity to increase the sensitivity considerably and to measure currents of the order of some mA.

To observe the relaxation oscillations the circuit represented in Fig. 1 was used Such a circuit makes it possible to observe relaxation oscillations in the case of S-shaped characteristic of the diode. Back voltage U_s and parameters $D1$, $R1$, $C1$ determine the repetition frequency which may change from some kHz to some MHz ($C1 = 0.03$ to 0.3 μF, $R1 = 1$ to 10 kΩ).

3. Experimental Results and Discussion

The typical pulse I–U characteristics are represented in Fig. 2. In the I–U characteristic section describing the diode breakdown two specific regions may be singled out: a) the region of the abrupt jump of the reverse voltage (1 to 7 V) at the NDR section and b) the region with a weak change of the reverse voltage where the diode has a small positive differential resistance (Fig. 2, curve 1). The pulse I–U characteristics were characterized by the critical current density j_{cr} and by the value of the reverse voltage jump ΔU at section a (Fig. 2).

To exclude the possibility of surface break-down I–U characteristics of some diodes were measured before and after surface treatment (chemical etching to the depth of 1 μm, dipping in the organic silicon compound K T-2). No changes of I–U characteristics were observed.

The presence of NDR on the I–U characteristics may be explained by the influence of the space charge at high current densities. Although this effect was described theoretically [13, 15] as early as 1966, its experimental study is at the initial stage.

In the present work the influence of the impurity concentration in the base N_0 and the base width x_0 on the value of the critical current was evaluated for the first time.

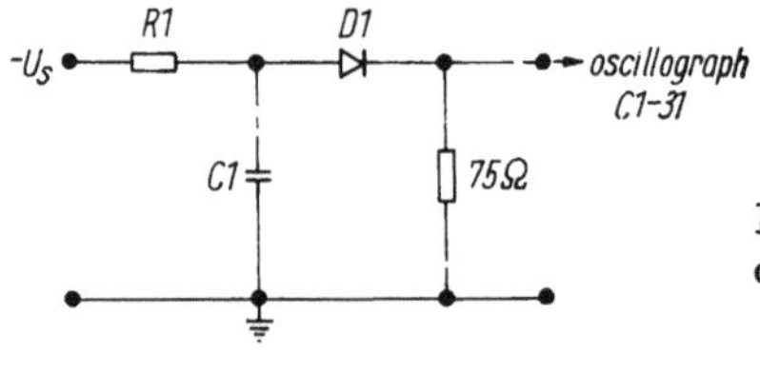

Fig. 1. The scheme for the investigation of the relaxation oscillation generation in p^+–n–n^+ diodes

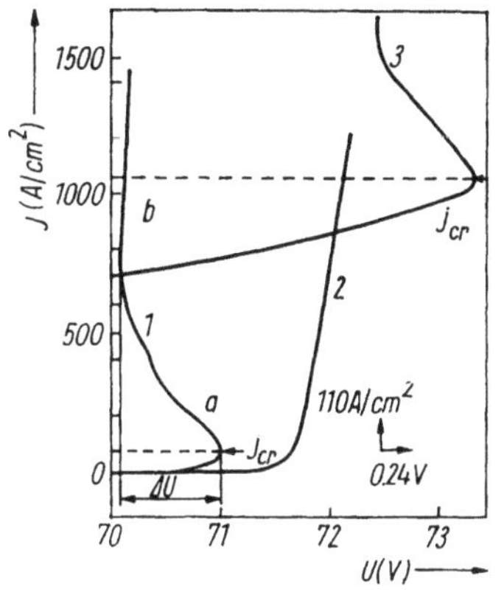

Fig. 2

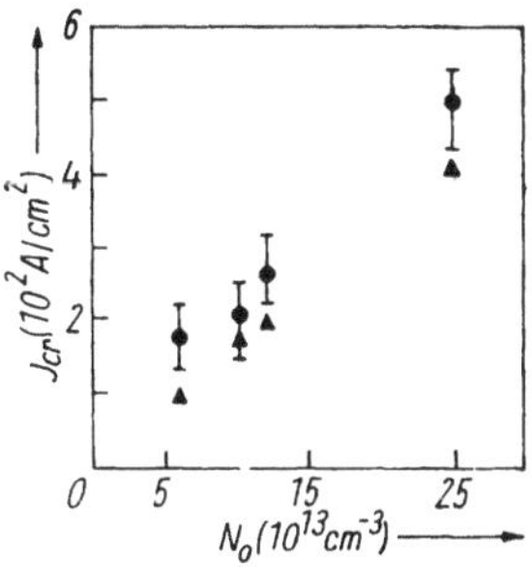

Fig. 3

Fig. 2. The pulse I–U characteristics of the diodes: (1) typical form of the pulse I–U characteristic; (2) pulse I–U characteristic in the case of small width of the base x_0; (3) pulse I–U characteristic in the case of considerably leakage current

Fig. 3. The dependence of the critical current j_{cr} on the impurity concentration in the base N_0: ● experimental, ▲ estimated values

It was established that the increase of the impurity concentration in the diode base leads to an increase of j_{cr}[2]) (Fig. 3). In this case the experimental values of j_{cr} agree with the estimated current density for a p–i–n diode obtaining according to the formula [12]

$$j_{cr} = N_0 q V_s ,$$

where q is the electron charge, V_s is the carrier drift velocity.

It is necessary to note that this formula is correct only for a Π-shaped field distribution in the depletion layer. A smoother change of the field at the edges leads to an increase of j_{cr} by some orders, as is shown in [15, 18]. That is why NDR is not observed for diodes with the base width of 1.8 μm in the current range under study (Fig. 2, curve 2), since in this case the tails of strongly doped areas, i.e. regions of smooth changes of the field, may occupy up to 0.5 μm from each side, i.e. a considerable part of the depletion layer.

The regularities, described above, are characteristic of NDR in p^+–n–n^+ diodes, caused by the space charge of the current carriers [15]. However, the changes of N_0 and x_0 in the studied diodes did not cause the changes of the NDR value predicted in [15], namely, the increase of NDR with the decrease of N_0 and decrease of NDR with the decrease of x_0. Besides we did not observe the smooth change of the voltage and NDR with increasing current as predicted by the theory. Thus, the change of the voltage in the real p^+–n–n^+ junctions occurred by a jump, the value of which was considerably smaller than expected change of the voltage and ranged usually from 2 to 10% of the break-down voltage (see Fig. 2).

As is known [14] the presence of NDR in the case of S-shaped characteristics leads to current filaments. The above-described deviations from the theory obtained for the case of a homogeneous distribution of the current over the p–n junction area are evidently caused by the formation of one or more longitudinal current "threads" [5].

[2]) In a number of cases the presence of a considerable leakage current also leads to an essential increase of j_{cr} (Fig. 2, curve 3).

Then range a in Fig. 2 corresponds to the current filament process, and the abrupt current increase in section b may be caused by the increase of the number of current "threads" and (or) by the increase of the area of each of them at constant current density and practically constant voltage.

Note that the static I–U characteristics measured in the constant current regime, were qualitively simliar to the pulse ones. The peculiarities of the static I–U characteristic are an essentially smaller critical current value compared to the puse I–U characteristic (from some tens to some hundreds of mA) and the absence of a correlation between impurity concentration in the base and critical current value. This may be explained by the fact that in the case of direct current a microplasma evolves and the current filament process develops only in the microplasma range. If one adopts the microplasma diameter to be approximately 1 μm, then to achieve the critical current density in the microplasma the value of the current through the diode must be of the order of 100 to 500 μA [16]. Since the experimental values of the critical current agree well with the latter, the assumption made above may be qualitatively correct.

When switching on the diode into the circuit shown in Fig. 1, the generation of relaxation oscillations was observed in the circuit at the breakdown of the diode (Fig. 4). Such oscillations appear if in the avalanche region of the static I–U characteristic sections with NDR occur [17]. The period of oscillations and the duration of the current pulses were dependent on the circuit parameters as well as on the sample parameters. The pulse duration for the given experiment and different samples was with in the limits $\tau_i = 10$ to 500 ns (Fig. 4a, b, c, d). It is evidently connected with the peculiarity of the current filament process in the diodes, and amplitude of the current pulses was proportional to the value of ΔU in the pulse I–U characteristic. For instance, for $\Delta U = 2$ V the amplitude of the current pulses achieved 50 mA, whereas the current supplied to the diode was 4 mA. One has to notice the following peculiarities of the relaxation oscillation regime in the diode: 1) the pulse current value is considerably larger than the mean current value; 2) small filament range where heat evolution takes place; 3) comparable values of the relaxation pulse duration and the heat relaxation time of the filament range. The peculiarities mentioned above lead to ta considerable pulse overheating of the filament range and are the reason

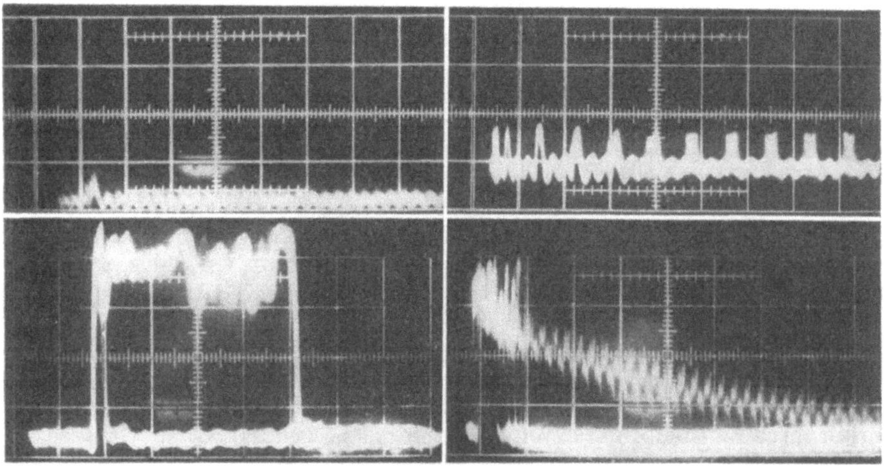

a b c d

Fig. 4. The relaxation oscillations in p^+–n–n^+ diode: current supply of the diodes is $I_S = 1$ mA; $x = 50$ ns/cm, $y = 13$ mA/cm (a, b, c); $x = 500$ ns/cm, $y = 13$ mA/cm (d) of the catastrophic diode failure. It is obvious that the overheating of the diode increases with the increase of energy supplied to the diode in the pulse time, and this energy is proportional to ΔU. Therefore, one may expect a decrease of the diode reliability with increasing ΔU. This phenomenon was observed in experiment.

4. Conclusions

It is experimentally established that the appearaence of NDR in silicon p^+–n–n^+ diodes is caused by the space charge of the current carriers. The presence of NDR leads to the process of current filament creation.

In the regime of constant voltage in the diodes relaxation oscillations (current impulses) were observed. As the circuit of the diode determines whether the given diode generates oscillations on the static curve possessing section with negative differential resistance, to avoid generation one has to use a working circuit with low stray capacitance.

References

[1] S. M. Sze and R. M. Ryder, Proc. IEEE **59**, 1140 (1971).
[2] R. A. Giblin, Electronics Letters **4**, 52 (1968).
[3] M. W. Muller, Appl. Phys. Letters **12**, 218 (1968).
[4] M. W. Muller, IEEE Trans. Electron. Devices **15**, 510 (1968).
[5] A. W. Carlson, Proc. IEEE **57**, 109 (1969).
[6] H. Egawa, IEEE Trans. Electron Devices **13**, 754 (1966).
[7] Yu. R. Nosov, Poluprovodnikovie impulsnie diodi, Izd. Sov. radio, Moscow 1965 (in Russian).
[8] V. K. Aladinskii, V. I. Doshin, A. S. Sushik, and A. M. Timerbulatov Radiotekhnika i Elektronika **18**, 342 (1973).
[9] V. I. Fineberg, Poluprov. Tekhnika i Mikroelektronika. **34**, 78 (1980).
[10] I. U. Zaizevskii, R. V. Konakova, Yu. A. Tkhorik, and V. I. Shakhovtsov, phys. stat. sol. (a) **53**, K153 (1979).
[11] W. J. Evans, D. L. Schafetter, R. L. Johnston, and R. L. Kev, J. appl. Phys., **42**, 799 (1971).
[12] A. S. Tager and V. M. Vald-Perlov, Lavinno-prolyotnie diodi i ikh primenenie v tekhnike SVCh, Izd. Sov. radio. Moscow 1968 (in Russian).
[13] B. Hofflinger, IEEE Trans. Electron Devices **13**, 151 (1966).
[14] J. B. Gunn, High Electric Field Effect in Semiconductors, in: Progress in Semiconductors, Vol. 2, Ed. A. F. Gibson, Heywood, London 1957.
[15] H. C. Bowers, IEEE Trans. Electron Devices **15**, 343 (1968).
[16] T. Ya. Puritis, I. E. Ozolinya, R. Ya. Krike, Ya. P. Murans, and Ya. K. Balodis, Microplasmennie yavleniya v kremnievom electronno-dirочhnom perekhode, v kn.: Fizika p-n perekhodov, Izd. Sinante, Riga 1966.
[17] A. W. Carlson, Proc. IEEE **56**, 1228 (1968).
[18] B. K. Petroff, Radiotekhnika i Elektronika **21**, 2365 (1976).
[19] A. V. Kardo-Sysoev, E. A. Panutin, and I. G. Tchashnikov, Fiz. Tekh. Poluprov. **10**, 1486 (1976).
[20] R. V. Konakova, V. I. Fineberg, and M. Yu. Filatov, Ukr. fiz. Zh. **25**, 1563 (1980).

(Received January 29, 1981)

phys. stat. sol. (a) 68, 45 (1981)

Subject classification: 10.2 and 11; 22.1.2

Institute of Nuclear Research, Warsaw[1])

The Effect of Dislocations on the Planar Dechanneling

II. Transparency Effect[2])

By

D. WIELUŃSKA, L. WIELUŃSKI, and A. TUROS

A computer simulation model is used to investigate the possibility of particle penetration through distorted atomic planes and subsequent rechanneling in the neighbouring channel. It is found that such processes significantly influence the dechanneling behaviour. The modification of the dechanneling cross-section is the strongest in the vicinity of the core of a dislocation. The amplitude and the phase dependence of this effect are also discussed.

Методом симуляции на эвм изучена возможность прохода частиц через исскаженную атомную плоскость и далнейшее их каналирование в соседнем канале. Установлено, что уменьшение эффективного сечения деканалирования самое сильное в непосредственной блискости дислокации. Обсуждаются величины амплитудных и фазовых эффектов.

1. Introduction

In the first part of this series, henceforth denoted Part I [1], we have described an analytical model for calculating dechanneling probabilities of particles in single crystals containing dislocations. This model based on the harmonic approximation of the interplanar potential reproduces well the principal features of the dechanneling cross-section such as energy or lattice spacings dependence. However, the absolute values of the calculated cross-sections are significantly higher than experimental results [2, 3] and analytical calculations based on the Quéré model [4]. Such a discrepancy can be attributed to the fact that the analytical model based on the continuum approximation of the interplanar potential disregards the detailed structure of atomic planes. On the other hand, this can be relevant to the dechanneling behaviour of particles approaching a distorted channel wall in the vicinity of a dislocation. In fact, if the amplitude of the particle oscillation is sufficiently large its trajectory is governed rather by binary collisions with individual atoms than by the continuum potential of the planes. Therefore the probability of a particle penetration through the atomic array of the curved channel wall and its possible rechanneling in the neighbouring planar channel should be taken into account. Such a process we termed transparency effect. This consideration will form a basic physical framework for a detailed analysis of particle trajectories in the vicinity of a dislocation.

2. Reflection and Penetration

The intention of these calculations was to investigate, with a computer, the conditions under which a particle can penetrate through the atomic plane and enter the neighbouring channel. One can expect that such an effect will be strongly influenced by the curvature of atomic planes.

[1]) ul. Hoża 69, 00-681 Warsaw, Poland.

[2]) Part I see phys. stat. sol. (a) 67, 413 (1981).

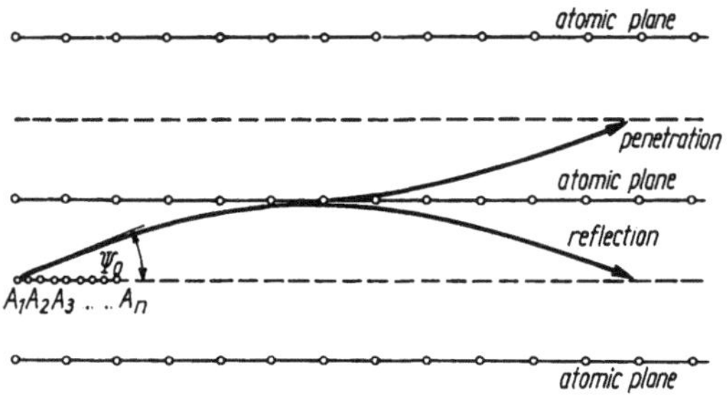

Fig. 1. Trajectories of reflected and penetrant particles

The computer program developed by Desalvo et al. [5] has been adapted for that purpose. The underlying assumptions are:

(i) the silicon unit cell is considered,

(ii) the interaction between an ion and lattice atoms is treated as a series of independent binary collisions,

(iii) the interatomic potential used was the Lindhard standard potential,

(iv) scattering angles are obtained in the momentum approximation,

(v) each trajectory is made up of a series of straight-line segments,

(vi) thermal vibrations of crystal atoms are simulated by a random choice of atomic displacements from a Gaussian distribution.

In order to avoid difficulties involved in the modelling of the distorted lattice the probability of particle penetration through a channel wall was calculated for a perfect crystal by following the trajectories of particles starting from the center of a planar channel with different total and transverse energies at starting points A_1, A_2, ..., A_n as shown in Fig. 1. The particle is considered as reflected if it arrives again in the center of the channel in which it was initially inserted. Penetration occurs when the atomic plane is crossed, i.e. the particle arrives in the center of the neighbouring channel. Following this procedure the probability distributions in terms of the transverse energy for both types of particles were determined. Such distributions have the form of narrow peaks of half- widths less than the critical value of the transverse energy for planar channeling. This is shown in Fig. 2.

The reflection and penetration probabilities have been calculated by integrating over both distributions and are given by

$$P_r = N_r/N_t \tag{1a}$$

and

$$P_p = N_p/N_t\,, \tag{1b}$$

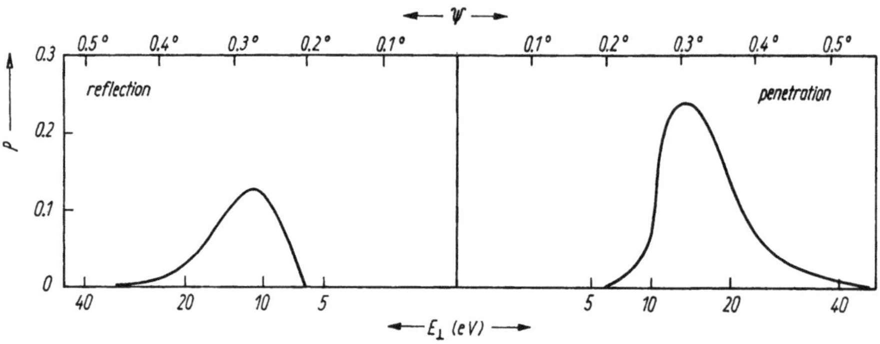

Fig. 2. Probability distributions for reflection and penetration of particles throughan atomic plane

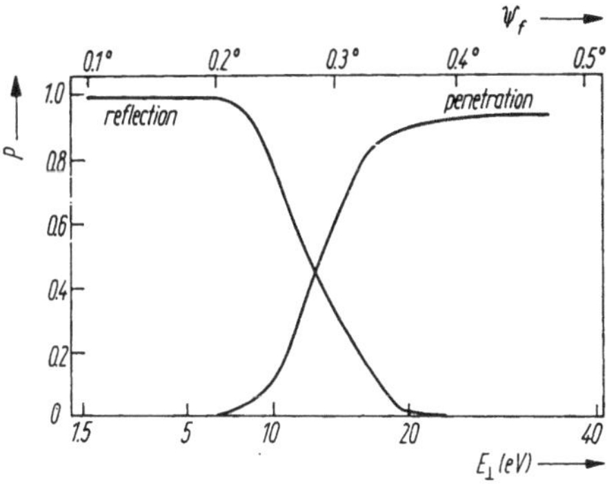

Fig. 3. Total probability of reflection and penetration as a function of the particle transverse energy

respectively, where N_r and N_p are total numbers of reflected and penetrating particles, respectively, N_t is the number of all the calculated trajectories.

Energy distributions and probability values are strongly influenced by the initial value of the transverse energy of a particle, $E_\perp$. Fig. 3 shows the variation of P_r and P_p with the starting value of $E_\perp$.

3. Dechanneling Probability

Once the reflection and penetration probabilities are known one can calculate, step by step, the trajectories of particles in the vicinity of a dislocation. Fig. 4 is a scheme of two of the trajectory types expected. The dashed line shows the trajectory of a particle which after passing the distorted region attained an amplitude larger than the critical one and became obviously dechanneled. The solid line represents the trajectory of a particle which has penetrated through the channel wall and, due to favourable circumstances, can oscillate with an amplitude less than the critical one. The calculation has been performed using the analytical model developed in Part I.

The particle was considered to be dechanneled if

$$A_f > A_c = d/2 - a_{TF} , \tag{2}$$

where A_f final amplitude of particles oscillation, A_c critical amplitude for channeling, d interplanar spacing, a_{TF} Thomas-Fermi screening distance.

In the following this criterion was replaced by a more sophisticated procedure.

(i) The elements of the particle trajectory $x(t)$ are selected for which

$$|x(t)| = A_c \quad \text{and} \quad \frac{d\,|x(t)|}{dt} > 0 .$$

In such points the particle approaches the channel wall with energy higher than the critical one, i.e. it can penetrate the atomic plane.

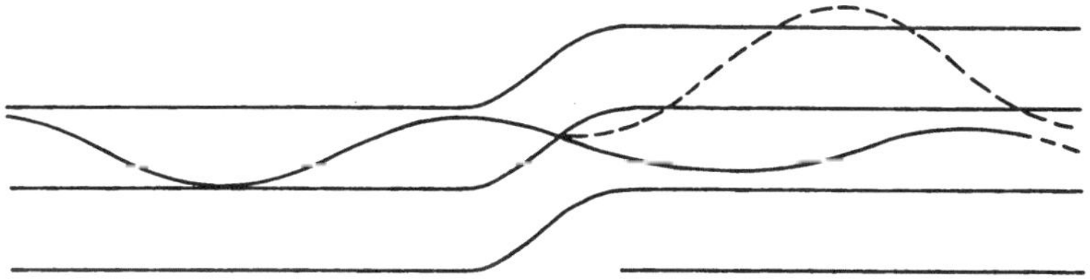

Fig. 4. Schematic representation of two types of particle trajectories in a planar channel in the vicinity of a dislocation: penetration (solid line) and reflection (dashed line)

(ii) From the points selected according to i) the impact point with the highest probability of the transparency effect can be found. It corresponds to the maximum value of the product

$$-\frac{\mathrm{d}x(t)}{\mathrm{d}t}\frac{\mathrm{d}u}{\mathrm{d}z},$$

where $\mathrm{d}u/\mathrm{d}z$ is the slope of the channel wall. The other points can be disregarded due to the following argument:

a) for small $\mathrm{d}x/\mathrm{d}t$, i.e. for small excess of the particle energy above the critical value the penetration probability is negligible,

b) penetration does not lead to rechanneling if it takes place in the region of a small channel slope,

c) penetration does not lead to rechanneling if $\mathrm{d}x/\mathrm{d}t$ and $\mathrm{d}u/\mathrm{d}z$ are of the same signs.

In the latter two cases particles which have penetrated the channel wall interact many times with other atomic planes. Since their energy is higher than the critical energy they are eventually dechanneled.

(iii) Once the impact point is selected the total energy of interaction $E = U(x) + E_\perp$ can be calculated. This allows a determination of the probabilities of penetration $P_\mathrm{p}(E)$ and reflextion $P_\mathrm{r}(E)$.

(iv) Starting from the selected impact point the penetration trajectory is calculated by means of (5) in Part I for a particle entering the neighbouring channel (see Fig. 4). The initial parameters of the trajectory are given by the assumption of energy and momentum conservation during the penetration through the channel wall.

(v) The final amplitudes A_f^r and A_f^p as well as the corresponding dechanneling probabilities Π_r and Π_p were calculated for both types of trajectories shown in Fig. 4. One gets

$$\Pi_\mathrm{r} = \begin{cases} 0 & \text{for} \quad A_\mathrm{f}^\mathrm{r} \leqq A_\mathrm{c}, \\ 1 & \text{for} \quad A_\mathrm{f}^\mathrm{r} > A_\mathrm{c} \end{cases} \tag{3a}$$

and

$$\Pi_\mathrm{p} = \begin{cases} 0 & \text{for} \quad A_\mathrm{f}^\mathrm{p} \leqq A_\mathrm{c}, \\ 1 & \text{for} \quad A_\mathrm{f}^\mathrm{p} > A_\mathrm{c}. \end{cases} \tag{3b}$$

The final dechanneling probability is given by

$$\Pi = P_\mathrm{p}\Pi_\mathrm{p} + P_\mathrm{r}\Pi_\mathrm{r}. \tag{4}$$

In order to get the dechanneling cross-section one has to perform integration over all the initial amplitudes, phases, and distances from the dislocation in a way similar to that described in Part I.

4. Results

Fig. 5 shows the results of calculations performed for 1 MeV α-particles in a {110} planar channel of Si single crystal distorted by the presence of an edge dislocation with magnitude of Burgers vector of 3.84 Å. The initial oscillation amplitude corresponds to $\eta_0 = A_0/A_\mathrm{c} = 0.5$. The dechanneling probability is plotted as a function of the distance from the dislocation r for different phases. One notes that the dechanneling probability $\Pi = 0$ for small r, the Π — value increases slowly as r increases until the level of about 10% is reached. Further increase of r produces a sharp rise of Π up to 100%. The dechanneling probability remains constant as long as $r < R_2$, where R_2

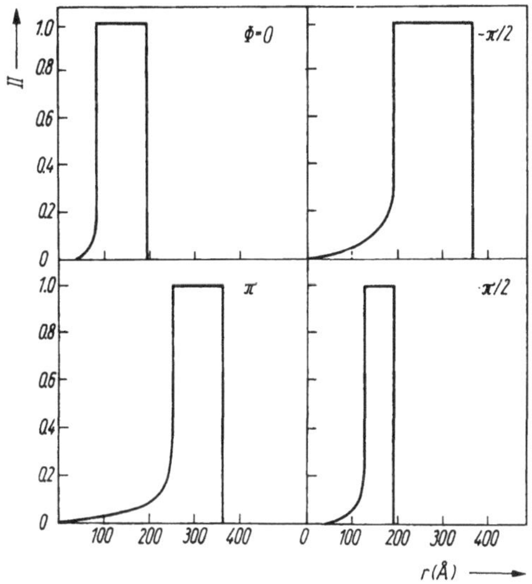

Fig. 5. Variation of the dechanneling probability with the distance from the dislocation for different phases

is the characteristic dechanneling radius as calculated using the analytical model described in Part I. For $r > R_2$ the dechanneling probability drops to zero.

Fig. 6 illustrates the typical phase dependence of the dechanneling probability shown in Fig. 5. There is no dechanneling for $r < R_3$ and $r > R_2$. In the region between R_1 and R_2 $\Pi = 100\%$ whereas between R_2 and R_3 the mean value of $\Pi = 10\%$.

It should be pointed out that the above results show that in the closest vicinity of a dislocation there exists a region in which the dechanneling probability is very low (less than 10%) or even zero. The reason for this behaviour is that penetration in the region of the highest curvature of atomic planes is easier and the transparency effect becomes the dominating process.

In the real channeling experiment all phases and distances from the dislocation are equally probable. Thus only the dechanneling probability averaged over all, available Φ and r can be of experimental significance. The result of the first step, i.e. averaging over Φ for different A_0, is shown in Fig. 7. The value of P_Φ is given by

$$P_\Phi = \frac{1}{2\pi} \int_{-\pi}^{\pi} P(\Phi)\, d\Phi \,. \tag{5}$$

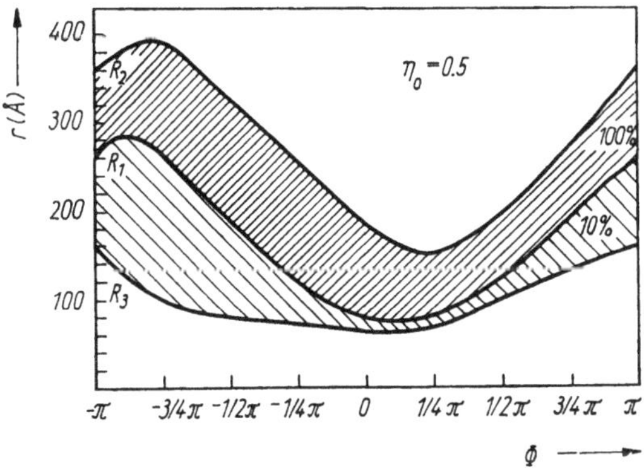

Fig. 6. Phase dependence of the dechanneling probability

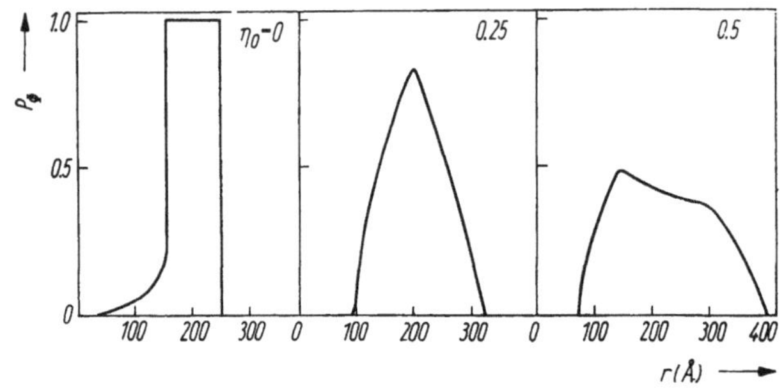

Fig. 7. Phase averaged dechanneling probability as a function of the distance from a dislocation for different η_0

For $A_0 = 0$ the particle moves in the center of the channel without performing oscillations. The important transparency occurs for $r < 160$ Å. For particles moving along oscillatory trajectories ($A_0 \neq 0$) the same effect is observed, however, for somewhat smaller r ($r < 100$ Å). The most important conclusion which can be drawn from Fig. 7 is that for sufficiently large A_0 the maximum dechanneling probability P_Φ is always lower than 100%. This is in contradiction with the model of an opaque dechanneling cylinder [4]. Averaging over r for a given initial amplitude A_0 gives the equivalent dechanneling diameter, λ, of an opaque dechanneling cylinder producing the same dechanneling as our analytical model with the transparency effect included.

Fig. 8 shows the variation of the equivalent dechanneling cylinder radius $\lambda/2$ with η_0.

Finally in order to obtain the value of λ which can be directly compared with the experiment one has to calculate $\bar{\lambda}$ which is given by

$$\bar{\lambda} = \frac{1}{A_c} \int_0^{A_c} \lambda\,(A)\,\mathrm{d}A\,. \tag{6}$$

Table 1 lists the available dechanneling cross-sections for planar channeling of 2 MeV α-particles in silicon single crystals containing edge dislocations. It can be seen that the agreement between the experimental data of Wieluński et al. [2] and the

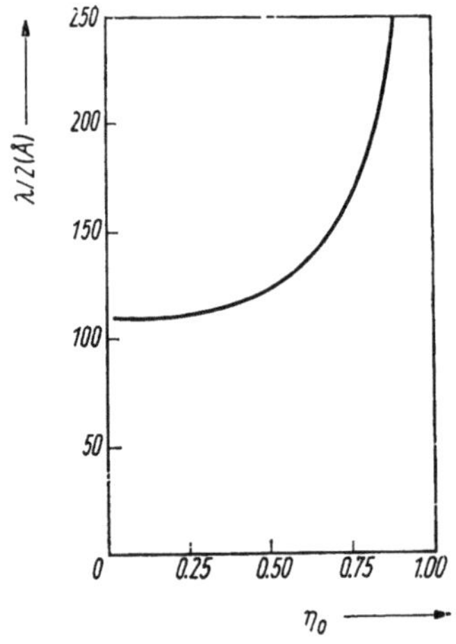

Fig. 8. Radius of the equivalent dechanneling cylinder as a function of η_0

Table 1

plane	dechanneling radius (Å)			
	experiment [2]	Quéré	harmonic approximation	this work
100	185 ± 30	121	333	145
110	168 ± 23	147	243	125
111	110 ± 23	121	250	

calculated ones based on the Quéré model and this work is satisfactory. Also the inadequacy of the harmonic oscillator model without transparency effect is evident.

5. Conclusions

The major results of the presented analysis of planar channeling in single crystals containing dislocations are as follows:

a) Comparison of calculated results with available experimental data gives good qualitative agreement indicating the validity of the model.

b) Oscillatory motion of channeled particles leads to a phase and amplitude dependence of the dechanneling cross-section.

c) The penetration of channeled ions through the curved planes gives rise to a group of ions which are rechanneled in the neighbouring channel. We have termed this transparency effect.

d) The transparency effect is a dominant factor governing the dechanneling behaviour of ions passing the dislocation at a distances less than $\lambda_Q/2$. λ_Q is the diameter of the dechanneling cylinder in the Quéré model [4]. It means that the dechanneling cylinder is in fact partially transparent and an important part of the dechanneled flux is due to dechanneling at distances greater than $\lambda_Q/2$. Nevertheless the mean value of the dechanneling cross-section coincides well with $\lambda_Q/2$.

References

[1] D. Wieluńska, L. Wieluński, and A. Turos, phys. stat. sol. (a) **67**, 413 (1981).

[2] L. Wieluński, D. Wieluńska, G. Della Mea, and A. Turos, Nuclear Instrum. and Methods **168**, 323 (1978).

[3] S. T. Picraux, D. M. Follstaedt, P. Baeri, S. U. Campisano, G. Foti, and E. Rimini, Proc. Internat. Conf. Ion Beam Modification of Materials, Budapest 1978 (p. 1863).

[4] Y. Quéré, J. nuclear Mater. **53**, 262 (1974).

[5] A. Desalvo, R. Rosa, and F. Zignani, CNR LAMEL Bologna, Relazione Interna No. 7, April 1971.

(Received June 1, 1981)

phys. stat. sol. (a) 68, 53 (1981)

Subject classification: 10.2; 22.1.2

Institute of Solid State Physics, Academy of Sciences of the USSR, Chernogolovka[1])

On the Real Structure of Monocrystalline Silicon near Dislocation Slip Planes

By

I. E. BONDARENKO, V. G. EREMENKO, B. YA. FARBER, V. I. NIKITENKO, and E. B. YAKIMOV

The properties of regions swept by a moving dislocation in silicon crystals are experimentally studied. The peculiarities of forming traces of various types behind dislocations observed by electron microscopy and chemical etching are investigated. These traces are found to arise as result of point defect redistribution by moving dislocations and change in their structural state in a volume swept by a dislocation. It is shown that the processes mentioned enable the slip plane to assume properties of a specific two-dimensional defect responsible for the appearance of the new mechanism of a charge carrier mobility anisotropy in the plastically deformed crystals.

Экспериментально изучены свойства областей кристалла Si, обметенных движущимися дислокациями. Выявлены особенности возникновения различного типа следов за дислокациями, наблюдаемых с помощью электронной микроскопии и химического травления. Установлено, что они появляются в результате перераспределения точечных дефектов движущимися дислокациями и изменения их структурного состояния в пройденном дислокациями объёме. Показано, что вследствие протекания указанных процессов плоскость скольжения дислокации приобретает свойства специфического квазидвумерного дефекта, который обусловливает появление нового механизма анизотропии подвижности носителей заряда в пластически деформированных кристаллах.

1. Introduction

It is known that interaction of dislocations with point defects of the crystal lattice plays a significant role in the formation of physical properties of plastically deformed crystals. As a result of such interaction an essential change in the state of impurities and intrinsic point defects takes place in the dislocation core and in the adjacent volume, various atmospheres of point defects are formed (such as Cottrell atmosphere, Snoek atmosphere, Suzuki segregation, etc.) [1].

Up to now while analysing numerous effects [2 to 6] of the change in mechanical, electrical, and other properties of crystals, which are due to dislocation–point defect interaction, only the above-mentioned changes in the state of the dislocation core and its atmosphere were taken into account.

However, the interaction of dislocations with point defects must also lead to a change in the state and properties of the crystal volume adjacent to the dislocation slip plane. Indeed the investigation of dislocation motion in real crystals has shown that in many cases an ideal contact between gliding layers is not kept. The change in the properties of the slip plane can be revealed experimentally by observing different types of traces with the help of chemical etching [7, 8] or transmission electron microscopy [9]. In most cases these traces were explained by the formation of clusters of intrinsic point defects during the motion of a dislocation with jogs [7], prismatic

[1]) 142432 Chernogolovka, Moscow district, USSR.

dislocation loops, which were left in the slip plane [10], or incomplete shear owing to oxide film on the crystal surface [9]. But recently microscopic segregations were observed along the slip plane trace on the crystal surface [11] as well as the effect of anomalous anisotropy of the charge carrier mobility [12], which cannot be explained on the basis of the above-mentioned mechanisms without taking into account the change of the point defect state in the volume adjacent to the slip plane.

The present paper is devoted to the investigation of the effects which are determined by the change of the point defect state in a crystal due to dislocation motion. The changes of both surface and bulk crystal properties as well as the dependence of the observed effects on impurity content and thermal treatment have been investigated. The nature of the observed effects is analysed.

2. Experimental Procedure

The starting materials were dislocation-free Si crystals grown by the Czochralski and floating zone methods, and doped with phosphorus up to 4×10^{19} to 1×10^{21} m^{-3} and with B, Ga, Al, Sb in the concentration range from 10^{21} m^{-3} up to their solubility limit in Si.

These ingots were cut into the rectangular samples of two kinds. The first ones with the edge orientation ⟨110⟩, ⟨110⟩, ⟨001⟩ and ⟨110⟩, ⟨112⟩, ⟨111⟩ and dimensions $30 \times 3.5 \times 3.5$ mm^3 and $30 \times 3.5 \times 2$ mm^3, respectively, were intended for the four-point bending around the ⟨110⟩ and ⟨112⟩ axes. Plastic deformation was performed at temperatures from 823 to 1023 K and stresses from 10 to 40 MN/m^2.

The samples of the second kind were thin slices $3 \times 3 \times 0.2$ mm^3 in size with various orientations of the wide face. They were deformed by indentation with a Vickers type indenter under the load from 5×10^{-2} to 10^{-1} N and at temperatures from 298 to 973 K according to the experimental procedure described previously in [4, 5, 13].

The standard Sirtl solution was used for dislocation etching and 1 HF : 7 HNO_3 mixture for polishing and preparing foils for the electron microscopy.

The dislocation structure of deformed crystals was examined with the aid of a JEM-150 electron microscope at 150 kV. The state of crystal surface was studied in an optical microscope Neophot-2 and with the help of microinterferometer with He–Ne laser. Electrical contacts were welded using gold wire [4].

3. Experimental Results and Discussion

3.1 Investigation of traces left by moving dislocations

Electron microscopic investigation of the dislocation rosette structure in silicon single crystals doped with Sb, B, Al, and Ga have revealed the traces behind 60° dislocations crossing a deformed surface. The contrast features depended on the impurity concentration. As a rule, the traces in lightly doped crystals were revealed as black-white unbroken lines (Fig. 1), similar to the glide traces in metals [9]. In the latter case they were explained by the existence of an oxide film which protects the surface against the slip penetration. In principle, the traces in lightly doped crystals might be of the same nature, all the more so they disappeared under electron beam irradiation in a time from a few seconds to 1 min depending on the beam intensity. However, the oxide film is unlikely to inhibit the shear at the deformation temperature ($T = 500$ to 800 °C) and at the same time a weak heating under the beam is enough to form the step.

Fig. 1. The traces behind 60° dislocations on the Si (111) surface revealed by transmission electron microscopy

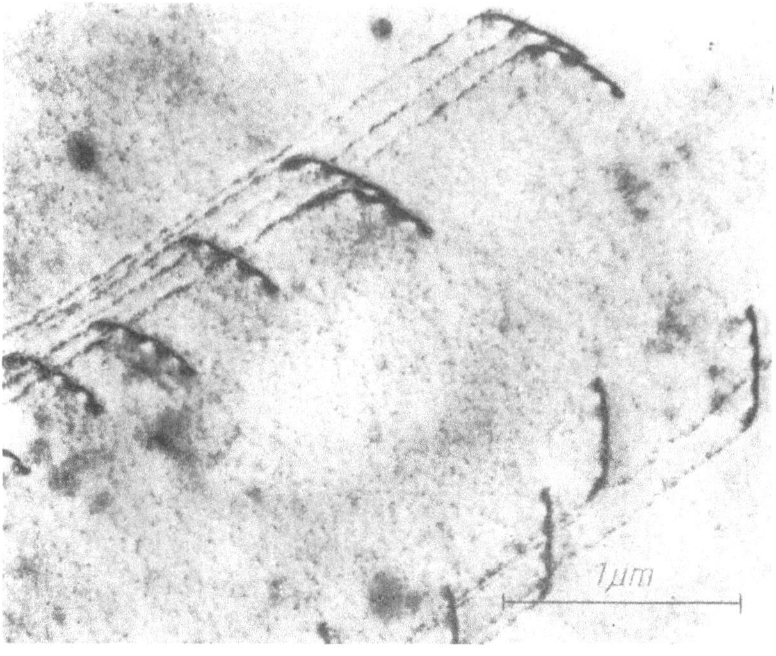

Fig. 2. Electron microphotograph of the traces on the silicon (111) surface doped with Sb ($N_{Sb} \approx \approx 6 \times 10^{24}$ m^{-3}) deformed by indentation at 600 °C and load 0.01 N

The traces behind dislocations are of a quite different appearance in strongly doped crystals, with impurity concentration near the solubility limit. In such crystals rows of discrete dotted traces were left on the deformed crystal surface by moving dislocation half-loops (Fig. 2). Unlike lightly doped crystals these traces were stable enough and did not change their size and form even at the highest (850 °C) annealing temperature "in situ". The features of the electron microscope image of traces did not depend neither on the deformation temperature T_D within the range from 550 to 800 °C, nor on the orientation of the crystal surface deformed. Annealing of the low temperature indentations ($T_D \approx 250$ °C) in an electron microscope at 850 °C permitted us to conclude that the traces start to form immediately at the moment of intersection of the dislocation with the surface.

The analysis of the contrast gave evidences that these traces are neither little dislocation loops nor short parts of screw dislocations. As it is seen, for example, in Fig. 1 the image of the traces behind dislocations considerably differs from that of the dislocation A lying just under a surface. The formation of traces is connected with impurity atoms and intrinsic point defects, which are collected by dislocations and taken out to the surface owing to the pipe diffusion. In strongly doped crystals their concentration near the surface can exceed a critical value required for the nucleation of a certain phase. In lightly doped crystals the traces have most probably the same nature, and the low stability of traces in such crystals may arise from a low concentration of impurities in the segregates on the surface. As a result these segregates are not stable and vanish due to the diffusion stimulated by the electron beam.

It should be pointed out that in some lightly doped crystals the stable traces were also observed. This means that microsegregates on the surface contain not only the main doping impurities but also electrically inactive ones. A sink of point defects on the surface owing to the pipe diffusion gives rise to the formation of a steady-state

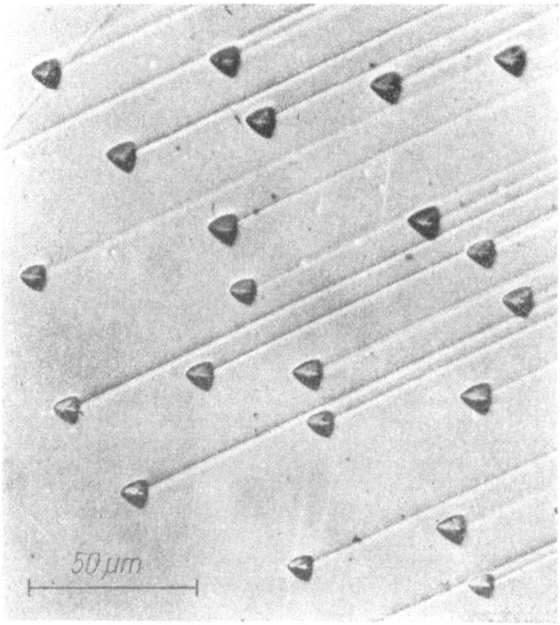

Fig. 3. Etch pattern of deformed n-Si crystal. Etch pits of dislocations and traces along slip planes are observed. Surface (111)

impurity atmosphere on the dislocation. In closed dislocation loops the formation of a steady-state atmosphere is likely embarrassed and this leads to a distinction in the mobility between them and dislocation half-loops crossing the surface [14].

The traces observed by electron microscopy on the crystal surface are formed only behind high speed dislocations under conditions of indentation. At low dislocation velocities ($v \leqq 10^{-5}$ m/s) the traces were not observed. In this case the pipe diffusion is likely to become inefficient, for example, because of a decreasing concentration of intrinsic point defects generated by moving dislocations due to the formation of impurity complexes in the dislocation core [5, 6]. However, in such conditions specific traces are revealed by etching behind the dislocations (Fig. 3). Similar traces were observed in all crystals investigated, independent of doping level, impurity type, and growing conditions. Step by step polishing and next etching showed that the traces are preserved up to the deepest part of the dislocation loops. The investigations carried out revealed a number of regularities in the formation of such traces. As a rule, the most intensive, continuous traces are observed in lightly doped Czochralski grown crystals. In floating-zone crystals which are purer with respect to electrically non-active impurities, as well as in strongly doped crystals the traces are less intensive and often discontinuous. The traces behind individual dislocations, generated from a scratch under four-point bending of the crystal, usually arise only at a some distance from the source. In some cases the double traces are revealed behind the dislocations having moved far away from the scratch.

The trace morphology was investigated by scanning electron microscopy and micro-interferometry. These traces behind the dislocations turned out to be extended ridges (Fig. 4).

For the given crystal and under the concrete etching conditions the ridge height (the trace intensity) is the same (within the experimental error) for all individual dislocations and increases with the number of dislocations (N) moving in the neighbouring glide planes (Fig. 4), becoming saturated at $N \approx 5$ to 10.

The observed etching effects evidence that the chemical potential in a crystal region adjacent to the acting slip plane changes significantly and this gives rise to the distinc-

a  b

Fig. 4. Microinterferometric image of the trace shape on the (111) surface of lightly doped n-type silicon. $\lambda = 0.63$ µm. a) Trace behind single dislocation, b) trace after moving of four dislocations tion in the etching rates of the region and of the host material. As it was noted the character of traces is somewhat different for crystals with different doping levels.

In addition, the dislocation velocity during deformation has a decisive influence on the formation of the etching traces. So, for example, during deformation at 700 °C the traces are observed only behind dislocations moving with the velocity $v \leq 10^{-5}$ m/s. The traces did not appear at much higher velocities of the dislocation.

These facts show that the formation of etching traces is connected with a change in the impurity content of the region, adjacent to the part of the slip plane swept by the dislocation because the dislocation velocity (at a constant temperature) determines the effective radius of dislocation interaction with the impurity atoms. The character of etching traces also depends on the state of intrinsic point defects in this part of the crystal. It results from a slight distinction in the trace intensities observed, as a rule, on extended and compressed sides of the sample under four-point bending as well as from a difference in traces behind the screw and the 60° dislocations and from a change in the nature and intensity of the traces in the course successive polishing of the crystal.

The presented experimental results indicate that in the region of the crystal adjacent to the dislocation slip plane complexes consisting of impurity atoms and intrinsic point defects are formed. These complexes are stable enough, because the etching traces disappear only after one hour annealing or deformation at high temperatures (above 900 °C). Sometimes, in the depth of the crystal a fine structure of traces is observed. This structure may result from the formation of large impurity clusters. But the transmission electron microscope investigations have not revealed any segregations, small dislocation loops, or stacking faults. Only a weak diffusion contrast was observed which quickly disappeared under electron beam irradiation.

It should be noted that etching traces similar to the discussed ones were observed in other materials, too [15]. We observed these traces behind the dislocations in n-type Ge crystals which were deformed by four-point bending at 420 °C and etched in a H_2O_2:HF:H_2O = 1:1:4 mixture.

3.2 The effect of anomalous anisotropy of electron mobility

The change of the state of the crystal volume swept by a dislocation also gives rise to a significant change in its electrical properties. The effect of anomalous anisotropy of the mobility of electrons moving along and across the parallel slip planes is the most striking illustration of such a change [12].

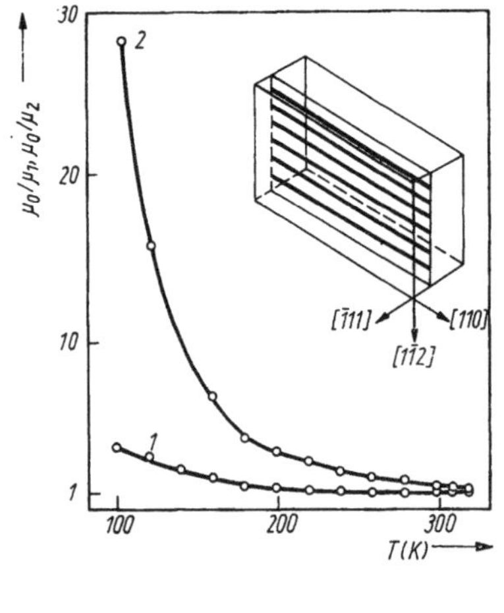

Fig. 5. The temperature dependence of anisotropy of the electron mobility in n-Si crystal with low dislocation density ($\approx 4 \times 10^9$ m^{-2}). μ_0 the mobility at current direction along dislocation, μ_1 current lines are parallel to slip plane but across dislocations, μ_2 current lines are perpendicular to slip plane and dislocation. (1) μ_0/μ_1, (2) μ_0/μ_2

It is known [16] that charge carrier scattering on dislocations is anisotropic. The experimentally determined difference between the values of charge carrier mobility in directions parallel and perpendicular to dislocation lines is usually taken as a measure of this anisotropy. The temperature dependence of the anisotropy coefficients K_1 and K_2 which are the ratios of mobility values in directions parallel to dislocation lines (μ_0) and perpendicular to them, but parallel or perpendicular to the slip plane (μ_1 and μ_2, respectively) is shown in Fig. 5. As is seen the anisotropy coefficients, i.e., the electron mobilities along and across the slip plane, differ approximately by an order of magnitude. The value of the carrier mobility in the case when the current lines were directed along the dislocation lines (μ_0) or perpendicular to them (μ_1), but parallel to the slip plane were practically the same as in the starting crystal. This effect was investigated on plastically deformed samples at temperatures 600 to 650 °C up to the dislocation density $N_D \leqq 10^{10}$ m^{-2}. The dislocation effect on the mobility in this case ($T \geqq 85$ K) is neglected because of the low dislocation density. Moreover, the anisotropy coefficient in the crystal with equal density of dislocations was observed to be the greater the farther away from neutral plane the sample was cut, i.e., the more slip planes swept by the dislocation it contained. It is important to point out that the magnitude of this effect is determined by the impurity content of the starting crystal. The anisotropy coefficient in low-oxygen floating-zone crystals was in accordance with values predicted by the theory [16].

The described effect cannot be explained by any mechanisms of scattering by the dislocations without taking into account a disturbance of the crystal volume adjacent to the slip plane. In the conditions of experiment the slip plane is assumed to play a new role of a quasi-two-dimensional defect consisting of the changed point defects and their complexes. In our opinion, the most probable mechanism of the formation of such defects consists in the following. While gliding through the crystal the dislocation gathers impurity atoms. As a result a Cottrell atmosphere is formed and the absorption of impurity atoms at stacking fault ribbons may take place as well, provided the dislocations move in a dissociated state. So the local concentration of impurities and intrinsic point defects increases and there appear conditions for reactions of complex formation. The impurity complexes formed cannot move with a dislocation because of their smaller diffusion coefficient than that of impurity atoms and are left in the slip plane behind the dislcocation. A few types of centres are formed simultaneously, which possess different electrical properties and thermal stability. The charged centres stipulating the anomalous anisotropy of the electron mobility dissociate or transform into more stable ones during thermal treatment at 700 °C as well as during plastic deformation at temperatures $T_D \geqq 700$ °C. But at the same time there are more stable centres in the slip plane at the temperatures of deformation or annealing up to 900 °C. At higher temperatures all traces disappear, i.e. there takes place a thermal dissolution of all complexes.

4. Summary

The presented data provide the convincing evidence that the properties of the crystal region through which the dislocations passed, differ essentially from the starting ones. In addition to the dislocations introduced qualitatively new two-dimensional defects adjacent to the slip plane arise in the plastically deformed crystal. These regions exert an essential influence on electrical and chemical properties of the crystal and its impurity distribution. The formation of these defects is determined by the redistribution and the change of the point defect state stimulated by a moving dislocation. Therefore, both should be taken into account in studying the properties of plastically deformed semiconductor crystals and for understanding the effects of dislocations on semiconductor devices.

References

[1] J. Hirth and J. Lothe, Theory of Dislocations, McGraw-Hill Publ. Co., New York 1968.
[2] J. Friedel, Dislocations, Pergamon Press, London 1964.
[3] K. Ahlborn, J. Physique **40**, C6-185 (1979).
[4] V. G. Eremenko, V. I. Nikitenko, and E. B. Yakimov, Zh. eksper. teor. Fiz. **73**, 1129 (1977).
[5] V. G. Eremenko, V. I. Nikitenko, E. B. Yakimov, and N. A. Yarykin, Fiz. Tekh. Poluprov. **12**, 273 (1978).
[6] I. E. Bondarenko, V. G. Eremenko, V. I. Nikitenko, and E. B. Yakimov, phys. stat. sol. (a) **60**, 341 (1980).
[7] W. C. Dash, J. appl. Phys. **29**, 705 (1958).
[8] V. N. Erofeev, V. I. Nikitenko, V. I. Polovinkina, and E. V. Suvorov, Kristallografiya **16**, 190 (1971).
[9] P. B. Hirsch, A. Howie, R. B. Nicholson, D. W. Pashley, and M. J. Whelan, Electron Microscopy of Thin Crystals, Butterworths, London 1965.
[10] W. G. Johnston and J. J. Gilman, J. appl. Phys. **31**, 632 (1960).
[11] I. E. Bondarenko, V. G. Eremenko, and V. I. Nikitenko, Dokl. Akad. Nauk SSSR **229**, 1087 (1976).
[12] V. G. Eremenko, V. I. Nikitenko, and E. B. Yakimov, Zh. eksper. teor. Fiz., Pisma **26**, 72 (1977).
[13] V. G. Eremenko and V. I. Nikitenko, phys. stat. sol. (a) **14**, 317 (1972).
[14] L. S. Milevskii and I. L. Smolskii, Fiz. tverd. Tela **16**, 1028 (1974).
[15] L. M. Dedukh, Yu. P. Kabanov, and V. I. Nikitenko, Fiz. tverd. Tela **18**, 2352 (1976).
[16] W. T. Read, Phil. Mag. **46**, 111 (1955).

(Received July 9, 1981)

phys. stat. sol. (a) 68, 61 (1981)

Subject classification: 14.3.4; 22.1.1

S. V. Starodubtsev Physico-Technical Institute, Academy of Sciences of the Uzbekian SSR (a) and V. I. Lenin State University, Tashkent[1]) *(b)*

On the Theory of Carrier Exclusion in "Pure" Intrinsic Semiconductors

By

D. A. Aronov (a), R. Mamatkulov (b), and V. V. Rubinov (a)

Theoretical results are given concerning exclusion in intrinsic semiconductors with non-injecting contacts under the conditions where the intrinsic concentration of charge carriers exceeds by far the impurity carrier concentration. Unlike other works, this paper considers exclusion not at low but at high level of free carrier depletion in the basic semiconductors. The concentration of carriers of both types near contacts as functions of current, distribution of carrier concentrations and field in a crystal, current and voltage dependences of the depletion region width, current–voltage characteristic, and specific values of voltage in its various regions, and their dependences on the intensity of "intrinsic" light and temperature are studied. Good agreement is shown between the theoretical and experimental results for "high-purity" intrinsic germanium.

В работе изложены результаты теоретического изучения эксклюзии в полупроводниках с собственной проводимостью и неинжектирующими контактами в условиях, когда собственная концентрация носителей тока намного превышает примесную. В отличие от других работ по эксклюзии рассмотрение проведено не при низком, а при высоком уровне обеднения базового полупроводника свободными носителями. Найдены и исследованы приконтактные концентрации обоих типов носителей как функции тока, распределения концентрации носителей и поля в кристалле, зависимость ширины области обеднения от тока или напряжения, вольтамперная характеристика и характерные значения напряжения на различных её участках и зависимость их от интенсивности ,,собственного" света и температуры. Показано хорошее согласие результатов теории и эксперимента в случае ,,высокочистого" собственного германия.

1. Introduction

Gibson [1] in 1954 was the first to observe in germanium the phenomenon of carrier exclusion in n^+–n or p^+–p (h–l) junctions with low-doped (l-) region conductance approximating the intrinsic value ($n_0 \approx p_0 \approx n_i$). Some laws related to this effect for germanium were experimentally and to some extent theoretically studied by Arthur et al. [2], Bray [3], Low [4], Many et al. [5], Burgess [6], as well as by Read [7] in 1955 and 1956. Later works by Lade and Jordan [8], Prokhorov et al. [9], and Kassur [10] also dealt with exclusion in germanium. In the above papers, however, the phenomenon of exclusion was considered in a rather simplified aspect and mainly qualitatively; it was analysed quantitatively only at low level of mobile carrier depletion in semiconductors, i.e. with only a slight deviation from thermodynamical equilibrium. At the same time, experimental data obtained by the same authors show that in most cases, and especially in germanium with resistivity $\varrho_0 \approx 40$ to $50\ \Omega$ cm at $T = 300$ K, the exclusion level was very high (80 to 85% majority carriers were removed), and that the effect should be quantitatively described as a considerable disturbance of thermal equilibrium. The only paper reporting on such an attempt is that by Vikulin

---

[1]) 700060 Tashkent, USSR.

et al. [11]. However, [11] considers only some dark characteristics of exclusion such as the distribution of carrier concentration, depletion layer length, its resistance and voltage drop across it, dark current–voltage (I–U) characteristic in the first ohmic and sublinear regions, as well as carrier concentrations near contacts as functions of current. The authors of [11] did not consider a number of problems such as the role of residual impurity concentration N_{im} in exclusion in "pure" semiconductors, field distribution features in a crystal, the influence of light and temperature on exclusion and its characteristics, determination of specific voltage values for the whole I–U characteristic and, not only for its sublinear region, the possibility of observing and features of strong exclusion in compensated materials with conductance approximating its intrinsic value or exceeding it by far, etc.

Since a considerable carrier depletion in the bulk of semiconductors improves the characteristics at high temperatures and makes the material a "quasi-impurity" one, a study more comprehensive than in [11] of exclusion in "pure" crystals is of interest. This paper contains more correct theoretical results related not only to dark but also to illumination characteristics of h–l (for definiteness, n^+–n) structures at great deviations in the concentration of non-equilibrium carriers from its thermodynamically equilibrium value ($n, p \ll n_i$, $|\Delta p| \approx |\Delta n| > N_{\text{im}}$) for materials like "high-purity" germanium ($N_{\text{im}} \approx 10^9$ to $10^{12}\ \text{cm}^{-3}$) whose intrinsic conductance exceeds by far the impurity value ($n_0 \approx p_0 \approx n_i \gg N_{\text{im}}$). The case of compensated semiconductors such as silicon with cadmium, zinc, or gold was considered by us in [12 to 23].

2. Calculation of the Distribution of Carrier Concentrations in the Base and Their Values at an Exclusion Contact

To solve the problem suppose that "intrinsic" light being uniformly absorbed in the bulk is incident on the structure in the direction perpendicular to the current j (axis x), and that exclusion causes carrier recombination to be less essential than carrier generation. In this case the condition $D_a\, \mathrm{d}^2p/\mathrm{d}x^2 \ll g + (p_0/\tau)$ is satisfied because the exclusion effect results from the action of field. Taking into account that electron and hole lifetimes in intrinsic material are both equal to [24]

$$\tau = \tau_0 \frac{n_0 + p_0}{n_0 + p_0 + \Delta p} = \tau_0 \frac{2n_i}{n_i + p}$$

we write the bipolar continuity equation for holes in the form [25]

$$\frac{bj}{q} \frac{N_D}{[(b+1)\,p + bN_D]^2} \frac{\mathrm{d}p}{\mathrm{d}x} - \frac{p + p_0(1 + 2G)}{2\tau_0} = 0\,. \tag{1}$$

Here and above, D_a is the ambipolar diffusion factor, g the carrier photogeneration rate in the bulk, p_0 the equilibrium hole concentration in the base, $G \equiv g\tau_0/p_0$ and τ_0 are, respectively, the photocarrier concentration reduced to the equilibrium value and carrier lifetime outside the depletion layer ($0 \leq x \leq x_1$). The quasi-neutrality condition used in deriving (1) is $n \approx p + N_D$ where N_D is the concentration of residual shallow donors in the n-base.

Integrating (1) we obtain

$$x = \frac{2bj\tau_0 N_D}{(b+1)\,qn_i\,(1+2G)} \left[\frac{1}{(b+1)\,p(0) + bN_D} - \frac{1}{(b+1)\,p(x) + bN_D} + \right.$$
$$\left. + \frac{1}{n_i(1+2G)} \ln \left\{\frac{(b+1)\,p(0) + bN_D}{(b+1)\,p(x) + bN_D} \frac{p(x) + n_i(1+2G)}{p(0) + n_i(1+2G)}\right\}\right]. \tag{2}$$

Taking into account that $p = n_i(1 + G)$ outside the depletion layer and that (as it is shown below) $p(0) \ll \big(b/(b + 1)\big) N_D \ll n_i$, we find that the depletion region length

$$x_1 \approx \frac{2j\tau_0}{(b + 1)\, qn_i\, (1 + 2G)}. \tag{3}$$

To calculate the values of $p(0)$ and $n(0) \approx p(0) + N_D$ we take into account that the hole and electron currents near the contact in the n^+- and n-regions are determined by diffusion and drift, respectively. Assuming that the n^+–n junction region is thin so that the generation–recombination current in the space charge could be neglected, we have the following equation for the hole current into the base at the point $x = 0$ [26]:

$$j_p(-0) \approx -qD_p \left.\frac{dp}{dx}\right|_{x=-0} = -\frac{qD_p}{l_p^*}\,[p(-0) - p_0^+]\,, \tag{4}$$

where p_0^+ is the equilibrium hole concentration in the n^+-region,

$$l_p^* = l_p^+ \frac{\coth \tilde{w}^+ + s_p^* \exp \psi_M^+}{1 + s_p^* \exp \psi_M^+ \coth \tilde{w}^+}$$

is the effective hole diffusion length in the n^+-region, $l_p^+ \equiv \sqrt{D_p\tau_p^+}$; $\tilde{w}^+ \equiv w^+/l_p^+$ the dimensionless length of the n^+-region, ψ_M^+ the external voltage drop across the n^+-region metal contact divided by kT/q, and s_p^* is the effective contact surface recombination rate [26]. Using Boltzmann statistics for carriers of both types on both sides of the n^+–n junction barrier, viz.

$$\left.\begin{aligned} p(-0) &= p(+0) \exp\left[-\frac{q}{kT}\,(U_{n\text{-}n^+} + U_K)\right], & p_0^+ &= p_0 \exp\left(-\frac{qU_K}{kT}\right), \\ n(-0) &\approx n_0^+ \approx n(+0) \exp\left[\frac{q}{kT}\,(U_{n\text{-}n^+} + U_K)\right], & n_0^+ &\approx n_0 \exp\left(\frac{qU_K}{kT}\right), \end{aligned}\right\} \tag{5}$$

we represent (4) in the form

$$j_p(0) = -j_s^{n^+} \left[\frac{p(0)}{p_0} \exp\left(-\frac{qU_{n\text{-}n^+}}{kT}\right) - 1\right], \tag{6}$$

where $U_{n\text{-}n^+}$ and U_K are the voltage drop across the n^+–n junction and the contact potential difference, and $j_s^{n^+} = (qD_p p_0/l_p^*) \exp(-qU_K/kT)$ is the hole saturation current in the n^+-layer. Since (6) is essentially the hole extraction current, the hole current should tend to saturation with increasing external voltage, and not be proportional to the concentration $\Delta p(0)$ as it follows from equation (14) in [11].

The drift component of the electron current at the point $x = 0$ is

$$j_n(0) = q\mu_n n(0)\, E(0) = bj\, \frac{p(0) + N_D}{(b + 1)\, p(0) + bN_D}. \tag{7}$$

Taking into account that $j = j_p(0) + j_n(0)$ and that the inequalities $j \gg (b + 1)\, j_s^{n^+}$, $p(0) \ll p_0 \exp(qU_{n\text{-}n^+}/kT)$ are satisfied at a high level of depletion, we determine from (5) to (7) that

$$p(0) \approx bN_D \frac{j_s^{n+}}{j - (b + 1)\, j_s^{n+}} \approx bN_D \frac{j_s^{n+}}{j} \ll N_D\,, \tag{8}$$

$$n(0) = n_0 \exp\left(-\frac{qU_{n\text{-}n^+}}{kT}\right) = N_D \frac{j - j_s^{n+}}{j - (b + 1)\, j_s^{n+}} \approx N_D\,. \tag{9}$$

It is evident that both carrier concentrations near the contact drop with increasing current, $p(0)$ being inversely proportional to current and tending to zero at $j \to \infty$, and $n(0)$ approaching its saturation value equal to N_D; the voltage drop across the contact also reaches saturation $U^{sat}_{n-n^+} = (kT/q) \ln (n_i/N_D)$ acquiring greater values for purer material, i.e. the greater is the ratio n_i/N_D. For instance, $U^{sat}_{n-n^+} = 4kT/q = 0.1$ V for Ge at $T = 300$ K ($n_i = 2.5 \times 10^{13}$ cm^{-3}) and $N_D = 5 \times 10^{11}$ cm^{-3} [11]. Consequently, the necessary condition for essential charge carrier depletion near the contact is that the voltage drop across the n$^+$–n junction should be at least $2.3kT/q \approx 57.5$ mV, i.e. it should amount to tens of millivolts and not to 10^2 to 10^3 μV as it is stated in [10].

As regards (8) and (9), it should be noted that: (i) equation (9) follows from a more general formula obtained in [12, 13] for high-resistance compensated materials whose conductance exceeds by far the intrinsic value. Indeed, we have $n(0) = n_0\delta$ (where $\delta = 1 - p_0\tau_{n0}/n_0\tau_{p0}$) instead of (9) for samples with low ($p_{t0} \approx n_0$) or, vice versa, high ($p_{t0} \gg n_0$) hole concentrations on deep levels; assuming that $\tau_{n0} = \tau_{p0} = \tau_0$ we have $\delta = (n_0 - p_0)/n_0 = N_D/n_0$ and $n(0) = N_D$; (ii) it follows from equal recombination rates for electrons and holes at $\tau_n = \tau_p$ that $p - p_0 = n - n_0$. It may seem that an approximate equation $n \approx p$ follows from the condition $n_0 \approx p_0 \approx n_i$ valid for intrinsic material. But as a matter of fact, it is not so because $p \approx n - N_D$. Since $n(0) \approx N_D$ (this results from the fact that the impurity electrons are not subject to exclusion whereas intrinsic carriers can be completely removed by it), in agreement with (8), we obtain $p(0) \ll N_D$. The fact that $p(0)$ tends to zero with increasing current corresponds to a quite insignificant hole inflow current from the n$^+$-region to the n-region as compared with the hole removal current ($bj_s^{n+} \ll j$). (iii) Since $j_p(0) = j - j_n(0)$, (8) and (9) can also be obtained from the continuity condition for the hole current through the contact on the assumption that the hole inflow and removal currents on the left and right of the plane $x = 0$ are equal [12, 13].

3. Current–Voltage Characteristic for the n$^+$–n Junction

It follows from (2) and (3) for the field distribution in the exclusion layer that

$$E \approx \frac{j}{q\mu_p[(b+1)\,p(x) + bN_D]} \approx E(0) - [E(0) - E(\infty)]\,\frac{x}{x_1}, \tag{10}$$

where

$$E(0) \approx \frac{b+1}{b}\,(1+G)\,\frac{n_i}{N_D}\,E(\infty) \approx \frac{j}{q\mu_n N_D}, \tag{11}$$

$$x_1 \approx \frac{2b}{b+1}\,\frac{\mu_p\tau_0}{1+2G}\,\frac{N_D}{n_i}\,[E(0) - E(\infty)]\,. \tag{12}$$

Consequently, the field decreases linearly with the coordinate from the contact $x = 0$ to the value of $E(\infty) \approx E(x_1) \ll E(0)$ in the bulk of the semiconductor, i.e. exclusion divides the sample into layers of relatively weak ($x_1 \leqq x \leqq L$) and strong ($0 \leqq x \leqq x_1$) field. It also follows from (10) to (12) that the lifetime τ_0 can be determined from the equations

$$\tau_0 = \frac{x_{1D}}{2\mu_p}\,\frac{E_d(0)/E_d(\infty)}{E_d(0) - E_d(\infty)}, \qquad \tau_0 = \frac{x_{1l}}{\mu_p}\,\frac{E_l(0)/E_l(\infty)}{E_l(0) - E_l(\infty)}, \tag{13}$$

if x_1 is known from the experimental results, and the field distribution in the crystal in darkness ($G = 0$) and with sufficient illumination ($G \gg 1$) is measured. However, τ_0 can be determined sufficiently accurately only if the hole inflow current j_s^{n+} is small as compared to the full current j. This is accomplished by sufficiently doping the n^+-layer of such a length that $\tilde{w}^+ \geqq 2.3$. For instance, the experimental results for Ge given in [11] show that at $T = 293$ K, $D_p = 45$ cm²/s, $n_0^+ = 6 \times 10^{14}$ ($p_0^+ = 9.3 \times 10^{11}$ cm^{-3}), and $\tau_p^+ = 1$ µs ($l_p^+ = 67$ µm), the current $j_s^{n+} = 1$ mA/cm² while the current density j in the sublinear region of the current–voltage characteristic exceeds this value (e.g., $j = 15$ mA/cm² at a voltage of $U = 5$ V).

Integrating the field E over intervals of 0 to x_1 and x_1 to L we find from (10) to (12) for the I–U characteristic that

$$j = \frac{bqN_D L}{2\tau_0} \frac{1 + 2G}{1 + G} \left(\sqrt{1 + \frac{U}{U_{cut}^{(1)}}} - 1\right), \tag{14}$$

where the cut-off voltage in the I–U characteristic plotted in $(j, \sqrt{U})$ coordinates is equal to

$$U_{cut}^{(1)} = \tfrac{1}{4} U_{ex} = \frac{kT}{q} \frac{bN_D}{(b + 1)\, n_i} \left(\frac{L}{2l_d}\right)^2 \frac{1 + 2G}{(1 + G)^2}. \tag{15}$$

According to (14), (15), the I–U characteristic is linear at low and sublinear,

$$j = \frac{qD_p}{l_d} \sqrt{N_D n_i} \sqrt{b(b + 1)(1 + 2G) \frac{qU}{kT}}, \tag{16}$$

at high voltages, and the lower threshold voltage for the sublinear region U_{ex}, in the same way as the depletion layer width x_1, decreases with increasing light intensity and acquires the smaller values, the lower is the impurity concentration; in other words, the lower N_D, the lower is the voltage at which essential exclusion occurs. It should be noted that the I–U characteristic for compensated semiconductors is analogous in its form to (14), (16), and its sublinear region at $u \equiv n_0/p_{t0} = 1$ and $u \ll 1$ may be expressed in the form [15, 19]

$$j_{com} = \frac{bqD_p}{l_d} n_0 \sqrt{\delta} \sqrt{(1 + G) \frac{2qU}{kT}} \quad \left(G \equiv \frac{g\tau_{n0}}{n_0}\right). \tag{17}$$

An equation for intrinsic material ($n_0 \approx p_0 \approx n_i$, $\tau_{p0} = \tau_{n0} = \tau_0$, $\delta = N_D/n_i$) following from the above one differs from (16) in the factor $f = \sqrt{\frac{2b}{b + 1} \frac{1 + G}{1 + 2G}}$ which is about unity at any illumination (for germanium, in particular, $f = 0.82$ to 1.15).

In compliance with (14), (15), a parabolically sublinear I–U characteristic can easier be observed in materials of low N_D and great n_1, and the voltage U_{ex} can be decreased by decreasing the base lengh L and increasing the diffusion length $l_d \equiv \sqrt{D_p \tau_0}$. According to [11], for instance, $U_{ex}^d = 0.275$ V for Ge at $G = 0$, $N_D = 5 \times 10^{11}$, $n_i = 1.7 \times 10^{13}$ cm^{-3}, $\tau_0 = 10$ µs, and $L = 5$ mm, whereas for Si with Cd under the same conditions and $\tau_{p0} = 10$ µs, $\delta = 0.1$ this voltage $U_{ex}^d = 12.5$ V [19] which is essentially higher. In compliance with (3), (16), the depletion region becomes wider with increasing applied field and extends over the whole base at a voltage satisfying the condition $x_1(U_{kr}) = L$. A further increase in the field results in majority carrier concentration in the whole exclusion-affected base approaching its value $n(0) \approx N_D$ near the contact. In this case the I–U characteristic has a second linear region described by the equation

$$j_{lin}^{(2)} = q\mu_n N_D \frac{U + U_{cut}^{(2)}}{L}, \tag{18}$$

where the voltage $U_{cut}^{(2)}$ corresponds to negative U because $j = 0$ at $U = -U_{cut}^{(2)}$. We calculate it from the condition of intersecting extensions of both linear regions in the I–U characteristic, i.e. by equating the current $j_{lin}^{(1)} = q\mu_p(b + 1)\, n_i(1 + G)\, U/L$ to current (18) at $U = U_{int}$. Experiment [2] shows that the value of U_{int} called "exclusion voltage" by the authors of [2] is independent of light intensity and equals $U_{int} = 0.3$ V for the investigated Ge samples with given L and τ_0. This enables us to put down

$$U_{cut}^{(2)} \approx U_{int} \frac{(b + 1)\, n_i}{bN_D} (1 + G) \gg U_{int} . \tag{19}$$

It is evident that the cut-off voltage exceeds U_{int} by far and rises with increasing crystal illumination and temperature. The current $j_{lin}^{(2)}$ increases with increasing light intensity and temperature only owing to the variation in $U_{cut}^{(2)}$ with G and T. Thus, the second linear region of the I–U characteristic caused by exclusion differs from the first (non-exclusion) one in the following:

1. The voltage $U_{cut}^{(2)} \neq 0$. 2. Its slope is proportional to the impurity concentration N_D and independent of light intensity and temperature (a slight temperature dependence is caused by a variation in mobility). 3. Because of exclusion, the lifetime τ is twice as long as τ_0 in darkness, and drops with increasing G as we show below. In fact, taking into account that $\Delta p = \Delta n = g\tau - n_0$ we find from (1) that

$$\tau \approx \frac{n_i}{2g} \left(\sqrt{1 + \frac{8g\tau_0}{n_i}} - 1 \right) \approx \begin{cases} 2\tau_0 & \text{at } 8g\tau_0 \ll n_i , \\ \sqrt{(2/g) n_i \tau_0} & \text{at } 8g\tau_0 \gg n_i , \end{cases} \tag{20}$$

i.e. τ is constant at low light intensities and drops inversely with the square root of g at high ones.

According to (16), (18), and (19), the transition voltage between the sublinear and second linear regions of the I–U characteristic is determined from the equation

$$U^2 - \left[\frac{kT}{q} \left(\frac{L}{l_d} \right)^2 \frac{(b + 1)\, n_i}{bN_D} (1 + 2G) - 2U_{cut}^{(2)} \right] U + U_{cut}^{(2)2} = 0 . \tag{21}$$

For long-base structures with the second term in the brackets of (21) exceeding the first one by far, we find from (21) that

$$U_{lin}^{(2)} \approx \frac{kT}{q} \left(\frac{L}{l_d} \right)^2 \frac{(b + 1)\, n_i}{bN_D} (1 + 2G) \gg U_{ex} . \tag{22}$$

Consequently, the voltage range of the sublinear region of the I–U characteristic can be quite considerable; it becomes wider with rising light intensity because U_{ex} decreases with increasing G. At the above values of parameters [11], $U_{ex}^{d} = 0.27$ and $U_{lin}^{(2)} = 700$ V.

Substituting (16) into (3) we obtain the following equation for the depletion layer width as a function of voltage:

$$x_1 \approx 2l_d \sqrt{\frac{bN_D}{(b + 1)\, n_i(1 + G)} \frac{qU}{kT}} , \tag{23}$$

i.e. $x_1 \sim U^{1/2}$ and drops with increasing G as $G^{-1/2}$. According to [11], for instance, $x_1 = 5l_d = 1.1$ mm for Ge at $T = 300$ K, $U = 5$ V and $\tau_0 = 10$ µs. In compliance with (16), photocurrent and photoelectric current gain are determined from the equations

$$j_{ph} = j - j_d = \frac{qD_p}{l_d} \sqrt{N_D n_i} \sqrt{b(b+1) \frac{qU}{kT}} \left(\sqrt{1+2G} - 1\right), \tag{24}$$

$$K_j \equiv \frac{j_{ph}}{qgL} = \frac{l_d}{L} \sqrt{b(b+1) \frac{N_D}{n_i} \frac{qU}{kT}} \frac{\sqrt{1+2G} - 1}{G}, \tag{25}$$

i.e., as for compensated semiconductors [15, 19], the values under investigation are proportional to the square root of voltage; the photocurrent is linearly dependent on illumination at low light intensities ($2G < 1$) and sublinearly dependent on it at high light intensities ($2G > 1$). At the same time the gain K_j is independent of G at low and decreases $\sim G^{-1/2}$ at high illumination intensity. Estimating the value of K_j at a low excitation level and at the parameter values of [11] used above we find that the gain $K_j = 3$ at $U = 25$ V and $L = 1$ mm, i.e. it is essentially lower for intrinsic materials where τ_0 is relatively short ($\tau_0 = 10$ to 100 μs) than for compensated semiconductors (for Si with Zn or Cd, $\tau_{n0} \approx 10^{-4}$ to 10^{-1} s and $K_j \approx 10^2$ to 10^4 [21]).

4. Comparison between Theory and Experiment

A comparison between the theoretical results obtained by us and experimental data for Ge [2, 11] shows good qualitative and partly quantitative agreement between them. Indeed, the experimental results, in agreement with the equations found, show that the I–U characteristic is linear at very low voltages where the voltage drop across the depletion layer is small as compared with its value for the part of the base not affected by depletion and at a voltage equal to U_{ex} which is quite low (lower than 0.3 V); then, if $N_D \ll n_i$ and the ratio L/l_d is not too great, it becomes parabolically sublinear ($j \sim \sqrt{U}$), and after that it passes into the second linear region whose slope is essentially less than that of the first one and not so dependent on temperature and light intensity. As in theory, the cut-off voltage $U_{cut}^{(2)}$ and the lower threshold voltage of the second linear region $U_{lin}^{(2)}$ rise with temperature and light intensity, the current rise with increasing g being caused not by a steeper slope of the illumination I–U characteristic but by a higher $U_{cut}^{(2)}$. Physically, this results from a greater rate of carrier thermo- and photogeneration, and for this reason stronger external fields are necessary for decreasing the carrier concentration to the level corresponding to impurity conductance. In this connection it may be indicated that the comparison between theoretical and experimental data in [11] is not completely correct. In [11], the I–U characteristic at $T = 77$ K was linear with a slope of $R_1 = 10/0.5 \times 10^{-3} = 20$ kΩ, at $T = 293$ K, $\varrho = 68$ Ω cm, i.e. $n_i = 1.7 \times 10^{13}$ cm^{-3}, and the I–U characteristic included two linear regions separated by a sublinear region $j_d \sim U^{1/2}$, so that $R_{lin}^{(2)} = 4.2 R_{lin}^{(1)} = 12.7$ kΩ and thence $L/S = 44$ cm^{-1}, $\varrho_{lin}^{(2)} = 300$ Ω cm. The latter values correspond to impurity conductance with $N_D = 6 \times 10^{12}$ cm^{-3} which value is an order of magnitude higher than the value of $N_D = 5 \times 10^{11}$ cm^{-3} given in [11]. At $T = 77$ K, $\varrho = 2 \times 10^4/44 = 500$ Ω cm; taking into account that $\mu_n = 10^4$ cm^2/Vs we obtain $N_D = 2 \times 10^{12}$ cm^{-3} in agreement with the value obtained by us. In the experiment described in [11], $j_s^{n^+} = 1$ mA/cm^2 as well and, according to (8), at $U = 7.5$ V ($I = 0.75$ mA, $j = 22.5$ mA/cm^2, $T = 293$ K, and $N_D = 4 \times 10^{12}$) we have $p(0) = 4 \times 10^{11}$ cm^{-3} and $Z_0 \equiv |\Delta p(0)|/p_0 = 0.98$ which noticeably differs from the value $Z_0 = 0.85$ given in [11]; in this case, according to (9), $n(0) = 4.4 \times 10^{12}$, i.e. $n(0) - N_D = 4 \times 10^{11}$ cm^{-3} and $|\Delta n(0)|/n_0 = 0.8$. Consequently, the concentration of remaining carriers is low and caused not by injection from the metallic part of the n$^+$–n contact as it is stated in [11], but by the presence of noticeable impurity conductance and hole inflow from the n^+-layer into the n-region. It should be noted that even at $Z_0 = 0.85$ there is no reason for the above statement because it is impossible to observe, after the linear and sublinear regions, a second linear region in the I–U characteristic whose slope is by far less than that of the first one. Injection after exclusion will more likely result in a superlinear or even S-shaped I–U characteristic than in a linear one. Also essential is the fact that the depletion region extends over the whole base, and carrier concentrations are everywhere equal to their values at the point $x = 0$.

References

[1] A. F. Gibson, Physica (Utrecht) **20**, 1058 (1954).
[2] J. B. Arthur, W. Berdsley, M. A. Brown, and A. F. Gibson, Proc. Phys. Soc. **B68**, 43 (1955).
[3] R. Bray, Phys. Rev. **100**, 1047 (1955).
[4] G. G. E. Low, Proc. Phys. Soc. **B68**, 310 (1955).
[5] A. Many, D. Gerlich, and E. Harnik, Proc. Phys. Soc. **B68**, 970 (1955).
[6] R. E. Burgess, Proc. Phys. Soc. **B68**, 793 (1955).
[7] W. T. Read, Bell Syst. tech. J. **35**, 1239 (1956).
[8] R. W. Lade and A. G. Jordan, J. Electronics and Control **13**, 23 (1962).
[9] E. D. Prokhorov, N. A. Shekhovtsov, and A. D. Prokhorov, Radiotekhnika i Elektronika **9**, 2014, 2174 (1964).
[10] A. Kassur, phys. stat. sol. (a) **16**, 507 (1973).
[11] I. M. Vikulin, Yu. A. Zaporozhchenko, L. F. Vikulina, and M. A. Glauberman, Radiotekhnika i Elektronika **19**, 2123 (1974).
[12] E. I. Adirovich, D. A. Aronov, and P. I. Knigin, Symp. Problemy dielektricheskoi elektroniki, Vol. 1, Izd. FAN UzSSR, Tashkent 1974 (p. 5).
[13] E. I. Adirovich, D. A. Aronov, P. I. Knigin, and Yu. S. Korolev, Fiz. Tekh. Poluprov. **8**, 2405 (1974).
[14] D. A. Aronov, I. Z. Karimova, and P. I. Knigin, Fiz. Tekh. Poluprov. **10**, 906 (1976).
[15] Kh. M. Abduraimov, D. A. Aronov, P. I. Knigin, Yu. S. Korolev, and S. V. Torosyan, Symp. Fizicheskie yavleniya v poluprovodnikovykh strukturakh s glubokimi urovnyami i optoelektronika, Izd. FAN UzSSR, Tashkent 1977 (p. 24).
[16] D. A. Aronov, P. I. Knigin, Yu. S. Korolev, and S. V. Torosyan, Fiz. Tekh. Poluprov. **11**, 2233 (1977).
[17] D. A. Aronov and R. Mamatkulov, Izv. Akad. Nauk UzSSR, Ser. fiz. mat. **23**, No. 1, 36 (1979).
[18] D. A. Aronov, E. P. Kotov, and R. Mamatkulov, Dokl. Akad. Nauk UzSSR **23**, No. 1, 28 (1979).
[19] D. A. Aronov, P. I. Knigin, and Yu. S. Korolev, Fiz. Tekh. Poluprov. **13**, 774 (1979).
[20] Kh. M. Abduraimov, D. A. Aronov, P. I. Knigin, and Yu. S. Korolev, Deposited at Cen. Sci. Res. Inst. Elektronika, No. DE-2578/1978. Abstract in: Fiz. Tekh. Poluprov. **13**, 1872 (1979).
[21] D. A. Aronov, P. I. Knigin, and Yu. S. Korolev, Fiz. Tekh. Poluprov. **14**, 1039 (1980).
[22] D. A. Aronov, Izv. Akad. Nauk UzSSR, Ser. fiz. mat. **24**, No. 6, 37 (1980).
[23] D. A. Aronov and R. Mamatkulov, Deposited at Cen. Sci. Res. Inst. Elektronika, No. PA-3067/1980. Abstract in: Fiz. Tekh. Poluprov. **15**, 1045 (1981).
[24] W. Shockley and W. T. Read, Phys. Rev. **87**, 835 (1952).
[25] W. van Roosbroeck, Bell Syst. tech. J. **29**, 560 (1950); **91**, 282 (1953).
[26] D. A. Aronov and G. M. Avakyants, Izv. Akad. Nauk UzSSR, Ser. fiz. mat. **3**, No. 6, 27 (1959).

(Received May 18, 1981)

phys. stat. sol. (a) 68, 69 (1981)

Subject classification: 11; 22.1.2

Institute for High Energy Physics, Academy of Sciences of the Kazakh SSR, Alma-Ata[1])

Range Distribution Characteristics of H^+ and He^{++} Ions with Energies (5 to 2) × 10^3 keV Implanted into Silicon

By

A. F. AKKERMAN and S. A. AKKERMAN

Detailed results of Monte-Carlo calculations on the depth distributions of H^+ and He^{++} ions with energies (5 to 2) × 10^3 keV implanted in silicon are presented. The energy dependences of mean projected and total ranges, standart deviations and skewnesses are given among them. The Pearson's approximation of the calculated distributions is discussed.

Ausführliche Ergebnisse werden mitgeteilt von Monte-Carlo-Rechnungen über die Tiefenverteilungen von H^+- und He^{++}-Ionen, die mit Energien von (5 bis 2) × 10^3 keV in Silizium implantiert werden. Die Energieabhängigkeiten der mittleren projizierten und der Gesamtreichweiten, Standardabweichungen und Winkelabweichungen werden angegeben. Die Pearson-Näherung der berechneten Verteilungen wird diskutiert.

1. Introduction

At present there are few experimental works on the measurement of range distributions for H^+ and He^{++} ions with energies up to 300 keV implanted into silicon [1 to 4]. For higher energies the only experiment [5] is known, where the mean projected ranges for H^+ ions with energy 0.8 to 2 MeV were determined. Quite recently the energy range up to 10 keV was reinvestigated in [4]. Really there are no experimental data on helium ion range distributions in silicon. In spite of the good agreement between experiments concerning the energy dependence of the mean projected range $\bar{R}_p(E)$ for H^+ ions there is considerable scatter in data characterizing the shape of distributions: the second, the third, and the fourth central moments and the associated quantities: standard deviation $-\sigma_p = \mu_2^{1/2}$, skewness $-S = \mu_3/\sigma_p^3$, and kurtosis $-k = \mu_4/\sigma_p^4$. This is partly due to the fact that most of the experimental methods of distribution function determination are indirect, they contain strong errors, especially in the distribution "tails". The theoretical methods are based on the solution of the kinetic transport equation [6 to 8]. Extended tabulated data for distribution characteristics and for various ion-target combination are obtained in [6, 7]. In the latter papers relations for the stopping power in inelastic and elastic collisions deduced from the atomic statistical model, were used. It is clear therefore that in case of the lightest ions the calculations contain a systematic error, whose apriori appreciation is difficult. Difficulties in the theory arise also for the reestablishment of the distribution function using a limited number of moments — as is well known, this procedure is not always correct. Compared with theoretical methods Monte-Carlo calculations [9 to 11] have definite advantages. They use realistic models of particle transport and more precise theoretical or experimental data of the elementary interaction process. Obtained distributions and their characteristics can be directly compared with the experiment.

[1]) 480082 Alma-Ata 82, USSR.

In this paper we report results of Monte-Carlo calculations on the distribution profiles of hydrogen and helium ions with energy up to 2 MeV implanted into silicon.

2. The Computational Model

Unlike the individual collision model used in [9], the collision grouping model as in [10, 11] was applied in this paper. The particle state (energy and angle) at the end of each step of random walk is sampled from two distributions: 1. energy loss distributions in inelastic processes and 2. angular distributions of multiple elastic scattering. As the first one we use Bohr distribution with the Lindhard standard deviation determined more precisely for hydrogen from the experimental data [12]. In [10] we discussed the validity of the indicated distribution. The data of [13] were used for the stopping power of H^+ in Si, while for He^{++} the computed values based on the compilation [14] by fitting of three-parameter Brice formula with $Z^B = 1.229$, $a = 0.446$, $n = 3.276$. The angular distribution functions $f(\theta)$ in multiple scattering are often deduced from the well-known tables [15] obtained for the Thomas-Fermi (T-F) potential. But they are not always convenient to use, because in the Monte-Carlo calculations it is necessary to carry out double interpolation in angle and target thickness. Besides, recently it was shown that the application of the T-F potential resulted in higher stopping powers in elastic collisions and in greater values of particle backscattering coefficients compared those obtained from the experiment. We have received $f(\theta)$ by summing up the Goudsmith-Sounderson series [16]. The algorithm with a short-range potential appearing in $f(\theta)$ calculations reduces to the following. Beginning from an angle $\theta_0 = \arcsin 8/\varepsilon$ (ε particle energy in Lindhard units) $f(\theta)$ corresponds to the differential Rutherford cross-section $d\sigma/d\Omega(\theta)$. For $\theta < \theta_0$, $f(\theta)$ is calculated by the Lindhard cross-section as shown in [16]. The composite distribution thus obtained is compared in Fig. 1 with that extracted from [15] and calculated by the same method, but for the short-range potential [17]. The usage of such distributions ensures correct values for the backscattering coefficients R_N and R_E (number and energy rates of backscattered particles). It is necessary to note that the application of the Lindhard-Nilsen $f(\theta)$ distribution, analysed in [18], does not provide correct R_N and R_E values. This can be seen from Monte-Carlo calculations [19], where for silicon at $E = 2.5$ MeV the value $R_N = 0.09$ was obtained, while the real value is $R_N \cong 0$. The use of this distribution must inevitably bring about a shift of the range distribution with respect to the experimental one and consequently overrates σ_p. Our elaborated ITRAN code permits to receive the whole of information on implanted atom distributions including its first central moments, the particle spectra at various depths in the silicon target, as well as the R_N and R_E values, the partition of energy transfer in elastic and inelastic processes, and the defect rate depth distribution. Usually 7×10^3 to 10^4 particle trajectories were simulated in each calculational run spending (0.2 to 0.4) s per trajectory.

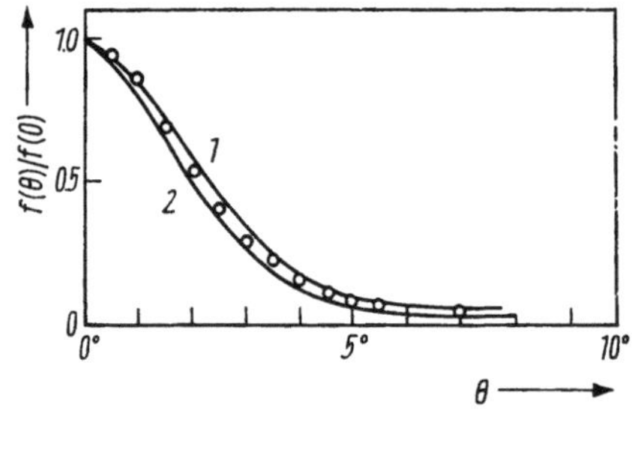

Fig. 1. Angular distribution of 100 keV helium ions penetrating a 700 Å silicon film. (1) as calculated in [15], (2) calculated for the potential [17], ●, ○ approximation used in our calculation

3. Results of Computations

The results of computations and the experimental data [4] of H^+ ions with energy up to 90 keV implanted into Si are compared in Fig. 2. One can see a good agreement until $E \leqq 30$ keV. However, a systematic discrepancy for $E = 50$ to 90 keV is observed in the region of maximum ionization energy losses. In this region the experimental data reviewed in [4] show a considerable scatter. The calculated energy dependence of the H^+ and He^{++} mean projected ranges $\bar{R}_p$ (in μm) to a precision of 5% may be described by the empirical formula

$$\bar{R}_p\,(\mu\text{m}) = a + bE + cE^2 \quad (E \text{ in keV})\,. \tag{1}$$

The parameter values entering into (1) for various particles and energy ranges are given in Table 1.

Table 1

Parameter values a, b, and c given in formula (1)

	$E \leqq 150$ keV		$150 < E \leqq 1000$ keV		$E > 1000$ keV	
	H^+	He^{++}	H^+	He^{++}	H^+	He^{++}
a	3.02×10^{-2}	-8.03×10^{-3}	1.13×10^{-2}	3.20×10^{-1}	1.25	1.09
b	8.65×10^{-3}	6.48×10^{-3}	7.11×10^{-3}	2.95×10^{-3}	8.12×10^{-3}	1.51×10^{-3}
c	-1.90×10^{-6}	-8.89×10^{-6}	9.17×10^{-6}	-9.20×10^{-8}	6.92×10^{-6}	5.80×10^{-7}

The obtained values for $\bar{R}_p$ are in good agreement with the compilation of experimental data in [4]. In Fig. 3 the energy dependence of the standard deviation $\sigma_p(E)$ together with the experimental data [4] and the theoretical calculations [7] are shown. In the same figure the $\sigma_t(E)$ dependence for the total range (along the trajectory) is given. It can be seen from this figure that our calculations sufficiently clearly indicate a plateau in the $\sigma_p(E)$ dependence in the region of maximum energy loss for H^+ and He^{++} ions (see Fig. 5). An indication of the plateau for H^+ ions was pointed out in [4]. The theoretical calculations [6, 7] compared to our data agree worse with the experimental data which is caused by the usage of the T-F potential. In our previous

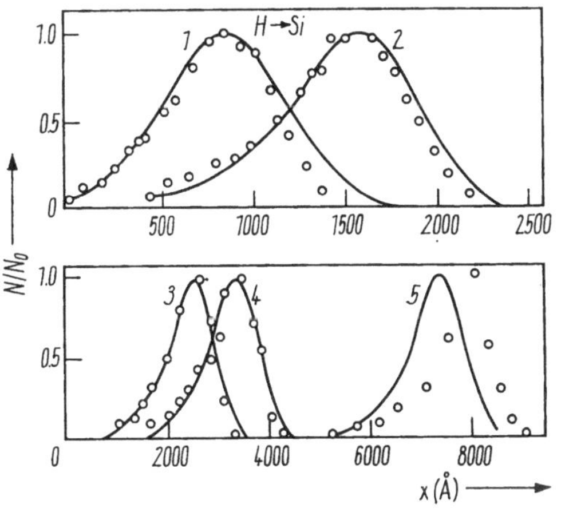

Fig. 2. H^+ ion depth distribution in Si. Curves 1 to 5 correspond to energies of 5, 10, 20, 30, and 90 keV, respectively (experimental data [4]), circles our calculations

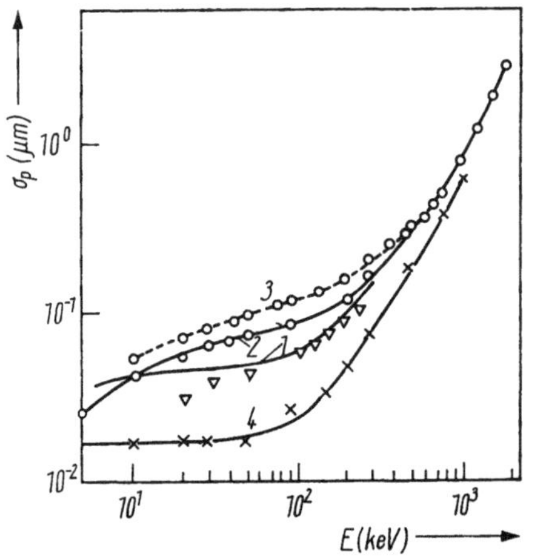

Fig. 3. Energy dependence of σ_p for H^+ ions; (1) experimental data [4], (2) our data, (3) calculations [7], △ experimental data [3], (4) our data for σ_t

paper [10] higher values of σ_p were obtained while the same potential was used. The calculated and experimental dependences $s(E)$ for H^+ ions are shown in Fig. 4. The experiment [4] gives nearly constant and small absolute values of skewness and kurtosis. The calculated values differ considerably from the experimental ones, although on the whole our data agree better with the experiment. The main reason of this discrepancy is the great errors arising in the calculations of the moments, but significant uncertainties arise also in the experiment as noted in [4].

The dependences $\sigma_p(E)$, $\sigma_t(E)$, and $s(E)$ for He^{++} ions are given in Fig. 5. We note that in this case a better agreement of σ_p with the theoretical calculation [6, 7] than for hydrogen is obtained. The $R(E)$ and $N_0(E)$ dependences for the total mean ranges and maximum concentration (at the distribution peak) per unit flux (1 cm^{-2}) for H^+ and He^{++} ions are shown in Fig. 6.

It is impossible to consider the problem of analytical representation of the concentration profiles to be solved. Edgeworth's expansion used in [6] requires a great number of moments, which cannot be found with sufficient accuracy neither in calculation

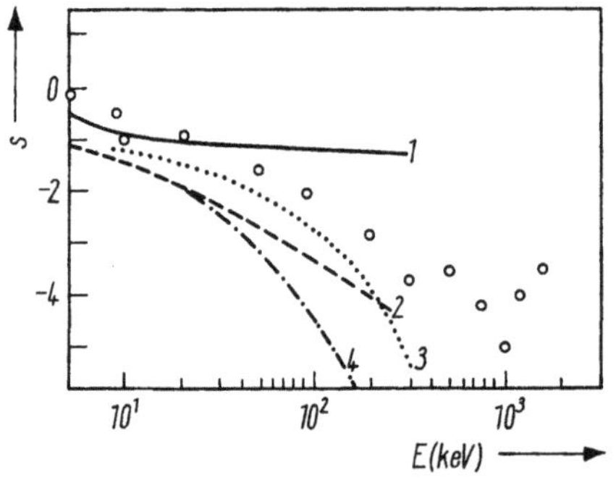

Fig. 4

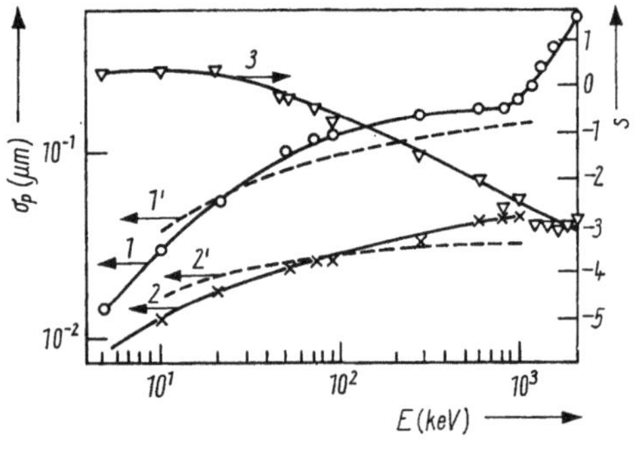

Fig. 5

Fig. 4. Energy dependence of skewness for H^+ ions; (1) experimental data [4], ○ our data, (2), (3), (4) as calculated in [6 to 8], respectively

Fig. 5. Energy dependences of σ_p, s, and σ_t for He^{++} ions; (1), (2) $\sigma_p(E)$, $\sigma_t(E)$ (our data), (1′), (2′) as calculated in [6], (3) $s(E)$ (our data)

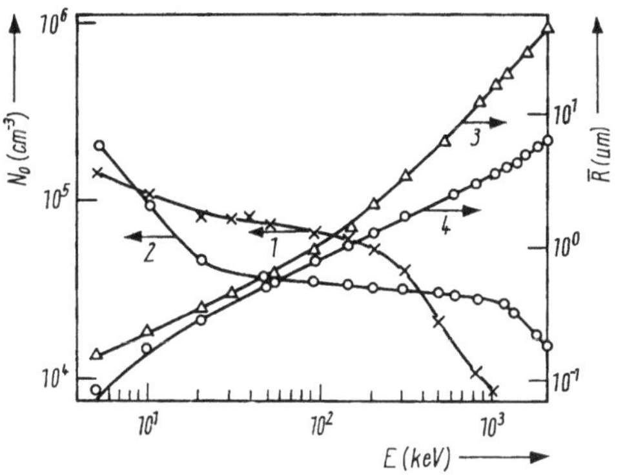

Fig. 6. Calculated energy dependences for the maximum of the depth distributions $N_0(E)$ ((1) H^+, (2) He^{++}) and total ranges $R(E)$ ((3) H^+, (4) He^{++})

nor in the experiment. The Pearson distribution f_p used in [7] is simpler and needs only three moments

$$f_p = k\left(1 + \frac{x - R}{a}\right)^{-q} \exp\left(-\nu \operatorname{arc\,tg} \frac{x - R}{a}\right), \tag{2}$$

where

$$R = \bar{R}_p + \frac{a\nu}{2q - 2}.$$

If the condition $s^2 < 32$ is fulfilled the application of (2) will be possible. Moreover, the usage of (2) assumes great values of kurtosis contrary to those deduced from the experiment [4]. However, Pearson's distribution may be used for approximately describing the concentration profiles of implanted hydrogen and helium ions with energy up to 300 and 1000 keV, respectively. The energy dependence of the parameters k, a, ν, q entering into (2) were obtained by a minimizing programme and are represented in Fig. 7. The accuracy is $\approx 10\%$ in the region of the peak distribution but gets worse for such depths x where N/N_0 becomes less than 0.01 (see Fig. 8, 9). In Fig. 8 and 9 the calculated concentration profiles for H^+ and He^{++} for energies

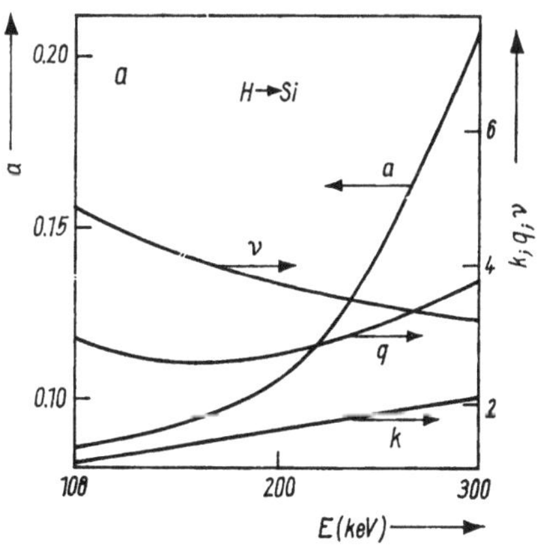

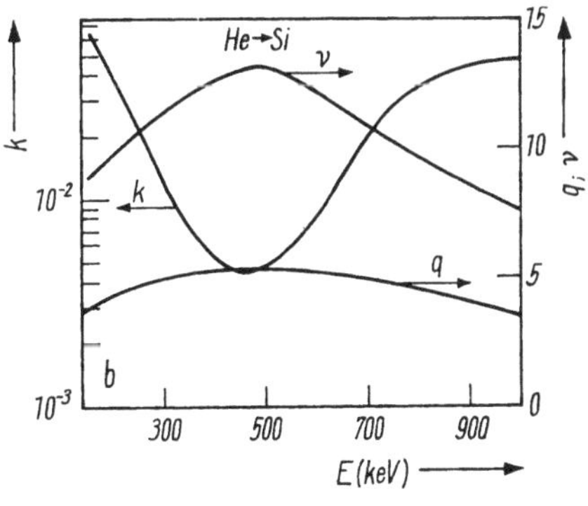

Fig. 7. Energy dependences of parameters entering in Pearson's formula (2). a) H^+–Si, b) He^{++}–Si $a = 0.2383$

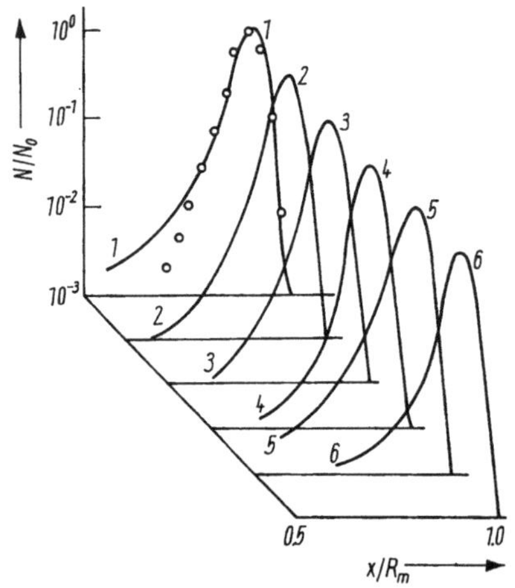

Fig. 8

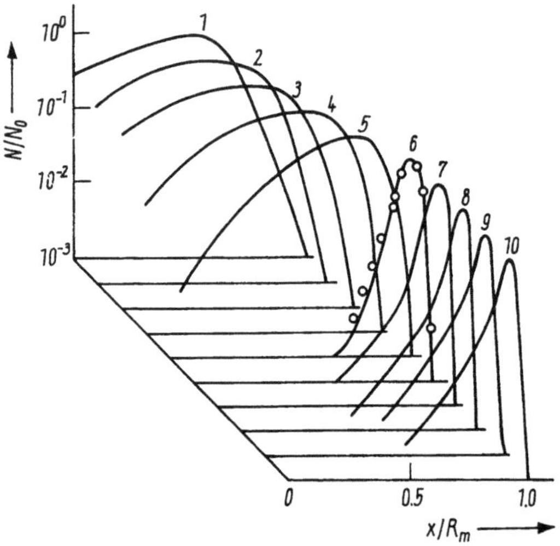

Fig. 9

Fig. 8. H^+ depth distribution of reduced concentration profiles. (1) to (6) correspond to energies of 300, 500, 800, 1000, 1500, and 2000 keV, respectively. Circles on curve 1 are calculated by formula (2)

Fig. 9. He^{++} depth distribution of reduced concentration profiles. (1) to (10) correspond to energies of 5, 10, 20, 50, 90, 1000, 1200, 1500, 1800, 2000 keV, respectively. Circles on curve 6 are calculated by formula (2)

outside the energy interval indicated above are shown. The R_m values may be approximated by the empirical relations

$$R_m\,(\mu\text{m}) = \begin{cases} 1.538 \times 10^{-1} + 8.278 \times 10^{-3}E + 9.828 \times 10^{-6}E^2; \\ \qquad H^+ \quad E \geqq 100\ \text{keV}, \\ 2.875 \times 10^{-2} + 1.206 \times 10^{-2}E - 3.429 \times 10^{-5}E^2; \\ \qquad He^{++}\ E \leqq 150\ \text{keV}, \\ 7.966 \times 10^{-1} + 2.446 \times 10^{-3}E + 4.053 \times 10^{-7}E^2; \\ \qquad He^{++}\ 150 < E \leqq 2000 E\ \text{in keV}. \end{cases} \tag{3}$$

4. Conclusion

We have demonstrated by the random walk model, using multiple angular distributions computed for a short-range potential approximation, that concentration profiles of hydrogen and helium ions implanted into silicon may be calculated with satisfactory accuracy.

However, the question on the probable interaction between hydrogen and helium ions with defects introduced remains open. It is unknown of present whether the calculated distribution is considerably altered if the defect distribution is included, especially for helium ions. An experimental investigation of the concentration profiles of helium ions is of interest. The experiment could give an additional confirmation of the calculational model of multiple scattering.

References

[1] E. Ligeon and A. Guivarch, Radiat. Eff. 27, 129 (1976).
[2] J. Mittenbacher and K. Gärtner, Internat. Conf. Ion Beam Modification of Materials, Vol. 3, Budapest 1978 (p. 1661).
[3] W. K. Chu, R. H. Kastl, R. F. Lever, S. Mader, and B. J. Masters, Phys. Rev. B **16**, 3851 (1977).
[4] F. J. Demond, S. Kalbitzer, H. Mannsperger, and G. Müller, Nuclear Instrum and Methods **168**, 69 (1980).
[5] A. Marcinkowski, H. Rzewuski, and Z. Werner, Nuclear Instrum. and Methods 57, 338 (1967).
[6] J. F. Gibbons, W. S. Johnson, and S. Maylorie, Projected Range Statistics, 2nd ed., Dowden, Hutchincon, and Ross, Inc. Stroudsburg, Pa, 1975.
[7] A. F. Burenkov, F. F. Komarow, M. A. Kumakhov, and M. M. Themkin, Tablitsy parametrov prostanstvennogo raspredelenija ionoimplantirovanikh primesy, Izd. BGU V. I. Lenina, Minsk 1980.
[8] K. B. Winterbon, Radiat. Eff. **13**, 215 (1972).
[9] D. A. Thomson and J. E. Robinson, Nuclear Instrum. and Methods **132**, 261 (1976).
[10] A. F. Akkerman, Fiz. Tekh. Poluprov **11**, 2005 (1977).
[11] A. F. Akkerman, Proc. Internat. Conf. Ion Beam Modification of Materials, Vol. 1, Budapest 1978, (p. 45).
[12] R. A. Langley and D. K. Brice, Phys. Rev. B **18**, 4673 (1978).
[13] H. H. Andersen and J. F. Zigler, Hydrogen Stopping Powers and Ranges in All Elements, Vol. 3, Pergamon, New York 1977.
[14] J. F. Ziegler and W. K. Chu, Atomic Data and Nuclear Data Tables **13**, 463 (1974).
[15] P. Sigmund and K. B. Winterborn, Nuclear Instrum. and Methods **119**, 541 (1974).
[16] A. F. Akkerman and S. A. Akkerman, Izv. Akad. Nauk Kazakh SSR, Ser. fiz. – mat. 2, 31 (1977).
[17] S. Kalbitzer and H. Oetzmann, Proc. Conf. Ion Beam Modification Materials, Vol. 1, Budapest 1978 (p. 3).
[18] A. Desalvo and R. Rosa, phys. stat. sol. (b) **69**, 71 (1975).
[19] J. Tatarkiewicz, Nuclear Instrum. and Methods **146**, 447 (1973).

(Received June 9, 1981)

phys. stat. sol. (a) 68, 77 (1981)

Subject classification: 10.2; 20.1; 20.3; 22.5.2

Vinogradov Institute of Geochemistry, Academy of Sciences of the USSR, Sibirian Branch, Irkutsk[1])

Magnesium Color Centers at 3.5 and 5.0 eV in Lithium Fluoride

By

E. A. RADYABOV and A. I. NEPOMNYACHIKH

Centers at 3.5 and 5.0 eV are created by X-irradiation of LiF:Mg or by optical bleaching of $Mg_c^+V_a^+V_c^-$ (at 4.0 eV) centers in LiF:Mg at 80 K. Both centers have [100] and [110] absorption dichroism. The thermal bleaching of centers at 3.5 eV is related with a glow peak at 150 K which obeys second-order kinetics. The energy of the absorption band of $Mg_c^+V_c^-$-centers in LiF is calculated to be about 3.5 eV. The absorption band shape and the oscillator strength of the centers at 5.0 eV and F-centers are almost equal. The color centers at 3.5 eV seem to be $Mg_c^+V_c^-$-centers and those at 5.0 eV seem to be F-centers near $Mg_c^{2+}V_c^-$ dipoles.

Центры при 3,5 и 5,0 eV образуются в процессе рентгеновского облучения LiF:Mg или путем оптического обесцвечивания $Mg_c^+V_a^+V_c^-$ (при 4,0 eV)-центров при 80 K. Обе полосы поглощения характеризуются как [100] так и [110] дихроизмом. Термообесцвечивание центров при 3,5 eV сопровождается термостимулированной люминесценцией с максимумом при 150 K, имеющей кинетику второго порядка. Согласно оценкам максимум полосы поглощения $Mg_c^+V_c^-$-центров в LiF должен находится около 3,5 eV. Центры при 5,0 eV и F-центры в LiF:Mg имеют близкую форму полосы поглощения и одинаковые силы осциллятора. Предполагается, что центры при 3,5 eV представляют собой $Mg_c^+V_c^-$-центры, а центры при 5,0 eV представляют собой F-центры около диполей $Mg_c^{2+}V_c^-$.

1. Introduction

Magnesium color centers in alkali halides are good electron traps [1 to 3]. This property of magnesium color centers in lithium fluoride is widely used in thermoluminescence dosimetry [4]. Some authors propose that magnesium color centers in LiF with absorption band at 4.0 eV are Z_2-centers and centers at 5.5 eV are Z_3-centers [5 to 7]. In a previous paper we have studied the thermal and optical bleaching of color centers at 4.0 eV from 85 to 440 K and found that the conversion of centers at 4.0 and 5.5 eV does not correspond to the Z_2–Z_3 center conversion in other alkali halides. We propose that centers at 4.0 eV are $Mg_c^+V_a^+V_c^-$-centers and those at 5.5 eV are magnesium atoms at anion site associated with cation vacancies [8]. The optical bleaching of the centers at 4.0 eV at liquid nitrogen temperatures leads to a creation of new centers at 3.5 and 5.0 eV [5, 6, 8], the nature of which is still unknown. The studies of centers at 5.0 eV are difficult due to the neighbourhood of the absorption bands of these centers and those of F-centers (at 5.1 eV at 80 K). We have studied LiF crystals with a large magnesium concentration (700 ppm) and there is no F-band in our crystals after X-irradiation at 85 K and warming up to 400 K. The present work deals with the nature of magnesium centers at 3.5 and 5.0 eV in LiF:Mg crystals.

---

[1]) 1a Favorskii Street, 664033 Irkutsk, USSR.

2. Results and Discussion

When the temperature is rising, the centers at 3.5 eV are bleaching at about 150 K (Fig. 1). Centers at 5.0 eV have three ranges of thermal bleaching at about 150, 230, and 320 K. Thermal bleaching of centers at 3.5 and 5.0 eV leads to an almost perfect recovery of $Mg_c^+V_a^+V_c^-$-centers. One can conclude the centers at 3.5 and 5.0 eV to be magnesium electron-trapping centers.

A pile of six fused silica plates was used to obtain absorption spectra with polarized light. Exposure of LiF:Mg to [001] polarized light at about 4.0 eV results in the preferential bleaching of $Mg_c^+V_a^+V_c^-$-centers having a definite orientation and in the creation of centers at 3.5 and 5.0 eV with the same orientation (Fig. 2). Curves on Fig. 2b show the [100] anisotropic absorption spectra which were obtained by subtracting the spectra taken with [010] polarized light from that taken with [001] light. By the

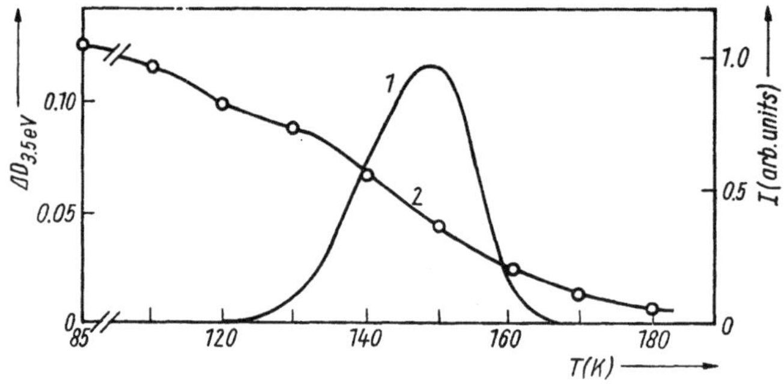

Fig. 1. TSL curve (1) and bleaching curve at 3.5 eV (2) of LiF:Mg crystal after optical bleaching of $Mg_c^+V_a^+V_c^-$-centers at 85 K

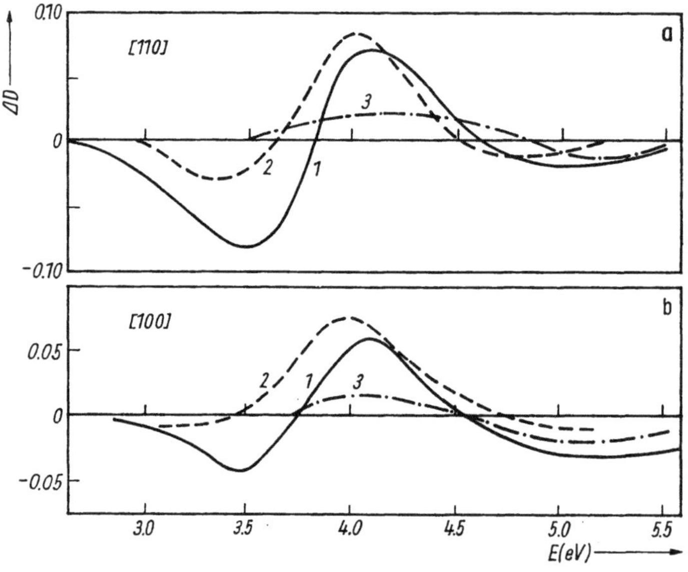

Fig. 2. Anisotropic absorption spectra of LiF:Mg crystal with $Mg_c^+V_a^+V_c^-$-centers after exposure to polarized light at (1) 4.14 eV, (2) 4.0 eV at 80 K and (3) 4.14 eV at 160 K

same way we obtained the [110] anisotropic spectra (Fig. 2a). These spectra coupled with [100] anisotropy show that all three centers are [110] polarized. We mark that the exposure of LiF:Mg with the centers at 3.5 and 5.0 eV to polarized light at 3.5 or 5.0 eV leads to a very small dichroism in these bands.

2.1 Color centers at 3.5 eV

The bleaching of these centers is related with thermostimulated luminescence (TSL) (Fig. 1) and with the recovery of $Mg_c^+V_a^+V_c^-$. One can find the order of TSL kinetics from the relations:

$$\ln (I/\int I \mathrm{d}T) = \text{const} - E/kT \quad \text{for first order,}$$

$$\ln \left(I/(\int I \mathrm{d}T)^2\right) = \text{const} - E/kT \quad \text{for second order.}$$

The TSL at about 150 K obeys a second-order kinetics (Fig. 3).

Color centers at 3.5 eV are created by X-irradiation of LiF:Mg crystals at 85 K too, but the absorption of these centers lies under the large absorption of V_k-centers (at 3.57 eV [9]). The V_k-centers in LiF are bleaching when the crystal is warming up to 145 K. After that the glow curve peak at 150 K is separated from the V_k-peak and the TSL at about 150 K obeys a second-order kinetics, too (Fig. 3). The two absorption bands in X-irradiated LiF:Mg bleached in the range 145 to 170 K (Fig. 4). One band is the absorption band of the centers at 3.5 eV and the other one lies within the range of V_k absorption and is probably a V_F band (in literature there are no data about the shape of the V_F band in LiF). Consequently, heating releases electrons from centers at 3.5 eV and they may be retrapped or recombine with hole (V_F, X_3^-) centers.

Two centers with Mg^+ ions are created in NaCl crystals after X-irradiation and F-bleaching at room temperature: Mg_c^+- and $Mg_c^+V_a^+V_c^-$-centers [1]. The calculation concludes that the absorption band energy of these centers in alkali halides has to be shifted to longer wavelength than the absorption energy of the free Mg^+ ions [10]. The shift is inversely proportional to the lattice constant [11], i.e.

$$E = \text{const}/d\,,$$

where E energy shift, d lattice constant. The energy of the $3s^1$–$3p^1$ transition of the free Mg^+ ion is 4.43 eV [12]. We calculated the absorption band energy of Mg_c^+ and

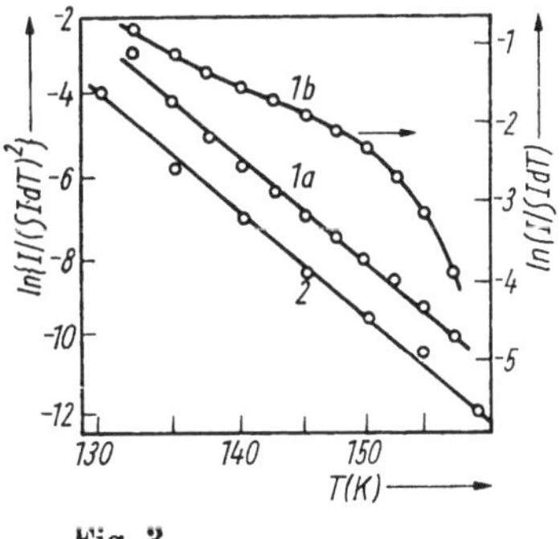

Fig. 3

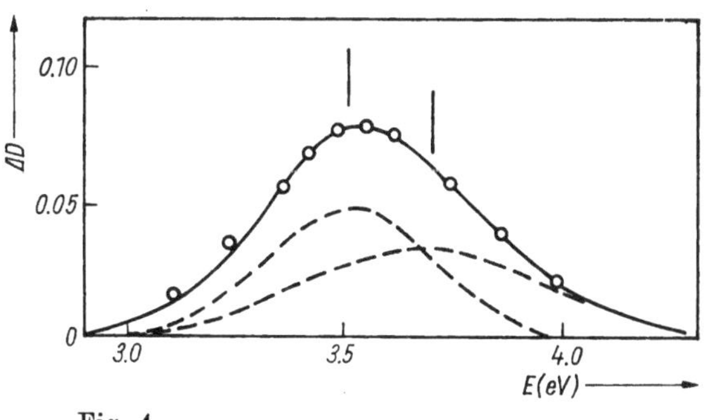

Fig. 4

Fig. 3. Determination of the order of TSL kinetics in the peak at 150 K, after bleaching of $Mg_c^+V_a^+V_c^-$-centers (1a and 1b) at 85 K and (2) after X-irradiation of LiF:Mg at 85 K

Fig. 4. Absorption band of the centers which are bleaching in the range 145 to 170 K in X-rayed LiF:Mg crystals. Dashed curves show the resolved bands

$Mg_c^+V_a^+V_c^-$ in LiF and KCl crystals on the base of the absorption band energy of these centers in NaCl crystal (Table 1). One can see that the calculated and experimental

Table 1

Energies of absorption bands of centers with Mg^+ ions in alkali halides

centers	NaCl			KCl		LiF	
	absorption band (eV)	ΔE (eV)	const (10^6 eV m)	absorption band (eV)			
				calc.	exp.	calc.	exp.
Mg_c^+	3.76 [1]	0.68	1.93	3.83	3.8 [11]	3.49	
$Mg_c^+V_a^+V_c^-$	4.14 [1]	0.30	0.85	4.17		4.01	4.0 [8]

data for the known centers are in good agreement. The calculated absorption band peak of Mg_c^+-centers in LiF lies at 3.5 eV. The Sn_c^+ electron transitions in KCl are shifted by cation vacancies in nearest neighbour position by about 0.1 eV [13]. Consequently $Mg_c^+V_c^-$-centers in LiF must have the absorption band near 3.5 eV. This center has [110] orientation. We can conclude that centers at 3.5 eV seem to be Mg_c^+ ions, connected with cation vacancies in nearest neighbour positions.

2.2 Color centers at 5.0 eV

When the $Mg_c^+V_a^+V_c^-$-centers are bleaching at 85 K, electrons do not enter the conduction band, because there is no photoconductivity of X-rayed LiF:Mg crystals with 4.0 eV light at room temperatures [14]. Consequently, electrons of excited levels of $Mg_c^+V_a^+V_c^-$-centers are trapped by levels of other centers due to a tunneling process. Electrons may be trapped by anion vacancies from $Mg_c^+V_a^+V_c^-$-centers and this creates F-centers near the $Mg_c^{2+}V_c^-$ dipoles. There are also some other results which support this conclusion:

1. When excited by polarized light $Mg_c^+V_a^+V_c^-$-centers are converted into centers at 5.0 eV with the same orientations (see Fig. 2).

2. The absorption band at 5.0 eV is near the F-band. Experiments have shown the band at 5.0 eV to be asymmetric (Fig. 5). The high energy side of this band is about 20% wider than the low energy side. A similar asymmetry is found for the F-band [3, 6].

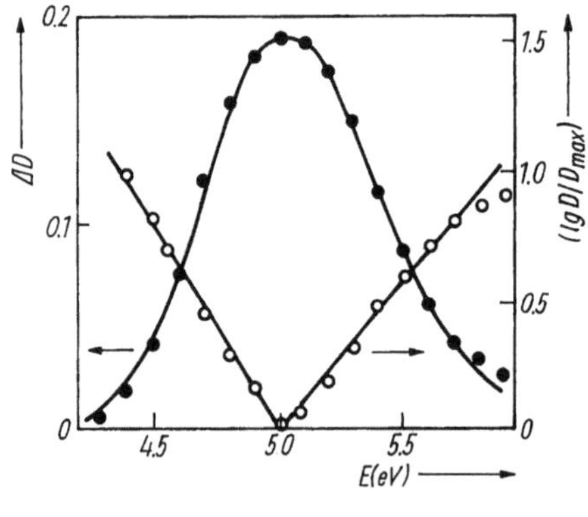

Fig. 5. Experimental shape (points) and theoretical shape (line) of the absorption bands at 5.0 eV

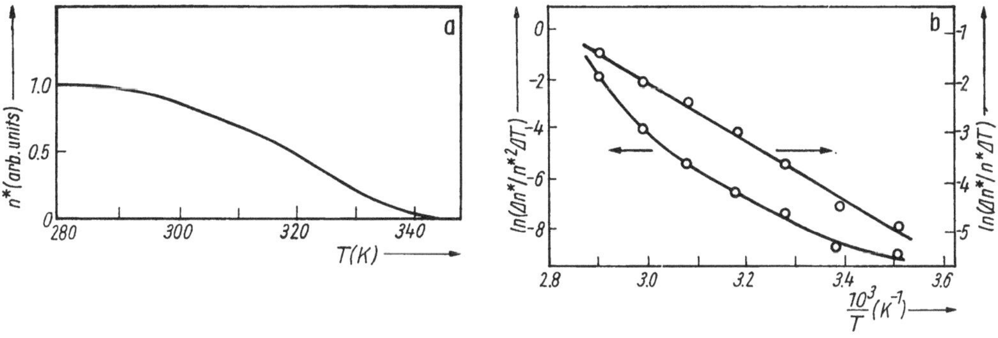

Fig. 6. a) Thermal bleaching of the centers at 5.0 eV at about 320 K and b) determination of the order of bleaching kinetics

3. We determined the relative oscillator strength $f(4.0\ \text{eV})/f(5.0\ \text{eV}) = 1.1 \pm 0.1$ in LiF:Mg crystals (where f is the oscillator strength of the corresponding bands). The relative oscillator strength $f(3.8\ \text{eV})/f(F) = 1.19$ in KCl:Mg crystals (where the band at 3.8 eV is the absorption band of Mg_c^+-centers) [11]. Because $f(F)$ in LiF is almost equal to $f(F)$ in KCl [15] the oscillator strengths of the centers at 5.0 eV and F-centers differ only slightly. It is well known that the oscillator strength of F-centers does only weakly depend on the surroundings (for example Z_1-centers [15]).

4. When the temperature rises the centers at 5.0 eV are bleaching and converting into $Mg_c^+V_a^+V_c^-$-centers. The bleaching at about 320 K obeys a first-order of kinetic (Fig. 6). Electrons are probably tunneling from the F-center levels to the levels of neighbouring Mg^{2+} ions.

Consequently centers at 5.0 eV seem to be F-centers near $Mg_c^{2+}V_c^-$ dipoles.

References

[1] A. Watterich and R. Voszka, Acta phys. (Acad. Sci.) Hungar. **33**, 323 (1973).
[2] R. Voszka and A. Watterich, phys. stat. sol. (b) **55**, 787 (1973).
[3] M. R. Mauhygh, R. W. Christy, and N. M. Johnson, J. appl. Phys. **41**, 2968 (1970).
[4] T. G. Stoebe and S. Watanabe, phys. stat. sol. (a) **29**, 11 (1975).
[5] R. Nink and H.-J. Kos, phys. stat. sol. (a) **35**, 121 (1976).
[6] S. K. Mehta, J. F. Merklin, and H. J. Donnert, phys. stat. sol. (a) **44**, 679 (1977).
[7] J. Mort, Phys. Letters (Netherlands) **21**, 124 (1966).
[8] A. I. Nepomnyachikh and E. A. Radyabov, Optika i Spektroskopiya **48**, 273 (1980).
[9] C. J. Delbecq, W. Hayes, and P. H. Yuster, Phys. Rev. **121**, 1043 (1961).
[10] K. Raksányi and A. Watterich, Symp. Thermodynamics, Spectral and Structural Features of Interaction between Molecules, Excited Molecules or Ions, Wepion (Belgium) 1974 (p. 232).
[11] A. Watterich and K. Raksányi, phys. stat. sol. (b) **80**, 105 (1977).
[12] A. R. Striganov and A. S. Sventitskii, Tablitsi spektralnikh linii, Atomizdat Nauka 1966 (p. 250).
[13] C. J. Delbecq, R. Hartford, D. Shoemaker, and P. Yuster, Phys. Rev. B **13**, 3631 (1976).
[14] W. E. Nelson, J. F. Merklin, and R. S. Lee, Nuclear Instrum. Methods **133**, 387 (1976).
[15] D. Y. Smith and D. L. Dexter, Progr. Optics **10**, 167 (1974).

(Received July 7, 1981)

phys. stat. sol. (a) 68, 83 (1981)

Subject classification: 14.3.2 and 14.3.4; 14.3.3

Department of Physics, Semiconductors, and Microelectronics, Voronezh State University[1])

Field Effect in Semiconductor Critical-Thickness Layers

By

V. F. SYNOROV

Theoretical electrostatic characteristics of structures based on metals, insulators, and semiconductor layers with thickness comparable to or less than the Debye screening radius (critical-thickness layers) and semiconductors of semi-infinite geometry are studied. Some peculiarities of electric field screening in semiconducting critical-thickness layers are revealed on the basis of computer calculations of potential and charge distributions as well as differential capacitances as a function of applied voltage. The obtained characteristics are compared with characteristics of analogous semiconducting structures with semi-infinite geometry.

Рассматриваются теоретические электростатические характеристики структур на основе металлов, диэлектриков, полупроводниковых слоев толщиной сравнимой или меньшей дебаевского радиуса экранирования (критической толщины) и полупроводников полубесконечной геометрии. На основании машинных расчетов распределений потенциала и заряда, а также дифференциальной емкости структур в зависимости от приложенного напряжения выявлен ряд специфических особенностей экранирования електрического поля в полупроводниковых слоях критической толщины. Получены характеристики сравниваются с характеристиками аналогичных структур с полупроводниками полубесконечной геометрии.

1. Introduction

The field effect of the surface charge carrier concentration influenced by the normal to the surface electric field in a semiconductor is generally treated with respect to the model of a semi-infinite semiconductor, whose thickness considerably exceeds the Debye screening radius ($d > L_D$ [1, 2]). This approach is well grounded for traditionally applied semiconductors with relatively high concentration of free charge carriers, such as, for instance, germanium and silicon. When investigating sufficiently thin semiconductor layers as well as semiconductor material with low charge carrier concentration the requirement of semiconductor semi-infinity may be disturbed. In this case the charge carrier concentration changes under the field effect throughout the entire semiconductor thickness, i.e. the thickness of the semiconductor becomes an additional parameter, affecting its characteristics. Further on, the semiconductor layers whose thickness is comparable to or less than their Debye screening length would be called semiconductor critical-thickness layers.

As it has been shown in [3, 4] manifestation of typical dielectric properties such as linear potential distribution and capacitance with respect to the electric field screening is characteristic of very thin semiconductor layers ($d \ll L_D$).

With increasing electric field strength there can be accomplished a transition of the semiconductor from "thin" to "thick" [5] in which the electric field is practically completely screened near the surface. It is as well possible to change the relation between the semiconductor thickness and its screening length by varying the free

[1]) University sg. 1, 394693 Voronezh, USSR.

charge carrier concentration under external effects as light or change of temperature. An essential change of electric field screening conditions may be used in the creation of new functional microelectronic devices. For these purposes one requires a detailed analysis of electric field screening conditions in semiconductor critical-thickness layers.

The problem of contact potential distribution for critical-thickness semiconductors in the absence of an external electric field has been treated in [6, 7]; spatial potential and charge distributions for semiconductors under symmetrical boundary conditions have been analyzed in [8, 9]; the effect of semiconductor thickness on its characteristics in case of fixed arbitrary potentials at its boundaries has been studied in [10, 11]. In practice, however, symmetrical boundary conditions can be realized only in a particular case for a definite state of a structure. An investigation of the problem of fixed arbitrary potential values for a separate layer gives no information about the "critical-thickness layer" behaviour in real structures due to the fact that in this case the potentials at the boundary appear to be mutually dependent.

In the general case the potential distribution caused by the field effect in semiconductor "critical-thickness layers" is determined either by the electric field strength value or by the surface potential at one of the layer boundaries, by the layer thickness, and by electric field screening conditions at the opposite boundary. Electric field screening conditions would be essentially different for semiconductor critical-thickness layer interfaces (S′) with metal (M), semiconductor (S), or insulator (I) due to qualitatively different electric field and potential distributions at the surface of these materials.

Therefore, with the purpose of analyzing the characteristics of real systems with semiconductor "critical-thickness layers" it is supposed appropriately to investigate the models of different specific multilayer structures including metals, insulators, and semi-infinite semiconductors.

2. Technique of Calculation of Electrostatic Structure Parameters

The investigation of models of different multilayer structures was carried out under the following assumption [12 to 14]: the semiconductors are non-degenerate and in the state of thermodynamic quasi-equilibrium with external electric fields; the energy band bending at distances of the order of a de Broglie wave do not exceed kT/q; space charge region dimensions are less than the diffusive length; the geometry of the system under consideration allows to restrict oneself to the solution of a one-dimensional model; impurity centres are completely ionized; free charge carriers are described by Boltzmann statistics. Calculation methods consist in the solution of the Poisson equation for the layer S′ with respect to its finite thickness and for layers of other materials with account of the requirement on continuity of the normal components of the electrostatic induction vector at the interface. For the structure with an insulator the solution presents no difficulties, the solution for a semi-infinite semiconductor has been described in [1, 2]; contact conditions for different boundaries have also been referred to in a number of articles: M–I, I–S [12], S_1–S_2 [13], M–S′ [6, 7]. The Poisson equation for the layer S′ in dimensionless variables may be written as

$$\frac{d^2\psi}{d\xi^2} = \exp\psi - 1 - \lambda_1^{-2}[\exp(-\psi) - 1]\,, \tag{1}$$

where

$$\psi \equiv \frac{q\varphi}{kT}, \qquad \xi \equiv \frac{x}{L_1}, \qquad \lambda_1^2 \equiv \frac{n_{10}}{p_{10}}, \qquad L_1 \equiv \left(\frac{\varepsilon_1\varepsilon_0 kT}{2q^2 n_{10}}\right)^{1/2}.$$

L_1 is the Debye screening length; φ the variation of the electrostatic potential in the layer S′ caused by the electric field, x the coordinate coinciding with electric field direction; n_{10}, p_{10} are, respectively, equilibrium concentrations of electrons and holes in S′ in the absence of electric fields (in case of flat bands), ε_1 and ε_0 the relative permittivities of S′ and vacuum; k is the Boltzmann constant; T the temperature; q the electron charge.

Integrating (1) between two planes with potentials ψ_1 and ψ_2 yields

$$\left(\frac{d\psi}{d\xi}\right)^2_{\psi_1} - \left(\frac{d\psi}{d\xi}\right)^2_{\psi_2} = \exp\psi_1 - \exp\psi_2 - \psi_1 + \psi_2 +$$

$$+ \lambda_1^{-2}[\exp(-\psi_1) - \exp(-\psi_2) + \psi_1 - \psi_2] \equiv f(\psi_1, \psi_2, \lambda_1) . \tag{2}$$

$d\psi/d\xi$ is the dimensionless electric field strength $E = d\psi/d\xi$. If ψ_2 corresponds to one of the boundaries of the layer S′ ($\psi_2 = \psi_{S2}$) and ψ_1 to an arbitrary plane with potential ψ the electric field strength at this plane could be expressed as

$$E(\psi) = \pm [E^2(\psi_{S2}) + f(\psi_1, \psi_{S2}, \lambda_1)]^{1/2} , \tag{3}$$

where the sign is chosen in accordance with the accepted direction of the x-axis regarding that $\psi > 0$ corresponds to electron accumulation.

Integrating (3) from one interface of the layer S′($\mathscr{E}_d$) $\equiv d/L_1$ where d is the thickness of layer S′ up to an arbitrary point with the coordinate $\mathscr{E}$ we obtain the dependence between the potential and the coordinate as

$$\mathscr{E} = \mathscr{E}_d - \left| \int\limits_{\psi}^{\psi_{S2}} [E^2(\psi_{S2}) + f(\psi, \psi_{S2}, \lambda_1)]^{-1/2}\, d\psi \right| . \tag{4}$$

Integrating (3) over the entire layer S′ thickness we find a relation between potentials at both boundaries, the strength at one of the boundaries, and the layer thickness,

$$\mathscr{E}_d = \left| \int\limits_{\psi_{S1}}^{\psi_{S2}} [E^2(\psi_{S2}) + f(\psi_1, \psi_{S2}, \lambda_1)]^{-1/2}\, d\psi \right| . \tag{5}$$

Now in order to bring the problem of the state of layer S′ at a fixed thickness to the problem depending on one parameter characterizing the state of the whole system, it is necessary to regard specific conditions at the boundaries of layer S′.

Let us assume that the layer S′ at the boundary S2 contacts with a semifinite semiconductor S so that the interface S_1–S is an abrupt heterojunction [13]. For a semi-infinite semiconductor the electric field strength at the surface is uniquely defined by the potential at the surface. Then, with respect to [1, 13] we may write

$$E^2(\psi_{S2}) = \frac{\varepsilon_2^2 L_1^2}{\varepsilon_1^2 L_2^2} f[(\psi_{S2} - \psi_{12}); 0; \lambda_2] , \tag{6}$$

where

$$L_2 \equiv \left(\frac{\varepsilon_2\varepsilon_0 kT}{2q^2 n_{20}}\right)^{1/2} , \qquad \lambda_2^2 = \frac{n_{20}}{p_{20}} .$$

ε_2, n_{20}, p_{20} are the relative permittivity and equilibrium charge carrier concentrations in the semiconductor S; ψ_{12} is the difference of work functions of semiconductors S′ and S in dimensionless units.

Substitution of (6) into (4) and (5) gives the potential distribution along the coordinate at each value of ψ_{S2} and ψ_{S1} at the given layer thickness. Substituting ψ_{S1} into (3) we define the strength at the external surface of layer S′ ($E(\psi_{S1})$).

If the heterojunction S′–S is a component of an MIS′S structure, the voltage (U) on the structure could be expressed as

$$U = \frac{kT}{q}\left(\psi_{M1} + \psi_{S1} + \frac{\varepsilon_1 h}{\varepsilon_i L_1} E(\psi_{S1})\right), \tag{7}$$

where ε_1 and h are relative permittivity and insulator layer thickness, respectively.

Thus, taking ψ_{S2} as an independent parameter we may calculate the potential distribution in structures with the heterojunction S′–S. Characteristic of an M–S′ contact is the potential constancy in the layer S′ at the interface with the metal ψ_{S2}, this potential being equal to the difference of work functions of materials M and S′. Then the dependence between $E(\psi_{S2})$ and ψ_{S1} may be defined from (5), $E(\psi_{S1})$ from (3).

The voltage on the structure $M_1S'IM_2$ is written as

$$U = \frac{kT}{g}\left[\frac{\varepsilon_1 h}{\varepsilon_i L_1} E(\psi_{S1}) - (\psi_{S1} - \psi_{S2}) + \psi_{M12}\right], \tag{8}$$

where ψ_{M12} is the difference of work functions of metals M_1 and M_2 in dimensionless units.

For a structure of MS′IS type the voltage is given as

$$U = \frac{kT}{g}\left[\frac{\varepsilon_1 h}{\varepsilon_i L_1} E(\psi_{S1}) - (\psi_{S1} - \psi_{S2}) + \psi_{SS} + \psi_{MS}\right], \tag{9}$$

where ψ_{SS} is the potential at the insulator boundary in the semi-infinite semiconductor S, ψ_{MS} the difference of work functions of M and S.

The charge Q in the layers of each material or in several layers was calculated according to the Gauss theorem,

$$Q = \varepsilon_0 \varepsilon (E_{S1} - E_{S2}), \tag{10}$$

where ε and E_{S1}, E_{S2} are permittivities and strength of electric fields at the layer boundaries.

The dependence of the differential capacitance of the structure (C) on the applied voltage is computed by the formula

$$C = \left|\frac{dQ}{dU}\right|. \tag{11}$$

As long as the given equations cannot be solved analytically in the general case, the solution was carried out by interpolation methods using computers at specific values of the system parameters under the assumption that all metals have equal work functions.

For the purpose of demonstration of the electric field screening peculiarities in structures with critical-thickness layers the characteristics of analogous structures with layer thicknesses $d \gg L_D$ (semi-infinite semiconductors) have been computed in the absence of free charge carriers in the layer S′ (layer I′).

3. Basic Results

3.1 Structures with heterojunction

Characteristic of structures of MIS′S type is the displacement of the effective localization centre from semiconductor S to layer S′ with increasing external voltage [14]. In the low-voltage region the electric field is distributed in the layer S′ as well as in the insulator, the total induced charge being localized in the semiconductor S. With

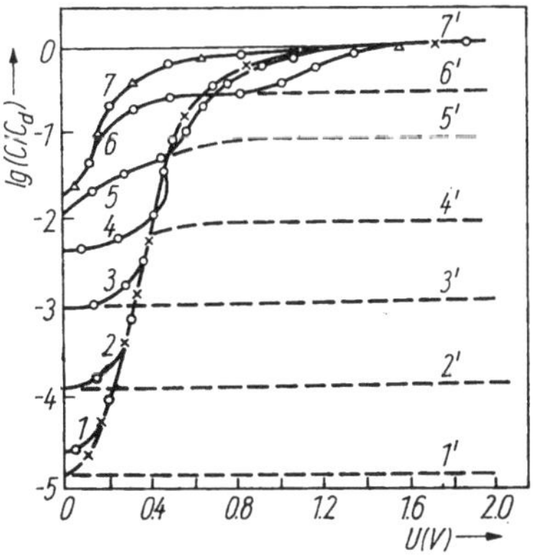

Fig. 1. Capacitance–voltage (C–U) characteristics of the structures: ○ MIS′S; - - - - MI′IS; × MIS*; △ MIS (1), (1′) $d/L_D = 1$; (2), (2′) 10^{-1}; (3) (3′) 10^{-2}; (4), (4′) 10^{-3}; (5), (5′) 10^{-4}; (6), (6′) 10^{-5}; (7), (7′) 10^{-6}

increasing voltage the charge carrier concentration throughout the entire thickness of layer S′ grows, the effective screening length consequently reduces and becomes less than the layer thickness.

The potential and the electric field strength at the interface S′–S do not change with further voltage increase, the charge in the layer S′ increases abruptly, while the charge in the semiconductor S remains constant. Thus, at different voltage levels it is possible to monitor each space charge in each semiconductor separately.

Fig. 1 shows equilibrium capacitance–voltage characteristics (C–U) of MIS′S-type structures computed for the following parameters of the system: equilibrium free charge carrier concentration in semiconductors S′ and S, respectively, $n_{01} = 10^{10}$ m^{-3} and $n_{02} = 1.5 \times 10^{16}$ m^{-3}; $n_0 = p_0$; semiconductor thickness $h = 10^{-7}$ m; permittivities of the insulator and semiconductors S′ and S, respectively, $\varepsilon_i = 3.8$, $\varepsilon_1 = 10$, $\varepsilon_2 = 12$, $T = 300$ K. Comparing these characteristics with those of corresponding structures with double insulator layer (MII′S), structures with a semi-infinite semiconductor of material S′(MIS*), and usual MIS structures one should point out the following facts: in the low-voltage region the characteristics of MIS′S structures are identical with those of MII′S structures, i.e. in the layer S′ there occurs the effect of capacitance constancy previously described in [3]. With increasing voltage the relation between the charges in S′ and S changes, and owing to the decrease of the effective screening length, the layer S′ becomes analogous to a semi-infinite semiconductor. With the thickness of layer S′ diminishing the voltage region of the capacitance constancy effect increases. With further diminishing the thickness of the layer S′ the C–U curve of an MIS′S structure coincides with the characteristics of an MIS structure, i.e. the critical-thickness layer influence is negligible. Thus, by means of varying the structure parameters it is possible to obtain a series of qualitatively different characteristics. The model of the MIS′S structure can be also applied for calculating the electrostatic characteristics of MIS structures with surface-doped semiconductors in which the functions of layer S′ are performed by a semiconductor surface layer with impurity concentrations distinct from the bulk concentration.

3.2 Structure with M–S′ contact

The calculation of the charge distribution in an MS′ system shows that in the regimes of weak accumulation and depletion of majority carriers and at the same surface potential values (φ_S) the charge in the layer S′ ($Q_{S'}$) is less than the charge in the semi-infinite semiconductor (Q_S, Fig. 2). This can be accounted for by the fact that the space charge region depth, when $d < L_1$, is limited by the layer thickness. Resulting

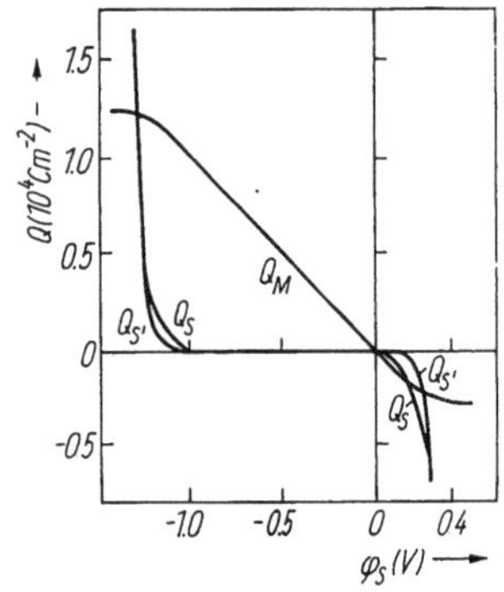

Fig. 2. Charge dependences in the layer S′ and in the semi-infinite semiconductor on the surfaces potential. $n_0 = 10^{17}\ m^{-3}$, $n_0/n_i = 10^7$, $\varepsilon_1 = 10$, $T = 300$ K, $d = 8.5 \times 10^{-7}$ m, $d/L_D = 0.1$

from this, the inversion in S′ occurs at larger surface potential values than in the semi-infinite semiconductor, which can be called an "inversion delay effect" distinct from the effect of "enhanced inversion" described in [8, 9] and characteristic of critical-thickness layers under symmetrical boundary conditions. With strong accumulation or inversion the dependence $Q_{S'}(\varphi_S)$ is the same as in a semi-infinite semiconductor. In the range of Q_S variation from weak accumulation to weak inversion the electric field screening by the layer S′ is negligible, the entire charge induced in the system MS′ is practically localized in the metal (Q_M) and changes linearly with the surface potential. At high charge carrier concentrations in S′ the charge value reaches saturation and further charge increase in the system is due to the layer S′.

Fig. 3 shows that capacitance–voltage characteristics of MS′IM structures differ from the characteristics of analogous MIS* structures with semi-infinite semiconductors by the existence of the constant capacitance section in the regimes of depletion and weak accumulation of majority carriers in the layer S′. The capacitance value in this case is greater than the minimum capacitance value of the structure MIS* due to the limitation of the space charge region depth by the thickness of layer S′ and is determined by the geometrical capacitance of the layer S′I. With increasing external voltage in regimes of accumulation or inversion in the layer S′ the structure capacitance increases abruptly and approaches the capacitance of the insulator layer. The high-frequency C–U characteristics of MS′IM structure calculated under the assumption that the testing signal period is considerably smaller than the relaxation time of minority carriers, but larger than the majority carrier relaxation time, possess a constant capacitance value in the region corresponding to the inversion in S′. The sign of majority carriers in the layers of semiconductors with wide energy band gap can be determined from the form of this characteristic.

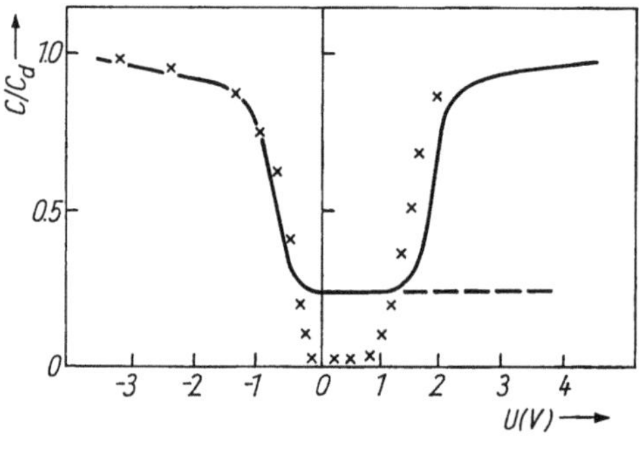

Fig. 3. Capacitance–voltage characteristics of the structures MS′IM (———) and MIS* (- - - -). × HF curve, $n_0 = 10^{17}\ m^{-3}$, $n_0/n_i = 10^7$, $d = 8.5 \times 10^{-7}$ m, $h = 10^{-7}$ m, $\mathscr{E}_d = 3.8$

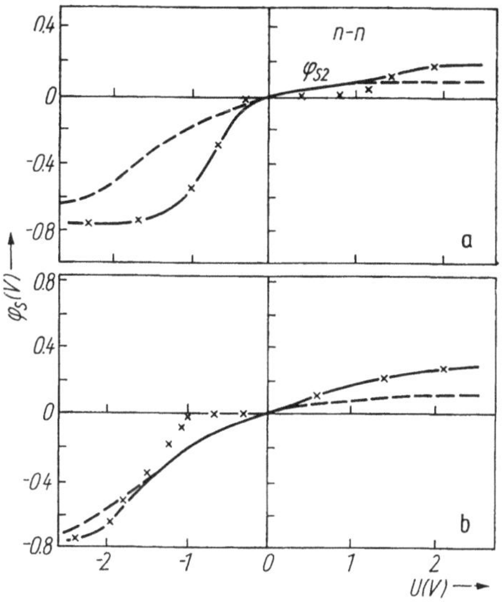

Fig. 4. The dependence of surface potential in semiconductors S on the applied voltage a) in isotype structures (—— MS′IS, - - - - MI′IS, × S*IS; $n_{01} = 10^{17}\ m^{-3}$, $n_{02} = 10^{21}\ m^{-3}$, $\lambda_1 = 10^7$, $\lambda_2 = 6.7 \times 10^4$, $d = 8.5 \times 10^{-7}$ m ($d/L_{D1} = 0.1$, $h = 2 \times 10^{-8}$ m) and b) in non-isotype MS′IS structures, in MI′IS structures with two semiconductors of semi-infinite geometry ($p_{01} = 10^{16}\ m^{-3}$, $n_{02} = 10^{21}\ m^{-3}$, $d/L'_{D1} = 0.1$, $h = 2 \times 10^{-8}$ m)

The existence of two space charge regions, separated by an insulator layer is characteristic of MS′IS-type structures. It should be noted that there exists an essential difference between the characteristics of structures with non-isotype and isotype semiconductors, caused by the fact that in non-isotype structures the space charge region state (accumulation or depletion) in both semiconductors is the same while in isotype structures it is reverse.

Fig. 4 shows an example of the surface potential dependence on the external voltage in semiconductor S for isotype and non-isotype structures MS′IS, S*IS, and MI′IS. In case of weak accumulation or depletion in the layer S these dependences for MS′IS structures coincide with the characteristics of structures with a double insulator layer; when accumulation or inversion is strong they are analogous to the characteristics of structures with two semi-infinite semiconductors. It is obvious that in the semiconductor S of an MS′IS structure the surface potential dependence on voltage is steeper than in case of M′IIS structures. This phenomenon can possibly be used for increasing the effectiveness of space charge control in devices based on MIS structures by partially or completely substituting the insulator layer by semiconductors with wide energy gap. In order to increase the characteristic steepness in the semiconductor S we should utilize non-isotype structures in the accumulation regime, isotype structures in the regimes of depletion and inversion.

Fig. 5 shows the reflection of the surface potential variation in the layer S caused by the increase of free charge carrier concentration in case when electron–hole pairs in

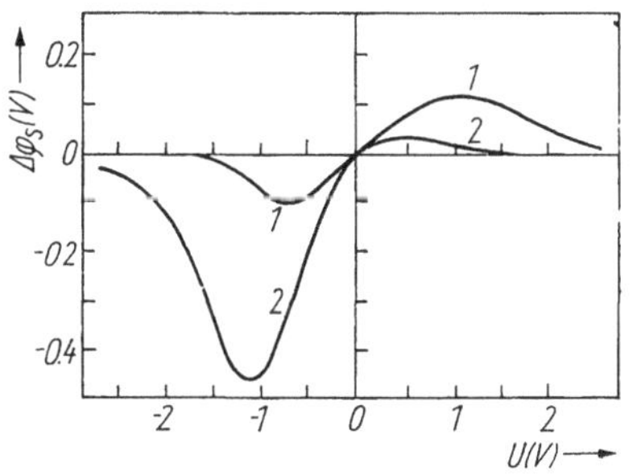

Fig. 5. Surface potential variation in semiconductor S of isotype (curve 1) and non-isotype (curve 2) MS′IS structures under the effect of light from the intrinsic absorption region. (1) n–n, (2) p–n

S′ are generated by light. The maximum surface potential variation in a semiconductor substrate under the effect of light is reached in the voltage region corresponding to the majority carrier depletion in the layer S′ for isotype as well as for non-isotype structures. However, the absolute value of this variation is considerably greater for non-isotype structures. So, for instance, at a voltage of 1 V on the isotype MS′IS structure the exposure to light causes a transition of the space charge region in the semiconductor S from weak accumulation to strong inversion ($\Delta\varphi_S = 0.45$ V) which can be applied in light-controlled MIS structures.

References

[1] C. G. B. Garrett and W. H. Brattain, Phys. Rev. **99**, 376 (1955).
[2] R. H. Kingston and S. T. Neustadter, J. appl. Phys. **26**, 718 (1955).
[3] B. I. Sysoev and V. F. Synorov, Fiz. Tekh. Poluprov. **6**, 1856 (1972).
[4] B. I. Sysoev, V. F. Synorov, A. N. Likholet, and L. A. Bityutskaya, Trudi mezhdunarodnogo simpozium, Fizika tonkykh plenok, Warsaw 1975 (p. 242).
[5] V. E. Lashkarev, Izv. Akad. Nauk SSSR, Ser. fiz. **16**, 203 (1952).
[6] V. B. Sandomirskii and V. P. Smylga, Fiz. tverd. Tela **1**, 307 (1959).
[7] A. I. Gubanov and S. U. Davydov, Fiz. Tekh. Poluprov. **5**, 369 (1971).
[8] L. S. Gasanov, Radiotekhnika i Elektronika **12**, 1524 (1967).
[9] L. S. Gasanov and V. I. Stafeev, Fiz. Tekh. Poluprov. **2**, 426 (1968).
[10] Z. Zerhot and V. Sneidar, J. Physique **20**, 903 (1970).
[11] Z. Zerhot and V. Sneidar, Czech. J. Phys. **B21**, 1114 (1971).
[12] V. F. Synorov, A. N. Likholet, and B. I. Sysoev, phys. stat. sol. (a) **33**, K161 (1976).
[13] V. F. Synorov, A. N. Likholet, and B. I. Sysoev, Mikroelektronika **6**, 56 (1977).
[14] B. I. Sysoev, A. N. Likholet, V. F. Synorov, and A. N. Rovinskii, Mikroelektronika **6**, 454 (1977).

(Received July 20, 1981)

phys. stat. sol. (a) 68, 91 (1981)

Subject classification: 10.1; 21.1

Department of Physics, University of Alberta, Edmonton[1])

Experimental Evidence of a Breakdown of the Column Approximation in Calculations of Contrast in Electron Microscope Images of Stacking Faults

By

R. PEREZ, K. Z. BOTROS[2]), and S. S. SHEININ

Experimental evidence is presented which indicates that calculations of diffraction contrast in electron microscope images obtained under weak beam diffraction conditions can be in considerable error if the column approximation is employed. This evidence is based on a comparison of experimental densitometer traces obtained from stacking faults in f.c.c. cobalt with theoretical non-column and column approximation calculations. A theoretical explanation of the significant differences between non-column and column approximation profiles is also given and some interpretive errors which can arise if the column approximation is employed are discussed.

Es wird experimentell nachgewiesen, daß die Berechnung des Beugungskontrastes in elektronenmikroskopischen Bildern unter der Bedingung schwacher Strahlbeugung mit beträchtlichen Fehlern behaftet sein kann, wenn die Säulennäherung verwendet wird. Dies wird anhand eines Vergleichs zwischen experimentellen Densitometerkurven von Stapelfehlern in kfz.-Kobalt mit theoretischen Rechnungen mit und ohne Verwendung der Säulenapproximation gezeigt. Eine theoretische Erklärung für die beobachteten erheblichen Unterschiede zwischen den mit beiden Approximationen berechneten Profilen wird angegeben und es werden einige Interpretationsfehler diskutiert, die bei Verwendung der Säulenapproximation auftreten können.

1. Introduction

The column approximation has been widely used in dynamical theory calculations of the contrast exhibited in electron microscope images of lattice defects. The validity of this approximation for weak beam diffraction conditions has been brought into question in theoretical calculations of dislocation profiles by the work of Howie and Sworn [1], Humphreys and Drummond [2], and Lewis and Villigrana [3]. In a recent paper Jap and Sheinin [4] have also shown that the column approximation may not be valid in calculating weak beam images of stacking faults. On the basis of theoretical considerations, therefore, there seems to be strong evidence that, under certain circumstances, the column approximation is not valid. Direct experimental evidence for a breakdown of the column approximation has, however, not been presented in the literature. The purpose of this paper is to present such evidence. The approach is based on a comparison of experimental densitometer traces obtained from images of stacking faults in crystals of known thickness with theoretical non-column and column approximation profiles.

---

[1]) Edmonton, Alberta, T6G 2J1, Canada.

[2]) Present address: Department of Physics, University of Kuwait, Kuwait.

2. Experimental Procedure

The experimental results to be presented in this paper were obtained from images of stacking faults in f.c.c. cobalt. In order to carry out a comparison of non-column and column approximation calculations with experiment, the following information about the experimental stacking fault images must first be obtained: the crystal thickness in which the fault lies, the phase angle, α, of the fault ($\alpha = 2\pi \boldsymbol{g} \cdot \boldsymbol{R}$ where $\boldsymbol{g}$ is the lowest order reflection in the systematic set excited and $\boldsymbol{R}$ is the displacement vector of the fault), the edges of the image which correspond to the intersections of the fault with the top and bottom surfaces of the crystal, and the angle of inclination of the fault plane with the crystal surfaces. In order to determine the crystal thickness in which the fault lies, wedge crystal specimens have been prepared. Since the wedge angles of the crystals were unknown, absolute values of crystal thickness could not be determined. However, for purposes of comparing experimental profiles with theoretical calculations, only the crystal thickness expressed in terms of extinction distance need be known. This can be easily determined from the thickness contours in the wedge crystal immediately adjacent to the image of the fault (see, for example, Fig. 1). The phase angle α and the edges of the image which correspond to the intersections of the fault with the top and bottom surfaces of the crystal have been determined from bright and dark field images recorded under strong beam diffraction conditions (i.e. with the deviation parameter S_g set close to zero) [5]. Finally the angle of inclination of the fault plane was determined from the projected width of the fault along the bottom surface of the crystal as obtained from the stacking fault image and the crystal thickness at this width (determined in extinction distances by the method described above and then multiplied by the theoretical extinction distance). The

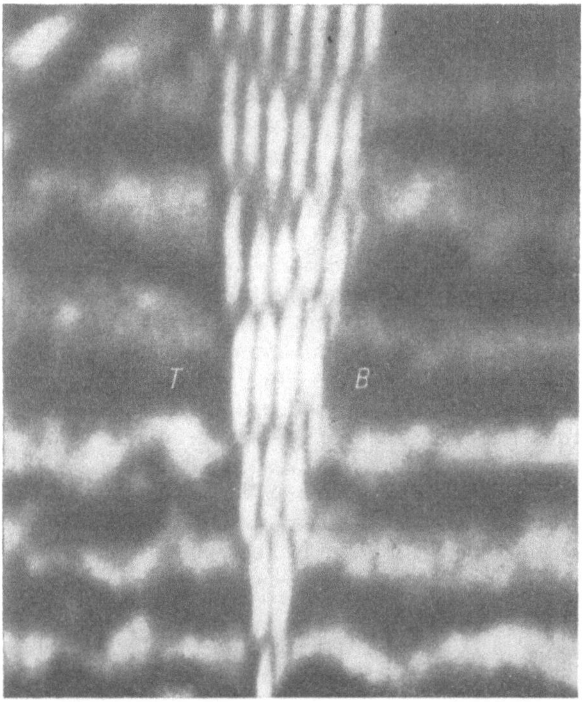

Fig. 1. Dark field image of an $\alpha = -2\pi/3$ stacking fault in f.c.c. cobalt obtained under weak beam diffraction conditions. The systematic set excited was (111) and $S_{111} = 9.0 \times 10^{-3}$ Å^{-1}. The intersections of the fault with top and bottom surfaces are indicated by T and B, respectively. The fault is inclined at an angle of 62° with respect to the foil surfaces

value so obtained was then checked by determining the angle of inclination from the orientation of the specimen, obtained from the diffraction pattern, and by assuming that the fault lies on a {111} plane.

3. Comparison of Theory with Experimental Results

Experimental stacking fault images have been obtained by first orienting the crystal so that (111) systematic set was excited (orientations were chosen so that the effects of non-systematic reflections were minimized). The specimen was then tilted to weak beam diffraction conditions, and a dark field image was recorded (see, for example, Fig. 1). Densitometer traces were then obtained from these images, an example of which is shown in Fig. 2. Using the methods referred to in Section 2, it was found that the fault shown in Fig. 1 had a value of $\alpha = -2\pi/3$. The most significant feature to note in the experimental profile in Fig. 2 is that there is a dark fringe at the side of the image corresponding to the intersection of the fault with the top surface while at the side of the image corresponding to the intersection of the fault with the bottom surface a bright fringe is obtained.

In order to compare the experimental results described above with theory, calculations of stacking fault profiles for both the non-column and column approximation cases were carried out. The calculations based on the column approximation were carried out by using the standard approach described extensively in the literature [5] while the non-column calculations employed the theory presented by Jap and Sheinin [4]. The results of the column approximation calculations are shown in Fig. 3a and indicate that a bright fringe should be obtained on both sides of the image. This is clearly in qualitative disagreement with the experimental profile in Fig. 2. The results of the non-column approximation calculation are shown in Fig. 3b and indicate that a dark fringe should be obtained at the side of the image corresponding to the intersection of the fault with the top surface while the other side should show a bright fringe. Clearly, therefore, this profile is in excellent agreement with experiment.

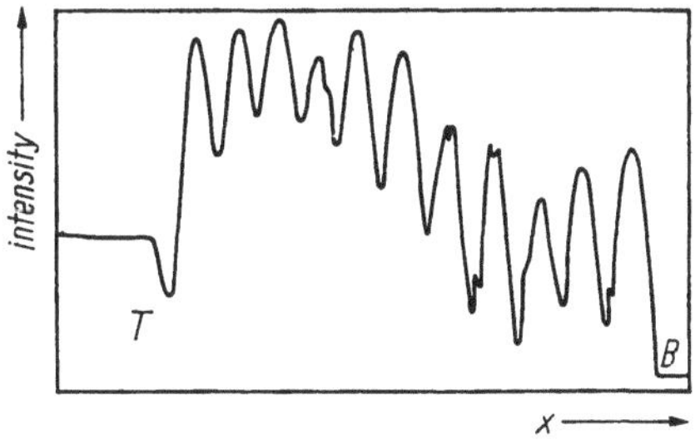

Fig. 2

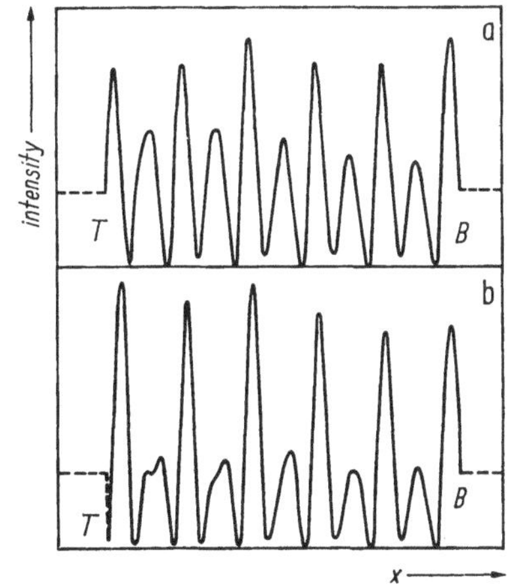

Fig. 3

Fig. 2. Experimental densitometer trace obtained from the image shown in Fig. 1 at a crystal thickness of $11\frac{1}{2}$ extinction distances. The intersections of the fault with the top and bottom surfaces are indicatet by T and B, respectively

Fig. 3. Theoretical stacking fault profiles corresponding to the experimental profile in Fig. 2; a) is based on the column approximation, and b) is based on the non-column theory. The intersection of the fault with the top and bottom surfaces are indicated by T and B, respectively

It is important to note the following point with regard to the non-column profile in Fig. 3b. The dark fringe at the side of this profile corresponding to the intersection of the fault with the top surface arises from a discontinuity in the diffracted beam intensity as given by the non-column stacking fault calculation and the background intensity. It should be stressed that this dark fringe was in fact observed experimentally. This result is, of course, in complete contradiction to calculations based on the assumption that the column approximation is valid which always predicts that the diffracted beam intensity at the edges of the profile will be continuous with background. An explanation of this significant difference between a theory based on the column approximation and one that is not is given in the next section.

4. Explanation of Differences between Column and Non-Column Profiles

Stacking fault profiles based on the column approximation always give diffracted beam intensities equal to the background intensity at positions in the profile corresponding to the intersections of the fault with top and bottom surfaces. Non-column and column approximation profiles would appear to agree on the side of the profile corresponding to the intersection of the fault with the bottom surface (see Fig. 3 and [4]). As mentioned in Section 3, however, these intensities can be quite different on the side of profile corresponding to the intersection of the fault with the top surface. In order to understand this behaviour, it is useful to obtain the expressions for the diffracted beam amplitudes in the non-column and column approximation cases in comparable form.

This can be done in the column approximation case by using an expression given by Sheinin and Botros [6] expressed in the following form:

$$\varphi_g(t_1, t) = \sum_j \sum_m \chi_C^{(j,m)} C_g^{(j)} \exp\left[2\pi i(\gamma^{(m)} - \gamma^{(j)})\, t_1\right] \exp\left[2\pi i \gamma^{(j)} t\right]; \tag{1a}$$

$$\boldsymbol{\chi}_C^{(j,m)} = C_0^{(m)} C_T \{\exp(2\pi i \boldsymbol{g} \cdot \boldsymbol{R})\}\, \boldsymbol{C}_g^{(m)}, \tag{1b}$$

where $\boldsymbol{\chi}_C^{(j,m)}$ is a column vector containing the coefficient $\chi_C^{(j,m)}$ ($j = 1, N$ where N is the number of beams taken into account). t_1 is the depth of the fault, t is the crystal thickness, $C_g^{(i)}$ is a Fourier coefficient appearing in the expression for the Bloch wave $b^{(i)} = \sum_g C_g^{(i)} \exp\left[2\pi i(\boldsymbol{k}^{(i)} + \boldsymbol{g}) \cdot \boldsymbol{r}\right]$, $\boldsymbol{C}_g^{(i)}$ is a column vector whose elements are $C_g^{(i)}$, C is a square matrix whose columns are $\boldsymbol{C}_g^{(i)}\, t (i = 1, N)$, the subscript T indicates the transpose of a matrix and $\{\exp(2\pi i \boldsymbol{g} \cdot \boldsymbol{R})\}$ is a diagonal matrix whose elements are $\exp(2\pi i \boldsymbol{g} \cdot \boldsymbol{R})$. The superscript (j, m) refers to a Bloch wave arising from interbranch scattering from dispersion surface branch m to j (see Fig. 4a, noting that in the column case $\theta = 0$ and the wave points (j, m) and (j, j) coincide). From (1a), the diffracted beam amplitudes at positions in the profile corresponding to the intersections of the fault with the top ($t_1 = 0$) and bottom ($t_1 = t$) surfaces are

$$\varphi_g(t) = \sum_j \sum_m \chi_C^{(j,m)} C_g^{(j)} \exp(2\pi i \gamma^{(j)} t) \tag{2}$$

and

$$\varphi_g(t) = \sum_j \sum_m \chi_C^{(j,m)} C_g^{(j)} \exp(2\pi i \gamma^{(m)} t), \tag{3}$$

respectively. The corresponding non-column expressions can be obtained from the general expression for diffracted beam amplitudes for this case [4],

$$\varphi_g(\boldsymbol{r}_e) = \sum_m \sum_j \chi_{NC}^{(j,m)} C_g^{(j,m)} \exp\left[2\pi i \boldsymbol{\gamma}^{(j,m)} \cdot \boldsymbol{r}_e\right]; \tag{4a}$$

$$\boldsymbol{\chi}_{NC}^{(j,m)} = C_0^{(m)} [C^{(m)}]^{-1} \{\exp 2\pi i \boldsymbol{g} \cdot \boldsymbol{R}\}\, \boldsymbol{C}_g^{(m)}, \tag{4b}$$

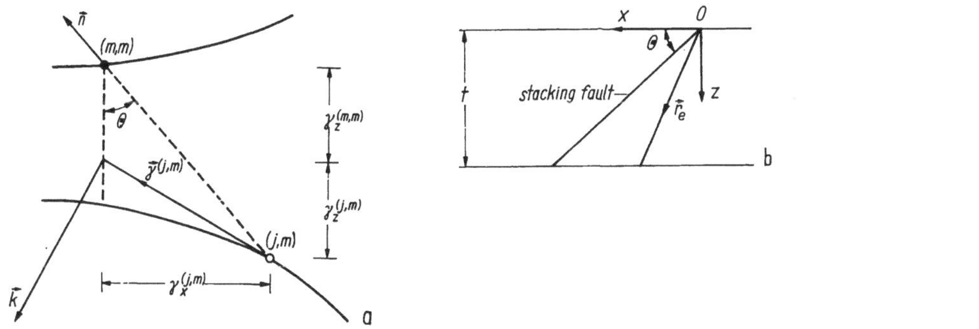

Fig. 4. a) Schematic diagram in reciprocal space showing the wave points associated with Bloch waves excited below the stacking fault in b). The wave points (m, m) and (j, m) represent intrabranch scattering of Bloch wave m and interbranch scattering from Bloch waves m to j, respectively. $\boldsymbol{k}$ is the electron wave vector after correction for the mean lattice potential. b) Schematic diagram in real space showing a stacking fault in a crystal of thickness t. 0 is the origin of the coordinate system and $\boldsymbol{r}_e$ is a position vector terminating on the exit surface

where $\boldsymbol{\chi}_{NC}^{(j,m)}$ is a column vector containing the coefficients $\chi_{NC}^{(j,m)}$ $(j = 1, N)$. $C_g^{(j,m)}$ is a Fourier coefficient in the expression for a Bloch wave excited below the fault by interbranch scattering from dispersion surface branch m to j, $b^{(j,m)} = \sum_g C_g^{(j,m)} \exp[2\pi i(\boldsymbol{k}^{(j,m)} + \boldsymbol{g}) \cdot \boldsymbol{r}]$, C^m is a square matrix whose columns are $\boldsymbol{C}_g^{(j,m)}$ $(j = 1, N)$, and $\boldsymbol{C}_g^{(j,m)}$ is a column vector whose elements are $C_g^{(j,m)}$. The parameters $\gamma^{(j,m)}$ and $\boldsymbol{r}_e$ are defined in Fig. 4a and b. Using (4a) and the fact that $\gamma^{(m)} = \gamma_z^{(m,m)}$ gives $\gamma_x^{(j,m)} = (\gamma^{(m)} - \gamma_z^{(j,m)}) \tan\theta$ (see Fig. 4a). The diffracted beam amplitudes at positions in the profile corresponding to the intersections of the fault with the top $(x = 0, z = t)$ and bottom $(x = t/\cos\theta, z = t)$ surfaces are therefore

$$\varphi_g(t) = \sum_j \sum_m \chi_{NC}^{(j,m)} C_g^{(j,m)} \exp[2\pi i \gamma_z^{(j,m)} t] \tag{5}$$

and

$$\varphi_g(t) = \sum_j \sum_m \chi_{NC}^{(j,m)} C_g^{(j,m)} \exp[2\pi i \gamma^{(m)} t]\,. \tag{6}$$

It is first of interest to compare the non-column and column approximation expressions for the diffracted beam amplitude at the intersection of the fault with the bottom surface by comparing (3) and (6). It is clear from these equations that the phase angles in each case are the same and any differences in diffracted beam amplitude must therefore arise from differences in the amplitude terms $\chi_{NC}^{(j,m)} C_g^{(j,m)}$ and $\chi_C^{(j,m)} C_g^{(j)}$. It turns out in fact that these amplitude terms are very nearly the same over a wide range of conditions. Thus, for example, under weak beam diffraction conditions $C_g^{(j,m)}$ and $C_g^{(j)}$ typically differ by less than about 5% for angles of inclination of the fault, θ, up to about 75°. Under these circumstances $C^{(m)}$ in (4b) is approximately equal to C in (1b). Since C is an orthogonal matrix, $[C^{(m)}]^{-1} \approx C_T$ and therefore $\chi_{NC}^{(j,m)} \approx \chi_C^{(j,m)}$. The result that diffracted beam intensities at the intersection of the fault with the bottom surface are very nearly equal in the non-column and column approximation cases is therefore explained. The reason that these intensities are different at the position in the profile corresponding to the intersection of the fault with the top surface can now also be understood by comparing (2) and (5). The amplitude terms, as discussed above, will be approximately the same for a wide range of conditions. The phase angles $(2\pi\gamma_z^{(j,m)} t)$ and $(2\pi\gamma^{(j)} t)$ can, however, be significantly different, even though the differences in the values of $\gamma_z^{(j,m)}$ and $\gamma^{(j)}$ may be relatively small. This can be seen by taking typical values of t and $\gamma^{(j)}$ of 1000 Å and 5 ×

$\times 10^{-3}$ Å^{-1}, respectively. These values give a difference in the above phase angles of π for only a 5% difference between $\gamma_z^{(j,m)}$ and $\gamma^{(j)}$ with the result that the diffracted beam amplitudes in the non-column and column approximation cases can be significantly different.

5. Discussion

The differences between non-column and column approximation calculations of stacking fault profiles discussed in Section 4 can lead to serious interpretive errors if the analysis is based on the column approximation profile. Thus, for example, column approximation stacking fault profiles obtained for weak beam diffraction conditions are symmetrical for both $\alpha = \pm 2\pi/3$ faults in the sense that both edge fringes exhibit the same contrast. In Section 3 it was shown that the experimentally observed profile is in fact asymmetrical. Thus, interpretation based on the column approximation could lead the electron microscopist to suspect that the defect imaged does not have the usual values of displacement vector of the type $\frac{1}{6}$ [112] and is therefore not one of the commonly occurring intrinsic or extrinsic faults in f.c.c. materials. The fact is, of course, that the fault does have a value of $\alpha = -2\pi/3$ and the asymmetry arises from the fact that the column approximation fails. It is clear from these results that great care should be exercised in using the column approximation to calculate weak beam images of stacking faults.

Acknowledgements

The authors are grateful to NSERC, Canada and CONACYT, Mexico for financial support.

References

[1] A. HOWIE and C. H. SWORN, Phil. Mag. **22**, 861 (1970).
[2] C. J. HUMPHREYS and R. A. DRUMMOND, VI. European Congr. Electron Microscopy, Jerusalem 1976 (p. 142).
[3] A. L. LEWIS and R. E. VILLIGRANA, Acta. cryst. **A35**, 276 (1979).
[4] B. K. JAP and S. S. SHEININ, phys. stat. sol. (b) **105**, 137 (1981).
[5] P. B. HIRSCH, A. HOWIE, R. B. NICHOLSON, D. W. PASHLEY, and M. J. WHELAN, Electron Microscopy of Thin Crystals, Robert E. Kreiger Publ. Co., New York 1977.
[6] S. SHEININ and K. Z. BOTROS, phys. stat. sol. (a) **13**, 585 (1972).

(Received June 16, 1981)

phys. stat. sol. (a) 68, 97 (1981)

Subject classification: 18.3; 22.8.2

Solid State Laboratory, Department of Chemistry, Indian Institute of Technology, Bombay[1])

Effect of Cation Distribution on the Properties of Some Magnesium–Nickel Ferrites

By

K. SESHAN, A. L. SHASHIMOHAN, D. K. CHAKRABARTY[2]), and A. B. BISWAS†

Monophasic magnesium–nickel ferrites with varying amounts of magnesium and nickel are prepared and subjected to different heat treatments. The cation distribution in these compounds is calculated from magnetisation values assuming a collinear spin arrangement. The amount of Mg^{2+} on the A sites in magnesium rich compounds is controlled by heat treatment of the solid. The observed structural, electric, and magnetic properties are very well explained by the obtained cation distribution.

Einphasige Magnesium–Nickel-Ferrite mit unterschiedlichen Konzentrationen von Magnesium und Nickel werden hergestellt und diversen Wärmebehandlungen unterworfen. Die Kationenverteilung in diesen Verbindungen wird aus der Magnetisierung berechnet, wobei kollineare Spin-Anordnung angenommen wird. Der Anteil von Mg^{2+} auf A-Plätzen in Magnesium-reichen Verbindungen wird durch die Wärmebehandlung kontrolliert. Die beobachteten strukturellen, elektrischen und magnetischen Eigenschaften werden sehr gut durch die erhaltene Kationenverteilung erklärt.

1. Introduction

It is well known that the structural, magnetic, and electric properties of spinel ferrites may suitably be modified by introduction of known impurities and by controlled preparative conditions (e.g. heat treatment etc.) [1]. Magnesium ferrite is very sensitive to heat treatment due to the high diffusibility of Mg^{2+} ions whereas cation distribution in nickel ferrite does not show any temperature dependence because Ni^{2+} has a strong octahedral site preference. A solid solution of the two, however, is expected to show changes in cation distribution, and hence properties as a function of heating schedule. Due to the strong octahedral site preference of Ni^{2+} ions the competition to occupy the tetrahedral sites would be between the Mg^{2+} and Fe^{3+} ions. Further, the magnetic Ni^{2+} ions in this series are being substituted for the non-magnetic Mg^{2+} ions. It would be interesting to study the structural, magnetic, and electric properties of $Ni_xMg_{1-x}Fe_2O_4$ in light of the above considerations.

2. Experimental

The ferrite compositions were prepared starting with NiO, MgO, and Fe_2O_3 of AR grade. The powders were mixed thoroughly and pre-fired at 1000 K for 24 h. They were pressed to disc-shaped pellets and finally sintered at 1570 K for 3 h. One set of samples was quenched to room temperature whereas another was cooled in the furnace slowly at the rate of about 60 K h^{-1}. The formation of monophasic ferrites was checked by recording the X-ray powder diffractograms using CoK_α in a Phillips X-ray diffractometer (PW 1010). The magnetic hysteresis loop properties were measured on an

[1]) Bombay 400076, India.

[2]) To whom all correspondance should be addressed.

ac hysteresis loop tracer [2] and the Curie temperatures by measuring the variation of initial susceptibility with temperature [3]. The electrical resistivity and the Seebeck coefficient were measured using the standard two-probe method.

For measuring magnetisation as a function of temperature the samples were equilibrated by dipping them in a solvent maintained at that temperature and mesuring the magnetisation using the method cited above.

3. Results and Discussion

The X-ray powder diffraction patterns of the compounds show that they are all monophasic cubic spinels. The lattice parameters of the compounds are given in Table 1.

Table 1
Properties of the system $Ni_xMg_{1-x}Fe_2O_4$

x	annealed samples				quenched samples			
	a (nm)	$4\pi M_s$ (10^{-4} T) 0 K*)	T_C (K)	ϱ at 500 K (10^5 Ω cm)	a (nm)	$4\pi M_s$ (10^{-4} T) 0 K*)	T_C (K)	ϱ at 500 K (10^5 Ω cm)
0.0	0.8372	1755	653	8.69	0.8400	3807	564	9.53
0.2	0.8350	1700	688	1.58	0.8368	3468	673	12.5
0.4	0.8362	1952	733	1.99	0.8354	3418	720	4.56
0.5	0.8341	2091	753	0.99	0.8351	3066	751	1.38
0.6	0.8355	2288	778	2.18	0.8350	3116	774	5.00
0.8	0.8333	2886	803	0.95	0.8332	3264	795	5.00
1.0	0.8320	3378	848	0.50	0.8320	3382	850	1.14

*) By extrapolation.

The cation distributions in these compounds have been calculated using the magnetisation values ($4\pi M_s$) at 0 K obtained by extrapolating the $4\pi M_s$ versus temperature plots (Fig. 1). A collinear spin arrangement was assumed as the ferromagnetic resonance studies of these compounds did not show any indication of canting [4]. It has been further assumed that Ni^{2+} ions occupy B sites only because of the octahedral site preference energies (P) of the three cations involved [5] (Mg^{2+}: −20.93, Fe^{3+}: −55.68, Ni^{2+}: 37.68 kJ/mol).

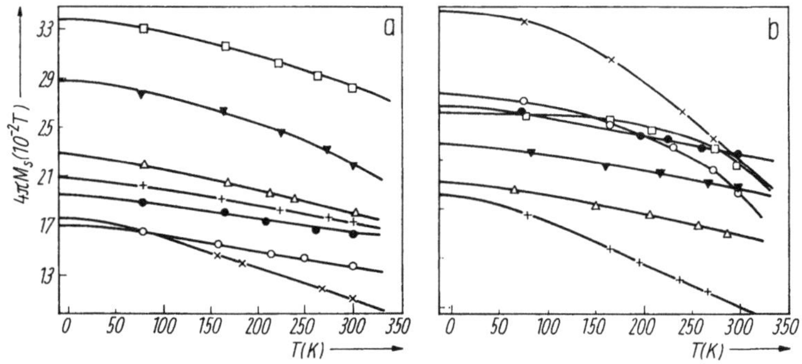

Fig. 1. Variation of magnetisation with temperature for $Ni_xMg_{1-x}Fe_2O_4$ a) for annealed and b) for quenched samples. × x = 0.0, ○ 0.2, ● 0.4, + 0.5, △ 0.6, ▼ 0.8, □ 1.0

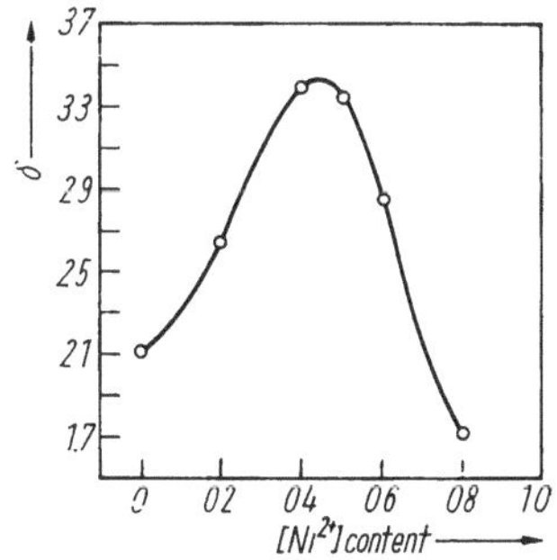

Fig. 2. Plot of δ versus $[Ni^{2+}]$ content for the system $Ni_xMg_{1-x}Fe_2O_4$

The cation distribution then can be written as

$$Mg^{2+}_{a(1-x)}Fe^{3+}_{1-a(1-x)}[Ni^{2+}_x Mg^{2+}_{(1-a)(1-x)} Fe^{3+}_{1+a(1-x)}] O^{2-}_4 ,$$

where

$$a = \frac{n_B - x\mu_{Ni^{2+}} - (1-x)\,\mu_{Mg^{2+}}}{2(1-x)\,[\mu_{Fe^{3+}} - \mu_{Mg^{2+}}]} .$$

The value of $\mu_{Ni^{2+}}$, $\mu_{Fe^{3+}}$, and $\mu_{Mg^{2+}}$ are taken as 2.3, 5.0, and $0.0\mu_B$, respectively, and n_B is the experimental value of magnetic moment of the sample at 0 K. The cation distributions thus calculated are presented in Table 2.

Table 2

Cation distributions in the system $Ni_xMg_{1-x}Fe_2O_4$

composition x	annealed samples	quenched samples
0.0	$Mg_{0.120}Fe_{0.880}[Mg_{0.880}Fe_{1.120}]O_4$	$Mg_{0.225}Fe_{0.745}[Mg_{0.745}Fe_{1.255}]O_4$
0.2	$Mg_{0.072}Fe_{0.928}[Mg_{0.728}Ni_{0.2}Fe_{1.072}]O_4$	$Mg_{0.190}Fe_{0.810}[Mg_{0.610}Ni_{0.2}Fe_{1.190}]O_4$
0.4	$Mg_{0.047}Fe_{0.953}[Mg_{0.553}Ni_{0.4}Fe_{1.047}]O_4$	$Mg_{0.159}Fe_{0.841}[Mg_{0.441}Ni_{0.4}Fe_{1.159}]O_4$
0.5	$Mg_{0.034}Fe_{0.966}[Mg_{0.466}Ni_{0.5}Fe_{1.034}]O_4$	$Mg_{0.114}Fe_{0.886}[Mg_{0.386}Ni_{0.5}Fe_{1.114}]O_4$
0.6	$Mg_{0.035}Fe_{0.965}[Mg_{0.365}Ni_{0.6}Fe_{1.035}]O_4$	$Mg_{0.100}Fe_{0.900}[Mg_{0.300}Ni_{0.6}Fe_{1.100}]O_4$
0.8	$Mg_{0.035}Fe_{0.965}[Mg_{0.165}Ni_{0.8}Fe_{1.035}]O_4$	$Mg_{0.060}Fe_{0.940}[Mg_{0.140}Ni_{0.8}Fe_{1.060}]O_4$
1.0	$Fe[NiFe]O_4$	$Fe[NiFe]O_4$

It can be observed that $NiFe_2O_4$ does not undergo any change in cation distribution with heat treatment. This is because of the strong B site preference of Ni^{2+} ion. However, the cation ordering in the other compositions is sensitive to heat treatment the effect being maximum for $MgFe_2O_4$. To understand the competitive effects of Ni^{2+} addition and heat treatment, the factor $\delta = Mg^{2+}_A$ (quenched sample)/Mg^{2+}_A (annealed sample) has been plotted against x in Fig. 2. The quenched compounds have higher amounts of Mg^{2+} on the A sites. It is clear from Fig. 2 that between $x = 0.0$ and 0.5 the distribution of Mg^{2+} on the A and B sites is strongly dependent on the heat treatment. If this was not so, δ versus x plot would result in a straight line. However, beyond $x = 0.5$ the factor decreases sharply due to the migration of Mg^{2+} ions to the A site leading to a similar cation distribution for the quenched and annealed samples, near $NiFe_2O_4$, δ becomes unity.

The lattice parameter increases linearly from nickel ferrite to magnesium ferrite following Vegard's law. The increase of lattice parameter on quenching is due to the

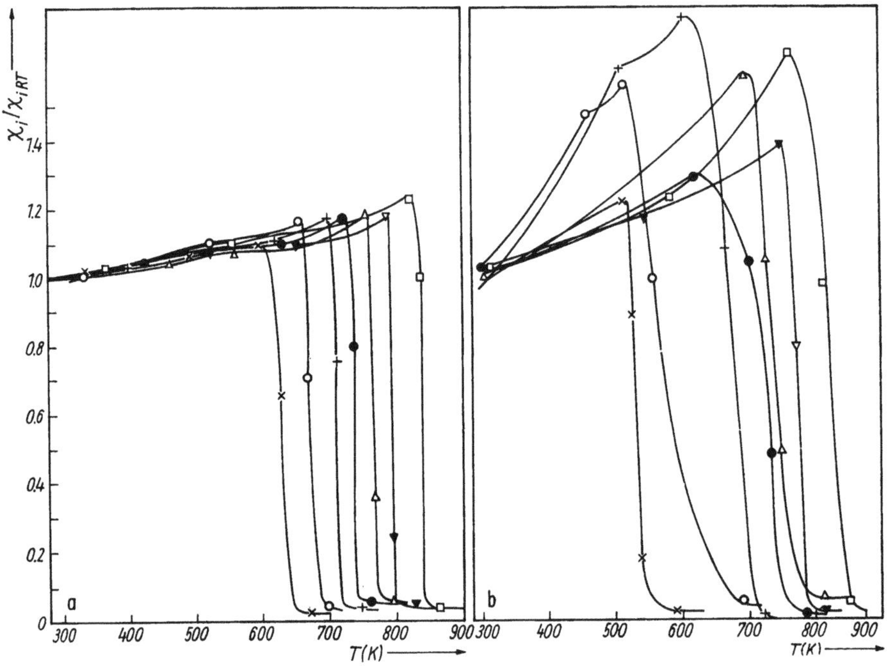

Fig. 3. Thermal variation of initial susceptibility for $Ni_xMg_{1-x}Fe_2O_4$ a) for annealed and b) for quenched samples. × $x = 0.0$, ○ 0.2, ● 0.4, + 0.5, △ 0.6, ▼ 0.8, □ 1.0

fact that at higher temperature the migration of Mg^{2+} ions to the A sites causes an increase in the average radius of the A sites. The lattice parameter of $NiFe_2O_4$ does not change with heat treatment.

The compounds all have low coercive force ($\approx$ 240 A/m) probably due to the absence of any anisotropic ions. The substitution of Ni^{2+} in $MgFe_2O_4$ increases the magnetisation. Similarly on quenching the higher amounts of Mg^{2+} ions found in the A sites displace an equal amount of Fe^{3+} ions to the B sites, also increasing the magnetisation values.

The thermal variation of initial susceptibility was recorded to find out the Curie temperature (T_C) of the samples. In the case of the annealed samples (Fig. 3a) the curves were typical of multidomain grains showing an increase before dropping to zero at T_C. The curves for the quenched compounds (Fig. 3b) show a large increase in the χ_i values before T_C. This is thought to be due to the strains developed between grains on quenching giving rise to a strain anisotropy. We did not observe the presence of any superparamagnetic clusters from our FMR studies [4] and Mössbauer spectra [6]. The increase of T_C from magnesium ferrite to nickel ferrite is due to the strengthening of A–B exchange interactions by the addition of Ni^{2+} ions. However, on quenching, the migration of Mg^{2+} ions to A sites reduces the A–B interactions and result in lower T_C values for quenched samples. Annealed and quenched $NiFe_2O_4$ show the same T_C value as expected.

While calculating the cation distributions in these compounds the presence of appreciable amounts of Fe^{2+} ions has been ruled out. This is evident from the high resistivities of the samples ($\varrho \approx 10^7$ to 10^8 Ω cm at room temperature), Seebeck coefficient values ($\approx$400 μV/K indicating less than 1% concentration of Fe^{2+} ions), and the Mössbauer somer shifts [6] (0.3 to 0.8 mm/s with respect to iron indicating the presence of Fe^{3+}

only) of the compounds. Any conductivity shown by these compounds is due to the hopping of electrons from Fe^{2+} ions, formed by the slight variations in preparative conditions, in the octahedral sites.

4. Conclusions

The cation distributions in the magnesium–nickel ferrites are in agreement with the site preference energies of the various cations involved. However, in compounds containing higher amounts of magnesium, cation ordering is predominantly controlled by the heat treatment. The structural, magnetic, and electric properties are consistant with the cation distributions.

References

[1] J. L. Snoek, Physica (Utrecht) **3**, 463 (1936).
[2] S. D. Likhite, C. Radhakrishnamurty, and P. W. Sahasrabudhe, Rev. sci. Instrum. **36** 1558 (1965).
[3] C. Radhakrishnamurty and P. W. Sahasrabudhe, J. Indian Geophys. Union **2**, 5 (1965).
[4] K. Seshan, M. J. Patni, and A. B. Biswas, Proc. Nuclear Phys., Solid State Phys. **22c**, 587 (1979).
[5] A. Miller, J. appl. Phys. **30**, 245 (1959).
[6] K. Seshan, Ph. D. Thesis, Indian Institute of Technology, Bombay 1980.

(Received March 12, 1981)

phys. stat. sol. (a) 67, 103 (1981)

Subject classification: 13.4; 14.3; 22.1.2

Mobil Tyco Solar Energy Corporation, Waltham[1])

Some Aspects of the Effect of Heat Treatment on the Minority Carrier Diffusion Length in Low Resistivity p-Type Silicon

By

C. T. Ho and F. V. Wald

Four different types of low resistivity p-type silicon, namely float-zoned and Czochralski-grown materials as well as EFG-grown ribbons containing high and low oxygen concentrations are subjected to heat treatments between 850 and 1050 °C, followed by slow cooling to 600 °C, after which they are air quenched. Subsequently, the response of their electron diffusion lengths to the flux of 1 μm photons is measured. The results show that in all materials a progressive destruction of the minority carrier diffusion length occurs as the heat treatment temperature increases. However, it appears that the dominant recombination centers in the oxygen rich materials differ from those present in the relatively oxygen free materials. It is concluded that a "gettering" effect based on oxygen occurs, which is not likely to be related to SiO_2 precipitation, however.

Es wird das Verhalten der Elektronendiffusionslänge in Bor-dotiertem Silizium als Funktion des Photonenflusses bei 1 μm Wellenlänge in Abhängigkeit von vorhergegangenen Wärmebehandlungen des Materials untersucht. Dazu werden vier Arten von Silizium, nämlich tiegelfrei gezogenes (FZ), tiegelgezogenes (CZ) und sauerstoffarme (OL) sowohl als auch sauerstoffreiche (OR) Bänder die im EFG-Verfahren hergestellt waren, bei Temperaturen zwischen 850 und 1050 °C vorgetempert und anschließend langsam auf 600 °C abgekühlt; danach werden sie an Luft abgeschreckt. Die Untersuchungen ergeben, daß bei allen Materialien eine Reduktion der Elektronendiffussionslänge eintritt, wenn bei höheren Temperaturen vorbehandelt wird. Allerdings scheinen sich die erzeugten Rekombinationszentren in den sauerstoffreichen Materialien von denen in den sauerstoffarmen zu unterscheiden. Es wird geschlossen, daß sich größere Komplexe zwischen Sauerstoff–Silizium und Verunreinigungen bilden, bevor es zur Präzipitation von SiO_2 kommt.

1. Introduction

Recently we have shown that the introduction of oxygen into EFG ribbon during growth can have profound consequences in altering the morphology of the resulting material and in improving its electronic properties [1]. We could also show that these improvements may be translated into clearly more efficient solar cells and that oxygen may be introduced both by use of quartz crucibles and by the introduction of suitable oxygen-bearing gases around the liquid meniscus during growth [2].

It has also become quite clear that it is the presence of oxygen in the ribbon that causes the previously observed "light enhancement effect" which manifests itself by an increase of the minority carrier diffusion length with the increasing photon flux. This may be interpreted as resulting from the filling of a number of closely spaced donor-like recombination states in the lower half of the gap as the electron Fermi level moves further upward with the increasing non-equilibrium electron population

[1]) 16 Hickory Drive, Waltham, Massachusetts 02254, USA.

injected into the material by the photogeneration process [3]. In a subsequent investigation, Pogany confirmed this interpretation, and also experimentally showed that such effects cannot be produced by one or a few simple impurity levels [4].

Thus, to us, the close spacing of these levels and their nearly Gaussian distribution through the gap has always suggested rather more complex structures for the centers which cause these effects and since they appear to be largely absent in oxygen-free material [2], oxygen must be implicated in their formation.

2. Experiments and Results

In this paper then, we present the results of a quantitative comparison of the changes in minority carrier diffusion length as a function of the flux of photons of wavelength 1 μm [5]. The characteristics of four different types of silicon materials, heat-treated through three differing sequences were compared. The materials were:

(i) Single-crystal float-zoned samples, boron doped to ≈ 1 Ω cm with a starting average electron diffusion length of ≈ 160 μm as measured by a surface photovoltage method [6].

This material also was analyzed for carbon and oxygen using standard infrared transmission techniques [7] and both of these elements were found to lie below a concentration of $\approx 1 \times 10^{16}$ and 2×10^{16} atoms cm^{-3}, respectively. These are the detection limits for our case, as we used relatively thin slices for the studies here.

(ii) Single-crystal Czochralski-grown slices, boron doped to ≈ 2 Ω cm with a starting diffusion length of ≈ 135 μm, a carbon content of 3×10^{17} atoms cm^{-3} and also containing 2×10^{18} atoms cm^{-3} of oxygen.

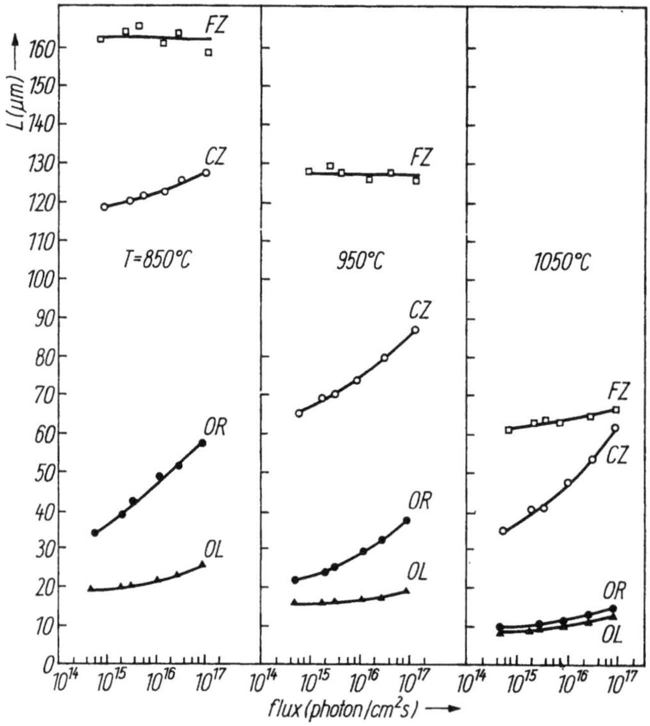

Fig. 1. The effect of pre-annealing temperature (see text) on the response of the electron diffusion length L to the flux of 1 μm photons, for various kinds of p-type silicon

(iii) 5 cm wide EFG ribbons, nominally 1 Ω cm boron doped, grown at ≈ 2 cm/min from a graphite crucible and using a graphite die. Oxygen was introduced into these ribbons by controlling the CO_2 concentration around the meniscus [1]. When no CO_2 was introduced into the gas ambient (the CO_2-off condition [2]), the oxygen concentration in this ribbon was found to be $\leqq 2 \times 10^{16}$ atoms cm^{-3}, i.e. below our detection limit, whereas carbon was found at $\approx 1 \times 10^{18}$ atoms cm^{-3}. For this material, which we designate "oxygen lean" (OL), the electron diffusion length averaged ≈ 34 μm.

(iv) Material from the same growth run, when CO_2 was added to the argon ambient around the meniscus, increased in diffusion length to ≈ 48 μm. This "oxygen rich" (OR) material now contained 4×10^{17} atoms cm^{-3} of oxygen and $\approx 1 \times 10^{18}$ atoms cm^{-3} of carbon. Of course, all ribbons contained the usual sets of defects inherent in EFG-grown silicon [8 to 10].

These four types of material, namely FZ (float zoned), CZ (Czochralski grown), OL (EFG ribbon grown under ambient conditions relatively free of oxygen), and OR (ribbon where CO_2 was introduced during growth), were all subjected to three annealing cycles in a phosphine containing atmosphere and at temperatures of 850, 950, and 1050 °C, using a standard diffusion tube furnace. The atmospheric and temperature conditions thus are similar to various diffusion sequences which one might use in solar cell preparation. However, in order to drive any solid-state reactions further to completion, the time at these three temperatures was extended to 90 min and in each case a slow cooling sequence was subsequently used which brought the slices down to 600 °C over 5 h, after which they were removed from the diffusion tube and air quenched.

The results of these experiments are all summarized in Fig. 1.

3. Discussion and Conclusions

It is quite clear from Fig. 1 that the heat treatment sequences employed here did have a significant effect on the final minority carrier lifetime of all the materials investigated, whether they contained high oxygen concentrations or not. We observe quite generally a progressive destruction of diffusion length as the pre-annealing temperature increases, irrespective of the fact that the subsequent slow cooling sequence exposes all materials for a significant length of time to temperatures around 600 to 800 °C.

This effect seems also not easily explained by assuming that impurity in-diffusion occurs from the annealing furnace or the atmosphere during the heat treatments. Since both the FZ and CZ materials were heat treated in the same furnace, they should have been exposed to the same impurity concentrations and should then be expected to show very similar diffusion lengths for the two higher temperature pre-anneals. That is however not the case, and one is therefore forced to assume that the observed decreases in diffusion length are due to internal rearrangements of recombination centers, i.e. centers which were not "active" for recombination in the as-grown material are now "active" after the heat treatment.

Such effects have been previously shown by Graff and Pieper [11] to exist in float-zoned material. They demonstrated that the lifetime of such crystals could be either increased or decreased in the temperature range between 600 and 900 °C, depending on the general defect concentration (dislocations, swirls, "A" and "B" clusters), and they associated the lifetime increases with an "internal gettering effect", caused by unspecified point defect complexes. It would then be possible to interpret the results of our experiments in this view, by saying that certain point defect complexes which may also incorporate bound impurities are progressively destroyed as the initial annealing temperature is increased, even though the cooling schedule is held relatively constant.

That effect clearly manifests itself (Fig. 1) in the float-zoned material and is quite total for the EFG material with its far higher defect and impurity contents. In fact, we have observed on occasion even more pronounced cases than those indicated in Fig. 1, where EFG material with an initial diffusion length of 70 to 100 μm was obtained in growth from graphite crucibles under oxygen-lean conditions, which was then reduced to a level of 20 to 25 μm during subsequent solar cell fabrication.

In other words, for all materials, whether they contain oxygen or not, in which pre-annealing and the slow cool are combined, a progressive destruction of the diffusion length with increased pre-annealing temperature is observed. Hence, it appears that during crystal growth of silicon at high temperatures a general "gettering" effect prevails, which is perhaps based on the "metal point defect" association model suggested by Graff and Pieper [11]. This association, however, appears to be quite unstable at lower temperatures and thus in all materials a reduction in lifetime occurs when they are heated for any duration in the temperature range between 1050 and 800 °C.

However, when oxygen is present in larger concentrations, this effect can apparently be counteracted somewhat, at least for heat treatments at intermediate temperatures. We conclude this of course from the fact that the diffusion length is now very sensitive to the photon flux, which indicates the presence of a set of quite different recombination centers in the oxygen-rich materials, which now clearly dominate the lifetime and which must have therefore been formed by an attraction between oxygen and those centers which dominate the lifetime in the oxygen-lean materials.

The cooling schedule used here indicates that the crystals are annealed for a considerable length of time at temperatures around 700 °C, but are quite quickly cooled through all temperatures below 600 °C so that we may assume that the present results are not due to effects associated with the well-known "thermal donor", whose effect is maximized at 450 °C. In fact, most investigations suggest that this "thermal donor" is destroyed at temperatures over 600 °C [12]. However other, more complex, donors [13] which are believed to take the form of larger Si_xO_y clusters and whose occurrence is maximized around 700 °C have recently been discussed [12, 14]. Thus we postulate that in our oxygen containing material these larger clusters play a role in the gettering of impurities. Such clusters of course can be viewed as part of a continuum which eventually leads to SiO_2 precipitation [15] and the sensitivity of such precipitation reactions to carbon content and pre-annealing temperature of silicon crystals has been well established recently [16, 17].

However, we do not believe that in this case the impurity gettering capability of the clusters is related to their eventual precipitation as SiO_2, which then generates dislocations that act as impurity sinks [18] since for "EFG"-grown ribbon with its relatively high defect contents [8 to 10] such an explanation seems hardly tenable. Yet our diffusion length measurements suggest that gettering by oxygen does in fact occur, and indeed is sensitive to the pre-annealing treatments which the material undergoes. It is then our postulate that larger Si_xO_y aggregates "getter" metallic impurities prior to the precipitation stage, by mechanisms not related to dislocations, but perhaps through coulombic attraction or physico-chemical forces of other kinds. However, it is quite possible that dislocations play a role in the initial nucleation of such aggregates.

Acknowledgements

The authors are grateful for the excellent work of Mr. J. Mathias on the diffusion length measurements, and the determinations of oxygen and carbon content which have been carried out by Ms. M. Cretella and Mr. G. Freedman, as well as for many valuable discussions with Dr. J. P. Kalejs. Also, part of the work was carried out as a special task of a project on EFG sheet growth under Jet Propulsion Laboratory contract No. 954355, Large Scale Solar Array (LSA) project, U.S. Department of Energy.

References

[1] J. P. Kalejs, M. C. Cretella, F. V. Wald, and B. Chalmers, ECS Meeting, St. Louis, May 1980 (Abstract No. 329); Proc. Symp. Electronic and Optical Properties of Polycrystalline or Impure Semiconductors and Novel Silicon Growth Methods, Ed. K. V. Ravi and O'Mara, The Electrochemical Society, Princeton 1980 (p. 242).

[2] B. Mackintosh, J. P. Kalejs, C. T. Ho, and F. V. Wald, III. E. C. Photovoltaic Solar Energy Conference, Ed. W. Palz, D. Reidel, Dordrecht 1981 (p. 553).

[3] C. T. Ho, R. O. Bell, and F. V. Wald, Appl. Phys. Letters **31**, 463 (1977).

[4] A. Pogany, XIV. IEEE Photovoltaic Specialists Conf. Record, IEEE, New York 1980 (p. 140).

[5] J. Ho and J. Mathias, IEEE Trans. Electron Devices **25**, 1332 (1978).

[6] R. O. Bell and G. M. Freedman, XIII. IEEE Photovoltaic Specialists Conf. Record. IEEE, New York 1978 (p. 89).

[7] ASTM Standard, F 121-79, and ASTM Standard, F 123-74, American Society for Testing and Materials, Philadelphia (PA) 1980.

[8] K. V. Ravi, J. Crystal Growth **39**, 1 (1977).

[9] K. Yang, G. H. Schwuttke, and T. F. Ciszek, J. Crystal Growth **50**, 301 (1980).

[10] C. V. H. N. Rao, M. C. Cretella, F. V. Wald, and K. V. Ravi, J. Crystal Growth **50**, 311 (1980).

[11] K. Graff and H. Pieper, J. electronic Mater. **4**, 281 (1975).

[12] V. Cazcarra and P. Zunino, J. appl. Phys. **51**, 4206 (1960).

[13] A. Kanamori and M. Kanamori, J. appl. Phys. **50**, 8095 (1980).

[14] D. Wruck and P. Gaworzewski, phys. stat. sol. (a) **56**, 557 (1979).

[15] M. Tajima, S. Kishino, M. Kanamori, and T. Lisaka, J. appl. Phys. **51**, 2247 (1980).

[16] S. Kishino, Y. Matsushita, and M. Kanamori, Appl. Phys. Letters **35**, 213 (1979).

[17] K. Yamamoto, S. Kishino, Y. Matsushita, and T. Lisuka, Appl. Phys. Letters **36**, 195 (1980).

[18] T. Y. Tan, E. E. Gardner, and W. K. Tice, Appl. Phys. Letters **30**, 175 (1977).

(Received June 4, 1981)

phys. stat. sol. (a) 68, 109 (1981)

Subject classification: 1.4 and 10.1; 4; 21.1

Institute of Industrial Science[1]) *(a) and*
Department of Metallurgy and Materials Science, Faculty of Engineering[2]) *(b),*
The University of Tokyo

Structure and Atomic Vibration of the $\Sigma = 7$ Tilt Boundary in α-Titanium

By

M. HASHIMOTO (a), Y. ISHIDA (a), R. YAMAMOTO (b), and M. DOYAMA (b)

The structure of the $\Sigma = 7$ symmetrical tilt boundary in titanium is simulated by molecular dynamics using the Lennard-Jones (12–6) potential. The boundary is characterized by trigonal prisms capped with two half-octahedra and tetrahedra. The local vibrational density of states (LDS) of each atom around the boundary is calculated by the recursion method. Some of the LDS's have resonant modes at the low-frequency side. The calculated results are compared with those of the bubble and colloid models.

Die Struktur der symmetrischen $\Sigma = 7$ Neigungskorngrenze im Titan wird molekular-dynamisch unter Verwendung des Lennard-Jones (12–6)-Potentials simuliert. Die Korngrenze wird durch trigonale Prismen charakterisiert, auf die zwei Halboktaeder und Tetraeder aufgesetzt sind. Die lokale Schwingungszustandsdichte (LDS) wird für jedes Atom an der Grenzfläche mit Hilfe der Rekursionsmethode berechnet. Einige der LDS haben Resonanzmoden auf der niederfrequenten Seite. Die berechneten Ergebnisse werden mit denen des Blasen- und Kolloid-Modells verglichen.

1. Introduction

The structures of coincidence-type tilt and twist boundaries in f.c.c. and b.c.c. metals have been simulated by many investigators using interaction potentials [1 to 7]. Using computer simulation and graphical methods based on hard-sphere models, Smith et al. [5 to 7] and Ashby et al. [8] recently showed that the grain boundary structures were characterized by the formation of polyhedral groups of atoms as found by Bernal [9] in assemblies of randomly packed hard spheres. Direct observations of the atomic configurations of the grain boundaries have been made by high-resolution electron microscopy [10]. By comparison between simulated and observed structures, we found that the simulated structure is generally in agreement with the observed results [11]. These studies have been restricted to boundaries in f.c.c. and b.c.c. metals. The structures of the boundaries in h.c.p. metals are also important. For example, it has been shown that the grain boundary sliding of ⟨10$\bar{1}$0⟩ tilt zinc bicrystals strongly depends on the misorientation [12].

The atomic vibration in the boundary is fundamental to understand the thermally activated phenomena, e.g. the diffusion, segregation, and structure transformation of the boundaries. The main results obtained so far for the atomic vibrations around the boundaries in aluminum [13] and gold [11] were that the resonant modes which gave the important contribution to the thermodynamic properties of the grain boundaries occurred in the low-frequency region, and disappeared at a few layers away from the boundary.

[1]) 7 Roppongi, Minato-ku, Tokyo 106, Japan.
[2]) Bunkyo-ku, Tokyo 113, Japan.

Since the h.c.p. lattice has crystalline anisotropy, a simple approximation such as a pairwise interatomic potential cannot be expected to be a good description of the bonding. Martin and Bacon [14] and Yamamoto et al. [15], however, showed that the elastic and vibrational properties of many h.c.p. metals with nearly ideal c/a ratio can be well fitted by Lennard-Jones (LJ) potentials. Minonishi et al. [16] also simulated the core structure of the edge dislocation in h.c.p. metals using LJ potentials. In order to study the grain boundary in h.c.p. metals we have chosen α-titanium ($c/a = 1.60$) with a nearly ideal c/a ratio ($= 1.63$) as an example. In the present paper, the structure of the $\Sigma = 7$, $\langle 0001 \rangle$ symmetrical tilt boundary was simulated by using the LJ potential of α-titanium, because its structure has a small coincidence lattice parameter and has been observed using two-dimensional models such as the bubble [17] and colloid [18] models. The local vibrational density of states (LDS) of each atom around this boundary was also calculated by the recursion method [19]. The calculated results were compared with those of the bubble and colloid models.

2. Method of Calculations

2.1 Interaction potential

The LJ potential which has the form

$$U(r) = \varepsilon_0[(d/r)^{12} - 2(d/r)^6] \tag{1}$$

was determined by the following conditions:

$$\left.\frac{\mathrm{d}E}{\mathrm{d}a}\right|_{a_{\text{obs}}} = 0 \quad \text{(keeping the } c/a \text{ ratio constant)} \tag{2}$$

and

$$\varepsilon_0 = 1.205 k_\mathrm{B} T_\mathrm{m} \,. \tag{3}$$

E, a_{obs}, and T_m are the total crystal energy, observed lattice constant along the a-axis, and the melting temperature of titanium, respectively. The potential depth ε_0 was determined in a similar manner as by Martin and Bacon [14]. The parameter d was computed from (2) to take into account the anisotropy of the h.c.p. lattice. The potential was truncated between the fourth- and fifth-nearest neighbours. The resulting parameters of the potential of α-titanium which were obtained by Yamamoto et al. [15] are $\varepsilon_0 = 0.20154$ (eV) and $d = 2.96292$ (A). This potential curve is shown in Fig. 1.

2.2 Computational technique

Periodic boundary conditions were used parallel to the boundary in order to reduce the size of the calculated zone. The number of atoms contained in the unit periodic zone is

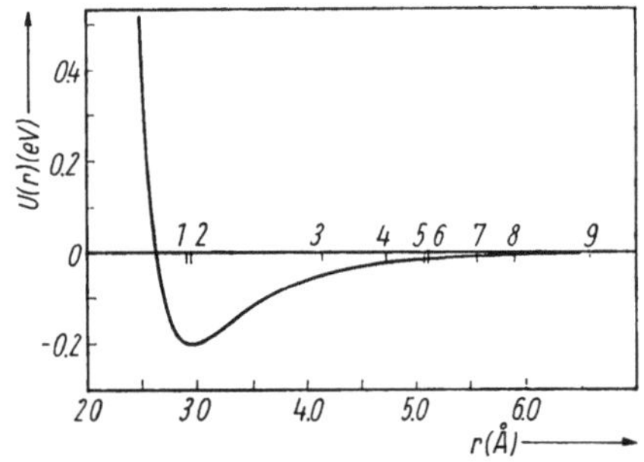

Fig. 1. Plot of the Lennard-Jones potential of α-titanium

58. The lattice coincidence model of the boundary was constructed in a computer as the initial configuration. The atomic configuration was relaxed in an iterative manner to achieve a minimum in the internal energy by the following procedures. First, the upper grain was transferred relative to the lower one by small increments parallel and perpendicular to the boundary without rotation. Since the lattice planes perpendicular to the tilt axis have an ... ABABA ... stacking sequence and the mirror symmetry on planes (0002) for the h.c.p. ⟨0001⟩ tilt boundary, the stable structures are consistent with this symmetry requirement (see [6]). Therefore, it was assumed that a relative translation parallel to the tilt axis between the both grains cannot occur. After each translation the boundary energy was calculated and the minimum energy site of the upper grain was determined. Second, individual atoms were displaced from the initial configuration by the molecular dynamic technique. In this technique, the equilibrium configuration can be obtained by damping out the kinetic energy of every atom as soon as the total kinetic energy of the relaxed zone reaches a relative maximum.

The LDS of each atom around the boundary was computed by using the relaxed structure which was calculated by the above technique. The LDS for atom 0 is given by

$$g_0(\omega) = -\frac{2\omega}{\pi} \sum_{\alpha} \operatorname{Im} G_{00}^{\alpha\alpha}(\omega^2 + i0^+) \qquad (\alpha = x, y, z)\,, \tag{4}$$

where the Green's function $G_{00}^{\alpha\alpha}(\omega^2)$ can be calculated as a continued fraction expansion by re-definition of the basis vectors of the hermitian matrix (see [19]). This technique is called the recursion method and was applied successfully to a variety of problems of the vibrational density of states in condensed matter [20, 21].

In order to calculate the LDS at or near the grain boundary we make use of a model system which contains a boundary in a spherical shape of about 1500 atoms. The calculation of the recursion coefficients (a_n, b_{n+1}) $n = 0, 1, 2, \dots, n_0$ has been made up to $n_0 = 14$. The vibrational spectrum, $g_0(\omega)$, is normalized here as

$$\int_0^{\omega_{\max}} g_0(\omega)\, d\omega = 3 \qquad \text{(modes per atom)}\,, \tag{5}$$

where $\omega_{\max}$ is the maximum frequency. The LDS reflects the local environment of each atom but reduces to the phonon frequency distribution function in the perfect crystal.

More details of the computational technique can be found in [13].

3. Results and Discussion

3.1 Structure

The simulated structure of the $\Sigma = 7$, $(1\bar{3}20)$ 38.21°/[0001] symmetrical tilt boundary in α-titanium is shown in Fig. 2. The energy of this boundary is estimated to be about 778 mJ/m². The width of the region formed by the characteristic configuration of the boundary is very large compared with those of the ⟨001⟩, and ⟨110⟩ symmetrical tilt boundary in f.c.c. and b.c.c. metals. If the (0002) plane marked by the symbol × is rotated about the tilt axis over 180° and is turned over, the atomic configuration of this plane becomes as that of the (0001) plane marked by the symbol ○. The structure of individual planes involves certain atom sites (atom E and E′ in Fig. 2) in near-coincidence positions. The planes formed by the coincidence sites on the individual planes (0001) and (0002) are separated as shown by full and dashed lines parallel

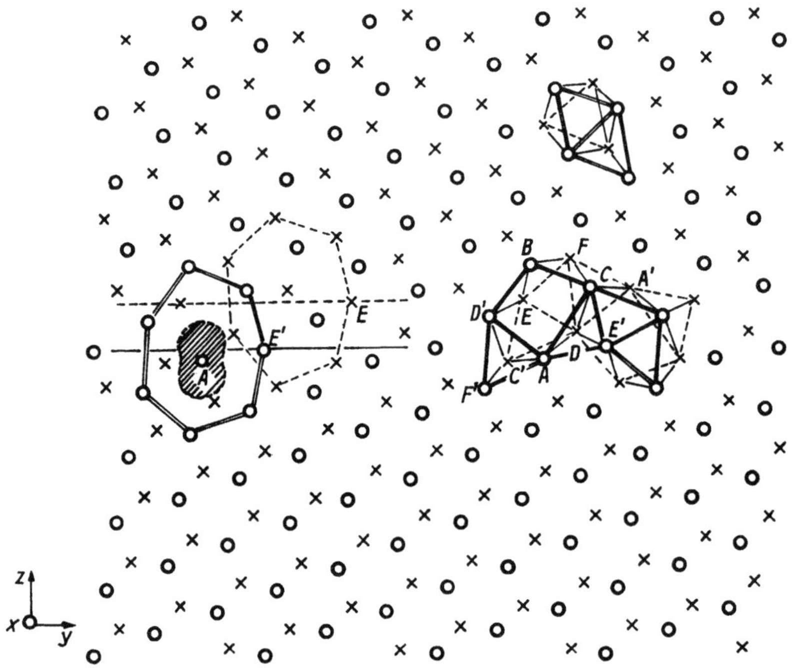

Fig. 2. Simulated structure of the $\Sigma = 7$, $(1\bar{3}20)$ 38.21°/[0001] symmetrical tilt boundary in α-titanium. The hatched region is explained in the text

to the boundary on the left hand in Fig. 2. An increase in the width of the grain boundary may be attributed to it.

In order to analyse the polyhedral groups of this boundary, the nearest neighbours of individual atoms were identified by setting an upper limit for the nearest- neighbour distance at the first approximation. The cut-off distance is somewhat arbitrary. Here the choice was made at $1.3a$. With this choice, the h.c.p. crystal consists of octahedra and tetrahedra. Configurations characterizing the $\Sigma = 7$ boundary are the distorted trigonal prism capped with two half-octahedra formed by atoms A, D, C, D′, E, C′, F, and B in Fig. 2 and two types of tetrahedra formed by atoms A, D′, F′, C′ and A, C, E′, D. The stacks of these polyhedra occur periodically along the boundary with a translation distance equal to the coincidence lattice parameter, $\sqrt{7}\,a$. A unit period contains two distorted trigonal prisms capped with two half-octahedra and four distorted tetrahedra. Trigonal prisms capped with three half-octahedra were frequently found at the boundaries in f.c.c. and b.c.c. metals, but are not found in the h.c.p. $\Sigma = 7$ tilt boundary.

In the two-dimensional representation, the distorted centred heptagon also appears at the boundary. Although this packing unit cannot form an orderly condensed pattern, the packing density is large. One of the authors (Y.I.) has suggested using the bubble model that even though the orientation relation may be the same, different structures can occur depending on the bubble size [17]. For the $\Sigma = 7$ boundary, three different structures appeared in the bubble model. The simulated structure almost agrees with one of the configurations formed in a raft of small bubbles. The minor difference exists, however, in the position of the central particle in the centred heptagon. In the present result this position slightly shifts from the central

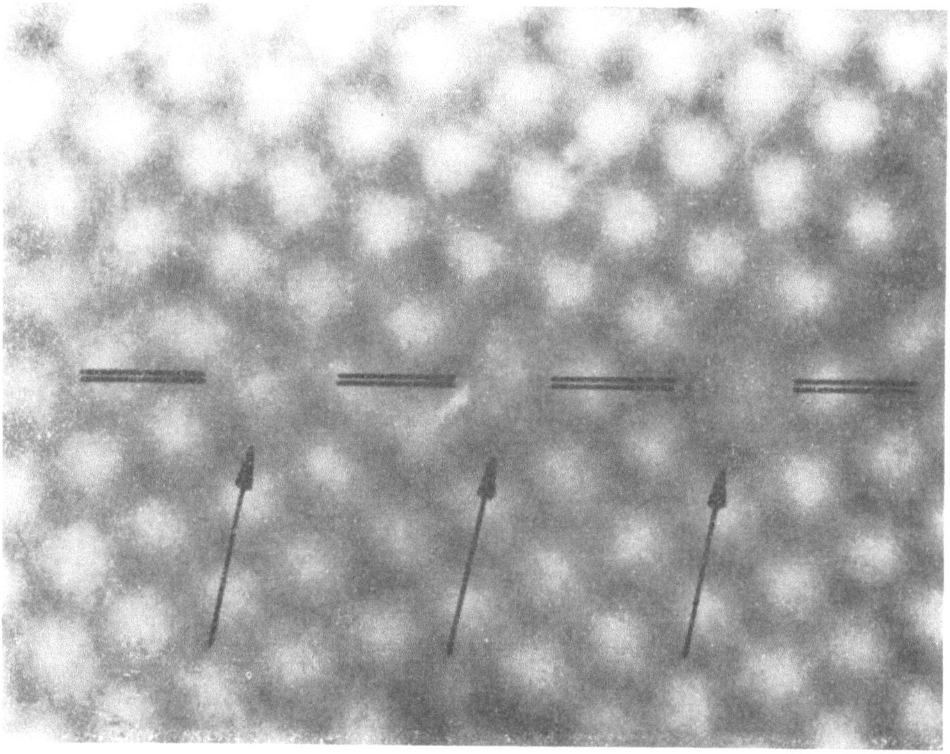

Fig. 3. Image of the $\Sigma = 7$ tilt boundary in the colloid crystal. The particles pointed by arrows are transferred up and down alternatively to each other

position of the heptagon, while it corresponds to the centre in the bubble model. The latter may be ascribed to the two-dimensional model formed by one layer of the bubble raft. Fig. 3 shows the $\Sigma = 7$ tilt boundary of the colloid model [16]. The position of the central particle is shifted from the centre of the heptagon, this site, however, is transferred up and down alternatively to each other. Therefore, the central particle may have two stable positions which are symmetric about the boundary plane.

3.2 Atomic vibration

The LDS's of atoms around the boundary are shown in Fig. 4. The characteristic feature is that the LDS's of atoms A (i.e. the central atom of the heptagon) and D have resonant modes in the lower frequency region. The LDS's of atoms E and C, on the other hand, have sharp peaks near the upper band edge. These features are in qualitative agreement with previous results of the $\Sigma = 5$ and 11 tilt boundaries in aluminum [13] and gold [11].

The Einstein frequency, ω_E, which can be obtained by

$$\omega_E = \frac{1}{3} \int_0^{\omega_{max}} \omega g(\omega)\, d\omega \tag{6}$$

gives us a simple image of the atomic vibration. The calculated values of ω_E of individual atoms are shown in Table 1. Those of atoms A and D are very small compared

Table 1

Einstein frequencies of individual atoms around the $\Sigma = 7$ boundary and of the bulk atom in titanium

atomic sites	A	B	C	D	E	F	bulk
$\omega_E/2\pi$ (10^{12} Hz)	4.12	6.19	5.65	4.41	6.78	6.06	6.08

with those of the other atoms. That of atom E is larger than that of the bulk. It may be considered that these atoms play an important role in the thermally activated phenomena of the boundary.

The component spectrum, $g_\alpha(\omega)$, of the LDS of atoms A and D is shown in Fig. 5, where $g_\alpha(\omega)$ is normalized to unity. It may be suggested that the characteristic resonant modes are mainly those of the vibration perpendicular to the boundary plane. The difference of the component spectra of atoms A and D exists in the y-component LDS. That of atom A has large peaks near the centre of the band, but that of atom D has also peaks in the low- and high-frequency regions.

The mean-square displacement, $\langle u_\alpha^2 \rangle$, along the α-direction is given by

$$\langle u_\alpha^2 \rangle = \frac{\hbar}{2M} \int_0^{\omega_{max}} \coth\left(\frac{\hbar\omega}{2k_B T}\right) \frac{g_\alpha(\omega)}{\omega} \, d\omega \tag{7}$$

using the α-component spectrum $g_\alpha(\omega)$ of the LDS, where M, k_B, and T are the atomic mass, the Boltzmann constant, and the absolute temperature, respectively. The temperature dependence of $\langle u^2 \rangle$ along three directions x, y, z, for the atoms A and D are

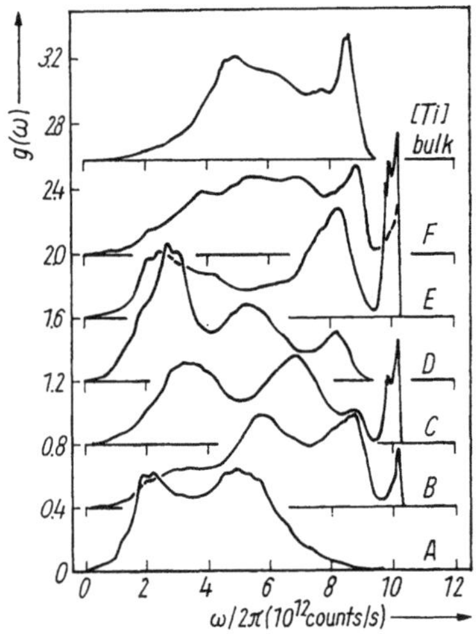

Fig. 4

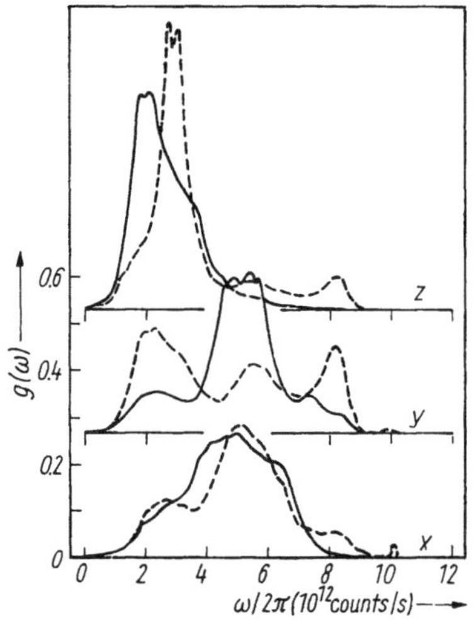

Fig. 5

Fig. 4. LDS's of individual atoms around the $\Sigma = 7$ tilt boundary and of the bulk atom in α-titanium. The alphabets indicate the atomic sites shown in Fig. 2

Fig. 5. Component spectra of LDS's of atoms A (———) and D (– – –). The x, y, z directions are shown in Fig. 2

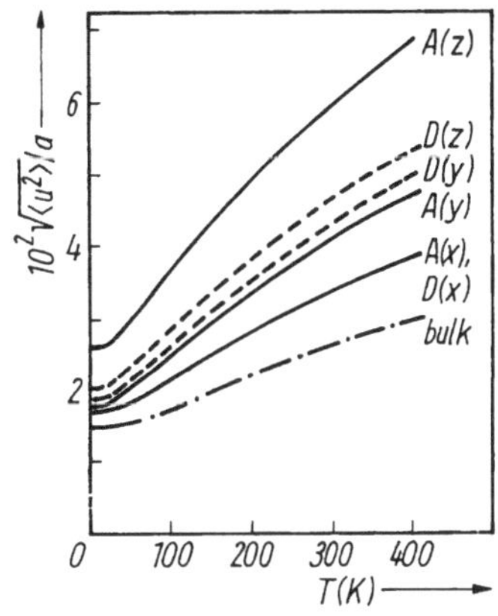

Fig. 6. Temperature dependence of the square root of the mean-square displacements of atoms A and D in units of the lattice parameter of a-axis

shown in Fig. 6. Atom A vibrates with a large amplitude perpendicular to the boundary. Atom D has not an anisotropic vibration such as that of atom A. Using an $\langle u^2 \rangle$ along the y- and z-directions, that of an arbitrary direction in the yz plane is obtained by

$$\langle u^2 \rangle = \langle u_y^2 \rangle \cos^2 \varphi + \langle u_z^2 \rangle \sin^2 \varphi , \tag{8}$$

where φ is the angle between the direction considered and the y-axis. The hatched region in the left-hand part in Fig. 2 shows $\sqrt{\langle u^2 \rangle}$ of individual directions of atom A at 300 K. The size of this region has been multiplied by ten. The shape is elongated perpendicular to the boundary, and almost agrees with the diffused particle image of the centre of the heptagon in the colloid model (shown in Fig. 3). It is expected from these facts that atom A can easily be displaced at the tetrahedral site formed with atoms D, E, and C′. If atom A is displaced from this site, the centre of the heptagon is transferred upward. Since such transfers of atoms are frequently expected at high temperatures, the structure at low temperatures would become unstable. Therefore, it is considered that the vibration of atoms with resonant modes induces structure transformations of the boundary at high temperatures.

4. Conclusion

In the present paper, the structure and atomic vibration of the $\Sigma = 7$ tilt boundary in titanium were calculated as the first approach, and it was verified that the computer simulations of the grain boundary in h.c.p. metals with nearly ideal c/a ratio are possible by using the LJ potentials. In order to understand the general features of the boundaries in the h.c.p. metals, however, studies of many other kinds of boundaries have to be done in the future.

Acknowledgements

The authors thank Mrs. S. Kagawa and M. Imafuku for their helpful discussions and technical assistance, and also I. Hashimoto for support in preparing the manuscript.

References

[1] G. Hasson, J. B. Guillot, B. Baroux, and C. Goux, phys. stat. sol. (a) **2**, 551 (1970).
[2] M. J. Weins, H. Gleiter, and B. Chalmers, J. appl. Phys. **42**, 2639 (1971).
[3] R. E. Dahl, J. R. Beeler, and R. D. Bourquin, Computer Phys. Commun. **2**, 301 (1971).
[4] P. D. Bristowe and A. G. Crocker, Phil. Mag. **31**, 503 (1975); Phil. Mag. A **38**, 487 (1978).

[5] D. A. SMITH, V. VITEK, and R. C. POND, Acta metall. **25**, 475 (1977).
[6] R. C. POND, D. A. SMITH, and V. VITEK, Acta metall. **27**, 235 (1979).
[7] V. VITEK, D. A. SMITH, and R. C. POND, Phil. Mag. A **41**, 649 (1980).
[8] M. F. ASHBY, F. SPAEPEN, and S. WILLIAMS, Acta metall. **26**, 1647 (1978).
[9] J. D. BERNAL, Sci. Amer. **203**, 125 (1960).
[10] H. ICHINOSE and Y. ISHIDA, Phil. Mag. A **43**, 1253 (1981).
[11] M. HASHIMOTO, H. ICHINOSE, Y. ISHIDA, R. YAMAMOTO, and M. DOYAMA, Japan. J. appl. Phys. **19**, 1045 (1980).
[12] T. WATANABE, M. YAMADA, S. SHIMA, and S. KARASHIMA, Phil. Mag. A **40**, 667 (1979).
[13] M. HASHIMOTO, Y. ISHIDA, R. YAMAMOTO, and M. DOYAMA, J. Phys. F **10**, 1109 (1980); Acta metall. **29**, 617 (1981).
[14] J. W. MARTIN and D. J. BACON, Proc. Intern. Conf. on Computer Simulation for Materials Applications, Ed. R. J. ARSENAULT, J. R. BEELER, Jr., and J. A. SIMMONS, Nat. Bur. Standards, Washington 1976 (p. 298).
[15] R. YAMAMOTO, S. KAGAWA, and M. DOYAMA, to be published.
[16] Y. MINONISHI, S. ISHIOKA, M. KOIWA, S. MOROZUMI, and M. YAMAGUCHI, Phil. Mag. A **43**, 1017 (1981).
[17] Y. ISHIDA, Grain Boundary Structure and Properties, Ed. G. A. CHADWICK and D. A. SMITH, Academic Press, New York/London 1976 (p. 93).
[18] Y. ISHIDA, S. OKAMOTO, and S. HACHISU, Acta metall. **26**, 651 (1978).
[19] R. HAYDOCK, V. HEINE, and M. J. KELLY, J. Phys. C **5**, 2845 (1972); **8**, 2591 (1975).
[20] R. YAMAMOTO, K. HAGA, and M. DOYAMA, J. Phys. Soc. Japan **48**, 341 (1980).
[21] Y. ISHII and T. FUGIWARA, J. Phys. F **10**, 2125 (1980).

(Received April 21, 1981)

phys. stat. sol. (a) 68, 117 (1981)

Subject classification: 14.4; 22.9

National Physical Laboratory, New Delhi[1])

Surface Charge Decay Characteristics in Polyvinyl Acetate Films

By

S. RADHAKRISHNAN, M. N. KAMALASANAN, and P. C. MEHENDRU

The charging and discharging characteristics of corona charged polyvinyl acetate films are investigated as a function of thickness. The saturation potential at constant corona field is found to vary with film thickness in a parabolic manner. The rise and decay transients of the surface charge are found to follow closely the theoretical expressions derived on the basis of space charge formation in the bulk of the polymer during corona charging.

Die Aufladungs- und Entladungscharakteristiken von Polyvinylazetatschichten, die in einer Koronaentladung aufgeladen werden, werden als Funktion der Filmdicke untersucht. Das Sättigungspotential bei konstantem Koronafeld variiert parabolisch mit der Filmdicke. Das An- und Abklingen der Oberflächenladung befolgt sehr gut die theoretischen Ausdrücke für den Aufbau von Raumladungen im Volumen des Polymers während der Koronaaufladung.

1. Introduction

In recent years both theoretical and experimental aspects of surface charge decay of insulators have attracted much attention because of the numerous applications of charged polymeric materials in electret microphones, electrophotography, etc. The insulators may be charged by various techniques, viz. triboelectric, exposure to radiations, electron beam, corona discharge, etc. Although many authors have dealt with the decay of surface charge in polymeric materials [1 to 6], practically no mention has been made on the charging behaviour of these materials. The charge distribution prior to the onset of discharge has hitherto been assumed by many authors to be of certain type such as localized surface charge, bulk space charge, etc., without any experimental evidence for the same. Thus, in order to understand the complete behaviour of corona-charged polymers, it is necessary to study both the charging as well as discharging characteristics. Our studies in this direction are reported in this paper.

2. Theory

In the simplest case an insulator when exposed to corona emission will be charged like a capacitor with the charge localised at the two opposite faces. However, there are typically ionizable centres in the bulk of the material such as deep neutral donors, defects, traps, etc. As soon as the charging process is initiated, the carriers that are thermally excited from these centres get swept away due to the field, resulting in the formation of bulk image charge. Thus the charge neutrality condition is met by compensating the surface charge through the creation of bulk space charge of opposite polarity in addition to the image charge at the base electrode. The variation of the maximum surface potential (U) as a function of layer thickness can be described by the Poisson equation as applied to a simple space charge model. This leads to the

[1]) Hillside Road, New Delhi 110012, India.

relation

$$\frac{\partial^2 U}{\partial x^2} = \frac{\varrho}{\varepsilon\varepsilon_0}, \tag{1}$$

where x is the distance along the film thickness, ϱ the space charge density, and ε the dielectric constant of the material. Assuming the space charge density to be constant it can easily be shown that

$$U = \frac{\varrho}{2\varepsilon\varepsilon_0} d^2 + \frac{\sigma^-}{\varepsilon\varepsilon_0} d, \tag{2}$$

where d is the thickness of the layer and σ^- the negative surface charge at the counter-electrode for positively charged films. If σ_{max} is the maximum positive surface charge the film can hold, at which the field across the film is E_{crit}, then we have from charge neutrality condition (mentioned earlier)

$$\sigma_{max} = \varepsilon\varepsilon_0 E_{crit} = \varrho d + \sigma^- . \tag{3}$$

Equation (2) can now be rewritten in terms of E_{crit} as

$$\left.\begin{aligned} U &= \frac{\varrho}{2\varepsilon\varepsilon_0} d^2 + dE_{crit} - \frac{\varrho}{\varepsilon\varepsilon_0} d^2 \\ \text{or}\quad U - \frac{\varepsilon\varepsilon_0 E_{crit}^2}{2\varrho} &= -\frac{\varrho}{2\varepsilon\varepsilon_0}\left(d - \frac{\varepsilon\varepsilon_0}{\varrho} E_{crit}\right)^2 . \end{aligned}\right\} \tag{4}$$

This equation suggests that the maximum potential which a film can retain depends on the thickness in a parabolic manner up to a certain thickness d_{crit} beyond which it remains constant. Thus at the critical thickness we have

$$d_{crit} = \frac{\varepsilon\varepsilon_0 E_{crit}}{\varrho} \quad \text{and} \quad U_{max} = \frac{E_{crit}^2 \varepsilon\varepsilon_0}{2\varrho} . \tag{5}$$

The theory of surface charge decay in polymers when conduction under SCLC conditions takes place has been developed by many authors [6 to 9]. The results of their calculations show that the surface potential U decays with time as

$$\left.\begin{aligned} U &= U_0 \{1 - [(\alpha + E_0\mu' t)/2d]\} \quad &\text{for} \quad t < t_r \\ \text{and}\quad U &= \frac{U_0 d}{2(\alpha + E_0 \mu' t)} \quad &\text{for} \quad t > t_r , \end{aligned}\right\} \tag{6}$$

where t_r is the transit time, U_0 the initial potential, E_0 the initial field developed across the sample, μ' the mobility of charge carrier, t the time, and α a parameter dependent on the space charge distribution.

Thus a complete picture can be obtained for the charging and discharging of polymers charged by corona technique through the study of surface potential behaviour with time as well as film thickness.

3. Experimental

Polyvinyl acetate (PVAc) obtained from Poly Sciences USA, and having a molecular weight of 17000, was cast into a film form on smooth aluminium substrates by isothermal evaporation technique at about 30 °C using methanolic solution containing different amounts of the solute. The different film thicknesses thus obtained ranged from 1 to 200 μm. The films were thoroughly dried in an oven at 40 °C prior to the studies on surface charge decay.

The charging and discharging characteristics for these films were investigated by the usual xerographic discharge technique except for the fact that no illumination was used in the measurement procedure. The films were exposed to corona emission from a wire source held at a dc potential of 7.0 kV for different durations and then put under a static probe (Keithley 2501) in order to measure the surface potential. The static detector was connected to a Keithley electrometer (610 C) the output of which was fed to a recorder so as to monitor the surface charge with time.

4. Results and Discusion

Curves 1 to 3 in Fig. 1 represent the charging characteristics for films of thicknesses 22, 30, and 50 μm, respectively. The films were exposed to corona emission at the same source potential (7.0 kV) but for different periods of time. It is seen that the surface potential increases rapidly in the beginning of the charging cycle and then gradually saturates to some value U_d depending on the film thickness. In order to investigate the thickness dependence of the saturation potential, films of various thicknesses were charged for 30 s at a corona potential of 7.0 kV. Fig. 2 shows a plot of saturation potential versus film thickness. It is interesting to note that the saturation potential is highly thickness-dependent for low film thicknesses ranging from 1 to 50 μm. However, for thicknesses above 50 μm the saturation potential is more or less independent of film thickness. In the case of space charge formation it has been shown earlier that the saturation potential increases with thickness in a parabolic manner according to (4). Thus one can plot $(U_s - U_d)^{1/2}$ versus d where U_s is the saturation potential for very large thickness and U_d is the saturation potential for small thicknesses. If one takes U_s to be the same as $E^2_{crit}\varepsilon\varepsilon_0/2\varrho$, the graph according to (4) is expected to be a linear one. Such a curve is shown in Fig. 3 and it is seen that the graph is indeed a linear one indicative of the space charge formation in bulk of the film.

The surface charge decay characteristics for the films charged as before is shown in Fig. 4. Curves 1 to 6 are for films of thickness of 3, 5, 18, 30, 50, and 80 μm, respectively. In order to investigate the exact time dependence of the surface charge the decay characteristics were analysed according to (6) using a curve fitting procedure. Various values of parameters such as α, μ', etc. giving the best fit for different film

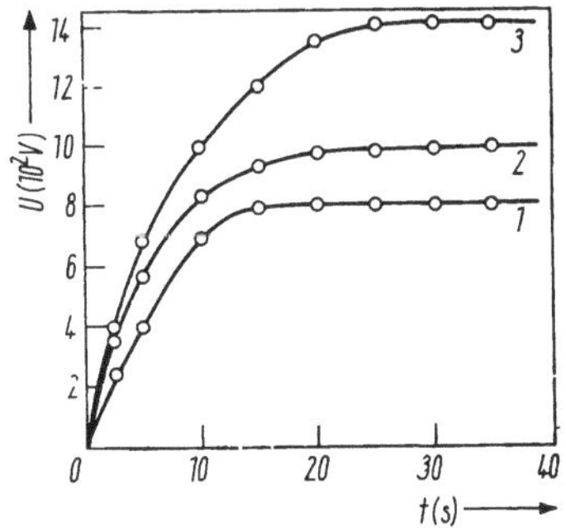

Fig. 1

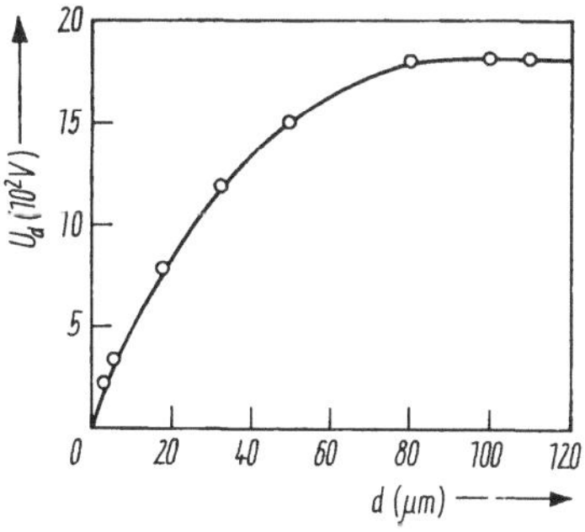

Fig. 2

Fig. 1. The charging characteristics of PVAc films at constant corona potential of 7.0 kV for three thicknesses, 22, 30, and 50 μm (curves 1, 2, 3, respectively)

Fig. 2. Thickness dependence of the saturation potential for films charged at 7.0 kV for 30 s

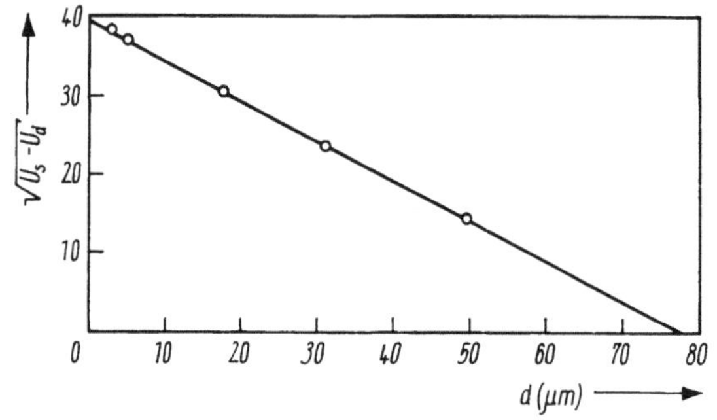

Fig. 3

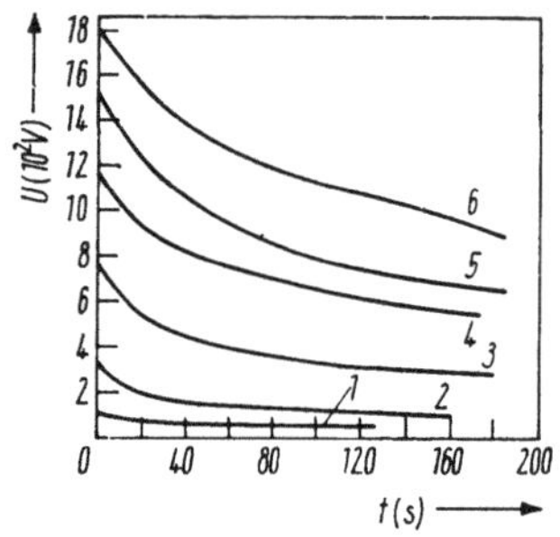

Fig. 4

Fig. 3. The plot of $(U_s - U_d)^{1/2}$ vs. d showing the parabolic nature as according to (4) of the text

Fig. 4. Surface charge decay characteristics for films with different thickness, curves 1 to 6 represent 3, 5, 18, 30, 50, and 80 µm, respectively

thicknesses are given in Table 1. It is thus clear that the surface charge decay follows (6), which is expected if the space charge formation were to take place in the bulk of the polymer film. The values of the mobility of charge carriers obtained from this analysis are also not far from those reported in polymeric materials. However, it seems that the mobility may be either thickness and/or field dependent in these films.

It may be of interest to mention here that the earlier studies reported on the saturation potential in corona-charged polymer films show no thickness dependence of U_d. This may be because the previous authors [10, 11] who have made some investigations in materials such as cellulose acetate, used very thick films ($d > 50$ µm). Such a thickness being far above the critical value (d_{crit}) gave no effect on the saturation potential. On the other hand, our studies on both charging and discharging characteristic of polyvinyl acetate films clearly show that the saturation potential is thickness dependent. Further, space charge formation plays an important role in the surface charge decay of corona-charged films and especially in thin PVAc films.

Acknowledgement

The authors wish to thank Dr. A. R. Verma, Director, National Physical Laboratory, for his keen interest and encouragement in this work.

Table 1

Charge decay analysis of PVAc films

sample thickness (10^{-4} cm)	U_0 (V)	α (10^{-4} cm)	μ' (10^{-10} cm²/Vs)
80	1850	79.6	4.1
50	1520	31.6	1.5
30	1200	22.3	1.26
18	750	11.8	0.85
5	300	2.7	0.18
3	100	2.2	0.39

References

[1] M. M. PERLMAN, Electrets and Related Electrostatic Charge Storage Phenomena, The Electrochem. Soc. USA, New York 1968 (p. 3).
[2] G. SAWA, D. LEE, and M. IEDA, Japan. J. appl. Phys. **14**, 634 (1975).
[3] D. K. DAVIES, J. Phys. D **5**, 162 (1972).
[4] H. J. WINTLE, J. appl. Phys. **41**, 4004 (1970).
[5] A. REISEV, M. W. BLOCK, and J. KNIGHT, Trans. Faraday Soc. **65**, 2168 (1969).
[6] H. KIESS, RCA Rev. **36**, 667 (1975).
[7] H. J. WINTLE, J. appl. Phys. **43**, 2927 (1972).
[8] L. R. WEISBERG, Solid State Devices, Inst. Phys. Conf. Ser. (London) **25**, 15 (1974).
[9] J. M. TAYLOR and T. P. T. WILLIAM, J. Phys. C **11**, 111 (1978).
[10] D. C. HOSTEREY, J. appl. Phys. **33**, 922 (1962).
[11] R. M. SCHAFFERT, Electrophotography, Focal Press, London 1965 (p. 243).

(Received March 4, 1981)

phys. stat. sol. (a) 68, 123 (1981)

Subject classification: 1.4 and 10

Institut für Physik, Max-Planck-Institut für Metallforschung, Stuttgart[1])

A Comment on the Kinematical Theory of Diffraction Contrast of Lattice Defects on Transmission Electron Microscope Images of Crystalline Specimens

By

M. WILKENS

The kinematical contrast integral, which is currently used for instance for the calculation of the weak-beam contrast of lattice defects on transmission electron microscope images, is reconsidered. It is shown that a term $\boldsymbol{s} \cdot \boldsymbol{R}$, which is normally neglected in deriving the integral, can be incorporated. This allows to take the effect of the local lattice dilation on the diffraction contrast into account. Some examples are discussed for which this additional contrast effect may be significant.

Bei der Begründung der Integraldarstellung des Beugungskontrastes in kinematischer Näherung, wie sie z. B. für die Berechnung des sogenannten weak-beam Kontrastes von Gitterfehlern auf elektronenmikroskopischen Durchstrahlungsaufnahmen benutzt wird, wurde bisher ein Term $\boldsymbol{s} \cdot \boldsymbol{R}$ vernachlässigt. Es wird gezeigt, daß dieser Term in das Integral aufgenommen werden kann und den Einfluß der lokalen Gitterdilation auf den Beugungskontrast berücksichtigt. Es werden Beispiele diskutiert, für die dieser zusätzliche Kontrastbeitrag von Bedeutung sein kann.

1. Introduction

The kinematical theory of diffraction contrast of lattice defects on transmission electron microscope (TEM) images of crystalline specimens was first established by Hirsch et al. [1, 2]. After the development of more sophisticated dynamical theories of diffraction contrast [2 to 4] the kinematical contrast theory is still used mainly for a qualitative or semi-quantitative interpretation of TEM images of lattice defects obtained under weak-beam conditions [5 to 7].

The kinematical contrast amplitude $A(x, y)$ is given by

$$A(x, y) = A_0 \int_0^t \exp 2\pi i[(\boldsymbol{g} \cdot \boldsymbol{R}(x, y, z) + sz] \, dz \,, \tag{1.1}$$

$\boldsymbol{g}$ diffraction vector = reciprocal lattice vector of the perfect reference lattice, $\boldsymbol{R}$ atomic displacement field vector of an atom at position $\boldsymbol{r} = [x, y, z]$ in components $\boldsymbol{R} = [R_x, R_y, R_z]$, s excitation error. The diffracted electron beam (wave vector $\boldsymbol{k}_g$) is parallel to the z-axis of a Cartesian coordinate system, t is the foil thickness. x, y are coordinates in the plane perpendicular to $\boldsymbol{k}_g$, just below the specimen. The prefactor A_0 ($A_0 = \pi/\xi_g$, ξ_g extinction length) will not be considered here.

One simple way in deriving equation (1.1) is to start from the kinematical scattering amplitude $F(\boldsymbol{\chi})$ in three-dimensional reciprocal space well-known from X-ray crystallography [1, 2]. For a crystal having a primitive unit cell $F(\boldsymbol{\chi})$ is given by

$$F(\boldsymbol{\chi}) = f \sum_n \exp 2\pi i\{(\boldsymbol{g} + \boldsymbol{s}) \cdot (\boldsymbol{r}_n + \boldsymbol{R}(\boldsymbol{r}_n))\} \,, \tag{1.2}$$

[1]) Heisenbergstr. 1, D-7000 Stuttgart, FRG.

where the sum goes over all crystal atoms at positions $\boldsymbol{r}_n$, f atomic scattering amplitude, and

$$\boldsymbol{\chi} = \boldsymbol{g} + \mathbf{s}\,, \tag{1.3}$$

where the deviation vector $\mathbf{s}$ may be represented in the frame of the already introduced Cartesian coordinate system in the component form

$$\mathbf{s} = (s_x, s_y, s_z)\,. \tag{1.4}$$

Following Hirsch et al. [1, 2] equation (1.1) is easily derived from (1.2) by omitting $\boldsymbol{g} \cdot \boldsymbol{r}_n$ in the exponent ($\boldsymbol{g} \cdot \boldsymbol{r}_n$ is the constant phase term common to all sum terms of (1.2)) and neglecting $\mathbf{s} \cdot \boldsymbol{R}$ as being small.

It is the aim of the present paper to consider in more detail the significance of the term $\mathbf{s} \cdot \boldsymbol{R}$. That this is not a trivial problem emerges immediately if one disregards for the moment the vectorial nature of $\mathbf{s}$ and $\boldsymbol{R}$. Setting $s = 0.2\ \text{nm}^{-1}$ as a typical value for the weak-beam imaging mode [5, 6] and assuming for R a value of the order of magnitude of a perfect Burgers vector in a typical metal, say, $R \approx 0.25$ nm, we obtain for $\mathbf{s} \cdot \boldsymbol{R}$ values of the order of 0.05; i.e. $2\pi s R$ expressed in degrees may be of the order of 15°. Such phase shifts are not a priori negligible.

2. The Significance of $(\mathbf{s} \cdot \boldsymbol{R})$

We omit $\boldsymbol{g} \cdot \boldsymbol{r}_n$ in the exponent of (1.2) (see above) and assume

$$|\mathbf{s}| \ll |\boldsymbol{g}|\,. \tag{2.1}$$

Then the exponent of (1.2) is a slowly varying function when going from one to the adjacent atom. Consequently, the summation in (1.2) can be replaced by an integral

$$F(\boldsymbol{\chi}) = \frac{f}{V_0} \int \exp 2\pi i(\boldsymbol{g} \cdot \boldsymbol{R}(\boldsymbol{r}) + \mathbf{s} \cdot \boldsymbol{R}(\boldsymbol{r}) + \mathbf{s} \cdot \boldsymbol{r})\, \mathrm{d}^3\boldsymbol{r}\,, \tag{2.2}$$

where the integration goes over the crystal volume (TEM specimen). V_0 is the volume of elementary cell.

Assuming a perfect electron microscope the amplitude of the diffraction pattern in the back focal plane of the objective lens is given by a cut of the (two-dimensional) Ewald sphere through the (three-dimensional) diffraction amplitude $F(\boldsymbol{\chi})$ of (2.2). We approximate the Ewald sphere by a tangent plane which is — because of $\boldsymbol{k}_g$ parallel to the z-axis — parallel to the s_x–s_y-plane. Then the kinematical contrast amplitude $A(x, y)$ just below the specimen is obtained by a two-dimensional Fourier transformation over the s_x–s_y-plane

$$A(x, y) = \int F(\boldsymbol{\chi}) \exp -2\pi i(s_x x + s_y y)\, \mathrm{d}s_x\, \mathrm{d}s_y\,, \tag{2.3}$$

where $s_z = \text{const}$ is the excitation error s in (1.1). However, instead of doing so, it is expedient to make a small detour: Instead of evaluating (2.3) we apply first a three-dimensional Fourier transformation leading to an auxiliary function $\tilde{F}(\boldsymbol{r})$,

$$\tilde{F}(\boldsymbol{r}) = \int F(\boldsymbol{\chi}) \exp -2\pi i(\mathbf{s} \cdot \boldsymbol{r})\, \mathrm{d}^3\mathbf{s}\,, \tag{2.4}$$

and subsequently a one-dimensional inverse Fourier transformation over the z-axis. The so obtained contrast amplitude will be denoted by $\hat{A}(x, y)$,

$$\hat{A}(x, y) = \int \tilde{F}(\boldsymbol{r}) \exp 2\pi i s_z z\, \mathrm{d}z\,. \tag{2.5}$$

For the purpose of the present paper we may assume that the atomic scattering amplitude f is independent of $\boldsymbol{\chi}$ and, consequently, independent of $\mathbf{s}$. Then (2.4) leads to a three-dimensional δ-function of a complicated argument. We prefer to express it in terms of a product of three one-dimensional δ-functions,

$$\tilde{F}(\boldsymbol{r}) = \frac{f}{V_0} \int \exp 2\pi i \boldsymbol{g} \cdot \boldsymbol{R}(\boldsymbol{r}')\, \delta(R_x(\boldsymbol{r}') + x' - x)\, \delta(R_y(\boldsymbol{r}') + y' - y)\, \delta(R_z(\boldsymbol{r}') + z' - z)\, dx'\, dy'\, dz' . \tag{2.6}$$

It is sufficient to consider the integration over one of these δ-functions. We refer to the first δ-function in (2.6) and assume that the components of $\boldsymbol{R}$ are diffentiable. Then

$$R_x(x', y', z') = R_x\big(x + (x' - x), y', z'\big) = R_x(x, y', z') + (x' - x)\, R_{x,x}(x, y', z') + + , \tag{2.7}$$

where the subscript beyond the comma denotes differentiation with respect to the indicated coordinate. Then

$$\int \delta(R_x(x', y', z') + x' - x)\, dx' = \int \delta[R_x(x, y', z') + (1 + R_{x,x}(x, y', z'))\, (x' - x)]\, dx' = \frac{1}{1 + R_{x,x}(x, y', z')} \tag{2.8}$$

with

$$x' = x - \frac{R_x}{1 + R_{x,x}} . \tag{2.9}$$

Proceeding with the two other δ-functions in the same way and assuming

$$R_{j,j} \ll 1; \quad j = x, y, z \tag{2.10}$$

and neglecting the small difference between x' and x in (2.9) and similarly for the other coordinates y and z we arrive at

$$\tilde{F}(\boldsymbol{r}) = \frac{f}{V_0} \exp 2\pi i \boldsymbol{g} \cdot \boldsymbol{R}(\boldsymbol{r}) \frac{1}{1 + \theta(\boldsymbol{r})} \tag{2.11}$$

with

$$\theta(\boldsymbol{r}) = R_{x,x}(\boldsymbol{r}) + R_{y,y}(\boldsymbol{r}) + R_{z,z}(\boldsymbol{r}) , \tag{2.12}$$

where $\theta(\boldsymbol{r})$ is the local volume dilation.

Now we proceed easily from (2.4) to (2.5) and end up with

$$\hat{A}(x, y) = \frac{f}{V_0} \int_0^t \frac{1}{1 + \theta(x, y, z)} \exp 2\pi i(\boldsymbol{g} \cdot \boldsymbol{R}(x, y, z) + sz)\, dz , \tag{2.13}$$

where the subscript z at s_z in (2.5) is now omitted in order to keep the analogy to the standard form of the kinematical contrast integral (1.1). Because in general $|\theta| \ll 1$ we may write instead of (2.13)

$$\hat{A}(x, y) = A(x, y) + A_c(x, y) \tag{2.14}$$

with $A(x, y)$ according to (1.1) with $A_0 = f/V_0$ and the "correction" term A_c given by

$$A_c(x, y) = -\frac{f}{V_0} \int_0^t \theta(x, y, z) \exp 2\pi i\big(\boldsymbol{g} \cdot \boldsymbol{R}(x, y, z) + sz\big)\, dz . \tag{2.15}$$

It is instructive to bring $\hat{A}(x, y)$ into a form which is more similar to $A(x, y)$. This is easily done by recalling the way in which the denominator $(1 + R_{x,x})$ on the right-hand side of (2.8) emerged from the term $s_x R_x$ in the exponent of (2.2). Going the corresponding steps backwards we find as an alternative form of $\hat{A}(x, y)$

$$\hat{A}(x, y) = A_0 \int_0^t \exp 2\pi i(\alpha_g + \alpha_s + sz) \, dz \,, \tag{2.16}$$

where α_g is the well-known phase term [2], commonly used,

$$\alpha_g = \boldsymbol{g} \cdot \boldsymbol{R}(x, y, z) \,, \tag{2.17}$$

and α_s is an additional phase term,

$$\alpha_s = sD(x, y, z) \tag{2.18}$$

with

$$D(x, y, z) = \int^z \theta(x, y, \zeta) \, d\zeta \tag{2.19}$$

being an "integral dilation" along the z-axis. The integration constant in (2.19) may be left open since it gives rise to an unimportant phase factor only.

3. Discussion

3.1 General remarks

Equations (2.13) to (2.15) show that retaining the term $\boldsymbol{s} \cdot \boldsymbol{R}$ in deriving the kinematical contrast integral gives rise in the denominator of the integrand to an additional term $(1 + \theta(\boldsymbol{r}))$ which reflects the local variation of the size of the elementary cell and therefore the local variation of the scattering power per unit volume expressed in terms of $f/V_0(1 + \theta(r))$. In this sense the term $\boldsymbol{s} \cdot \boldsymbol{R}$ represents a special case of "structure factor contrast". (The incorporation of structure factor contrast into current dynamical contrast theories has been considered by [8, 9].)

Obviously $A(x, y)$ (1.1) and $\hat{A}(x, y)$ (2.13) are identical if the volume dilation θ vanishes. This holds e.g. for screw dislocations if $\boldsymbol{R}$ is calculated by means of the linear theory of elasticity and for stacking faults if the shift vector is parallel to the fault plane (the case of stacking faults will be considered in some more detail in Section 3.2). In other cases where $\theta \neq 0$ it depends on the accuracy to be achieved whether or not the additional phase term α_s must be taken into account.

If alloy or impurity atoms segregate substitutionally close to the cores of dislocations they may modify the scattering power per unit volume even more effectively than the elastic volume dilation of lattice defects. The influence of such segregation effects on the diffraction contrast can be brought into a form similar to that by which θ enters into (2.13). Segregation at interstitial positions may have a more complicated effect on the diffraction contrast and will not be considered here.

3.2 Planar defects

We consider an intrinsic stacking fault with the fault normal $\boldsymbol{e}_n$,

$$\boldsymbol{e}_n = [\sin \varphi, 0, \cos \varphi] \tag{3.1}$$

and with a unit vector $\boldsymbol{e}_t$,

$$\boldsymbol{e}_t = [\cos \varphi, 0, -\sin \varphi] \tag{3.2}$$

along the trace of the stacking fault plane with the x–z plane. Further we use auxiliary coordinates ξ, ζ along the unit vectors $\boldsymbol{e}_t$ and $\boldsymbol{e}_n$, respectively. The displacement field

$\boldsymbol{R}$ as caused by the stacking fault is given by

$$\boldsymbol{R} = \boldsymbol{R}_{\mathrm{sf}} H(\zeta) \tag{3.3}$$

with

$$H(\zeta) = \begin{cases} 0 & \text{for} \quad \zeta < 0 \,, \\ 1 & \text{for} \quad \zeta > 0 \,, \end{cases} \tag{3.4}$$

and with

$$\boldsymbol{R}_{\mathrm{sf}} = R_{\mathrm{sf}}^{\parallel} \boldsymbol{e}_{\mathrm{t}} + R_{\mathrm{sf}}^{\perp} \boldsymbol{e}_{\mathrm{n}} \,. \tag{3.5}$$

Then, simple algebra leads to

$$D = \frac{1}{\cos \varphi} R_{\mathrm{sf}}^{\perp} H(\zeta) \tag{3.6}$$

or

$$\alpha_s = \begin{cases} 0 & \text{for } z \text{ above the fault plane,} \\ s \dfrac{1}{\cos \varphi} R_{\mathrm{sf}}^{\perp} & \text{for } z \text{ below the fault plane.} \end{cases}$$

Note that D in (3.6) is the projection of $R_{\mathrm{sf}}^{\perp} \boldsymbol{e}_{\mathrm{n}} H(\zeta)$ parallel to $\boldsymbol{e}_{\mathrm{t}}$ onto the z-axis.

If the phase shift term α as caused by the stacking fault is solely given by $\alpha_g = \boldsymbol{g} \cdot \boldsymbol{R}$, the "overall visibility" ([10]) of an intrinsic stacking fault should not change if the diffraction vector is changed from $\boldsymbol{g}$ to $-\boldsymbol{g}$. This holds irrespective of whether or not the shift vector $\boldsymbol{R}_{\mathrm{sf}}$ contains a component parallel to $\boldsymbol{e}_{\mathrm{n}}$. If, however, $R_{\mathrm{sf}}^{\perp} \neq 0$ (e.g. [11]) and α_s according to (3.7) has to be added to α_g, then even intrinsic faults may in principle show up a change of the "overall" fringe visibility ("contrast asymmetry") with changing the sign of $\boldsymbol{g}$ (e.g. changing the sign of α_g) but keeping the sign of s (and so the sign of α_s) unchanged. This conclusion revised the corresponding contrary statement in [10] where the term α_s was not yet incorporated. Further a contribution α_s to the α-values of the two successive intrinsic faults constituting an extrinsic stacking fault may influence the contrast asymmetry of extrinsic stacking faults as found experimentally and theoretically in [10]. The term α_s may also be important for a correct interpretation of the quasi-stacking fault contrast of coincidence boundaries arising when both crystals adjacent to the boundary are imaged with a common diffraction vector $\boldsymbol{g}$ and both crystals are shifted against each other by a "rigid-body translation" [12, 13]. If the latter contains a component perpendicular to the boundary [12], again, α_s is not zero and a contrast asymmetry when changing from $\boldsymbol{g}$ to $-\boldsymbol{g}$ may happen. Indeed, such asymmetries on weak-beam images of coincidence boundaries have been observed [13] and were so far not satisfactorily explained.

3.3 Final remarks

During the time when the kinematical contrast theory was first developed and experimentally tested [1, 2] rather small values of the excitation errors were applied, say $s \approx 1/20$ nm. At this stage neglecting the term $\boldsymbol{s} \cdot \boldsymbol{R}$ or its equivalent — the volume dilation θ — was presumably justified in all practical cases. However, with the introduction of the weak-beam imaging mode [5, 6] values of s up to 0.2 nm^{-1} and larger are used and — as a consequence — much finer details of the contrast figures become accessible and require interpretation. The calculation of the weak-beam contrast of specific lattices defects using (2.13) to (2.15) instead of (1.1), which is beyond the scope of the present paper, must show whether at a given stage of image resolution of the volume dilation θ is important.

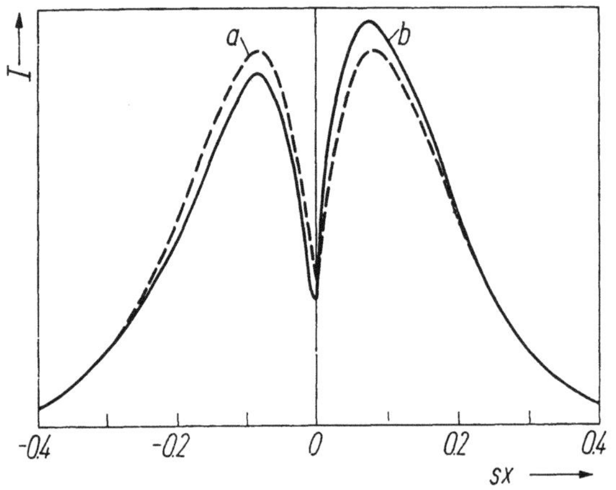

Fig. 1. Weak-beam contrast of an edge dislocation (background intensity subtracted). Burgers vector $a/2\langle 110\rangle$ parallel to the incident electron beam, $a = 0.562$ nm. Diffraction vector $\boldsymbol{g} = \{220\}$ perpendicular to the dislocation line, excitation parameter $s = 0.2\ \text{nm}^{-1}$. (a) Contrast calculated without the dilation term θ, equation (1.1). (b) Contrast calculated with the term θ included, equation (2.13). Contrast intensity I in arbitrary units

Preliminary calculations of the weak-beam contrast of an edge dislocation in Si reveal that for $\boldsymbol{g} \cdot \boldsymbol{b} = 0$ the double-peaked symmetrical contrast profile as it comes out of (1.1) becomes asymmetric if the θ-term is included, see Fig. 1. Such effects, although at present at the limit of experimental detectability, should be taken into account if details of the core structure of dislocations are to be derived from the fine structure of the weak-beam images of dislocations, e.g., [14].

Acknowledgements

The author is very grateful to the AAEC, Lucas Heights, Australia, in particular to Dr. P. M. Kelly, for supporting a three-months stay at the Materials Division of the AAEC from March to June 1981. The author is further indebted to Dr. D. J. H. Cockayne, Electron Microscope Unit of the University of Sydney for very helpful and fruitful discussions not only on the subject of the present paper.

References

[1] P. B. HIRSCH, A. HOWIE, and M. J. WHELAN, Phil. Trans. Roy. Soc. (London) **A252**, 499 (1960).
[2] P. B. HIRSCH, A. HOWIE, R. B. NICHOLSON, D. W. PASHLEY, and M. J. WHELAN, Electron Microscopy of Thin Crystals, Butterworths, London 1965.
[3] A. HOWIE and M. J. WHELAN, Proc. Roy. Soc. **A263**, 217 (1961).
[4] M. WILKENS, phys. stat. sol. **6**, 939 (1964).
[5] D. J. H. COCKAYNE, I. L. F. RAY, and M. J. WHELAN, Phil. Mag. **20**, 1265 (1969).
[6] D. J. H. COCKAYNE, Z. Naturf. **27a**, 452 (1972).
[7] D. J. H. COCKAYNE, in: Modern Diffraction and Imaging Techniques, Vol. I, Ed. S. AMELINCKX, R. GEVERS, and J. VAN LANDUYT, North-Holland Publ. Co., Amsterdam 1978 (p. 153).
[8] M. F. ASHBY and L. M. BROWN, Phil. Mag. **8**, 1083, 1649 (1963).
[9] M. WILKENS, M. J. JENKINS, and K.-H. KATERBAU, phys. stat. sol. (a) **39**, 103 (1977).
[10] H. FÖLL, C. B. CARTER, and M. WILKENS, phys. stat. sol. (a) **58**, 393 (1980).
[11] I. P. JONES and H. L. LORETTO, Proc. 6th Europ. Congr. Electron Microscopy, Jerusalem 1976 (p. 503).
[12] R. C. POND and V. VITEK, Proc. Roy. Soc. **A357**, 453 (1977).
[13] C. B. CARTER and H. FÖLL, Scripta metall. **12**, 1135 (1978).
[14] G. R. ANSTIS and D. J. H. COCKAYNE, Acta cryst. **A35**, 511 (1979).

(Received August 18, 1981)

phys. stat. sol. (a) 68, 129 (1981)

Subject classification: 10.2 and 14.4; 20.1; 22.8.1

I. Physikalisches Institut der Universität Gießen[1])

Dipole Centres in $CaMoO_4$ Crystals[2])

By

M. Böhm, S. Herrmann, and A. Scharmann

In $CaMoO_4$ crystals with oxygen deficiency thermally stimulated depolarization (TSD) currents are investigated in dependence on polarization time as well as on polarization temperature. The results can be well understood on the base of dipole orientation. Optical absorption measurements yield a band near 460 nm and another very broad one at around 850 nm, either band showing anisotropic absorption of plane polarized light. From molecular orbital theory the infrared band is assumed to be due to a transition within a $(MoO_4)^{3-}$ centre. The other absorption band possibly arises from oxygen vacancies which have trapped one or two electrons. The gradual disappearance of either band due to annealing procedure can be correlated to that of the TSD signal. Thus the proposed models of dipole centres involve absorbing defects, too.

In $CaMoO_4$-Kristallen mit Sauerstoffmangel werden thermisch stimulierte Depolarisationsströme (TSD) in Abhängigkeit von der Polarisationszeit sowie von der Polarisationstemperatur untersucht. Die Ergebnisse können gut auf der Grundlage der Dipolorientierung verstanden werden. Optische Absorptionsmessungen ergeben eine Bande nahe bei 460 nm und eine weitere sehr breite bei ungefähr 850 nm, die beide anisotrope Absorption von linear polarisiertem Licht zeigen. Nach der Molekülorbitaltheorie wird angenommen, daß die infrarote Bande durch einen Übergang innerhalb eines $(MoO_4)^{3-}$ Zentrums verursacht wird. Die andere Absorptionsbande stammt möglicherweise von Sauerstoffleerstellen, die ein oder zwei Elektronen eingefangen haben. Das allmähliche Verschwinden beider Banden infolge des Temperns kann mit dem des TSD-Signals korreliert werden. So enthalten die vorgeschlagenen Modelle der Dipolzentren auch absorbierende Defekte.

1. Introduction

Oxygen dominated phosphors with scheelite structure such as $CaMoO_4$ are of considerable interest concerning the solid state properties as well as the practical application. Most of the characteristic phenomena arise from defects which are generated or excited by X-, γ-, or particle radiation.

However, even without any irradiation some defects are already present and can be investigated by means of thermally stimulated conductivity [1]. In some cases the peak heights of the current glow curves can be monitored by irradiation of visible light [2]. This leads to the assumption that electronic carriers are seen to be responsible for the charge transport. On the other hand, permanently existing dipole complexes are to be assumed as well for the explanation of current glow peaks especially in $CaWO_3$ crystals with oxygen deficiency [3]. The correlation to the results of optical absorption measurements indicates that oxygen vacancies are likely included in these processes.

The aim of this work is to study similar dipole centres in $CaMoO_4$ crystals with the method of thermally stimulated depolarization (TSD) in more detail. In addition optical absorption measurement were performed in order to get some information on the structure of defects in question.

---

[1]) Heinrich-Buff-Ring 16, D-6300 Gießen, FRG.

[2]) Dedicated to Prof. Dr. D. Hahn, Braunschweig, on occasion of his 60th birthday.

2. Experimental Techniques and Results

The single crystals of calcium molybdate investigated in these studies have been grown in oxygen deficient atmosphere. The size of the samples was about $3 \times 4 \times \times 1$ mm³. Silver paint was used for making the electrodes. The samples are sandwiched between insulating foils of mica in form of a capacitor to avoid the formation of space charges due to ion accumulation or carrier injection.

The thermally stimulated depolarization measurements were performed using the ionic thermo-current technique [4, 5]. An external field of about 300 V/cm was applied to the crystal at room temperature for an interval of time. After polarization the temperature was reduced with the field still applied to the temperature of liquid air; the cooling rate was about —41 K/min. The field is then removed and the sample is warmed at a linear rate of about 24 K/min. During heating-up a depolarization current is detected in short-circuit conditions which peaks at about 155 K (Fig. 1).

This depolarization peak which exhibits typical non-symmetrical single band structure described by first-order kinetics [6, 7] is supposed to be due to reorientation of microscopic dipoles.

Some other measurements yield results which can be explained quite well in terms of this model. The effect of the polarization time t_p on the current intensity is illustrated in Fig. 2. The built-up of polarization at a fixed temperature can be described

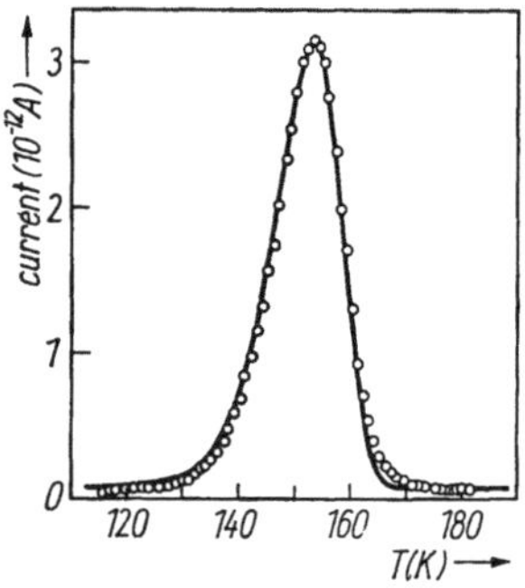

Fig. 1. Thermally stimulated depolarization current in short-circuit conditions by polarizing at room temperature (○) and calculated glow curve (———) with the activation energy $E = 0.33$ eV and the pre-exponential factor $\tau_0 = 2.4 \times 10^{-10}$ s

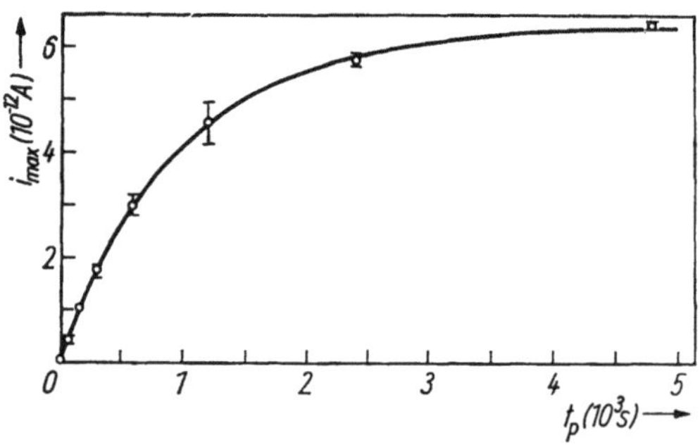

Fig. 2

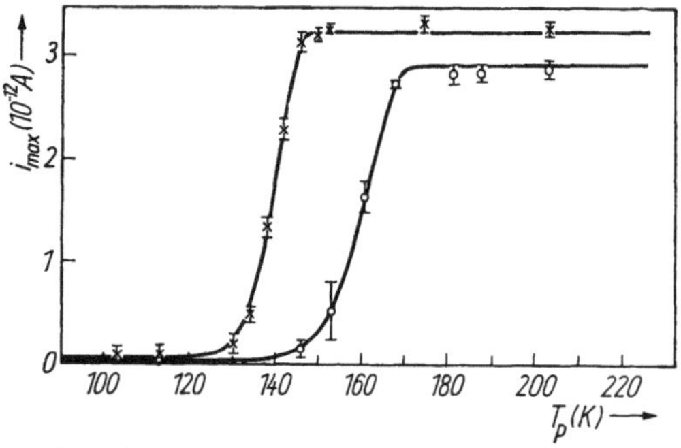

Fig. 3

Fig. 2. Peak current i_{max} of thermally stimulated depolarization in dependence on polarizing time (○); calculated curve (———)

Fig. 3. Peak current i_{max} of thermally stimulated depolarization vs. polarization temperature with an average polarizing time of 0 min (○) and 30 min (×); theoretical curve calculated from the model of a dipolar relaxation mechanism (———)

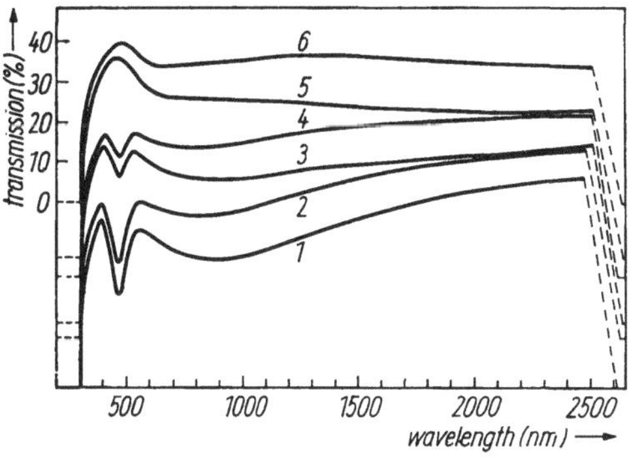

Fig. 4

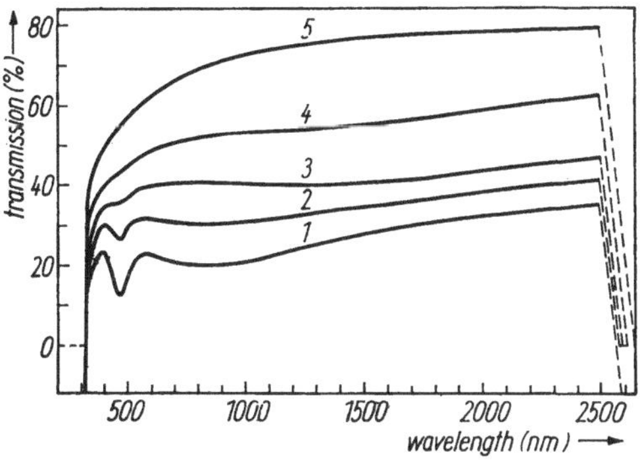

Fig. 5

Fig. 4. Transmission spectra at $T = 110$ K. (1) and (2) direction of propagation $\boldsymbol{k} \perp (110)$ plane and $\boldsymbol{k} \perp (010)$ plane, $\boldsymbol{E} \parallel c$-axis; (3) and (4) unpolarized light; (5) and (6) $\boldsymbol{E} \perp c$-axis

Fig. 5. Transmission spectra at $T = 300$ K with annealing time t_a as parameter ($T = 1320$ K, $\boldsymbol{E} \parallel c$-axis). (1) $t_a = 0$, (2) 10, (3) 330, (4) 1410, (5) 4530 min

by an exponential law and a saturation behaviour is found as is expected from relaxation processes.

The effect of the polarization temperature T_p is demonstrated in Fig. 3. The curve shifted to lower temperatures is obtained by polarizing the sample for 30 min and cooling down with the electric field applied.

The transmission measurements were performed at low temperatures by a double-beam photometer (Zeiss, DMR 21). The spectra depend on the direction of light propagation as well as on the polarization with respect to the c-axis. Two absorption bands can be observed when the sample is irradiated by plane polarized light with the E-vector parallel to the c-axis. Both the band at 460 nm and the broader one at around 850 nm show dichroic properties. The highest transmission is obtained by irradiation with light of which the E-vector is perpendicular to the c-axis (Fig. 4).

The effect of annealing in air at about 1320 K on transmission intensity is illustrated in Fig. 5. As a result the total transmission is found to be enhanced after each annealing procedure.

3. Discussion

The results of depolarization measurements cannot be explained by the occurrence of free charge carriers but rather by a single dipolar relaxation process [8]. The glow peak of thermally stimulated current in short circuit shows a non-symmetrical shape which is typical for kinetics with first-order conditions. Presuming a monomolecular mechanism at the reorientation during heating, i.e. non-interacting dipoles there are various approximative methods for evaluating the thermal activation energy E based on the shape of the glow curve. A value of about 0.33 eV is determined in this way (Fig. 1). The pre-exponential factor τ_0 is found to be 2.4×10^{-10} s.

The dependence of the polarization upon the polarizing time (Fig. 2) provides another method for evaluating the relaxation time. In the case of a single relaxation process the polarization P approaches its final value P_s exponentially with the relaxation time constant τ

$$P = P_s(T_p) \left\{1 - \exp\left[-\frac{t}{\tau(T_p)}\right]\right\}. \tag{1}$$

Assuming an activation energy of about 0.33 eV a pre-exponential factor of $8.5 \times \times 10^{-10}$ s is found from the experimental data. This value is well consistent with others determined from glow curve analysis.

An essential hint at the occurrence of dipolar reorientation processes is given by the dependence of the built-up of polarization on temperature (Fig. 3). The maximum current shows a steep decrease at polarization temperatures below the glow peak temperature. Besides, this decrease depends also on both the polarization time and the cooling rate. These phenomena can be ascribed to the increase of relaxation time and the decrease of jump probability, respectively. It is important to note that the position of the depolarization peak remains constant. This fact gives further evidence for the appearance of a dipolar relaxation process [9]. The theoretical curves are calculated from (1) neglecting the reciprocal temperature dependence of the steady-state polarization P_s [10]. From this the values of activation energy and pre-exponential factor are found to be 0.41 eV and 10^{-12} s.

The results of optical absorption measurements allow one to discuss the structure of defects involved in the reorientation processes. An infrared band can also be observed in other oxygen dominated phosphors with oxygen deficiency such as $CaWO_4$ [3] or $LiNbO_3$ [11]. This band is suggested to arise from a transition within the molecular states of the oxocomplex. From molecular orbital calculations the energetically highest occupied orbital is found to be a threefold degenerate t_1-state [12]. Assuming an additional electron to be trapped in the oxyanion the empty twofold degenerate molecular orbital 2e will be occupied and a $(MoO_4)^{3-}$ complex will be formed. A transition into the next higher lying $4t_2$ orbital is electric dipole allowed. The correlated energy roughly estimated from calculations corresponds to a wavelength of about 735 nm.

The symmetry T_d of the oxyanions is lowered to the site group symmetry D_{2d} by a slight compression of the tetrahedra along the *c*-axis [13]. As a consequence the orbital degeneracy is partially removed and the splitting of the molecular states 2e and $4t_2$ into a_1, b_1 and e, b_2 is expected to be about 300 to 600 cm^{-1} [14]. The allowed transitions are polarized either in z-direction (a_1–b_2) or in x,y-direction (a_1–e). However, the calculated degree of polarization is found too low and to be not consistent with the experimental results.

Let us consider one more distortion, due to an oxygen vacancy within the molybdate complex ion a new configuration is formed consisting of a single negative charged oxyanion $(MoO_3)^-$. The symmetry should then be further reduced to C_{3v} or C_s and z-polarized transitions are consequently anticipated to appear in the short-wavelength region of the wide absorption band. However, such absorption cannot be unambiguously detected by measurement because of the broadening possibly arising from some more transitions in this region. Nevertheless it seems likely that Mo^{5+} in connection with a $(MoO_4)^{3-}$ or $(MoO_3)^-$ complex is to be seen as the absorbing centre. Results of XPS investigations in amorphous WO_3 argue rather for transitions in the W^{5+} ion to be responsible for the long-wavelength band [15].

As for the 460 nm band the more pronounced anisotropy can no longer be explained by energy splitting due to crystal field effects. Since the observed absorption is almost completely z-polarized the splitting of energy levels and consequently the separation of z- and x,y-polarized bands would have to be assumed extremely large. One more way of discussing the strong anisotropic behaviour will be given by the theory of hybridization. The wave function of at least one of the states involved in the transition is rather suggested to have directional properties related to the axes of the crystal which can be classically represented by an electric dipole transition moment [16].

For a defect whose dipole moment makes an angle θ with the c-axis the anisotropic absorption ratio is given by

$$A = \frac{\sum f(\boldsymbol{E} \parallel \boldsymbol{c})}{\sum f(\boldsymbol{E} \perp \boldsymbol{c})} = 2 \cot^2 \theta \,, \tag{2}$$

where f is oscillator strength ($\approx$ square of electric dipole moment) [17]. The requirement of high anisotropy implies the presence of a dipole making a small angle with the c-axis. This can be fulfilled by the line joining two adjacent oxygen atoms of different tetrahedra. Similar ideas have been discussed in context with some other oxygen compounds [17 to 19]. The angle with the c-axis is found to be less than about 10° (Fig. 6).

For calculating the anisotropic ratio it is necessary to account for the eight possible dipole positions connecting five tetrahedra. As a result we again obtain (2) which does not state any dependence of the E-vector upon its direction in the (001) plane. The high A-value of about 64 estimated from (2) is qualitatively consistent with the experimental results. From this the origin of the absorption band might be due to a defect centre consisting of two neighbouring oxygen vacancies which have trapped one or two electrons similar to F^+- or F-centres in alkali-earth oxydes [20, 21].

The results of annealing measurements indicate that the decrease of absorption bands goes as fast as the TSD signal disappears. From this the absorption centres are supposed to be correlated to the dipole centres responsible for depolarization currents. Thus the structure of dipoles can be roughly described in a twofold manner. First we have two reduced molybdate complex ions and one more complex ion including an oxygen vacancy: $2(MoO_4)^{3-}$–$(MoO_3)^0$. The electric moment originates from two units of positive charge of the oxygen vacancy and two units of negative charge of two Mo^{5+} ions. Second the dipole might be formed by two reduced molybdate complex ions either of which involves an oxygen vacancy again: $(MoO_4)^{3-}$–$(MoO_3)^-$. The electric moment originates from one unit of negative charge of one Mo^{5+} ion and one unit of positive charge of the Mo^{5+}–oxygen vacancy complex. The density of the ab-

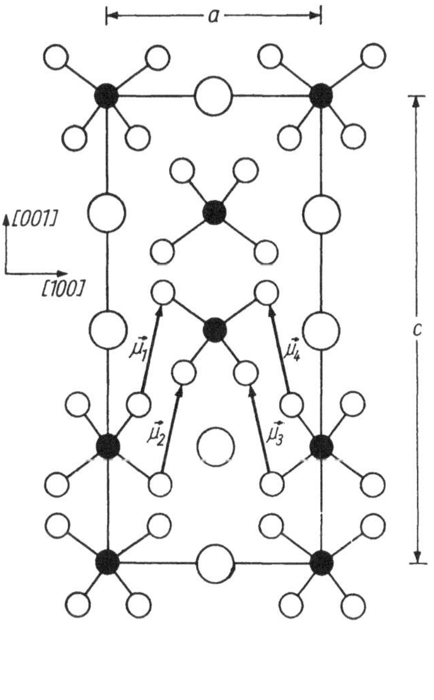

Fig. 6. Projection of unit cell on the (101) plane; dipole positions between adjacent oxygen sites

sorption centres evaluated from the Smakula equation ($N_{abs} \approx 1.5 \times 10^{22}$ m^{-3}) is found to be less than that of dipole centres determined from glow measurements ($N_{dip} \approx 7 \times 10^{23}$ m^{-3}). This leads to the assumption that not all defects are participating in the absorption process.

References

[1] M. Böhm, O. Erb, and A. Scharmann, phys. stat. sol. (a) **41**, 535 (1977).
[2] M. Böhm and A. Scharmann, phys. stat. sol. (a) **22**, 143 (1974).
[3] M. Böhm, A. Scharmann, and M. Thomae, Charge Storage, Charge Transport and Electrostatics with their Application, Elsevier Sci. Publ. Co., Amsterdam 1979 (p. 261).
[4] C. Bucci and R. Fieschi, Phys. Rev. Letters **12**, 16 (1964).
[5] C. Bucci, R. Fieschi, and G. Guidi, Phys. Rev. **148**, 816 (1966).
[6] M. Böhm and A. Scharmann, phys. stat. sol. (a) **4**, 99 (1971).
[7] M. Böhm and A. Scharmann, phys. stat. sol. (a) **5**, 563 (1971).
[8] A. Scharmann and M. Böhm, Proc. Internat. Conf. Defects in Insulating Crystals, Riga (USSR) 1981, to be published.
[9] A. Kessler, J. Electrochem. Soc. **123**, 1237 (1976).
[10] J. C. Manifacier, J. Gasiot, P. Parot, and J. P. Fillard, J. Phys. C **11**, 1011 (1978).
[11] W. Bollmann and M. Gernand, phys. stat. sol. (a) **9**, 301 (1972).
[12] R. Kebabcioclu and A. Müller, Chem. Phys. Letters **8**, 59 (1971).
[13] E. Gürmen, E. Daniels, and J. S. King, J. chem. Phys. **55**, 1093 (1971).
[14] R. Oeder, A. Scharmann, and D. Schwabe, J. Crystal Growth **49**, 349 (1980).
[15] G. Hollinger, Tran Minh Duc, and A. Deneuville, Phys. Rev. Letters **37**, 1546 (1976).
[16] E. W. J. Mitchell and E. G. S. Paige, Phil. Mag. **46**, 1353 (1955).
[17] E. W. J. Mitchell and J. D. Rigden, Phil. Mag. **2**, 941 (1957).
[18] E. W. J. Mitchell, J. D. Rigden, and P. D. Townsend, Phil. Mag. **5**, 1013 (1960).
[19] L. S. Welch, A. E. Hughes, and G. P. Pells, J. Physique **41**, C6-533 (1979).
[20] R. G. Bessent, B. C. Cavenett, and I. C. Hunter, J. Phys. Chem. Solids **29**, 1523 (1968).
[21] Y. Chen, J. L. Kolopus, and W. A. Sibley, Phys. Rev. **186**, 865 (1969).

(Received August 26, 1981)

phys. stat. sol. (a) 68, 135 (1981)

Subject classification: 14.1; 1.5; 21.1; 21.6

Fakultät für Physik der Universität Konstanz[1])

Electrical Resistivity of CuCo Solid Solutions at Low Temperature

By

H. HALLER, D. KORN, and G. ZIBOLD

Metastable CuCo solid solutions are produced by vapour condensation onto a substrate held at 5 K. Co solved in Cu causes a negative temperature coefficient of the electrical resistivity at low temperature (resistance anomaly). The resistance anomaly depends on the square of the Co concentration up to about 3 at% Co. This result agrees with the observation that single Co atoms have no localized magnetic moment in a non-magnetic host but get a moment by interaction. Annealing the CuCo films to e.g. $T_a = 390$ K leads to clustering of the Co atoms. With the Co atoms in magnetic clusters, the resistance anomaly varies linearly with Co concentration. The residual resistivity as a function of Co concentration follows straight lines always and the slope of these lines decreases with annealing.

Metastabile CuCo-Mischkristalle kann man durch Abschrecken aus der Dampfphase auf einer mit flüssigem Helium gekühlten Unterlage herstellen. Das im Cu gelöste Co bewirkt bei tiefen Temperaturen einen negativen Temperaturkoeffizienten des elektrischen Widerstandes (Widerstandsanomalie). Die Widerstandsanomalie wächst bis 3 at% Co quadratisch mit der Co-Konzentration. Dieses Ergebnis steht im Einklang mit der Beobachtung, daß einzelne Co-Atome in einem nichtmagnetischen Wirtsmetall kein magnetisches Moment haben, daß aber durch die Wechselwirkung zwischen den Co-Atomen ein magnetisches Moment entsteht. Wenn man die CuCo-Schichten bei $T_a = 390$ K tempert, so bilden sich Haufen von Co-Atomen. Die Widerstandsanomalie dieser Haufen hängt linear von der Konzentration ab. Der vom Co herrührende Anteil des Restwiderstandes ist proportional zur Co-Konzentration, wobei der Proportionalitätsfaktor mit dem Tempern abnimmt.

1. Introduction

Magnetic atoms in a nonmagnetic host cause a negative temperature coefficient at low temperature (resistance anomaly). This effect is theoretically described by Kondo's model [1]. The resistance anomaly increases linearly with the concentration of the magnetic component. This concentration dependence is valid if each magnetic atom has a localized magnetic moment. In opposition to that fact single Co atoms in a nonmagnetic host have no localized magnetic moment. We want to investigate the resistance anomaly for this case. Does a resistance anomaly exist and how does it depend on concentration? The problem has been regarded with respect to AuCo. Boucai et al. [2] found that single Co atoms in Au have no magnetic moment. The resistance anomaly as a function of Co concentration c varies as c^2 for AuCo solid solutions [3].

Concerning CuCo, the resistance anomaly of bulk alloys is already measured [4 to 7]. The investigation of the concentration dependence is difficult in the case of bulk samples, because Co precipitates [8]. The number and size of the precipitations depend on thermal treatment [9, 10].

[1]) Postfach 5560, D-7750 Konstanz, FRG.

In order to get solid solutions of CuCo we evaporate both components simultaneously on a substrate held at 5 K. The atoms randomly hit the substrate. Diffusion is suppressed at the temperature of 5 K. By this method [11, 12] a solid solution is produced with the Co atoms distributed at random. Within 0 to 15 at% Co the solid solutions are single phase and metastable as shown by Mader et al. [13] for CuCo films [14]. The films contain lattice defects. The solid solution exists up to a temperature higher then 150 K. Near 290 K diffusion sets in enabled by lattice vacancies which are present in the film. This means, films annealed to 290 K are no longer representative of a random solid solution and films annealed to 390 K are close to thermal equilibrium.

2. Experimental Method

A master alloy is produced by melting the two components together in a tungsten crucible under argon (99.999%) atmosphere. The master alloy is filed to small particles. Only bronze tools are used. The alloy particles are evaporated step by step from a hot tungsten ribbon onto a substrate of crystalline quartz optically polished. To avoid fractionation each particle contributes less than one atomic layer. During evaporation the vacuum amounts to 2×10^{-8} mbar. After condensation the quartz plate is warmed up from 5 K to an annealing temperature T_a using a rate of about 1.5 K/min. The time of waiting t at T_a is noted in Table 2.

The electrical resistance is measured by a conventional four-electrode method during and after condensation of the film within the same UHV cryostat (in situ measurement). The electrical resistivity ϱ results from the measured resistance R according to

$$\varrho = D \frac{B}{L} R \tag{1}$$

(thickness D, width $B = 1$ mm, and length $L = 10$ mm of the film). The absolute error of the resistivity is mainly determined by the error of the film thickness D. There are two methods to determine D. First, D is calculated from the temperature-dependent part of the resistance R_a of annealed films ($T_a = 390$ K)

$$D = \frac{\varrho_{Cu}(T_2) - \varrho_{Cu}(T_1)}{R_a(T_2) - R_a(T_1)} \frac{L}{B} \tag{2}$$

with $T_2 = 250$ and $T_1 = 100$ K. The application of this method [15] is possible because the annealed CuCo films consist of Cu with Co precipitations. The Co precipitations do not contribute to the temperature dependence of the resistivity between $T_1 = 100$ and $T_2 = 250$ K. This means we put the resistivity drop of annealed CuCo films equal to the resistivity drop of pure Cu between T_2 and T_1. This method has the advantage that the resistivity is independent of the geometry of the film since (1) and (2) yield

$$\varrho = \frac{\varrho_{Cu}(T_2) - \varrho_{Cu}(T_1)}{R_a(T_2) - R_a(T_1)} R\,. \tag{3}$$

The error of ϱ of this method amounts to 3%. Secondly we measure D by an optical method after the experiment. The second method leads to an error in ϱ of about 9%. Therefore, we apply (3) to determine the resistivity and use for $\varrho_{Cu}(T)$ the values of pure Cu. We check the so determined values of ϱ by using the optically measured thickness D and (1). The results agree within the stated errors.

Surface scattering is neglected [16] because the thickness $D = 500$ Å is larger than the electron mean free path of $l = 300$ Å for our films.

3. Results

Fig. 1 shows a typical example of the annealing behaviour of an alloy film produced by vapour deposition. The electrical resistivity decreases irreversibly during annealing up to T_a due to healing of lattice defects. Afterwards the resistivity is reversible below T_a. Only reversible data of the resistivity are described in the following (e.g. branch II is drawn in Fig. 2b on a larger scale).

3.1 Solid solutions (T_a = 80, 150 K)

Fig. 2 shows results for the annealing stage T_a = 80 K. The results are representative of CuCo solid solutions (Section 1). The curves show a resistance minimum shifting to higher temperature with rising Co concentration (numerical data in Table 1). The resistance minimum is due to superposition of the phonon part of the electrical resistivity dominating at higher temperature and a contribution of the Co atoms at low tem-

Table 1

Temperature (in K) of the resistance minimum of CuCo alloys as a function of Co concentration c and annealing temperature T_a

c (at%)	T_a (K)			
	80	150	290	390
0.0	—	—	—	—
0.21	—	8.5	15.0	15.5
0.51	9.5	9.5	20.5	20.5
0.98	14.4	14.5	25.5	25.5
1.5	19.5	21.5	30	30
2.0	26.5	27	35	33
2.6	33.5	34	37.5	34.5
3.1	40.0	43.5	42	36
4.1	52.5	56	46	39
6.1	66	67	49	40

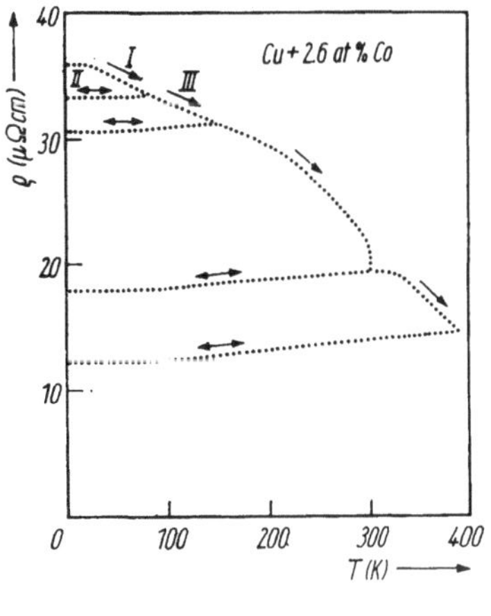

Fig. 1. Electrical resistivity of a CuCo film as a function of temperature. Substrate held at 5 K during condensation; different annealing stages (T_a) noted in Table 2

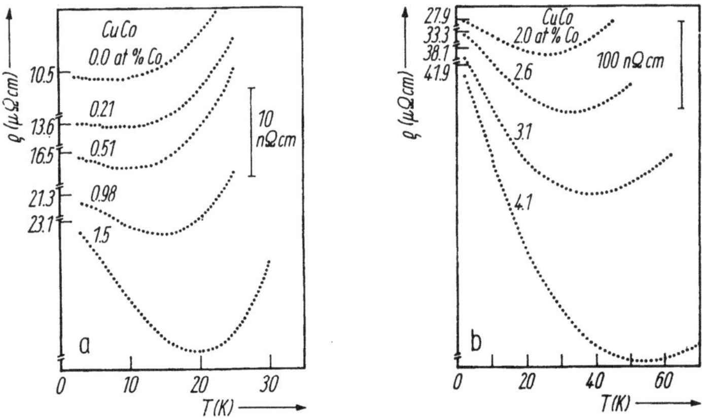

Fig. 2. Electrical resistivity of CuCo solid solutions as a function of temperature. Co concentration a) 0.0 to 1.5 and b) 2.0 to 4.1 at%; ordinate zero suppressed; $T_a = 80$ K

perature. The addition of Co causes an increase of the resistivity with falling temperature. We define as resistance anomaly the negative slope $(-\mathrm{d}\varrho/\mathrm{d}T)$ below the minimum. We take the resistance anomaly at 6 K because the phonon contribution is zero at this temperature within our accuracy (Fig. 2a: 0.0 at% Co). The resistance anomaly rises almost quadratically with the Co concentration (Fig. 4).

3.2 Annealing behaviour

To demonstrate the influence of annealing on the resistance anomaly a low and a high Co concentration are chosen in Fig. 3. The resistance anomaly grows with annealing in the case of 0.21 at% Co and decreases for Cu with 6.1 at% Co. In connection with the resistance anomaly the temperature of the resistance minimum rises with annealing for low Co concentration in Cu and falls with annealing in the cases of high Co concentration (Fig. 3, Table 1). Formally the reason for this annealing behaviour can be seen in Fig. 4. The resistance anomaly depends on Co concentration corresponding to

$$-\frac{\mathrm{d}\varrho}{\mathrm{d}T} = ac^{\gamma}\,.$$

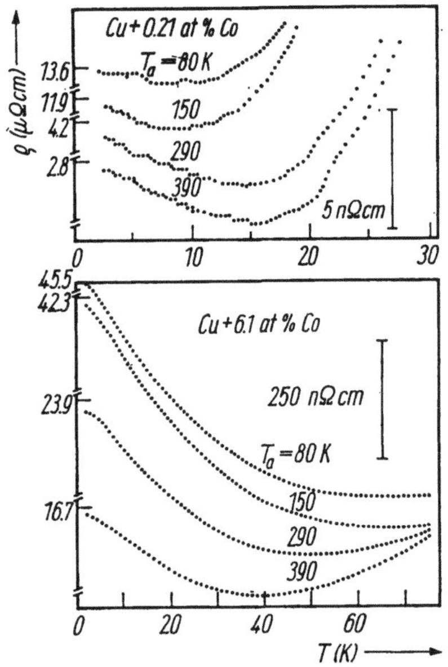

Fig. 3. Electrical resistivity of CuCo alloys as a function of temperature for different annealing stages T_a. Ordinate zero suppressed

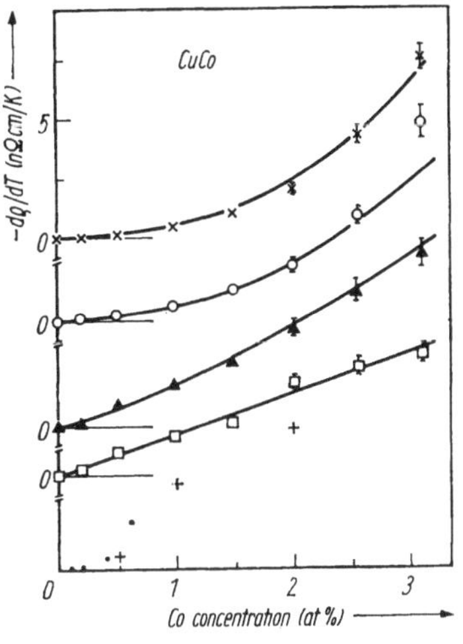

Fig. 4

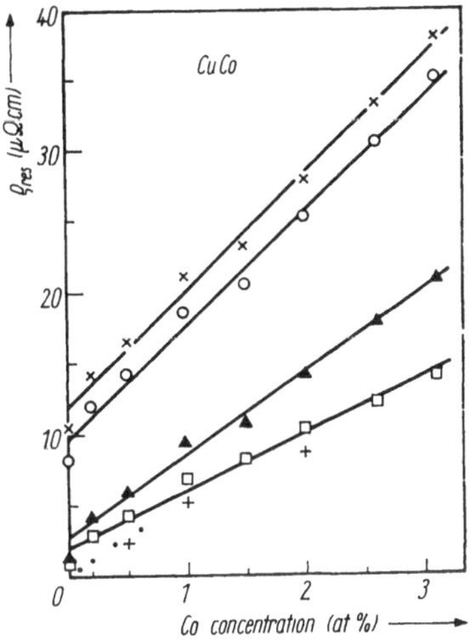

Fig. 5

Fig. 4. Resistance anomaly $-d\varrho/dT$ taken at 6 K as a function of Co concentration in Cu. Different annealing stages T_a of the films and data of bulk [6, 7] alloys; same scale of the ordinate; ordinate zero shifted. × $T_a = 80$, ○ 150, ▲ 290, □ 390 K; ●, + bulk

Fig. 5. Residual resistivity of CuCo alloys as a function of Co concentration. Data of films (different annealing stages T_a) and bulk [6, 7] samples. × $T_a = 80$, ○ 150, ▲ 290, □ 390 K; ●, + bulk

The coefficient a and the exponent γ for the drawn curves are given in Table 2. The resistance anomaly increases nearly quadratically with Co concentration for $T_a = 80$ and 150 K and depends linearly on Co concentration for $T_a = 390$ K. The curves $-d\varrho(c)/dT$ for $T_a = 80$ and 390 K intersect and this gives rise to the opposite annealing behaviour at low and high concentrations. In Fig. 4 values for the resistance anomaly of bulk samples are additionally introduced. The values of bulk samples and films are of comparable magnitude. (Further data for films with high Co concentration are shown in Table 3.)

The data for the residual resistivity are obtained by extrapolating the curves (e.g. Fig. 3) to $T = 0$ K. The last digit of the data (e.g. Table 3) is unaffected by taking the measured resistivity at 2 or 0 K. The residual resistivity as shown in Fig. 5 increases linearly with Co concentration for all annealing stages.

Table 2

CuCo alloys annealed at T_a for the time t

T_a (K)	t (min)	a (n Ωcm/K(at%)$^\gamma$)	γ	$\Delta\varrho_{res}/c$ (μΩcm/at%)
80	6	0.5	2.2	8.4
150	6	0.7	1.9	8.3
290	900	1.9	1.2	6.0
390	6	1.7	1.0	4.1

Resistance anomaly $-d\varrho/dT = ac^\gamma$. Normalized residual resistivity $\Delta\varrho_{res}/c$

Table 3

Resistance anomaly $-d\varrho/dT$ taken at 6 K and residual resistivity ϱ_{res} for 4.1 and 6.1 at% Co in Cu after annealing at T_a

c (at%)	T_a (K)	$-d\varrho/dT$ (n ΩcmK^{-1})	ϱ_{res} (μΩcm)
4.1	80	12.4	41.9
	150	13.9	38.7
	290	9.4	19.8
	390	6.1	14.4
6.1	80	15.0	45.5
	150	15.1	42.3
	290	11.1	23.9
	390	7.0	16.7

4. Discussion

4.1 Solid solution

The films annealed at 80 and 150 K are representative of solid solutions. The residual resistivity varies linearly with Co concentration. This means each Co atom contributes to the residual resistivity. The resistance anomaly increases with the square of the Co concentration. Accordingly the resistance anomaly is caused by the interaction between the Co atoms. This result agrees with the magnetic properties of bulk CuCo which indicate that single Co atoms have no magnetic moment in Cu [17]. The problem is similar to that of AuCo [12, 3]. In Au single Co atoms have no magnetic moment and the resistance anomaly depends on the square of the concentration showing that interactions between the Co atoms cause the resistance anomaly. By comparing AuCo [3] and CuCo solid solutions we find that the resistance anomaly is about four times larger for AuCo than for CuCo. The residual resistivity, however, has almost the same value in both cases.

4.2 Annealed films and bulk alloys

Annealing to 290 and 390 K allows diffusion towards positions given by thermodynamic equilibrium. According to the equilibrium diagram [8] Co precipitates in Cu. This means annealed films and bulk samples contain Co clusters larger than those clusters given by the statistics of random distribution.

With respect to the residual resistivity annealing from $T_a = 80$ to 150 K causes a parallel shift of the straight line. This means host lattice defects heal out but the Co atoms remain in solution. Annealing to 290 and 390 K diminishes the slope of the straight line. At these annealing stages ($T_a = 290$ and 390 K) clustering occurs. The Co atoms in clusters scatter the conduction electrons less than single Co atoms. The results of films annealed at 290 and 390 K are comparable with those of bulk alloys (Fig. 5).

The resistance anomaly depends on the interaction between the Co atoms. Therefore this property is very sensitive to the spatial distribution of the Co atoms in the host lattice. The resistance anomaly varies strongly with the thermal history of the sample. If all Co atoms interact within entities of the same average size, the number of entities grows linearly with concentration. In this case a linear dependence of the resistance anomaly upon Co concentration is expected. CuCo films annealed to 390 K (Fig. 4) show this behaviour.

References

[1] J. Kondo, Progr. theoret. Phys. (Kyoto) **32**, 37 (1964).
[2] E. Boucai, B. Lecoanet, J. Pilon, J. L. Tholence, and R. Tournier, Phys. Rev. B **3**, 3834 (1971).
[3] U. Henger and D. Korn, J. Phys. F, in the press.
[4] J. O. Linde, Proc. V. Internat. Conf. Low Temperature Physics and Chemistry, August 26 to 31, 1957, University of Wisconsin Press, Madison 1957 (p. 402).
[5] N. Bäcklund, J. Phys. Chem. Solids **7**, 94 (1958).
[6] I. S. Jacobs and R. W. Schmitt, Phys. Rev. **113**, 459 (1959).
[7] C. A. Domenicali and E. L. Christenson, J. appl. Phys. **32**, 11 (1961).
[8] M. Hansen, Constitution of Binary Alloys, McGraw-Hill Publ. Co., London 1958 (p. 470).
[9] S. Nasu, T. Shinjo, and Y. Nakamura, J. Phys. Soc. Japan **23**, 664 (1967).
[10] K. Krop, J. Mizia, and A. Korbel, Acta metall. **15**, 463 (1967).
[11] W. Buckel and R. Hilsch, Z. Phys. **146**, 27 (1956).
[12] D. Korn, D. Schilling, and G. Zibold, J. Phys. F **9**, 1111 (1979).
[13] S. Mader, H. Widmer, F. M. d'Heurle, and A. S. Nowick, Appl. Phys. Letters **3**, 201 (1963).
[14] E. Kneller, J. appl. Phys. **33**, 1355 (1962).
[15] W. Buckel and R. Hilsch, Z. Phys. **132**, 450 (1952).
[16] A. von Bassewitz and G. von Minnigerode, Z. Phys. **181**, 368 (1964).
[17] R. Tournier and A. Blandin, Phys. Rev. Letters **8**, 397 (1970).

(Received May 29, 1981)

phys. stat. sol. (a) 68, 143 (1981)

Subject classification: 10.2 and 14.3; 22.4.1

Institut für Angewandte Physik der Technischen Universität Clausthal, Clausthal-Zellerfeld[1]) *and SFB 126*

Anisotropic Conductivity of CdS after Plastic Deformation and Conduction along Dislocations

I. Macroscopic Measurements

By

G. Döding and R. Labusch

The anisotropy of the conductivity in deformed CdS is investigated using undoped and In–Cd-doped specimens. While the results for undoped specimens appear to be inconclusive because of very strong annealing effects during plastic deformation, the data of doped specimens indicate that the effect is due to the presence of highly compensated cylinders around screw dislocations. All observed properties can be explained by a percolation model for the current flow through the bulk material between these cylinders. A possible mechanism for their formation is suggested.

Die Anisotropie der Leitfähigkeit von verformtem CdS wird an undotierten und In–Cd-dotierten Proben untersucht. Während die Ergebnisse an undotierten Proben keine sicheren Schlüsse zulassen, weil die mit der Verformung verbundene Temperaturbehandlung die Leitfähigkeit sehr stark verändert, weisen die Ergebnisse der dotierten Proben auf die Existenz von stark kompensierten Zylindern an Schraubenversetzungen hin. Alle beobachteten Eigenschaften können im Rahmen eines Perkolationsmodells für den Stromfluß durch das Material zwischen diesen Zylindern erklärt werden, für deren Entstehung ein möglicher Mechanismus vorgeschlagen wird.

1. Introduction

An extraordinarily high anisotropy of the dc conductivity has been found in CdS after plastic deformation [1, 2]. This effect was attributed to a high density of straight parallel screw dislocations with Burgers vector $\langle 2\bar{1}\bar{1}0\rangle/3$ and it was suggested that their cores have a high conductivity and can be described as a pseudo one-dimensional metal.

In view of the translational symmetry of the dislocation core [3], combined with the observation of long straight screw segments [4], this explanation appears to be very plausible, but nevertheless further investigation of the effect seems to be justified for several reasons.

(i) The specimen size in the experiments was of the order of several millimeters while the length of straight screw segments seems to be at least an order of magnitude lower so that one cannot expect that the same screw dislocations are contacted at both ends of the specimen and branching into other types is more likely. Therefore, the question of bridging between separate dislocation segments arises.

(ii) The temperature during plastic deformation was rather high ($T \approx 600$ °C). This can lead to the formation of Frenkel defects which have a rather low enthalpy of formation [5]. In our own experiments, to be reported below, we found that even at appreciably lower temperatures the conductivity could be changed by orders of magnitude after a short annealing treatment alone. A clear distinction between dislocation effects and changes due to the heat treatment that is necessary for plastic deformation, is therefore desirable.

[1]) Arnold-Sommerfeld-Str. 6, D-3392 Clausthal-Zellerfeld, FRG.

In the present paper we report on macroscopic conductivity measurements from which a new alternative model of the anisotropy effect is derived. In another forthcoming paper this model is confirmed by local conductivity measurements at dislocations on a microscopic scale but, at the same time, strong evidence is found for the additional existence of one-dimensional conductivity in the dislocation core.

2. Experimental Methods

Most experiments to be reported here were done on single crystals grown by a slightly modified Piper-Polich sublimation method [6]. Some crystals were purchased from the Eagle Picher Co. The base material of our self-grown crystals was high-purity CdS powder provided by the Preussag Corporation. The results that appear to be most relevant were obtained from a series of crystals that had been doped by adding metallic In and metallic Cd to the base material during growth. After growth at 1000 °C the crystals were slowly cooled down to room temperature at a rate of less than 1 K/min. The grown-in dislocation density was revealed by an etchant that was developed by one of the authors [7] and turned out to be of the order of 10^2 cm^{-2}.

As a rule, our specimens were rectangular bars with the dimensions $3 \times 4 \times 5$ mm^3 and with surface orientations along the (0001), $(21\bar{3}0)$, and $(4\bar{5}10)$ direction, respectively. Electrical contacts were obtained by evaporating spots of In on two opposite surfaces that had been chemically etched and sputtered. On top of the In we deposited a layer of aluminum for better mechanical stability and attached the contact wires with silver epoxy. The current–voltage characteristics of these contacts were checked and turned out to be perfectly ohmic.

Conductivity measurements were carried out in a continuous flow helium cryostat with a commercial temperature control that allowed us to keep the temperature constant within 0.01 K. The absolute value of T was measured with a repeatedly gauged carbon resistor. The accuracy was better than 0.2 K over the entire range.

For the electrical measurements we used a constant potential source and a Keithley 602 electrometer to measure the current. After correction for the internal resistance of the Keithley this allows reliable measurements with better than 1% accuracy down to a conductivity of about 10^{-12} Ω^{-1} cm^{-1}. Below this range the settling time increased prohibitively and leak currents parallel to the specimen became a problem.

Dislocations were introduced by static compression along the $\langle 4\bar{5}10 \rangle$ axis which is favourable for single glide with a $\langle 2\bar{1}\bar{1}0 \rangle$ Burgers vector in the main slip system. The deformation apparatus was entirely made of quartz. The compression stamps were "greased" with graphite powder of spectral purity to reduce friction and to facilitate single glide.

At low temperatures CdS is very brittle. Therefore, it is desirable to deform at high temperatures. On the other hand, we wanted to avoid drastic changes of the electrical properties from the heat treatment alone. The temperature of 400 °C used in most of our experiments is a compromise which seems to work quite well in this conflicting situation, at least for the doped specimens. The standard deformation time was 10 min; warm up from and cooling down to room temperature took about 2 min each. Etch pit densities after plastic deformation on a surface perpendicular to the screw direction were between 10^6 and $> 10^7$ cm^{-2}, depending on the degree of deformation. As no quantitative correlation between etch pit density and true dislocation density has been established so far, we prefer to characterise the specimens by their relative change in length ε.

All treatments at elevated temperatures after growth were done in an argon atmosphere.

3. Experimental Results for Undoped Specimens

Fig. 1 shows the conductivity σ versus $1/T$ of an undoped specimen (Eagle Picher material) in the as-grown state and after a heat treatment of 10 min at 400 °C. Within the experimental errors the conductivity in three perpendicular directions was found to be isotropic. This figure demonstrates the drastic effect of a comparatively gentle heat treatment on a highly resistive material which turns from nearly intrinsic behaviour to a rather flat $\sigma(1/T)$, due to the introduction of unidentified "thermal defects".

Also shown in Fig. 1 are the conductivities of an undoped specimen after plastic deformation parallel and perpendicular to the line direction of screw dislocations in the main glide system. Obviously we have reproduced the anisotropy effect. The ratio $\sigma_{||}/\sigma_{\perp}$ between the conductivities becomes certainly higher than 10^7 at low temperatures.

Quantitatively, we notice that the temperature dependence of $\sigma_{\perp}$ is similar to σ in the undeformed specimen while $\sigma_{||}(1/T)$ is rather flat. Comparison with the as-grown specimen would therefore favour the model of highly conductive dislocation threads in a highly resistive matrix. However, if we compare with a specimen that underwent the same heat treatment as the deformed one, we would rather conclude that plastic deformation has somehow inhibited or compensated the effect of the simultaneous heat treatment for $\sigma_{\perp}$ while $\sigma_{||}$ looks similar to the conductivity of the undeformed sample with a reduced effective cross-section. This would suggest an alternative model in which $\sigma_{||}$ is attributed to the volume between dislocations. In summary it seems that, so far, our results are inconclusive and that specimens with better thermal stability are needed for further experiments.

For later use we have to mention here another result for specimens that had been grown in our laboratory under the same conditions as the doped ones but without intentional doping. Here we found a perfectly linear dependence of $\ln \sigma$ on $1/T$ (not shown here for brevity) between room temperature and 60 K, corresponding to an activation energy of 123 meV $\pm$ 1%. This value is in agreement with earlier estimates of the energy for double ionisation of Cd interstitials. Furthermore, the extrapolated value of σ at $1/T \to 0$ is extremely low ($10^{-3}\ \Omega^{-1}\text{cm}^{-1}$) which indicates strong compensation. The most plausible explanation of these features is that Cd interstitials and vacancies are the dominating defects and that their densities are almost equal because

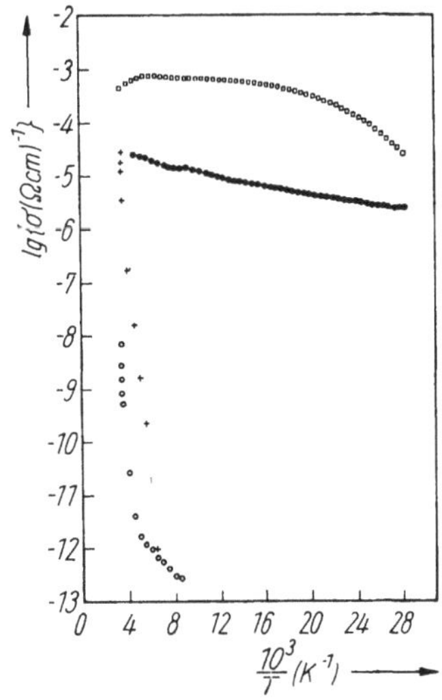

Fig. 1. Conductivities of undoped specimens before and after annealing and plastic deformation; ○ as-received; □ annealed for 10 min at 400 °C; ● after plastic deformation to $\varepsilon = 14\%$ at 400 °C during 10 min, current direction parallel to screw dislocations; + same as ● but current direction perpendicular to screws

our crystals were grown from stoichiometric material in a closed ampoule. We therefore assume that the level of doubly ionised Cd is at 123 meV and that its occurrence as an activation energy indicates strong compensation.

4. Doped Specimens Undeformed

In Fig. 2 we have plotted $\sigma(1/T)$ of an as-grown doped specimen and of the same specimen after annealing for 10 min at 400 °C. Further annealing for another 10 min at the same temperature changed the conductivity only by a few per cent which is within the experimental uncertainty because the contacts were removed and reapplied between the measurements. Fig. 3 shows the data of another specimen of the same type in the as-grown state, after annealing for 10 min at 450 °C and after further annealing for 1 h at 500 °C. Again the conductivity change after further annealing at 450 °C was negligible.

From a pragmatic point of view we could be satisfied with the statement that, although the material is not completely stable at 400 and 450 °C, it settles to a relatively stable state in a short time and that the changes are moderate enough to recognise the characteristic shape of the $\ln \sigma(1/T)$ curve after the treatment. We shall in fact draw on these features in our analysis of the $\ln \sigma(1/T)$ curves after plastic deformation.

On the other hand, a more detailed understanding of the material properties before plastic deformation may be useful. CdS that was annealed at 855 °C in the presence of metallic In and excess Cd has been investigated by Partain et al. [8]. They found that, although a concentration $[\text{In}] \approx 3.5 \times 10^{18}\ \text{cm}^{-3}$ of indium is incorporated in the material, the In atoms stay electrically inactive and only the number of interstitial Cd donors is increased to $\approx 10^{17}\ \text{cm}^{-3}$, compared with about 10^{15} in a control sample without In. The neutrality of the In was inferred only indirectly from the mobility $\mu(T)$ which corresponds to a high density of neutral impurities. From Hall measurements, not shown here for brevity, we obtain the same temperature dependence of the mobility, and from a comparison with an investigation by Boyn et al. [5] we are able to present an additional proof that the active donors in our material are in fact Cd interstitials. For specimens that had been annealed under controlled vapour pressures of Cd at 700 °C, these authors found the same dependence of $\ln \sigma(1/T)$ that is characteristic of our doped specimens: A high temperature plateau is followed by a thermally activated regime at lower temperatures. The activation energy in this regime is $Q = 50$ meV. Q was identified with the upper level of interstitial Cd which has two donor levels for single and double ionisation at 50 and $\geqq 120$ meV, respectively (according to our own measurements 123 meV). The upper level was assumed to be partly compensated by Cd vacancies.

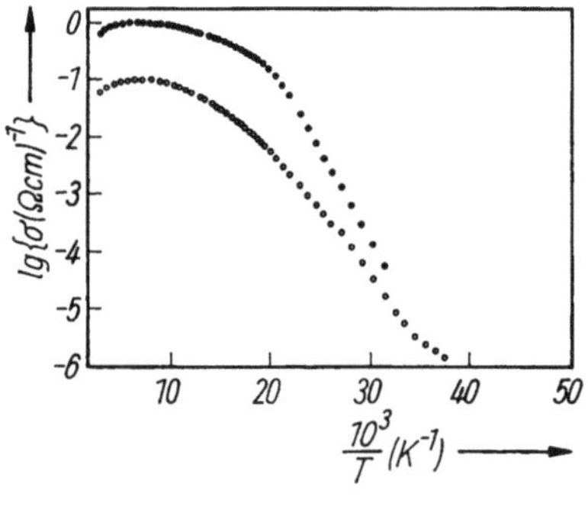

Fig. 2. Conductivities before (●) and after (○) annealing for 10 min at 400 °C of a specimen doped with In and Cd during growth

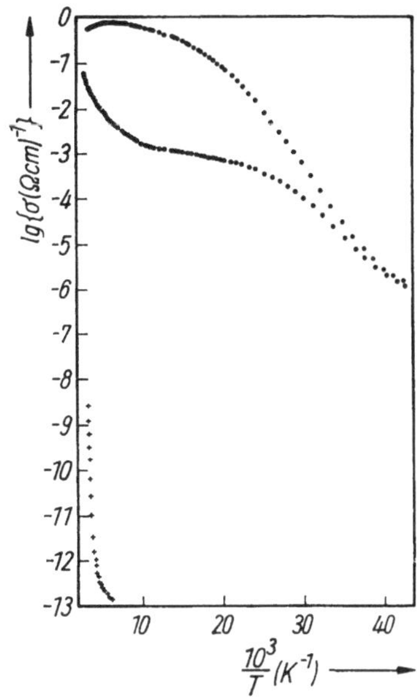

Fig. 3. Conductivities of a doped specimen; ○ as-grown, ● annealed for 10 min at 450 °C, + after further annealing for 1 h at 500 °C

Adopting this interpretation, we can determine the density $[Cd_{\bigcirc}]$ of interstitial Cd and estimate, from the degree of compensation, the density $[Cd_{\square}]$ of Cd vacancies. The result is $[Cd_{\bigcirc}] = 7 \times 10^{15}$ and $[Cd_{\square}] \approx 10^{13}\ cm^{-3}$ for the as-grown specimen of Fig. 2.

At about 25 K the $\ln \sigma(1/T)$ curves seem to bend around to a lower slope. At present we have no unambiguous explanation for this behaviour which has not been observed in previous investigations where the temperature range ended well above 25 K.

Obviously the interstitial density, $[Cd_{\bigcirc}]$ is reduced by annealing at 400 and 450 °C but this cannot be explained by out-diffusion from the specimen surface. From the known diffusion constant of interstitial Cd [5, 9] we obtain a diffusion length $\lambda = \sqrt{Dt}$ of approximately 5 μm at 400 °C which is less than the thickness of the surface layer that was removed by etching after the annealing treatment before new electrical contacts were applied. Therefore another mechanism must be found. Because of the high speed of the process an association with In impurities appears to be likely, but this question requires further investigation.

After annealing at 500 °C the material is very highly compensated with a position of the Fermi level well below the Cd levels. If we adopt an estimated value of 2.5 eV for the enthalpy of pair formation [5] we find that, with the given initial densities of Cd interstitials and vacancies, we must expect the creation of new vacancies at this temperature. Therefore the strong compensation is not due to a reduction of the Cd interstitial density alone but also to an increased density of Cd vacancies which have acceptor states near the middle of the gap [5]. The creation of other defects at 500 °C can of course not be excluded, but Cd vacancies appear most likely because of their low enthalpy of formation.

In summary, we conclude that deformation of doped specimens at 400 and even at 450 °C (but not higher!) should produce acceptable changes of the bulk conductivity between dislocations and that a distinction between dislocation effects and annealing effects during deformation should be possible.

5. Doped Specimens after Plastic Deformation

In Fig. 4 we have plotted the conductivities of four specimens that were deformed to $\varepsilon = 7.7$, 10.2, 15.5, and 26.7% at 400 °C. Each plot shows the undeformed specimen together with the curves after deformation in three different current directions. Con-

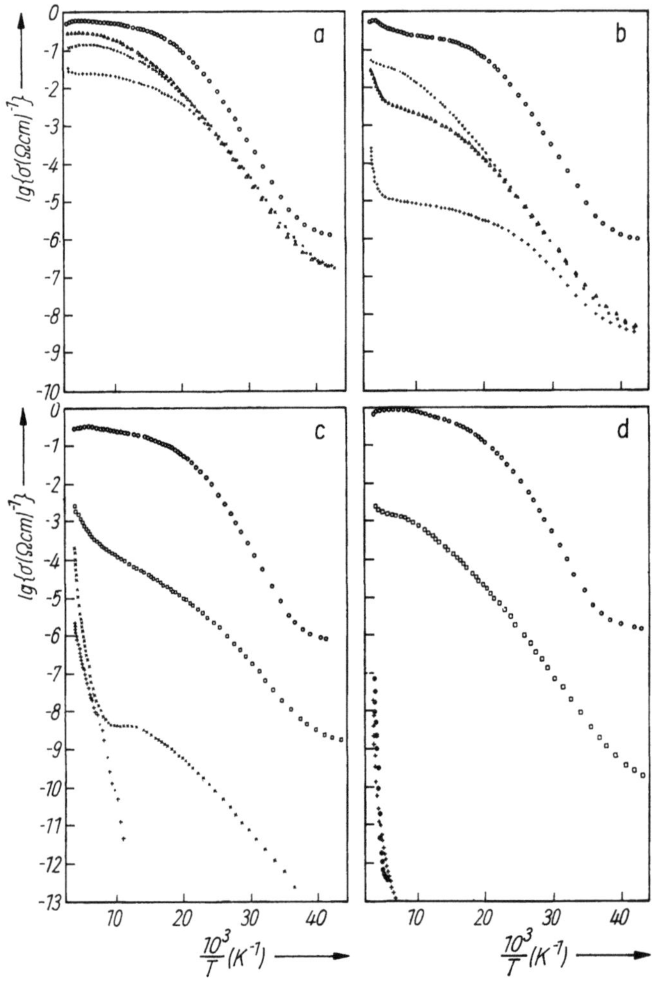

Fig. 4. Conductivities of doped specimens before and after various degrees of plastic deformation. a) $\varepsilon = 7.7$, b) 10.2, c) 15.5, d) 26.7%; ○ undeformed (isotropic), □ parallel to [21̄1̄0] (screw direction), + parallel to [0001] (perpendicular to screws), × parallel to [45̄10], △ parallel to [213̄0], ● parallel to [011̄0]

duction along the [21̄1̄0] direction yields $\sigma_{\parallel}$ (parallel to the screw direction) and conduction along [0001] yields $\sigma_{\perp}$ (perpendicular to screws). The latter was measured in all cases. In the weakly deformed specimens ($\varepsilon = 7.7$ and 10.2%) the other two directions are [213̄0] and [45̄10] which are both under approximately 45° to the screw direction. This was done for convencience although it is not ideal. On the other hand, the contact areas were large and their distance small enough to have a continuous path along the screw direction connecting both contacts. The cross-section of this path is a fraction of the specimen cross-section. Therefore, if $\sigma_{\parallel} \gg \sigma_{\perp}$, the conductivity along [213̄0] and [45̄10] is approximately proportional to $\sigma_{\parallel}$ with a proportionality factor smaller than unity. The other two specimens were cut along the [21̄1̄0] plane to yield the true value of $\sigma_{\parallel}$. The third direction was [45̄10] in one case. This gave a $\ln \sigma\,(1/T)$ curve between $\sigma_{\parallel}$ and $\sigma_{\perp}$ but similar in shape to $\sigma_{\parallel}$, as expected. In the other case it was the [01$\bar{1}$0] direction which is perpendicular to [2$\bar{1}\bar{1}$0] and, consequently, had almost the same $\ln \sigma(1/T)$ as [0001].

The following features of the experimental results are most relevant for further discussions and model considerations.

After plastic deformation $\sigma_{\parallel}$ is of the same order of magnitude or smaller than the conductivity σ_0 of the heat-treated but undeformed specimen. The shape of $\ln \sigma_{\parallel}(1/T)$ is very similar to that of $\ln \sigma_0(1/T)$ and only shifted to lower values with increasing ε. This suggests that the conduction mechanism is not qualitatively different either.

At low degrees of deformation $\sigma_{\perp}$ seems to exhibit the same properties, only that the ratio $\sigma_{\perp}/\sigma_0$ is smaller than $\sigma_{\parallel}/\sigma_0$. At higher degrees of deformation $\ln \sigma(1/T)$ is changed drastically to a qualitatively different dependence. There is no indication of a high temperature plateau and below room temperature the slope of $\ln \sigma_{\perp}(1/T)$ (see Fig. 4c) corresponds now to an activation energy of approximately (125 ± 5) meV which agrees well with the level of doubly ionised interstitial Cd at 123 meV, found in undoped specimens. This is indicative of a high degree of compensation which could be caused by a reduction of the Cd interstitial density and/or by an increase of the density of deep acceptors.

6. New Model of the Anisotropy Effect in Doped CdS

If we accept that $\sigma_{\parallel} \leqq \sigma_0$ and that the similarity in the shapes of $\sigma_{\parallel}$ and σ_0 is not accidental, $\sigma_{\parallel}$ must be associated with the bulk material between dislocations, and we cannot explain our results in terms of one-dimensional conduction along screw dislocations. Therefore we suggest the following alternative model for doped specimens: After plastic deformation in single glide, the specimens contain parallel straight screw dislocations, surrounded by cylinders in which the material is highly compensated and the local conductivity σ_c is thermally activated with an activation energy of ≈ 125 meV. Outside the cylinders the bulk conductivity σ_b is essentially the same as in the undeformed heat-treated material (Fig. 2), possibly with minor modifications due to point defects produced by moving dislocations. σ_c is much smaller than σ_b below room temperature.

If the cylinders are perfectly straight and no other obstacles to the electric current present, we expect $\sigma_{\parallel} = (1 - \alpha)\,\sigma_b + \alpha\sigma_c$ from this model, where α is the volume fraction that is occupied by the compensated cylinders. Below room temperature we have $\sigma_c \ll \sigma_b$ so that $\sigma_{\parallel} \approx (1 - \alpha)\,\sigma_b$. In reality the path between the cylinders will be frequently obstructed by screw dislocations of secondary slip systems, so that $\sigma_{\parallel}/\sigma_b$ can be much smaller than $1 - \alpha$. This effect is more pronounced in heavily deformed specimens because secondary slip systems are more and more activated with increasing ε.

Perpendicular to the cylinders we have a percolation problem. As long as the volume fraction of highly resistive material is small enough, there are percolation paths for the current through low-resistivity bulk material and the conductivity is again an appropriate average of σ_b and σ_c which is approximately proportional to σ_b because $\sigma_b \gg \sigma_c$. The proportionality constant is smaller than for $\sigma_{\parallel}$ because each percolation path is curved and has narrow spots. If, on the other hand, the volume fraction of the cylinders gets higher with increasing ε, all percolation paths are closed so that the current is forced to flow through compensated material in some places.

This explains the transition from $\sigma_{\perp} \sim \sigma_{\parallel}$ at low dislocation densities to a dependence $\sigma(T)$, at high dislocation densities, that is characteristic of the interior of the cylinders. A simulation experiment [10] shows that the probability to find a continuous percolation path through a two-dimensional random array of circles in a plane drops sharply to zero at $\pi R^2 \varrho \approx 0.4$, where R is the radius of the circles and ϱ their density. In our experiments the corresponding transition takes place between $\varepsilon = 10.2\%$ and 15.5% at an etch pit density on a surface perpendicular to the screw direction of about 10^6 cm^{-2}. Therefore $R \approx 4$ µm. The true dislocation density ϱ at the transition may differ appreciably from the value given here, but not by more than two orders of magnitude, so that our value of R which is proportional to $\varrho^{-1/2}$ should be a fair estimate.

7. Conclusions

Our percolation model of the anisotropy effect accounts for all experimental results in a semiquantitative way. For a full quantitative evaluation it would be necessary to know the dislocation arrangement as well as the distribution and type of all electrically active point defects in the compensated cylinders and in the bulk. These informations are hard to get and cannot be obtained at all from macroscopic averages. Measurements of the local conductivities near dislocations and elsewhere on a microscopic scale are necessary to prove or disprove the model. These will be presented in Part II of this investigation.

The phenomenon described here is not a singular one. Tubes, several microns in diameter, in which non-radiative recombination was enhanced and photoluminescence suppressed due to the presence of unknown defects, have also been observed at dislocations in GaAs [11]. The question arises by what mechanism the cylinders are in both materials and, possibly, in others.

In this respect it could be significant that the cylinder radius is considerably bigger than the range of any conceivable interaction between dislocations and point defects but is roughly equal to the diffusion length of interstitial Cd in CdS. This suggests that the cylinders are the result of an incomplete diffusion process of Cd towards screw dislocations. This hypothesis agrees with the observation that screw dislocations are very immobile compared with other types [4] so that the diffusion process is not disturbed by dislocation motion during plastic deformation. Furthermore, an interaction between point defects and dislocations is very likely and has been invoked in the interpretation of other experiments before [4, 12, 13]. We notice that the mechanism suggested here works only if the interaction is attractive. The effect of a repulsive interaction would be a diffusion of Cd away from dislocations and would result in a depleted cylinder whose radius is given by the requirement that the interaction potential must be greater than kT at the deformation temperature. An upper limit of the radius is therefore the Debye screening length [11] which, in our material, is only of the order of a few hundred Angstroms at 400 °C. This is too small to account for the observed effects. On the other hand, an attractive interaction would result in a Cd-rich inner cylinder of the same size, surrounded by a much larger depletion zone whose radius is given by the diffusion length. The absence of compensated cylinders at edge-type dislocations could then be explained if their interaction with interstitials is either very weak or repulsive. In summary, the mechanism of cylinder formation suggested here is plausible, but somewhat speculative and requires further experimental proof.

From our results we obtain no indication of one-dimensional conduction along screw dislocations but no disagreement with the existence of this mechanism either. Its contribution to $\sigma_{\parallel}$ could be comparatively small and manifest itself only in the details of the shape of $\sigma(1/T)$. It is also not possible to decide whether previous experiments could be interpreted in terms of our model because we are not sure whether or not and to what degree the bulk conductivity was changed after plastic deformation at high temperatures. In Part II of this investigation we shall present very convincing evidence, not only for the validity of our model in doped material but also for the existence of one-dimensional conduction along screw dislocations. Therefore both mechanisms have to be considered as possible alternatives and it has to be checked individually which one is predominant in different materials.

References

[1] C. Elbaum, Phys. Rev. Letters **32**, 376 (1974).
[2] P. Merchant and C. Elbaum, Phys. Rev. B **19**, 2992 (1979).
[3] R. R. Holmes and C. Elbaum, Phys. Rev. **173**, 803 (1968).
[4] Yu. A. Osipyan, V. F. Petrenko, G. K. Strukova, and I. I. Khodos, phys. stat. sol. (a) **57**, 477 (1980).
[5] R. Boyn, O. Goede, and S. Kuschnerus, phys. stat. sol. **12**, 57 (1965).
[6] W. W. Piper and S. J. Polich, J. appl. Phys. **32**, 1278 (1961).
[7] G. Döding, J. appl. Phys., submitted for publication.
[8] L. D. Partain, G. J. Sullivan, and C. E. Birchenall, J. appl. Phys. **50**, 551 (1979).
[9] K. W. Boer, R. Boyn, and O. Goede, phys. stat. sol. **3**, 1684 (1963).
[10] B. J. Last and D. J. Thouless, Phys. Rev. Letters **27**, 1719 (1971).
[11] K. Böhm and B. Fischer, J. appl. Phys. **50**, 5453 (1979).
[12] L. Carlsson and C. Svensson, J. appl. Phys. **41**, 1652 (1970).
[13] N. V. Klassen, Yu. A. Osipyan, and M. Sh. Shikhsaidov, Soviet Phys. – Solid State **18**, 922 (1976).

(Received July 14, 1981)

phys. stat. sol. (a) 68, 153 (1981)

Subject classification: 20.1; 22.2.4

Institut für Hochfrequenztechnik der Technischen Universität Braunschweig[1])

Near-Band-Gap Absorption of InGaAsP at 1.3 μm Wavelength

By

W. KOWALSKY, A. SCHLACHETZKI, and F. FIEDLER

InGaAsP, adapted in its composition for a band gap equivalent to a wavelength of 1.3 μm and lattice matched to InP, is grown by liquid-phase epitaxy. Transmission and reflectance measurements in the vicinity of the band gap are made, from which the refractive index of both InP and the quaternary compound are derived. Also the optical absorption coefficient α of the quaternary compound is obtained. Above the band gap, α can be fitted to the model of direct band-to-band transitions, allowing the precise determination of the band-gap energy. Below the band gap, α follows an exponential dependence on photon energy. Using material parameters for the quaternary compound, the role of phonon-assisted transitions is discussed in this spectral range.

InGaAsP, dessen Bandabstand über die Zusammensetzung als äquivalent zu 1,3 μm gewählt ist und das in seinem Gitter InP angepaßt ist, wird epitaktisch aus der flüssigen Phase gezogen. Aus Transmissions- und Reflexionsmessungen wird der Brechungsindex des InP und der quaternären Verbindung in der Nähe der Bandkante bestimmt. Ebenso wird der optische Absorptionskoeffizient α der quaternären Verbindung gewonnen. Oberhalb der Bandkante kann α durch direkte Band-Band-Übergänge beschrieben werden, wodurch eine genaue Bestimmung der Breite des verbotenen Bandes möglich ist. Unterhalb der Bandkante ist α exponentiell von der Photonenenergie abhängig. Für diesen Spektralbereich werden Übergänge unter Mitwirkung von Phononen diskutiert, wobei die Materialparameter der quaternären Legierung verwendet werden.

1. Introduction

The quaternary semiconductor InGaAsP has become of increasing prominence as a material for light sources and detectors in glass-fiber communication systems. InGaAsP affords the possibility of freely choosing the band gap by varying the composition in a range suitable for this application. Consequently the system InGaAsP was the subject of numerous investigations in recent years. Yet, there are a number of quantities on which relatively little is known. Among them are the optical properties, namely the refractive index n and the absorption coefficient α of light. Both quantities are important for the design of light sources as well as of photodetectors.

So far only few measurements of n were reported [1], whereas the measured α values [2, 3] did not agree well with each other and also not with the calculations [4]. Depending on the physical model used, the calculated α values themselves show some discrepancies [4, 5]. These calculations are carried out by using band parameter values obtained by suitable interpolation schemes starting from the binary corner compositions. A similar procedure is followed in the calculation of n [6].

In this paper we will concentrate on InGaAsP of a composition adapted for use in the wavelength range near 1.3 μm, where glass fibers exhibit a minimum of attenuation and zero dispersion. The material was grown by liquid-phase epitaxy (LPE),

[1]) Postfach: 3329, D-3300 Braunschweig, FRG.

lattice-matched to InP substrates. We will report on measurements of the refractive index n and of the absorption coefficient α in the spectral range around the fundamental absorption edge.

2. Experimental

The crystals were grown on InP substrates commercially obtained from Metals Research, England. The substrates of (100) orientation were either Sn-doped with a carrier concentration of 2×10^{18} cm^{-3} or Fe-doped with a resistivity of about 10^6 Ωcm. The etch-pit density was around 5×10^4 cm^{-2}. The investigated quaternary crystals were 1.3, 2.1, and 3.3 μm in thickness. They were grown by standard sliding-boat LPE in Pd-purified hydrogen. Shortly before growth by gradual cooling around 645 °C, the substrates were backmelted for 30 s in an In:P melt saturated for 600 °C. The nominally undoped films were n-type, measured after van der Pauw, with a carrier concentration of 3×10^{16} and 2.5×10^{16} cm^{-3} at 300 and 77 K, respectively. The corresponding mobilities were 4500 and 6900 cm^2/Vs.

Before the optical investigations, the InGaAsP/InP crystals were polished on Pellon discs (PAN W) from the substrate side to a thickness between 150 and 200 μm using a 0.4% Br-isopropanol solution in nitrogen atmosphere. Thus the effect of stray light was strongly reduced. Fig. 1a shows surface profiles of InP substrates as obtained from the manufacturer (upper trace) and after the additional polish (lower trace). The curves were measured with a mechanical profiler (Perthometer) with a stylus of 3 μm radius of curvature. The surface roughness of the substrate is well below 0.1 μm. Fig. 1b displays a surface profile of a quaternary film which appears mirror-like to the naked eye.

Optical transmission and reflectance were measured at normal or near-normal incidence with an angle of incidence less than 7°. The optical set-up was extremely simple in order to minimize the adverse influence of stray light. A current-stabilized tungsten lamp with intensity fluctuations of less than 0.5% was employed, followed by a monochromator (Bausch & Lomb) with a grating of 675 lines/mm and a wavelength range from 0.7 to 1.6 μm. The wavelength control was calibrated with a HeNe laser. Our measurements were performed at a dispersion of 10 nm. To avoid lines of higher order, a long-pass filter transmitting at wavelengths longer than 0.83 μm was used. Two bi-convex lenses focused the radiation from the monochromator slit on the sample surface and from the sample on a germanium photodiode of a sensitive area

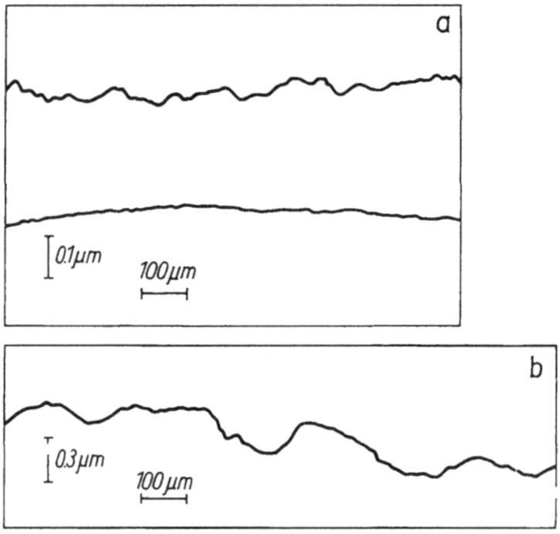

Fig. 1. a) Surface of an InP substrate as obtained from the manufacturer (upper trace) and after an additional polishing step with 0.4% Br-isopropanol (lower trace). b) Surface profile of an InGaAsP epitaxial layer as-grown

of 1 mm in diameter. The signal of the diode, which was operated in the photo-current mode, was amplified by an operational amplifier CA 3140 with a chopped high-impedance FET input stage. Thus small drift and low noise were achieved. The detection system was completed with a lock-in amplifier (PAR 186).

The linearity of the experimental set-up was very good, as controlled with an optical powermeter (Tektronix digital photometer J-16) at 1.0 μm. The scattered light, as indicated by the sum of the reflected and transmitted intensities in the spectral range of no absorption, was in most cases well below 1%. The conditions were worse for surface regions of samples of mediocre quality. The product of α and the thickness d of the samples was in the optimum range from 0.1 to 6.

3. Measurements and Their Evaluation

In evaluating the measured transmission and reflectance spectra, multireflections in the parallel-plate system film-substrate were taken into account. The reflected intensities were summed up neglecting interference effects. Because of thickness inhomogeneities and because of the limited dispersion of the monochromator, interference could neither be expected nor was observed. In order to make allowance for the loss due to stray light, the experimental error caused by this effect was reduced by attributing the loss in equal parts to the reflection and transmission data.

By making use of recursion formulas the measurements were evaluated to obtain the refractive index n and the absorption coefficient α. In Fig. 2, n in dependence on the wavelength λ (or the photon energy $\hbar\omega$ on the upper abscissa) is depicted. The band-gap energy is marked as $E_{g,InP}$ and $E_{g,Q}$ for InP and the quaternary alloy

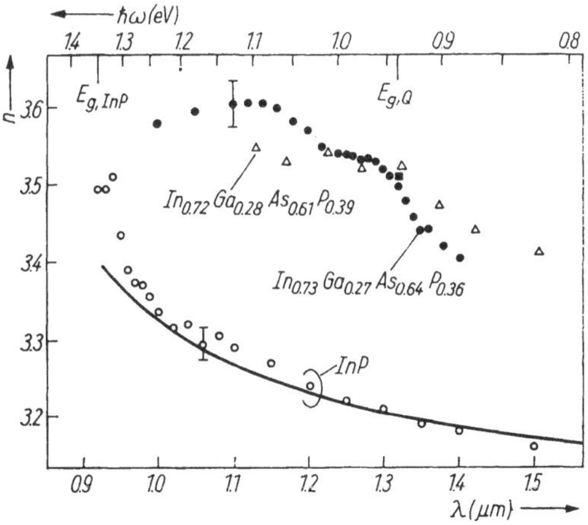

Fig. 2

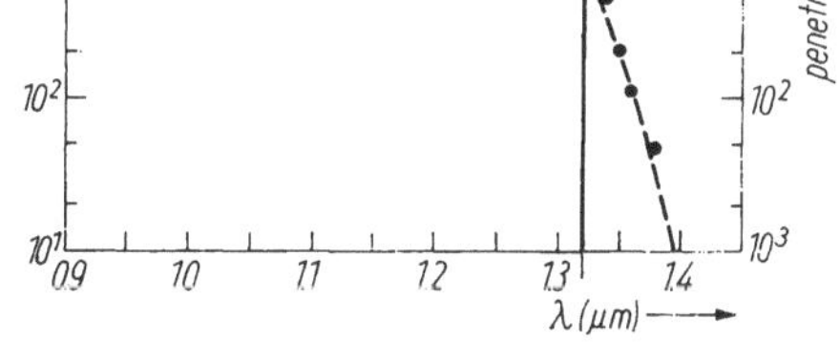

Fig. 3

Fig. 2. The refractive index n in dependence on the wavelength λ (or on the photon energy $\hbar\omega$, upper abscissa). ○ present measurements on InP; ● present measurements on the quaternary alloy InGaAsP; — measurements on prism-shaped InP by Pettit and Turner [7]; △ measurements on a quaternary alloy of similar composition as in the present case by Chandra et al. [1]; ■ interpolation for the band-gap value by Olson et al. [6]

Fig. 3. Absorption coefficient α in dependence on the wavelength λ for the quaternary alloy. ● present measurements ($In_{0.73}Ga_{0.27}As_{0.64}P_{0.36}$); ——— fit to the theory of direct band-to-band transitions; ····· forbidden transitions; ---- calculation by Dutta [4] for $In_{0.72}Ga_{0.28}As_{0.60}P_{0.40}$

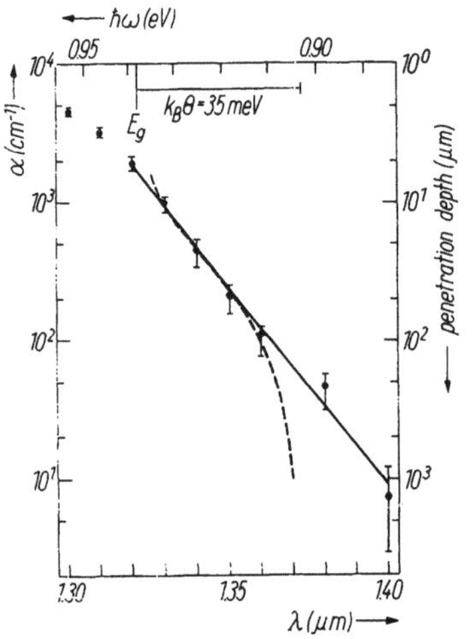

Fig. 4. The same measurements as in Fig. 3, but on an expanded abscissa. The straight line is a fit to the experimental points beyond the band-gap wavelength $\lambda_{g,Q}$. $k_B\theta$ marks the spectral range where phonon-assisted transitions might be expected. Dashed curve: after Dumke [13]

InGaAsP, respectively. The circles represent the experimental results for InP and the solid points apply to InGaAsP. The vertical bars indicate the experimental error which becomes somewhat larger near the band gap in the case of InP. Thus most of the measurements are accurate to within less than 2% for high-quality surface InP samples. Due to the limited perfection of the epitaxial InGaAsP layers the experimental error is about 3% in this case.

We noticed no difference in the refractive index for semi-insulating and highly n-type InP (up to 2×10^{18} cm^{-3}).

α versus λ for the quaternary alloy is given in Fig. 3. Again the solid points mark the experimental results. The experimental error below the band gap is almost negligible on the scale of Fig. 3. In the long-wavelength region the scatter of the experimental results is marked by vertical bars in Fig. 4.

4. Discussion

As to the refractive index, the most precise method of measurement is perhaps with samples in the form of prisms. The results of such measurements on InP [7] are shown as a solid curve in Fig. 2. The agreement with our measurements is very good. With respect to InGaAsP, we find reasonable accordance with recent experiments (triangles) [1] performed on an alloy of composition, which is close to the present one. Also shown in Fig. 2 as a square is the n value at the band-gap energy of the quaternary alloy, as determined by an interpolation scheme from the data of the corner compositions [6].

The absorption values in the short-wavelength range (Fig. 3), which are in agreement with recent ellipsometry measurements [8], are almost an order of magnitude larger than the results of experiments on material of comparable band-gap energy [2]. Applying the model of a direct semiconductor with allowed transitions from the valence band to the conduction band, the absorption should follow the relation [9]

$$\alpha\hbar\omega \sim (2\mu)^{3/2} (\hbar\omega - E_g)^{1/2}/n \,, \tag{1}$$

where μ is the reduced effective mass and E_g is the band-gap energy. The experimental points closely follow this square-root dependence leading to a band gap of the quaternary alloy of $E_{g,Q} = 0.938$ eV, corresponding to $\lambda_{g,Q} = 1.32$ μm. With this data we obtain a composition described by $x=0.30$ and $y=0.64$ for the system $In_{1-x}Ga_xAs_yP_{1-y}$ lattice-matched to InP [10]. This is reasonably close to the electron microprobe measurements which lead to $x = 0.27$ and $y = 0.64$.

The value of $E_{g,Q}$ together with the refractive-index step from the substrate to the quaternary film at the band gap, as obtained from Fig. 2, gives an indication of the lattice-parameter mismatch. From the calculations by Olsen et al. [6] we obtain about -0.3%.

The relation (1) is shown in Fig. 3 as a solid curve where the proportionality constant, containing the matrix element for transition, is used to adjust the absolute value of the curve. The good agreement of the experimental results with this simple model, which is also found with GaAs [11] and InSb [12], is surprising in view of the exciton absorption which can be expected to contribute considerably near the band edge.

The forbidden transitions denoted by the dotted curve in Fig. 3 clearly do not apply.

The dashed curve in Fig. 3 reproduces the results by Dutta [4] calculated for an electron concentration of 2×10^{17} cm^{-3} for a composition close to ours. This relatively high concentration causes band tailing towards longer wavelengths and a reduced absorption at shorter wavelengths. In this spectral range the effect of excitons can be expected, particularly for carrier concentrations as in the present case, which are below the shielding concentration of about 8×10^{16} cm^{-3} (see [9] p. 346). The shift of the band edge due to excitons is about 5 meV and thus negligible with the resolution of our apparatus.

The experimental results beyond the band-gap wavelength $\lambda_{g,Q}$ are redrawn in Fig. 4 with an expanded abscissa. The solid line is a fit to the experiments demonstrating that the absorption follows an exponential function in its steep part. The same behaviour was reported for other semiconductors [12] without a fully conclusive explanation of the reasons which might be envisaged. Among these reasons are shallow acceptors and crystal imperfections, whereas free-carrier absorption can be ruled out because of its λ^2 dependence.

We will discuss here phonon-assisted direct band-to-band transitions, as described by Dumke [13]. In this model only the phonons near the Γ-point of the Brillouin zone and of the highest energy, i.e., from the optical branch, can extend the absorption to any appreciable extent beyond $\lambda_{g,Q}$. Because of the small dispersion of the optical branch, their energy $k_B\theta$ (with Boltzmann's constant k_B) can be taken as essentially constant. We estimate $k_B\theta$ for our composition by employing Vegard's law with the parameters of the four binary semiconductors of the system InGaAsP. We find about 35 meV. Therefore the range can be marked in Fig. 4 where phonon-assisted transitions can be expected. Calculating the effective electron mass as 0.055 and the effective heavy-hole mass as 0.464 [4] leaves only one free parameter for non-polar scattering [3]. This parameter is fitted to the experimental points to obtain the dashed curve in Fig. 4.

5. Conclusion

We have measured the optical constant of $In_{1-x}Ga_xAs_yP_{1-y}$ of a composition denoted by $x = 0.27$ and $y = 0.64$. The crystals were epitaxially grown on InP substrates. The absorption in the short-wavelength range can be adequately described by direct band-to-band transitions leading to a band-gap energy of 0.938 eV. In the long-wavelength range the absorption follows closely an exponential dependence.

Acknowledgements

These investigations were financially supported by the Deutsche Forschungsgemeinschaft. The authors gratefully acknowledge the technical assistance by U. Heydecke in the crystal growth, and the microprobe measurements by A. Pöcker, Research

Institute of the German Post Office, Darmstadt. The authors wish to express their gratitude to Prof. Dr. F. R. Keßler and Dr. K. Dettmer, both of the Technical University Braunschweig, for extensive discussions.

References

[1] P. CHANDRA, L. A. COLDREN, and K. E. STREGE, Electronics Letters 17, 6 (1981).
[2] A. R. CLAWSON, W. Y. LUM, and H. H. WIEDER, J. Soc. Photo-Opt. Instrum. Engn. 132, 2 (1978).
[3] H. H. CASPER and H. H. WIEDER, Solid State Commun. 29, 403 (1979).
[4] N. K. DUTTA, J. appl. Phys. 51, 6095 (1980).
[5] T. TAKAGI, Japan. J. appl. Phys. 18, 2017 (1979).
[6] G. H. OLSON, T. Z. ZAMEROWSKI, R. T. SMITH, and E. P. BERTIN, J. Electronic Mat. 9, 977 (1980).
[7] G. D. PETTIT and W. J. TURNER, J. appl. Phys. 36, 2081 (1965).
[8] E. KUPHAL, private communication (1981).
[9] K. SEEGER, Semiconductor Physics, Springer, Wien/New York 1973 (p. 339).
[10] R. E. NAHORY, M. A. POLLACK, and W. D. JOHNSTON, Appl. Phys. Letters 37, 659 (1978).
[11] J. KUDMAN and T. SEIDEL, J. appl. Phys. 33, 771 (1962).
[12] E. J. JOHNSON, in: Semiconductors and Semimetals, Chap. 6, Ed. R. K. WILLARDSON and A. C. BEER, Vol. 3, Academic Press, 1967.
[13] W. P. DUMKE, Phys. Rev. 108, 1419 (1957).

(Received July 20, 1981)

phys. stat. sol. (a) 68, 159 (1981)

Subject classification: 10.1 and 18.2; 21.1.1

Institut für Physik, Max-Planck-Institut für Metallforschung, Stuttgart[1])

On Magnetic Domain Patterns and Dislocation Structures in Iron Single Crystals Deformed at 195 and 77 K

By

Y. UMAKOSHI[2]) and H. KRONMÜLLER

Magneto-optical investigations of domain patterns in plastically deformed iron single crystals are used for an analysis of dislocation structures and stress induced magnetic anisotropies. The observed laminar domains extending either parallel to the Burgers vector [111] or parallel to the edge dislocation lines [1$\bar{2}$1] of the primary glide system are attributed to dislocations sheets and dislocation walls. At large deformation screw dislocations of the conjugate glide system are activated which produce a laminar domain structure extending parallel to the [010] direction.

Magnetooptische Untersuchungen der Domänenstrukturen plastisch verformter Eiseneinkristalle werden zur Untersuchung der Versetzungsanordnung und der spannungsinduzierten magnetischen Anisotropie herangezogen. Die beobachteten streifenförmigen Domänen, die parallel zum Burgersvektor [111] oder parallel zur Linienrichtung [1$\bar{2}$1] der Stufenversetzungen des primären Gleitsystems verlaufen, werden Versetzungsschichten und Versetzungswänden zugeordnet. Bei großen Deformationen werden Schraubenversetzungen des konjugierten Gleitsystems erzeugt, die eine laminare Domänenstruktur parallel zur [010]-Richtung erzeugen.

1. Introduction

Many experimental studies of work-hardening of iron single crystals have shown that in the temperature range from 200 to 300 K the work-hardening behaviour of iron single crystals, oriented for single slip, changes from the three-stage work-hardening curves, which are similar to those of f.c.c. metals, to become parabolic in shape [1 to 5]. Direct observations by transmission electron microscopy have shown that with decreasing temperature the dislocation structure changes from arrays of edge dipole clusters and dislocation networks, aligned parallel to the primary glide planes, to continuously distributed long dislocations having small parts of jogs [6 to 8]. The difference in dislocation structures between low and intermediate temperatures is thought to be mainly dependent on the motion of screw dislocations in crystals. At intermediate temperatures screw dislocations can easily perform cross-slip resulting in the formation of edge dipoles and dislocation networks aligned parallel to the primary glide planes. These dislocation arrays may become sources of significant long-range internal stresses. At lower temperatures screw dislocations with jogged parts slip on the primary slip plane and the dislocations are uniformly distributed.

Previously, magnetic techniques have been developed and widely used for the study of internal stress in plastically deformed ferromagnetic metals [9 to 12]. More recently, several investigations of dislocation arrangements in plastically deformed iron single crystals have been performed using these magnetic techniques [13 to 16].

[1]) Heisenbergstr. 1, D-7000 Stuttgart 80, FRG.

[2]) On leave from Department of Materials Science and Engineering, Osaka University, Yamadakami, Suita, Osaka, Japan.

The present work was undertaken in order to obtain information on dislocation arrangements at various stages of deformation in iron single crystals deformed at 77 and 195 K; magnetic domain patterns have been observed using the magneto-optical Kerr effect.

2. Experimental Procedure

Iron single crystals with diameters of 5 mm were grown by the strain-anneal technique. After annealing in Pd-purified dry hydrogen at 1000 K for three weeks, the carbon content was less than 5 wt-ppm. The main impurity was nickel at a concentration of about 300 ppm. Tensile tests were performed at 195 and 77 K using an Instron type machine with a nominal strain rate of $3 \times 10^{-5}\,s^{-1}$. These temperatures were obtained by immersing the specimens in constant temperature baths. The crystal investigated at 77 K was pre-strained by 2% shear strain at 295 K in order to suppress heterogeneous deformation. The tensile tests were repeatedly interrupted after desired intervals of strain and, after unloading, cylindrical discs were cut from small parts of these specimens by spark cutting. Subsequently, the remaining crystal parts were deformed after chemically repolishing the surface. Slip traces were examined with an optical microscope using Nomarski interference contrast. Slip planes were determined by measuring the angle between the tensile axis and the tangent to the slip line for different rotations about the deformation axis. Slip directions were determined by noting the surface that contained ***b*** and therefore showed no slip markings. The change of orientation of the crystals after straining was determined by standard X-ray Laue back reflection technique. The stress axis is referred to the standard $\langle 001\rangle$–$\langle 011\rangle$–$\langle 111\rangle$ unit triangle. Discs for magnetic domain studies prepared from deformed specimens by spark machining were chemically polished in a solution of H_2O_2 (69%), HF (6%), and H_2O (25%) to remove the surface damage and subsequently electrolytically polished in a 10% perchloric acid-acetic acid solution, and then evaporated with a $\lambda/4$ film of ZnS for eliminating back reflection. The observation of magnetic domain patterns was carried out in the demagnetized state by means of the longitudinal optical Kerr effect where the spontaneous magnetization lies in the plane of incidence and also in the plane of the reflecting surface.

3. Experimental Results

3.1 Stress–strain behaviour and slip line morphology

The stress–strain curves and the initial orientations are given in Fig. 1. Crystal A deformed at 195 K shows a parabolic stress–strain curve with a yield point occurring at small strains and in addition the work-hardening rate has zero or a very small negative value at large strains after the shear stress passed through a maximum. Crystal B, tested at 77 K, was pre-strained at 295 K, because this depresses the yield strength at 77 K [17]. The pre-straining eliminated the yield point from the stress–strain curve, which may be described by a parabola with a high initial rate of work-hardening, followed by a small linear regime, and then a decrease of the work-hardening rate with continuing deformation.

Fig. 2 shows the results for the path of the tensile axis and the observed slip plane variation during deformation. For crystal A, the orientation of the tensile axis rotated towards the [111] primary slip direction which is almost exactly consistent with the calculated changes of the orientation assuming [111] single slip up to 10% strain. The observed slip plane was roughly the $(\bar{1}01)$ plane at the beginning of deformation and it approached from $(\bar{1}01)$ to the plane of maximum resolved shear stress (m.r.s.s.) with increasing deformation. For crystals deformed by more than 10% strain, slip

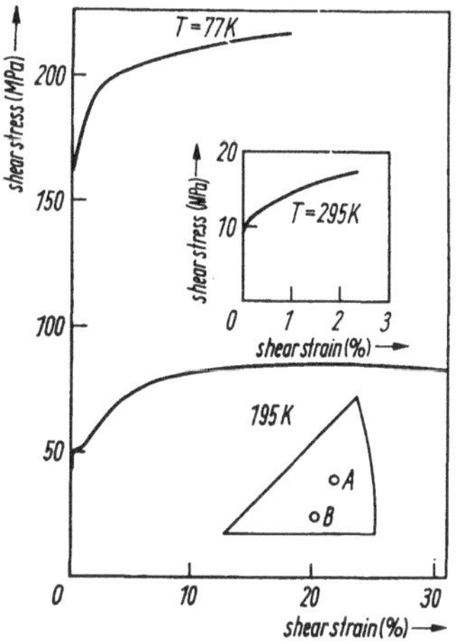

Fig. 1

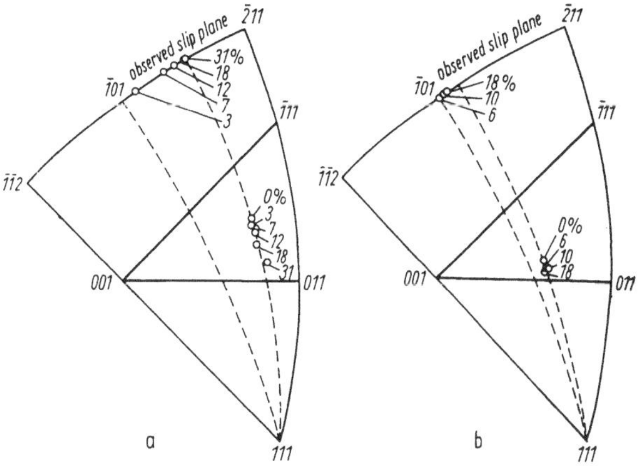

Fig. 2

Fig. 1. Stress–strain curves and initial orientation. The shear stress is calculated for the glide plane (1̄01) [111]

Fig. 2. Changes of the orientation of the tensile axis during deformation; a) crystal A deformed at 195 K, b) crystal B deformed at 77 K

follows the m.r.s.s. plane, on which the shear in [111] is in the anti-twinning sense. The orientation change of crystal B deformed at 77 K was smaller than that calculated because of the activation of conjugate slip. The observed slip plane in crystal B was almost exactly the (1̄01) plane regardless of the degree of deformation.

Fig. 3 and 4 show the slip traces observed on top faces of the single crystals deformed at 77 and 195 K, respectively. Top faces correspond to that part of the cylindrical surface to which primary edge dislocations are parallel. Typical fine and wavy slip traces of primary slip can be seen in Fig. 3. This nature may be interpreted on the basis of the composite slip on (1̄01) and (2̄11) planes because in this temperature range

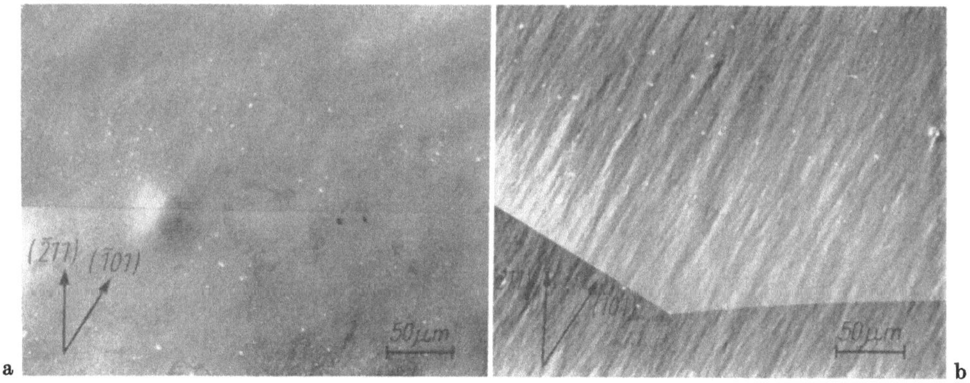

Fig. 3. Slip traces on the top face of crystal A deformed at 195 K, a) 3% shear strain, b) 18% shear strain

Fig. 4. Slip traces on the top face of crystal B deformed at 77 K, 18% shear strain

the screw dislocations can easily cross slip on another glide plane. With increasing deformation, the slip traces are found to become less fine and nearly parallel to the m.r.s.s. plane.

The slip traces on crystal B become increasingly coarse and somewhat straight when compared with those of crystal A. The slip bands in which the shear strain is concentrated can be observed on some regions of the surface. It may be possible that these slip bands are attributed to the interaction between screw dislocation and obstacles which consist of dislocation sheets produced during the pre-straining at 295 K [5]. According to Fig. 4 duplex slip of types [111] and [$\bar{1}$11] operates from the beginning of deformation although crystal B has an orientation where ($\bar{1}$01) [111] is more highly stressed than (101) [$\bar{1}$11].

a 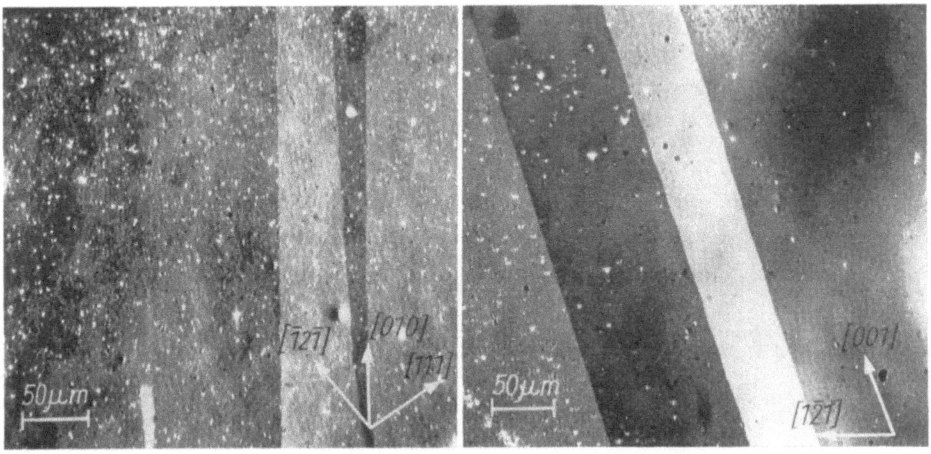 b

Fig. 5. Domain patterns of undeformed iron single crystals, a) ($\bar{1}$01), b) (210)

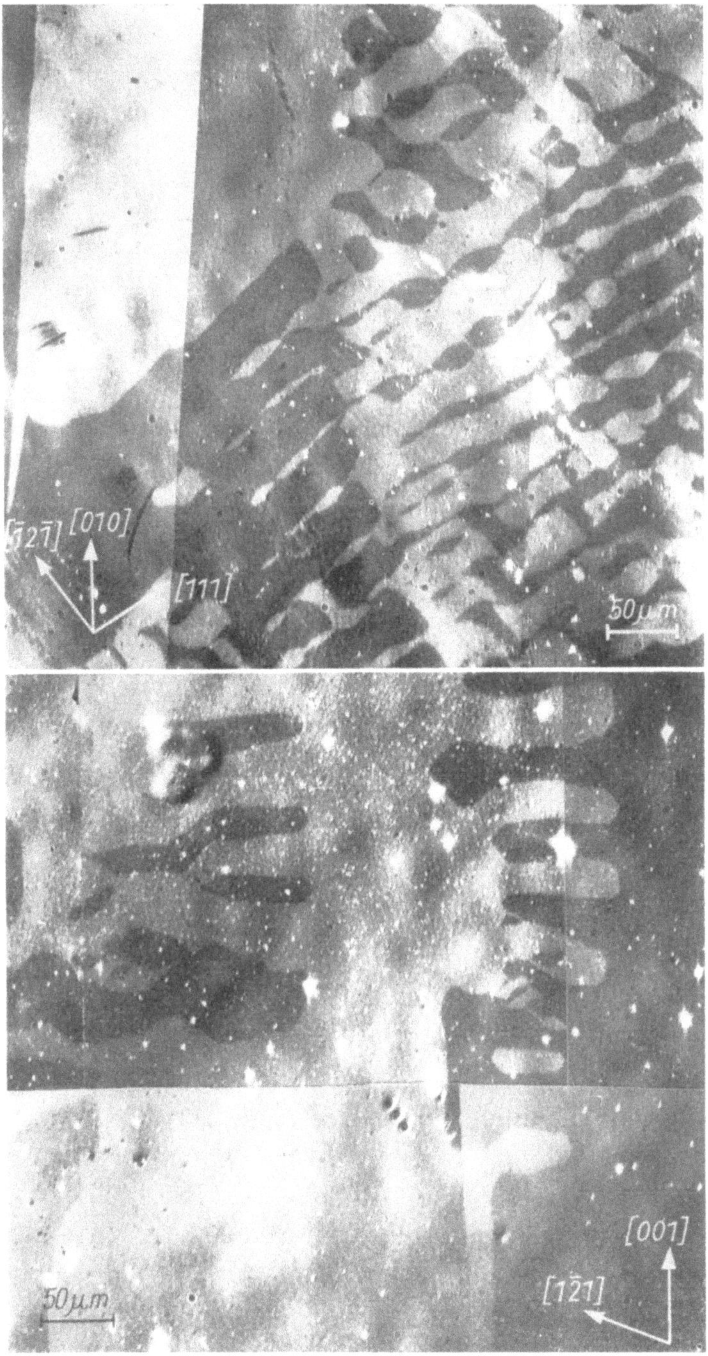

a

b

Fig. 6. Domain patterns of crystal A deformed at 195 K; a) 8% shear strain, (1̄01); b) 8% shear strain, (210)

3.2 Magnetic domain structure studied by means of the magneto-optical Kerr effect

We have observed the domain structure on the ($\bar{1}01$) primary slip plane and the (210) plane. On the ($\bar{1}01$) plane we have minima of the crystalline energy ([010] and [0$\bar{1}$0]), and the line direction [1$\bar{2}$1] and the Burgers vector [111] of primary dislocations lies within the ($\bar{1}01$) plane. The (210) plane contains two easy directions ([00$\bar{1}$] and [001]) as well as the [1$\bar{2}$1] direction.

3.2.1 Domain patterns in the undeformed crystal

In Fig. 5 we can observe the typical laminar two-phase domain structure lying parallel to the easy direction ([001] or [010]) on ($\bar{1}01$) and (210) planes. The lanceolate-shaped domains are due to a slight misorientation from the ($\bar{1}01$) plane.

3.2.2 Domain patterns in crystal A deformed at 195 K

As seen in Fig. 6 to 8, the laminar two-phase domain structure changes in the crystals with increasing internal stresses; this may be attributed mainly to dislocations induced by plastic deformation. At the beginning of deformation, the domain pattern of the

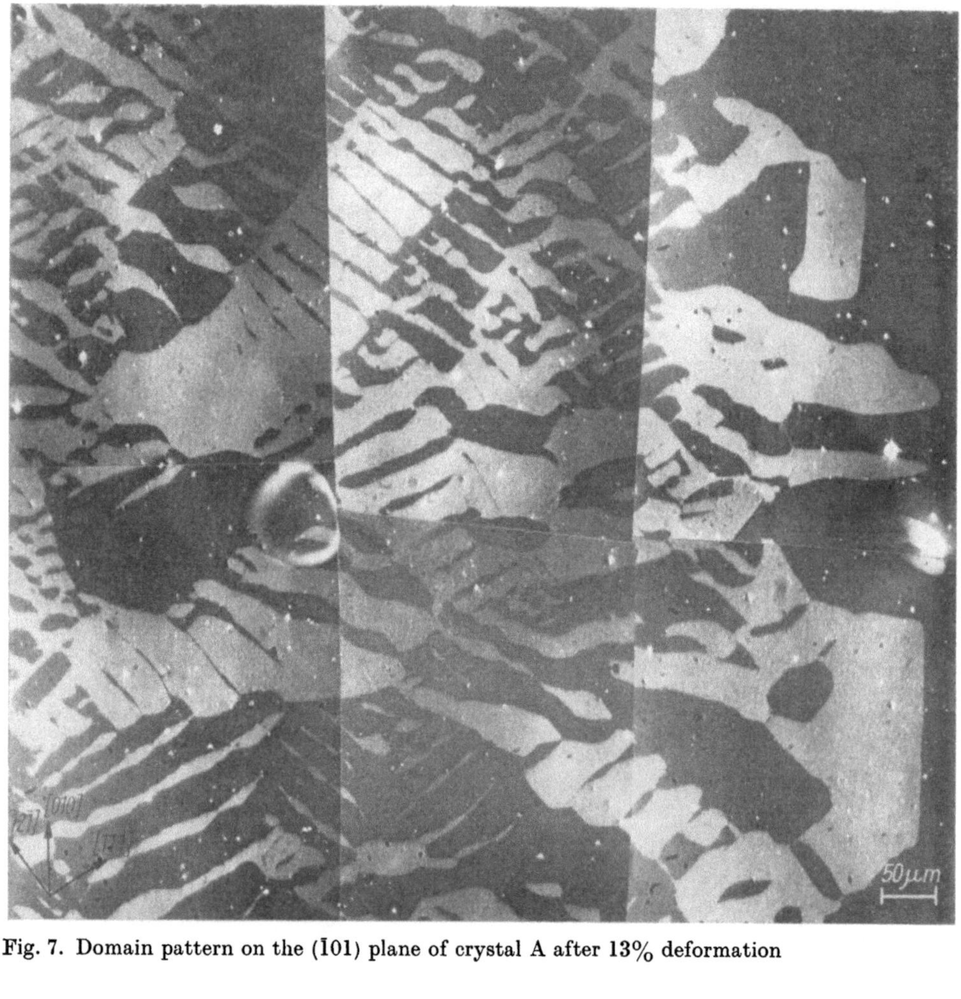

Fig. 7. Domain pattern on the ($\bar{1}01$) plane of crystal A after 13% deformation

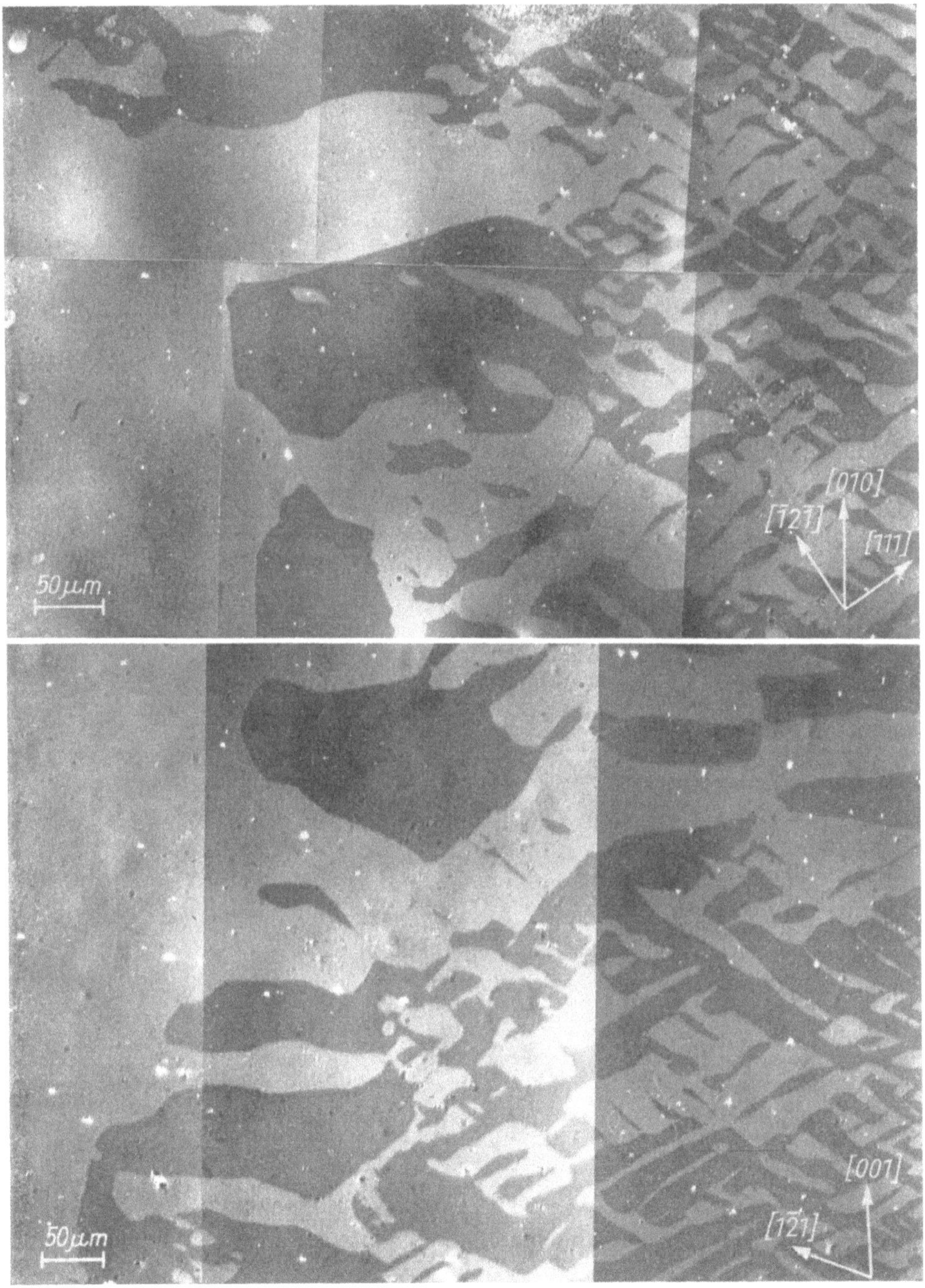

a

b

Fig. 8. Domain patterns of crystal A deformed at 195 K; a) 18% shear strain, $(\bar{1}01)$; b) 18% shear strain, (210)

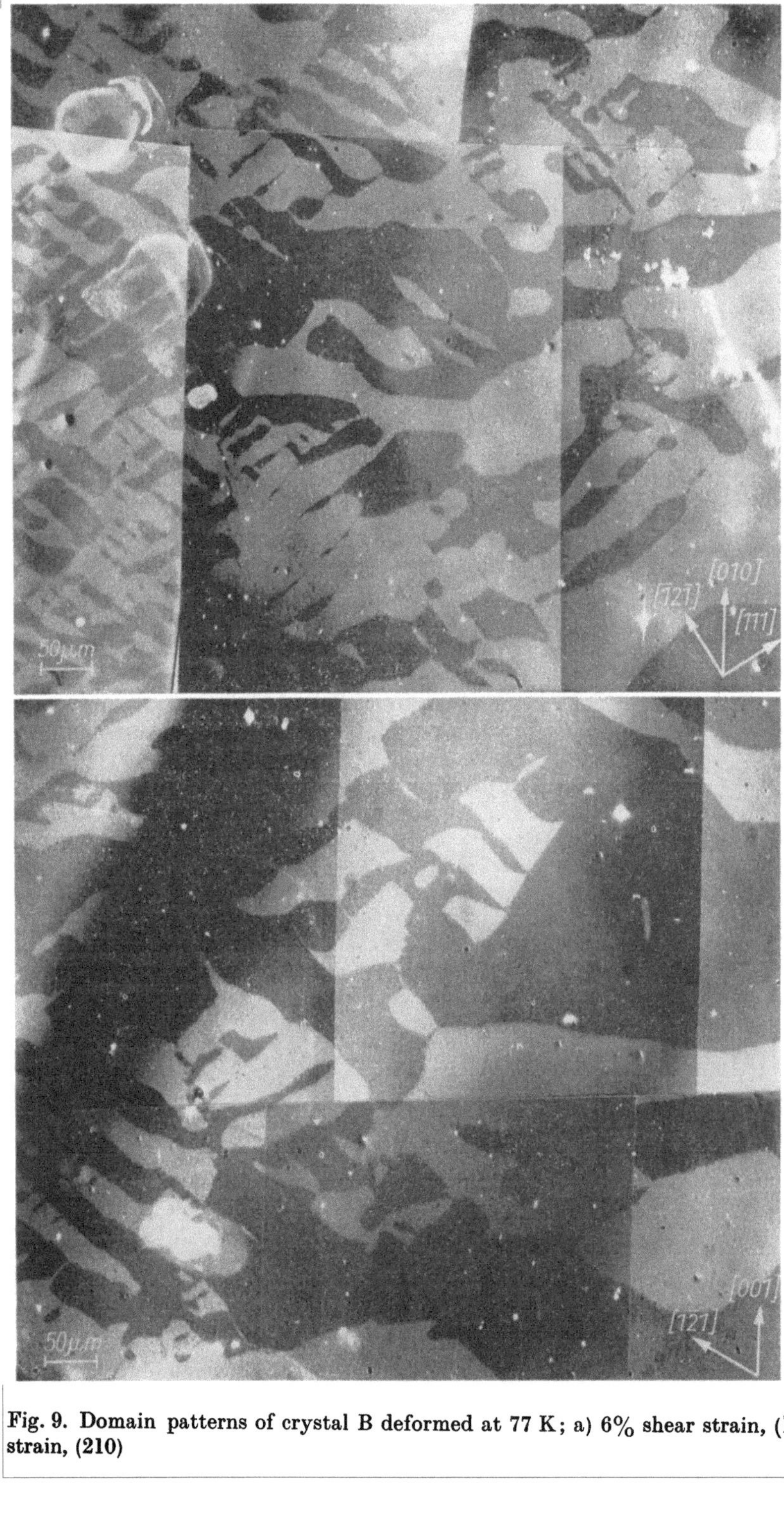

Fig. 9. Domain patterns of crystal B deformed at 77 K; a) 6% shear strain, (Ī01); b) 6% shear strain, (210)

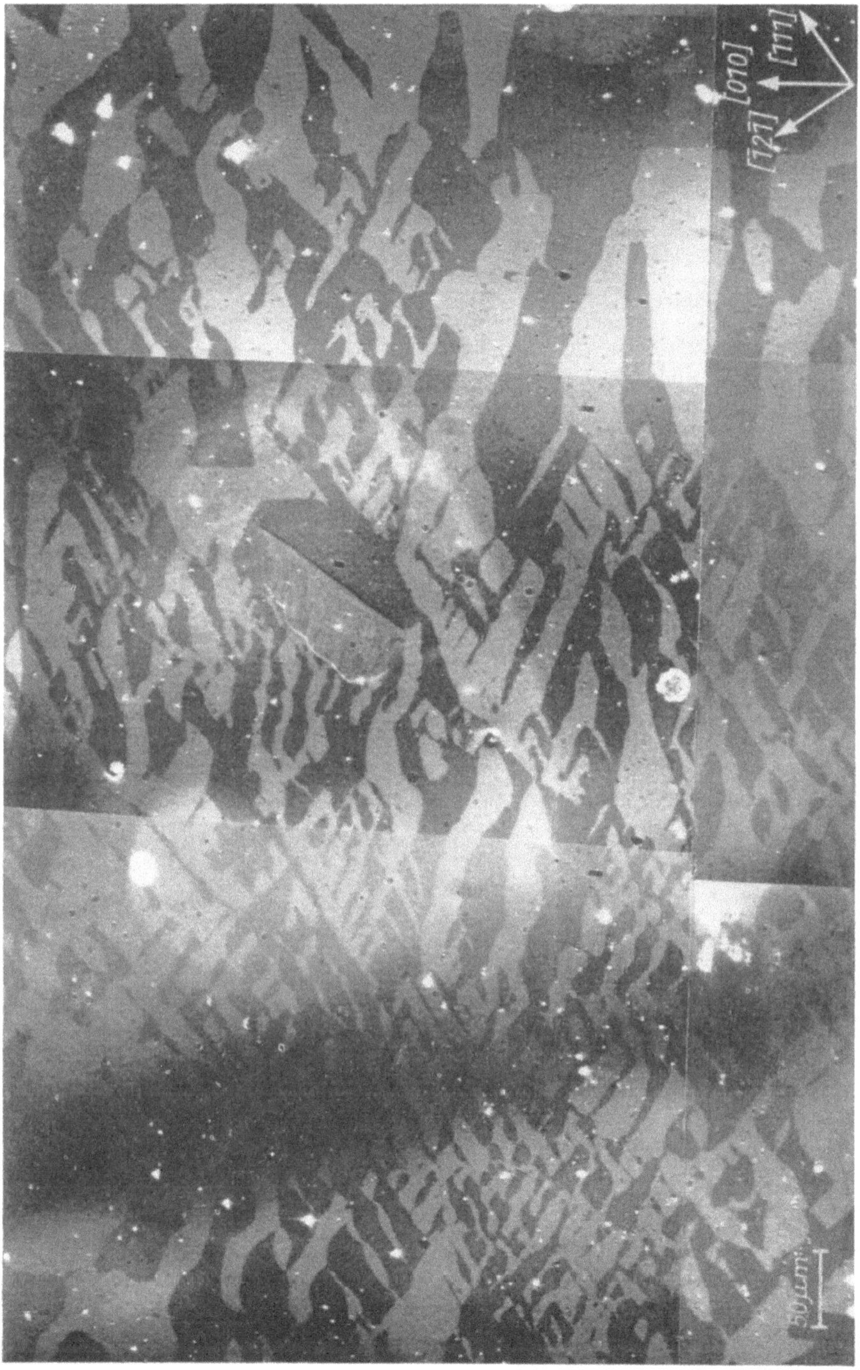

Fig. 10. Domain patterns on (Ī01) plane of crystal B deformed up to 10% shear strain

Fig. 11. Domain patterns of crystal B deformed at 77 K; a) 18% shear strain, ($\bar{1}$01); b) 18% shear strain, (210)

as-deformed specimen mainly consists of wide and long laminar domains oriented parallel to the [010] easy direction, and in addition the ribbon-shaped strips are distributed in some regions nearly parallel to the Burgers vector [111] of the primary edge dislocations (Fig. 6a). As shown in Fig. 6b, some small domains protruded from wide laminar phases can be observed (except such a ribbon-shaped strip) because of the difference in the induced anisotropy between ($\bar{1}$01) and (210) planes.

Fig. 7 and 8 show that the two-phase laminar phase on the $(\bar{1}01)$ primary slip plane is interrupted by two types of ribbon-shaped domain layers parallel to primary edge dislocations. The domain pattern on the $(\bar{2}10)$ plane is also interrupted by layers extending parallel to a $[1\bar{2}1]$ direction. Furthermore it should be noted that the ribbon-shaped stripes become much finer with increasing deformation.

3.2.3 Domain patterns in crystal B deformed at 77 K

Crystal B was pre-strained up to 2% at 295 K before deformation at 77 K. During pre-straining we cannot observe a great change in the two-phase laminar structure apart from the appearance of small disordering on local region owing to the induced edge dislocations. After a slight deformation at 77 K, the domain patterns change from laminar two-phase strips to the ribbon-shaped strips, which are similar to those on crystal A, aligning parallel to the directions $[1\bar{2}1]$ and $[111]$ (see Fig. 8). The size of the ribbon-shaped areas shows a wide-spread distribution. The domain patterns on (210) planes shown in Fig. 8b are not remarkable when compared with those on $(\bar{1}01)$ planes, however, they are found to lie parallel to the $[1\bar{2}1]$ direction. With increasing deformation, the ribbon-shaped strips become short and fine, and deviate from $[1\bar{2}1]$ and $[111]$ directions as shown in Fig. 9. The laminar-shaped phase becomes also much finer than those of weakly deformed specimens as demonstrated by Fig. 10.

Fig. 11 shows domain patterns for specimens deformed to 18% strain where the conjugate slip is mainly activated. The activation of conjugate slip transforms the ribbon-shaped strips into laminar phases parallel to the [010] easy direction, although laminar phases are shortened and rather narrow. These phases which appeared again cover almost the whole area of the specimen and, in addition, the traces of ribbon-shaped domains can be found in the isolated areas.

4. Discussion

It is well-known that a compromise between the strayfield energy, φ_s, and the Bloch wall energy, φ_γ, results in the formation of magnetic domains in a macroscopic specimen [18]. The spontaneous magnetization within the domains is aligned parallel to directions where the free energy has a minimum. In the undeformed crystal the direction of the spontaneous magnetization is parallel to crystallographic directions which correspond to minima of the crystalline energy

$$\Phi_K = K_1(\gamma_1^2\gamma_2^2 + \gamma_1^2\gamma_3^2 + \gamma_2^2\gamma_3^2) + K_2\gamma_1^2\gamma_2^2\gamma_3^2\,, \tag{1}$$

where γ_1, γ_2, and γ_3 are the direction cosines of the magnetization, and K_1 and K_2 are the crystalline energy constants. As seen from Fig. 12 in α-Fe the $\langle 100\rangle$ directions are magnetic easy directions. Thc typical laminar two-phase domain structures indicated in Fig. 5 can easily be understood from the anisotropy of the crystalline energy.

In crystals with internal stresses, the field energy is increased. The free energy difference between a homogeneously magnetized ideal crystal and the crystal with internal stress is given by

$$\Delta\Phi^{(i)} = \Phi_{ex} + \Phi_m + \Phi_K - \Phi_K^{(h)} + \Phi_{ms} - \Phi_{ms}^{(h)}\,, \tag{2}$$

where Φ_{ex} is the exchange energy, Φ_m the magnetoelastic coupling energy, Φ_K the crystalline energy, Φ_{ms} the magnetostatic energy. $\Phi_K^{(h)}$ and $\Phi_{ms}^{(h)}$ are the crystalline energy and the magnetostatic energy of the homogeneously magnetized ideal crystal, respectively. Referred to previous works [12, 19], we can obtain the following induced anisotropy, after taking into account the anisotropy of the magnetostriction of iron:

$$\Delta\Phi^{(i)} = AG^2b^2N(B + \ln R_0)\, F(\gamma_i)\,. \tag{3}$$

Here N is the dislocation density, G the shear modulus, b is the length of the Burgers vector, R_0 corresponds to the average distance between dislocations, and A and B are constants. $F(\gamma_i)$ denotes the anisotropy function which depends on the direction of the spontaneous magnetization and the stress tensor of the dislocation. From (3) it may be concluded that the induced anisotropy is remarkably influenced by the anisotropy of the stress tensor. The dislocation $(\boldsymbol{b}, \boldsymbol{n}, \boldsymbol{l})$ is introduced in the coordinate system where the z-axis is parallel to the line direction of the dislocation, $\boldsymbol{l}$ and the y-axis is parallel to the slip plane normal, $\boldsymbol{n}$. In this coordinate system the stress tensor of an edge dislocation has the following form:

$$\sigma = \begin{pmatrix} \sigma_{xx} & \sigma_{xy} & 0 \\ \sigma_{xy} & \sigma_{yy} & 0 \\ 0 & 0 & \sigma_{zz} \end{pmatrix}. \tag{4}$$

Under the assumption of statistically distributed dislocations we have calculated the induced anisotropy of primary and conjugate dislocations for a dislocation density of 10^9 cm^{-2} (cf. Fig. 13a and 13b). The induced anisotropy of primary screw and edge dislocations has a minimum near the direction of the Burgers vector. $\Delta\Phi^{(i)}$ must be superimposed on the crystalline energy $\Delta\Phi_K$ in order to interpret the domain structures in deformed crystals because the minimum of the *total free energy* determines the easy direction of magnetization. At the beginning of the deformation at 195 K the laminar domain structure was interrupted by ribbon-shaped layers extending parallel to the [111] Burgers vector of the primary dislocations. This interruption seems to be governed by the effect of edge dislocations rather than that of screw dislocations since the energy minima of $\Delta\Phi^{(i)}$ parallel to $\boldsymbol{b}$ and to the $\langle 100\rangle$ directions are found to be much lower for primary edge dislocations than for primary screw dislocations.

According to Fig. 12 and 13, we should consider the effect of dislocation groups which are composed of n dislocations and correspond to superdislocations with Burgers vector $n\boldsymbol{b}$ in order to dominate the large crystalline energy Φ_K by $\Delta\Phi^{(i)}$. The induced anisotropy, $\Delta\Phi_{nb}^{(i)}$ for dislocation groups is given by

$$\Delta\Phi_{nb}^{(i)} = AG^2b^2nN(B + \ln \sqrt{n}\, R_0)\, F(\gamma_i)\,. \tag{5}$$

With increasing deformation at 195 K, the ribbon-shaped domains are created parallel to the [111] Burgers vector as well as parallel to the [$1\bar{2}1$] dislocation line directions. This result cannot be explained only by the influence of the internal stresses produced

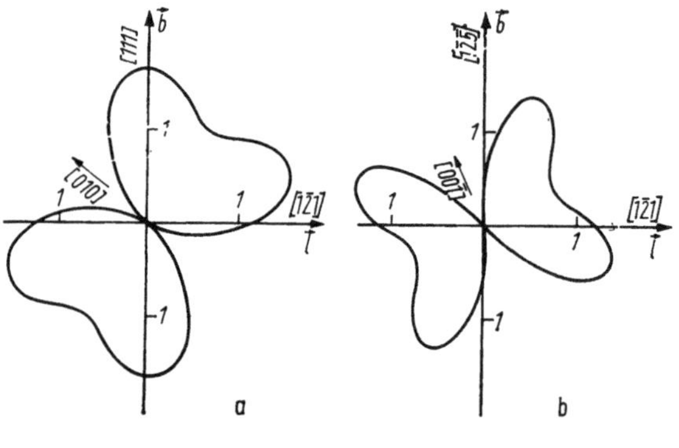

Fig. 12. Anisotropy of the crystalline energy (in units of 10^4 J/m^3) in the ($\bar{1}01$) plane. a) ($\bar{1}01$), b) (210)

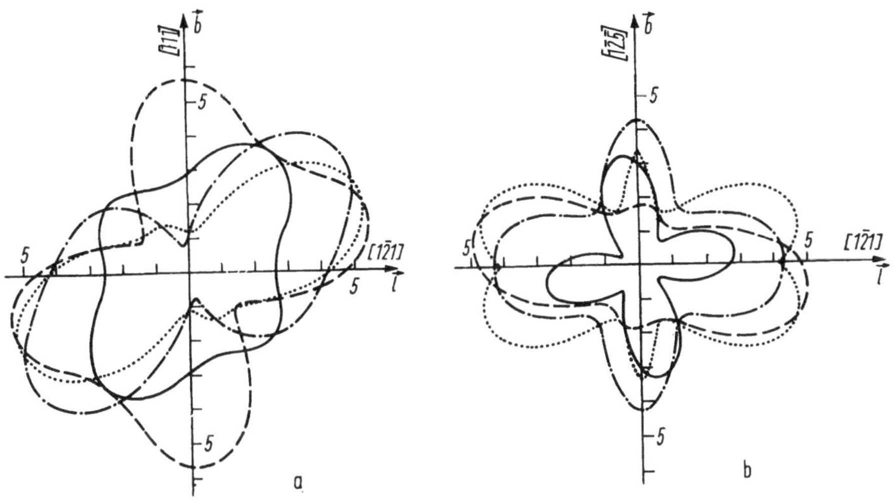

Fig. 13. Induced anisotropy energy $\Delta\Phi^{(i)}$ (in units of 10^2 J/m^3), calculated for a dislocation density of $N = 10^9$ cm^{-2}; a) ($\bar{1}$01), b) (210); ——— conjugated edge, - - - - conjugated screw, — · — primary edge, · · · · primary screw

with individual edge dislocations. At this temperature the screw dislocation can still easily cross slip on other planes and therefore the macroscopic slip plane corresponds to a plane between the ($\bar{1}$01) and the ($\bar{2}$11) planes as seen in Fig. 3. Dislocation sheets and/or dislocation walls composed of layered segments and dipoles seem to be created by the cross slip of primary and/or conjugate dislocations [5]. They are sources of long-range internal stresses. In the case where the dislocations develop extended dislocation sheets the stress tensor changes from (4) to

$$\sigma = \begin{pmatrix} \sigma_{xx} & 0 & 0 \\ 0 & \sigma_{yy} & 0 \\ 0 & 0 & 0 \end{pmatrix}. \tag{6}$$

From the anisotropy of this stress tensor we obtain an induced anisotropy $\Delta\Phi^{(i)}$ as shown in Fig. 14. This figure shows that the minima of $\Delta\Phi^{(i)}$ in the line direction $\boldsymbol{l}$

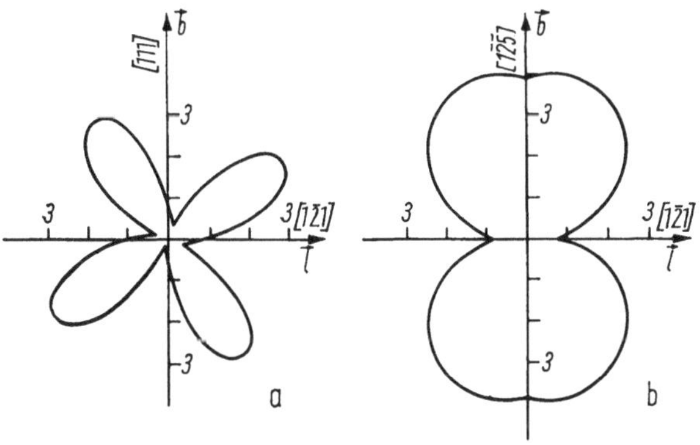

Fig. 14. Induced anisotropy energy $\Delta\Phi^{(i)}$ (in units of 10^2 J/m^3) of dislocation layers, calculated for a dislocation density of $N = 10^9$ cm^{-2}; a) ($\bar{1}$01), b) (210)

and in the direction of the Burgers vector ***b*** favour the formation of domain layers extending into the [111] and the [1$\bar{2}$1] directions. The ribbon-shaped domain structure extending not only parallel to [111] but also to [1$\bar{2}$1] suggests the presence of dislocation sheets and/or dislocation walls which are produced by cross slip of screw dislocations.

In the case of crystal B, deformed at 77 K, the mobility of edge dislocations is much larger than that of screw dislocations and, therefore, at the beginning of the deformation at 77 K the small portion of the primary edge dislocation created by pre-straining at 295 K can easily slip out of the crystal. However, the greater portion remains immobile belonging to dislocation tangles, sheets, and walls which is illustrated by the presence of ribbon-shaped domain structures. Although mainly screw dislocations are created with increasing deformation and the flow stress is determined by the motion of screw dislocations the dislocation sheets remain always in the same regions. With increasing deformation, the induced anisotropy of conjugate screw dislocations start to suppress the formerly induced anisotropy of primary dislocations which developed in the extended domain structure and, therefore, the two-phase domain structure appears again.

Acknowledgements

The authors are very grateful to Prof. A. Seeger for his interest in this paper, to Dr. H. Mughrabi for his valuable helps, and to Dr. H. Blythe for useful remarks on the manuscript. Y.U. would like to thank the Alexander von Humboldt Foundation for a visiting fellowship.

References

[1] H. BILGER, phys. stat. sol. **18**, 637 (1966).
[2] T. TAKEUCHI, R. HONDA, K. IWAYAMA, and T. TAOKA, Japan. J. appl. Phys. **6**, 1282 (1967).
[3] T. TAKEUCHI, J. Phys. Soc. Japan **26**, 355 (1969).
[4] B. ŠESTÁK and A. SEEGER, phys. stat. sol. (b) **43**, 433 (1971).
[5] V. NOVAK and B. ŠESTÁK, Proc. 4th Internat. Conf. Strength of Metals and Alloys, Vol. 1, Nancy 1976 (p. 156).
[6] S. IKEDA, J. Phys. Soc. Japan **27**, 1564 (1969).
[7] T. YAMASHITA and Y. TANEDA, Trans. Iron Steel Inst. Japan **10**, 1971 (1970).
[8] A. S. KEH, W. A. SPITZIG, and Y. NAKADA, Phil. Mag. **23**, 829 (1971).
[9] H. KRONMÜLLER, Internat. J. nondestr. Testing **3**, 315 (1972).
[10] G. SCHROEDER and H. KRONMÜLLER, phys. stat. sol. (a) **34**, 623 (1976).
[11] H. MUGHRABI, R. KÜTTERER, K. LUBITZ, and H. KRONMÜLLER, phys. stat. sol. (a) **38**, 261 (1976).
[12] H. KRONMÜLLER, J. appl. Phys. **38**, 1314 (1967).
[13] K. LUBITZ, Appl. Phys. **4**, 51 (1974).
[14] G. SCHROEDER and H. KRONMÜLLER, J. Magnetism magnetic Mater. **6**, 213 (1977).
[15] K. LUBITZ, Appl. Phys. **16**, 87 (1978).
[16] G. SCHROEDER, phys. stat. sol. (a) **47**, 145 (1978).
[17] W. A. SPITZIG and A. S. KEH, Acta metall. **18**, 611 (1970).
[18] H. KRONMÜLLER, AIP Conf. Proc. **10**, 1006 (1972).
[19] W. F. BROWN, Phys. Rev. **58**, 736 (1940).

(Received August 19, 1981)

phys. stat. sol. (a) 68, 173 (1981)

Subject classification: 1.2 and 14.4.1; 22.8.1

Instituto de Física "Gleb Wataghin", Universidade Estadual de Campinas[1])

On the Use of Powder Diffractometry in the Study of Phase Transitions

Case of $NaNO_2$

By

A. DA COSTA LAMAS[2]), S.-L. CHANG[3]), and S. CATICHA-ELLIS

The phase transition of $NaNO_2$ is investigated by using X-ray powder diffractometry. The temperature dependence of the lattice constants, peak intensities, and the order parameter of the ferroelectric phase, in the range of 27 to 200 °C, show the same results as obtained by using single-crystal diffractometry. These experimental findings imply that the X-ray powder method is suitable for the study of phase transition. The analysis of the powder diffraction profile would even improve the method.

Die strukturelle Phasenumwandlung von $NaNO_2$ wird mittels Pulverdiffraktometrie untersucht. Die Temperaturabhängigkeit der Gitterkonstanten, der Bragg-Intensität und des Ordnungsparameters der ferroelektrischen Phase im Bereich von 27 bis 200 °C ergibt die gleichen Resultate wie mit Diffraktometrie an Einkristallen. Daraus ergibt sich, daß die Pulvermethode für das Studium von Phasenübergängen geeignet ist. Eine Analyse der Intensitätsprofile würde sogar eine weitere Verbesserung bringen.

1. Introduction

The phase transition of sodium nitrite has been intensively studied for the past two decades on its diverse physical properties, such as dielectric, thermal, optical, and others [1 to 4]. The transition has been found of the type from ferroelectric phase, to a so-called FS phase [5], to sinusoidal antiferroelectric and then to paraelectric phase as temperature increases. This is due to the variation of polarization of the dipoles in the crystallographic a-planes along b-axis. The parameter of order–disorder as well as the lattice constants were usually determined by X-ray diffraction from single crystals [5 to 9]. Although the use of single crystal diffractometry has the advantage of revealing the satellite peaks in the antiferroelectric phase for this substance, it is limited by the fact that the intensity measurements can be carried out for a given reflection of a certain cut crystal. Investigation on various reflections, therefore, needs to have different samples prepared or have the same crystal cut differently. In addition, due to thermal expansion, the variation of lattice constant, which is frequently determined by dilatometric methods [10], cannot be estimated directly from X-ray diffraction by using a unique sample at a fixed crystal mounting. This paper is a tentative to study this subject by using the X-ray powder method, because then a simple powder sample is needed.

---

[1]) Campinas, S.P.-13100 Brazil.

[2]) Present address: Physics Department, Brandeis University, Walthan, Ma. 02154, USA.

[3]) On leave at Max-Planck-Institut für Festkörperforschung, D-7000 Stuttgart 80, FRG, from March 1981 to February 1982.

2. Experimental

$NaNO_2$ single crystals were grown using the Czochralsky technique [11], and then ground in powder form. A powder sample was prepared in the usual way to avoid preferred orientation and mounted on a Rigaku powder diffractometer with high temperature attachment. The furnace was controlled by a manual variac. The Fe–Ct thermocouple employed permits an adequate measurement of temperature in the range from 150 to 200 °C. A precision of ± 0.1 K in temperature was obtained. The Bragg angle θ and the peak intensities for various reflections were then measured at temperatures, ranging from 27 to 200 °C. CuK_α radiation was used throughout.

3. Results and Discussion

The diffracted intensity is proportional to

$$|F|^2 = A^2 + S^2(T)\, B^2 ,$$

where A and B are the real and imaginary parts of the structure factor F and S is the order parameter for the ferroelectric phase. The calculated values of A and B, listed in Table 1, were obtained with the positions of Na, N, and O as given by Tanisaki [6]. Reflections, (200) and (002), with $B = 0$ and (020) with $B \ll A$, were used to determine the lattice constants so that the influence of the satellite peaks on the peak position above the critical temperature T_0 ($\approx$ 163 °C) was negligible. The lattice constants a, b, and c, were plotted against temperature in Fig. 1 a, b, and c. Our measurements confirmed the known fact that a and b increase while c decreases as temperature increases. Discontinuities occurred at about 160.8, 162, and 163 °C. As expected, the lattice constant a was less affected by the temperature variation.

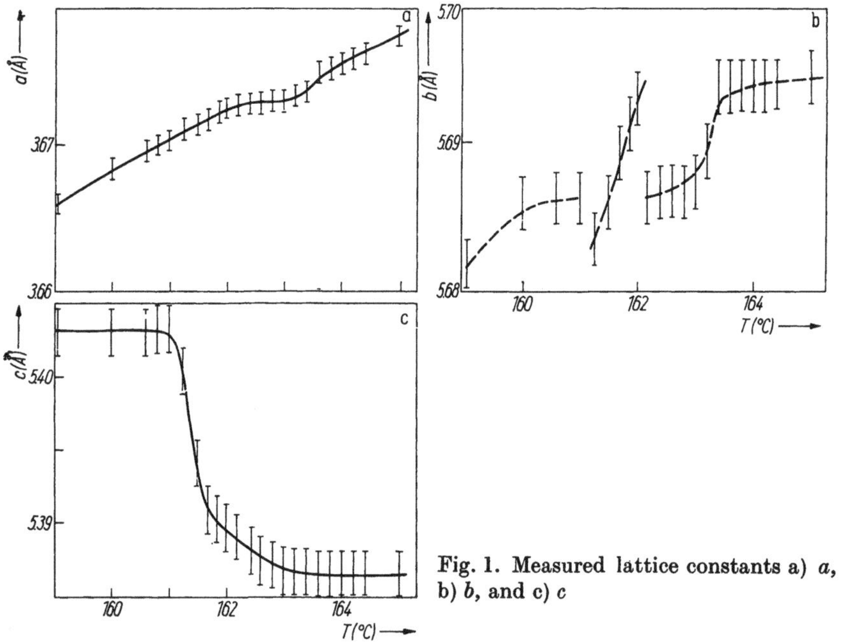

Fig. 1. Measured lattice constants a) a, b) b, and c) c

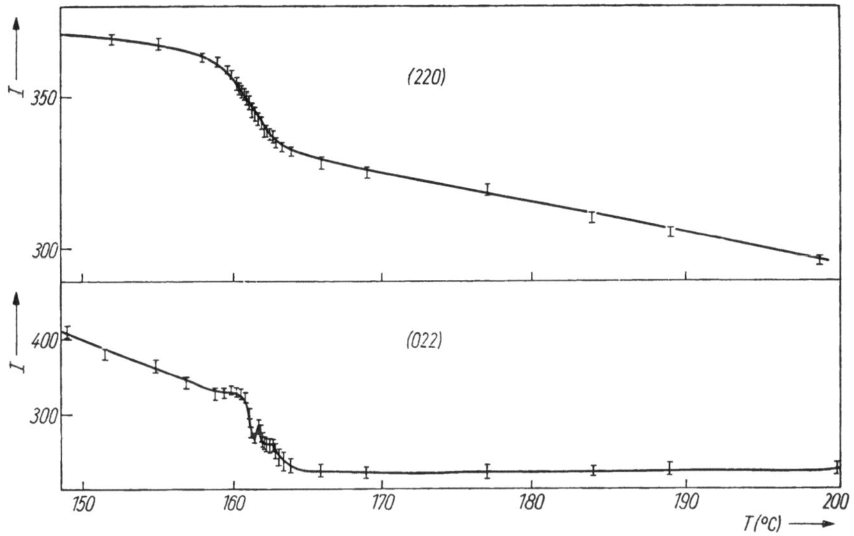

Fig. 2. Measured peak intensities vs. temperatures of (220) and (022) reflections for CuK_α radiation

The detailed changes of measured peak intensities, illustrated in Fig. 2 and 3, display a fine structure containing information on the transition mechanism. Reflection (101), with $B = 0$, showed only the ordinary thermal decay on the diffracted intensity. For the other reflections, phase transitions occurred within the range from 160.8 to 163 °C. The appearance of some kinks and dips is possibly due to the influence of satellite peaks, thermal diffused scattering, and temperature hysteresis effect.

The order parameter S of ferroelectric phase was obtained from the diffracted intensity measurement of (022), (123), (130), (042), and (222) reflections, of which $B \gg A$. The relation between S and T/T_0, with $T_0 = 163.4$ °C, is given in Fig. 4a. Detailed measurements in the temperature region close to T_0, shown in Fig. 4b, are seen to dis-

Table 1

Real A and imaginary B parts of the structure factors of the reflections involved of $NaNO_2$ at $T = 0$ °C

hkl	*A*	*B*
110	19.538	− 4.141
101	43.965	0.000
200	59.659	0.000
020	49.879	9.973
002	10.158	0.000
022	1.319	31.220
220	43.936	9.240
222	2.386	27.378
013	−31.311	6.673
130	2.810	−18.978
123	2.190	28.110
132	−33.382	9.393
042	−10.072	36.120

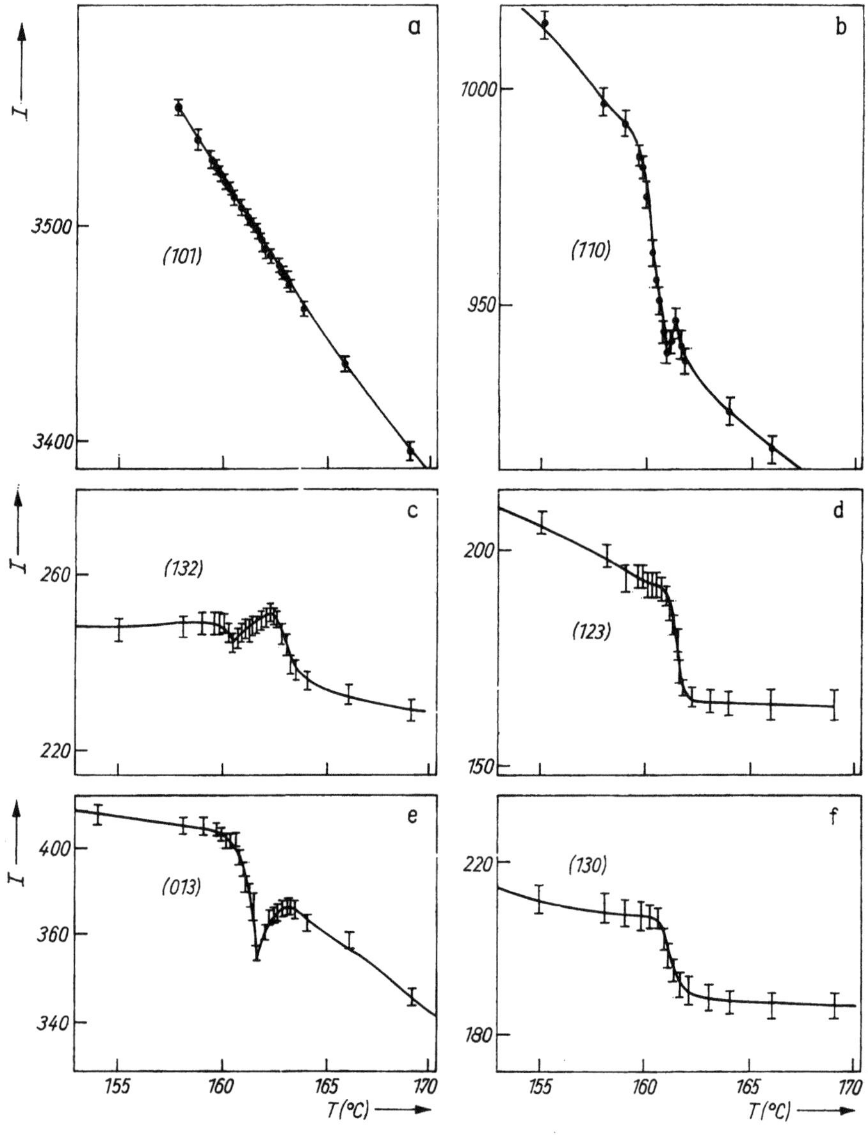

Fig. 3. Measured peak intensities vs. temperatures of a) (101), b) (110), c) (132), d) (123), e) (013), and f) (130) reflections for CuK_α radiation

play a fine structure; S decreased almost linearly with increasing temperature from 158 to 160.8 °C and then decreased rather rapidly up to about 162 °C. Above 162 °C a minimum was reached at about 163 °C and then S raised to a secondary maximum at about 163.4 °C and dropped afterwards quickly to zero at 164.4 °C. The dashed curve represents the contribution to the diffracted intensities from satellite peaks, which are probably responsible for the maximum at 163.4°. This curve resembles the intensity measurement obtained by Yamada et al. [12].

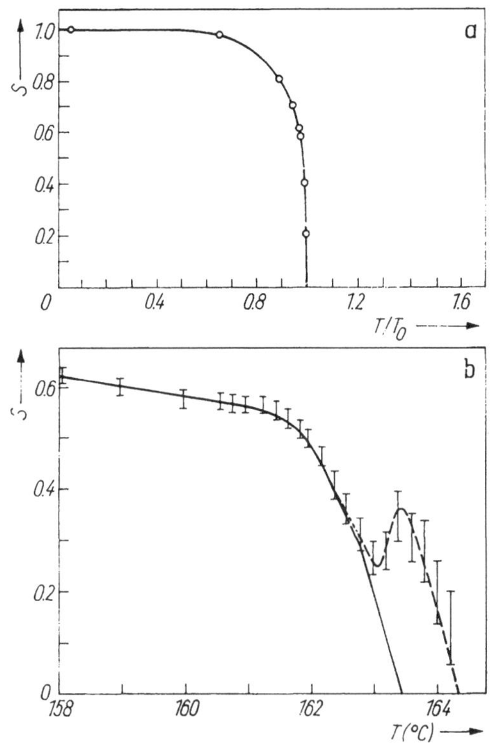

Fig. 4. Measured order parameter. a) S vs. T/T_0, b) S vs. T near the transition temperature

4. Conclusion

We have thus demonstrated that, using a commercially available powder diffractometer, one can observe the critical phenomena in $NaNO_2$, and certaintly in other crystals, for various reflections from a powder sample thus providing ample evidence of the complexity on the mechanism of the transformation. It is, of course, clear that this experimental set-up is not adequate to study the satellite intensity distribution since in the powder method the intensity measured corresponds to an annular surface in reciprocal space, i.e., one obtains the total intensity from this region and in general it is not likely that a small satellite peak included or overlapping this region will be detected by the usual methods. However, improvement on the resolution and evidence of this satellite peaks can possibly be achieved by using the powder profile fitting method [13].

Acknowledgements

The financial support from CNPq and FINEP are here gratefully acknowledged.

References

[1] Y. Takagi and K. Gesi, J. Phys. Soc. Japan **19**, 142 (1964).
[2] K. Hamano, J. Phys. Soc. Japan **19**, 945 (1964).
[3] S. Hoshino, J. Phys. Soc. Japan **19**, 140 (1964).
[4] M. Sakiyama, A. Kimoto, and S. Seki, J. Phys. Soc. Japan **20**, 2180 (1965).
[5] S. Hoshino and H. Motegi, Japan. J. appl. Phys. **6**, 708 (1967).

[6] S. TANISAKI, J. Phys. Soc. Japan **18**, 1181 (1963).
[7] M. CANUT and R. HOSEMANN, Acta cryst. **17**, 973 (1964).
[8] M. L. CANUT and J. MENDIOLA, phys. stat. sol. **5**, 313 (1964).
[9] Y. YAMADA and T. YAMADA, J. Phys. Soc. Japan **21**, 2167 (1966).
[10] N. MARUYAMA and S. SAWADA, J. Phys. Soc. Japan **20**, 811 (1965).
[11] J. R. CAUMO, R. LANDERS, and Z. P. ARGÜELLO, J. Crystal Growth **33**, 8 (1976).
[12] Y. YAMADA, I. SHIBUYA, and S. HOSHINO, J. Phys. Soc. Japan **18**, 1594 (1963).
[13] T. C. HUANG and W. PARRISH, Appl. Phys. Letters **27**, 123 (1975).

(Received May 21, 1981)

phys. stat. sol. (a) 68, 179 (1981)

Subject classification: 11; 1.1; 22.1.2

Institute of Nuclear Research, Świerk Research Establishment, Otwock[1])

X-Ray Diffraction Investigations of High Energy α-Particle Damage in Silicon

By

K. WIETESKA

Si single crystals irradiated by an inhomogeneous beam of α-particles are investigated by X-ray double crystal spectrometer and Lang camera methods. The interference fringe patterns on the topographs and subsidiary maxima for the low angle wing of the rocking curve are observed. A second system of fringes identified as a Pendellösung of equal inclination is observed on the topographs independently of the main fringes which are due to the bicrystal interference. It is found that the depth gradient of strain distribution depends mainly on the plane gradient of the α-particle density.

Mit der Zweikristall- und der Langschen Methode werden Si-Einkristalle untersucht, die mit einem inhomogenen Strahl von α-Teilchen bombardiert waren. Es werden Interferenzstreifen auf den Topogrammen als auch zusätzliche Maxima auf der Kleinwinkel-Flanke der Rockingkurve beobachtet. Unabhängig von den Bikristallinterferenzstreifen wird ein sekundäres System von Streifen beobachtet, das als Pendellösung gleicher Neigung identifiziert wird. Es wird gefunden, daß der Tiefenverteilungsgradient von Spannungen vor allem von dem lateralen Dichtegradienten der α-Teilchen abhängt.

1. Introduction

The aim of this study was to give a description of the damage created in the perfect crystal lattice of silicon during its bombardment by α-particles. Our investigation concerns the (111) surface of single crystals implanted by 4.1 to 5 MeV α-particles. The intensity of the beam was inhomogeneous with an average value of 2.5×10^{15} to 1×10^{16} cm^{-2}.

Previously Burgeat and Colella [1] reported that single crystals of silicon irradiated by a homogeneous beam of α-particles show a subsidiary peak in the double crystal rocking curve profile obtained for $(n, -n)$ setting. This peak was due to the increased lattice parameter in the thin layer shot through by α-particles. The results were in qualitative agreement with the dynamical theory of X-ray diffraction.

On the other hand, the investigations by Bonse et al. [2, 3] which were performed for silicon crystals implanted by an inhomogeneous beam of heavy ions showed that in the implantation process a bicrystal was created with a non-diffracting region between its two parts. Lang topographs obtained for such samples revealed characteristic interference fringes whose number N is defined by the following relation [2, 3]:

$$\frac{N}{\boldsymbol{h}\boldsymbol{z}} = \text{const.} \quad (1)$$

where $\boldsymbol{h}$ is the diffraction vector, $\boldsymbol{z}$ the unit vector normal to the bombarded area.

The intensity of the X-rays diffacted by such bicrystals with non-diffracting region is described by the following relations [2]:

[1]) 05-400 Otwock, Poland.

a) for the X-ray beam incident on the implanted side of the crystal (layer t)

$$I_0 = I \sim 1 + \sin(2\pi \boldsymbol{hc}) \sin(2\pi \Delta^{-1} t)\,, \tag{2}$$

b) for the X-ray beam incident on the opposite side (non-implanted) of the crystal

$$I_0' \sim 1 - \sin(2\pi \boldsymbol{hc}) \sin(2\pi \Delta^{-1} t)\,, \tag{3}$$

$$I_h' \sim 1 + \sin(2\pi \boldsymbol{hc}) \sin(2\pi \Delta^{-1} t)\,, \tag{4}$$

where I_0 intensity in the 0 direction, I_h intensity in the H direction, $\boldsymbol{c}$ thickness of the non-diffracting region, Δ extinction distance, t thickness of the layer shot through by α-particles.

2. Experimental Details

The plate-shaped samples with (111) surface were cut so that the angle of misorientation of each plate was not larger than 10 min of arc.

The circular area (7 mm in diameter) of each sample was implanted perpendicularly to its surface with α-particles using the cyclotron in the Institute of Nuclear Physics in Kraków. The dose of α-particles had its maximum in the central part of the beam. During the implantation process the sample was water-cooled. The time of the implantation process was about 4 h.

A double crystal spectrometer (DCS) of high precision and a Lang camera (L) were used for the X-ray topography. The DCS reflection topographs were taken with CuK_α radiation and the transmission DCS and L topographs with CuK_α and MoK_α radiations. In order to obtain a DCS topograph in an exactly defined position on the wing of the diffraction maximum a special mechanical device was used.

3. Results and Discussion

As we reported previously [4], the samples after implantation with a uniform beam of high energy α-particles show a uniform X-ray contrast for the bombarded area. The analysis of this contrast indicates that it is caused by an increase of the lattice parameter in the top layer in comparison to the lattice parameter of the bulk crystal. On the low-angle wing of rocking curves obtained for such samples one additional peak appeared. The Si single crystal samples investigated in the present study were bombarded by an inhomogeneous beam of α-particles. The interference fringes, similar to those obtained by Bonse et al. appeared on the DCS reflection topographs. The typical

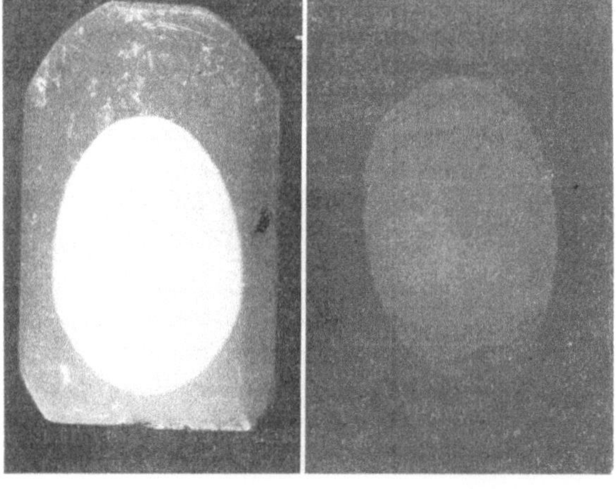

a b

Fig. 1. DCS topographs obtained for a crystal implanted by a) homogeneous and b) inhomogeneous beam of α-particles of similar dose and energy

X-ray contrasts obtained for samples bombarded by uniform and non-uniform beams of ions are presented in Fig. 1.

The symmetrical topographs taken for ($\bar{2}20$) and ($22\bar{4}$) maxima did not show any contrasts, but the asymmetrical topographs taken for the ($33\bar{3}$) and ($3\bar{1}1$) maxima in the directions H and 0 for both incident beams, on the bulk side and on the irradiated side, revealed interference fringes whose contrast changes in agreement with relations (2), (3), and (4), respectively. DCS Bragg topographs taken for different diffraction vectors $\boldsymbol{h}$ revealed that the number of interference fringes depends not only on the diffraction vector but also on the place on the rocking curve for which the topograph was taken. On the small-angle side of this curve the fringes are strong and more distinct but their number diminishes to zero when one proceeds in the direction to smaller angles θ on the rocking curve. For the peak and for the high-angle side of the peak the fringes are less distinct and their number reaches the maximum value and remains constant. Taking into account this maximum value of

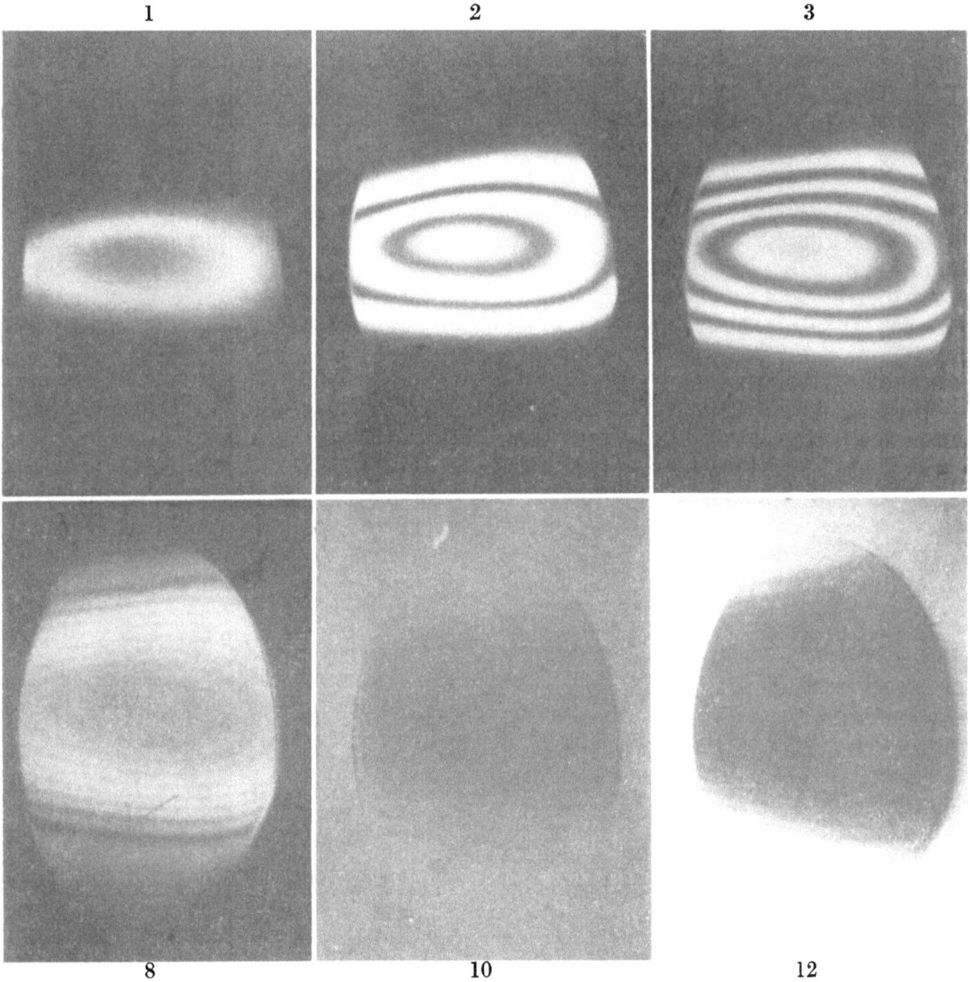

Fig. 2. Part of the series of topographs obtained for different positions on the rocking curve, symmetrical setting (511, −333), CuK_α radiation

fringes in the Bragg case and comparing it with the number of fringes obtained by Lang topography, relation (1) was confirmed. As an example, part of the series of topographs is presented in Fig. 2 and 3 shows the principle of the method. In Fig. 3a, the curve was measured using an X-ray beam diffracted from both implanted and non-implanted areas of the crystal. Accordingly, in Fig. 2 the topographs are denoted by proper numbers. The rocking curve taken for the implanted area (Fig. 3b) shows some additional maxima mainly on the low-angle side of the diffraction peak. The area (0.14 mm^2) reflecting the X-ray beam was approximately in the centre of the bombarded region. We can distinguish seven additional peaks on the low-angle side of the rocking curve. The lowest maximum is situated 35 s of arc from the main peak. The points corresponding to the topographs taken for curve a are marked on curve b.

It follows from the topographs that the number of fringes diminishes when one proceeds in the direction to smaller angles θ on the rocking curve. For a given spot on the implanted area we observe a sequence of minima and maxima of diffracted intensity brought about by a continuous change of the vector **c**, in other words, by a continuous change of the non-diffracting layer thickness. But the sample behaves like a dynamical monocrystal. Lattice damage caused by implantation reflects itself as an increase in the lattice parameter in the direction perpendicular to the crystal surface. This increase attains its maximum value when the penetration range of α-particles is reached. Then, for X-rays with angle of incidence close to θ_B no points on the dispersion surface are excited. Propagation of a wave field occurs if the angle of incidence is decreased. Clear and distinct interference fringes indicate that the damaged region is a perfect crystal. In this way the topographs should exhibit a number of fringes equal to that of subsidiary peaks occurring on the rocking curve taken for the spot implanted with maximum dose. In fact, such a conformity was observed. It seems that the diffraction situation in our case is similar to that observed by te Kaat and Schwuttke [5]. Going through the bulk crystal, the wave field penetrates the volume where the damage becomes increasingly stronger and adapts itself to the strain distribution. Consequently the wave points migrate on the dispersion surfaces. The radiation damage increases and the local gradient of deformation becomes so large that further propagation becomes impossible (point B in Fig. 4). Thus, the wave field is forced to separate into its components. Here, we can assume that the wave leaves the bulk region passing to the gap. After reaching point A the wave meets the same damage situation as in B. Now the wave field can reconstruct itself from the component waves. It will be able to propagate across the shot-through-layer.

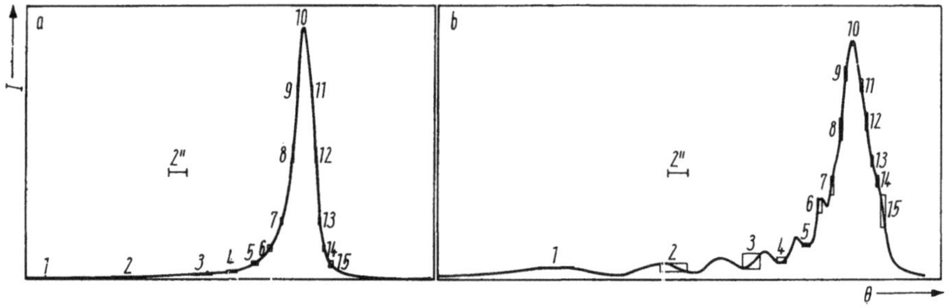

Fig. 3. Rocking curves observed for the same sample and for the same diffraction conditions as in Fig. 2. The numbers denote topographs from Fig. 2; a) curve measured with the X-ray beam diffracted both from implanted and non-implanted areas of the crystal, b) curve taken for the central part of the implanted area of the crystal

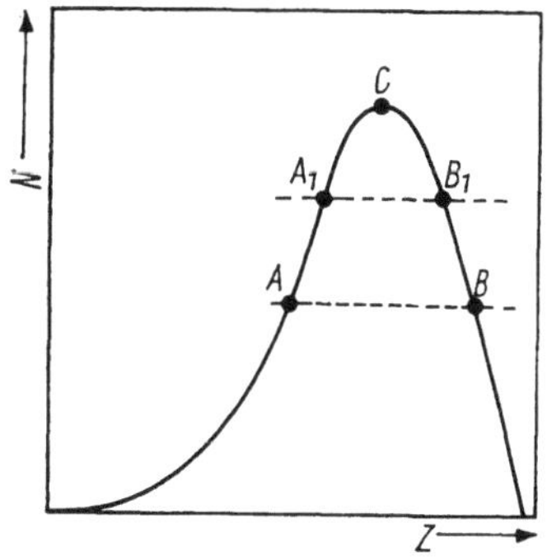

Fig. 4. Schematical drawing of the strain distribution for the implanted crystal. The point C corresponds the to range of ions (see text)

The wave fields excited in the bulk and in the thin layer are coherent and shifted in phase with respect to each other relative to the distance AB which is the vector **c**. Thus the situation is identical to the Bonse and Hart bicrystal model. On the topographs we obtain the interference fringes which are curves of equal ***hc***. From the above considerations it follows that for the Laue case the thickness of the non-diffracting layer will be maximum due to the transmission through the thick bulk crystal. For the Bragg case, a decrease of the angle of incidence of X-rays affects the diffraction in the top layer for smaller diffraction vector. We are able to obtain diffraction only near point C (Fig. 4) when we choose a proper angle θ and assume that the crystal is dynamical. This follows from Fig. 5. We can conclude that in the case of inhomogeneous implantation the strain fields strongly differ from the homogeneous implantation case and that extensive vertical gradients of lattice damage appear.

Sometimes one or more additional systems of fringes imposed on the standard bicrystal interference fringes are observed. A number of DCS topographs for different diffraction vectors and for different azimuths were obtained. Part of them was taken for (511, −333) setting and CuK_α radiation. They are displayed in Fig. 6. A similar phenomenon was observed by te Kaat and Schwuttke [5], but their fringes were more distinct than those found in the present study. However, on the presented topographs it is possible to observe a phase difference of approximately π between the secondary systems displayed in the topographs which were obtained at different points on the low-angle tail of the rocking curve. Such fringes appear due to bending of the top layer of the bombarded crystal. This are the so-called Pendellösung fringes of equal inclination. It was calculated from the interfringe distance that the radius of curvature ranges from 65 to 140 m.

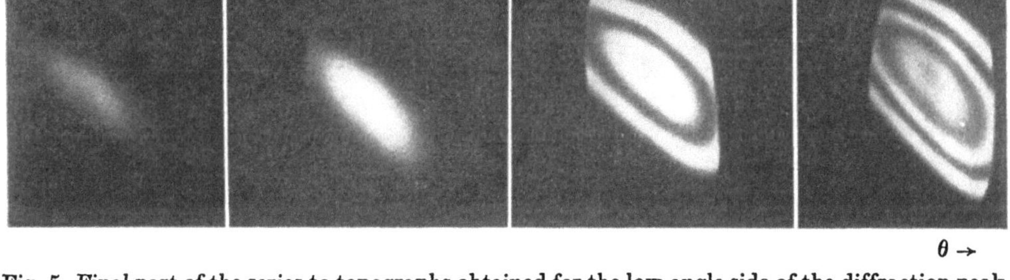

Fig. 5. Final part of the series to topographs obtained for the low-angle side of the diffraction peak. Asymmetrical arrangement (422, −422), CuK_α radiation

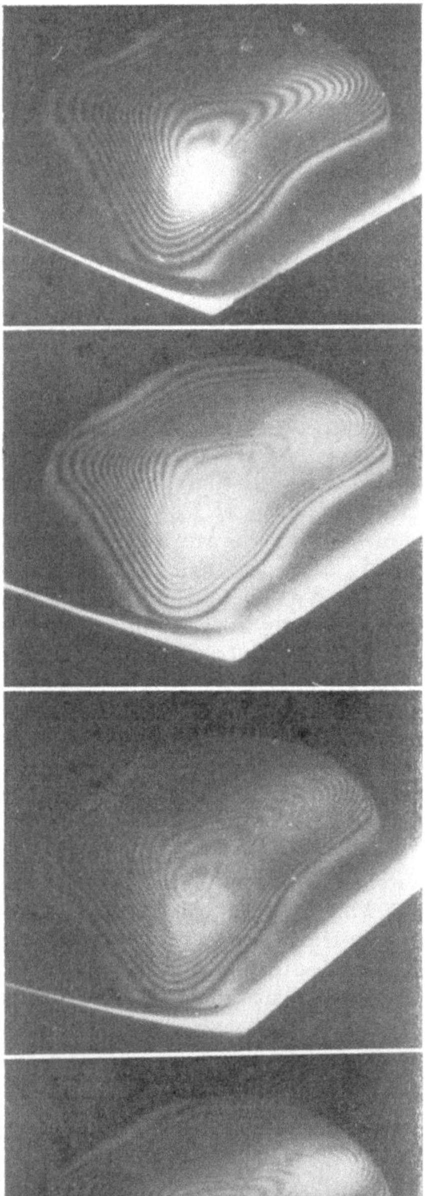

Fig. 6. Set of topographs showing the additional system of fringes. Note the phase difference π between the upper secondary systems

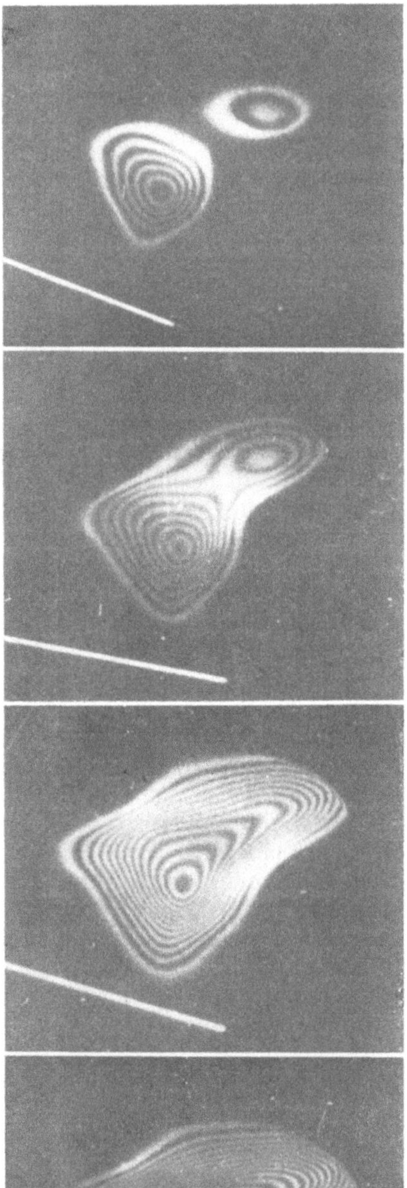

Fig. 7. Topographs taken for a crystal implanted by an asymmetrical beam of α-particles. The decomposition of a set of interference fringes between two systems is visible

At was mentioned above, the strain distribution in the crystal strongly depends on the dose distribution in the plane. Assuming that this strain distribution depends more on the plane gradient of the dose than on the magnitude of the dose one can expect to observe essential differences in the speeds of disappearance of the fringes with diminishing angle θ for different parts of the sample implanted by a beam with a strongly asymmetrical distribution of the ion density. Such an effect was observed and shown in Fig. 7. On topograph a) we see that the crystal was really implanted by an asymmetrical beam of α-particles. We observe on topograph b) that the number of fringes decreases from 15 to 12. The appearance of the second centre is visible on topograph c). This effect is caused by a decrease of the effective non-diffracting layer thickness occurring with different speeds when one proceeds towards smaller angles θ. Finally (Fig. 7d) two centres of the interference fringes separate. By comparing the four topographs displayed in Fig. 7 we can conclude that the two parts of the crystal implanted by the same dose but exhibiting different density gradients, show different vertical strain gradients.

Acknowledgements

The author wishes to thank Prof. J. Leciejewicz for his interest in this work and critical reading of the manuscript. Thanks are due also to Dr. Z. Szmid and Dr. S. Szarras for their support and useful discussions and to Mr. J. Krzewski for his technical assistance in the experiment.

References

[1] J. Burgeat and R. Colella, J. appl. Phys. **40**, 3505 (1969).
[2] U. Bonse and M. Hart, phys. stat. sol. **33**, 351 (1969).
[3] U. Bonse, M. Hart, and G. H. Schwuttke, phys. stat. sol. **33**, 361 (1969).
[4] S. Szarras, Z. Szmid, and K. Wieteska, Internat. Conf. Ion Implantation in Semiconductors Lublin 1974 (p. 281).
[5] E. H. TeKaat and G. H. Schwuttke, in: Advances in X-Ray Analysis, Ed. K. F. J. Heinrich, Vol. 15, Plenum Press, New York 1972 (p. 504).

(Received Juli 16, 1981)

phys. stat. sol. (a) 68, 187 (1981)

Subject classification: 11; 21; 22.1; 22.9

Institute of Nuclear Physics, Moscow State University[1])

Energy Straggling of Hydrogen and Helium Ions in Al, C, and Polystyrene at Energies of Tens and Hundreds keV per Nucleon

By

A. A. BEDNYAKOV, YU. V. BULGAKOV, V. S. NIKOLAEV, and V. L. CHERNOV

Energy loss and straggling measurements in thin aluminum, carbon, and polystyrene films are performed for H- and He-ions in the energy intervals ≈ 70 to 500 and ≈ 100 to 1200 keV, respectively. It is shown that the observed spread of the specific straggling experimental values for a given material may be significantly reduced by introducing corrections on nonuniformity of target thickness. Analysis of the results based on existing theories indicate that "charge-exchange" and "correlative" mechanisms of straggling play an important and even dominant (for He-ions) role at the energies near maximum of stopping power.

Energieverlust- und Straggling-Messungen werden an dünnen Aluminium-, Kohlenstoff- und Polysterenschichten mit H- und He-Ionen im Energieintervall ≈ 70 bis 500 bzw. ≈ 100 bis 1200 keV durchgeführt. Es wird gezeigt, daß die beobachtete Streuung der spezifischen experimentellen Stragglingwerte für ein gegebenes Material durch die Einführung von Korrekturen bezüglich der Nichtgleichförmigkeit der Targetdicken beträchtlich reduziert werden kann. Eine Analyse der Ergebnisse auf der Grundlage existierender Theorien zeigt, daß „Ladungsaustausch" und „Korrelationsmechanismen" des Straggling eine wesentliche und sogar dominierende Rolle (für He-Ionen) bei Energien in der Nähe des Maximums der Bremsleistung spielen.

1. Introduction

The wide application of ion beams with energies of tens and hundreds keV to study and alter the structure and properties of the surface layers of solids required a deeper and more exact knowledge of the process of stopping such particles in matter. It turns out important to know then not only the average energy loss $\overline{\Delta E}$, but also their straggling. Accordingly, there have recently been performed several investigations of the straggling of light ions in gaseous [1, 2] and solid [3 to 12] targets and attempts have been made to analyze the phenomenon more deeply [1, 13 to 15]. Some of these studies [1, 13] are devoted to a refinement of the traditional theory of straggling based on the model of independent binary collisions of ions with electrons of the stopping medium [16], other deal with the influence on straggling of additional mechanisms, viz., the charge fluctuations of the stopped ions [7, 14] and the correlation effect caused by the simultaneous interaction of the ion with more than one electron of the medium [15]. The results of the measurements [5, 7 to 10] indicate that these mechanisms contribute significantly to the straggling of light ions in the energy region corresponding to the maximum of the curve of specific energy losses. Yet, for a reliable comparison with the results of the theoretical estimates, the available experimental data are insufficient.

---

[1]) 117234 Moscow, USSR.

The present paper is an extension of the experimental study of the straggling of H and He ions with energies of tens and hundreds of keV per nucleon in solids, previous results having been published elsewhere [7, 9].

2. Experimental

The measurements were made using ions of hydrogen (H^+) and helium (He^+, He^{++}) in the energy regions from ≈ 70 to 500 and from 100 to 1200 keV, respectively. The ion beam was directed toward the target perpendicularly to its surface, the angular divergence of the beam being about 5×10^{-4} rad and the energy spread not greater than 0.5 keV. In order to ensure that the beam always hit the target at the same place, the last diaphragm of the collimator, which was ≈ 0.5 mm in diameter, was mounted directly on the casing of the target, about 0.5 mm above its surface. The particles that passed through the target were detected using a silicon surface-barrier detector and the standard spectrometric apparatus with a multichannel amplitude analyzer. The overall resolution of the spectrometer was 7 to 8 keV for protons and 10 to 13 keV for He ions. To minimize the contribution of elastic collisions to the energy straggling, particles were detected that had been scattered at angles not exceeding 0.5°.

As targets, free films of polystyrene $(C_8H_8)_n$, carbon and aluminum were used which were from ≈ 0.1 to 0.8 μm thick. Polystyrene films were produced from the polymer solution in dichlorethane on the surface of polished glass and were separated from it in distilled water. The remaining films were produced in vacuum by arc (carbon) or heat (aluminum) evaporation upon a thin layer of sodium chlorine, which had also been vacuum evaporated onto the surface of polished glass. The separation of the films from their substrate was also carried out in distilled water. For the experiment we selected films with "mirror" surfaces, without apparent defects when examined in a light microscope having a magnification of 56 ×. An analysis of the metallic films by electron microscopic methods has failed to reveal in them any noticeably preferred orientation of crystalline blocks and has shown that the penetration depth of surface roughness elements is several tens of angstroms.

The target thickness t was determined by comparing the average energy loss $\overline{\Delta E}$ of protons measured in the region from ≈ 300 to 500 keV with the data on the corresponding stopping powers dE/dt [17 to 19]. The error in the values of t thus obtained was 1 to 2% (disregarding the errors of measurement of dE/dt in the references cited).

In our experiment, the criteria of the "middle thickness target", viz. $\overline{\Delta E} \gg T_{max} = 4(m_e/M_1)\, E_0$ and $\Omega \ll E_0 - \overline{\Delta E}$, where $\overline{\Delta E}$ and Ω are the average values of energy loss and its straggling, m_e and M_1 the masses of the electron and ion, respectively, were fulfilled. In these conditions, the energy distributions of ions $I(E)$ are Gaussian [16]: $I = I_{max} \exp[-(E - E_m)^2/2\Omega^2]$. Indeed, the experimental distributions practically coincided with Gaussians in shape, a small "tail" on the side of the lowest energies being observed only in the spectra of helium ions of the lowest energy at $I \leqq 0.1 I_{max}$ thus indicating an increase in the contribution from elastic collisions. No distortions have been observed in the shape of the high-energy side ($E < E_m$) of the spectra, which might indicate the presence of channeled particles.

The processing of the experimental spectra consisted in their computer-aided approximation by the least-squares technique using the Gaussian functions (the "tailed" spectra were approximated in the main part $I \geqq 0.1 I_{max}$) and in the determination of the quantities E_m and Ω^2 in this way. The values of the energy loss of

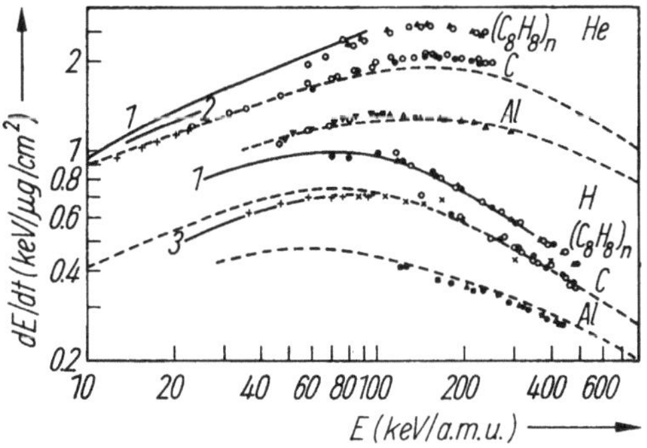

Fig. 1. Specific energy losses of protons and helium ions in carbon, polystyrene $(C_8H_8)_n$ and aluminum. Solid curves and symbols give experimental results: (1) [17]; (2) and (3) data of [11], 12], respectively; ×, + [10]; ●, ▼, ▲, ○, ■ present measurements with various targets; dashed curves calculated results of [20]

ions in the target $\overline{\Delta E}$ and its dispersion Ω were found as follows:

$$\overline{\Delta E}_{\text{exp}} = E_{\text{m0}} - E_{\text{m1}} \quad \text{and} \quad \Omega^2_{\text{exp}} = \Omega^2_1 - \Omega^2_0 \,, \tag{1}$$

where indices 0 and 1 refer to the spectra obtained with the primary beam and with the transmitted beam, respectively. The errors in the values of $\overline{\Delta E}_{\text{exp}}$ and Ω^2_{exp} were mainly determined by the statistical accuracy of measurements and by the relative values of the differences $E_{\text{m0}} - E_{\text{m1}}$ and $\Omega^2_1 - \Omega^2_0$. In the case of energy losses the errors were from ≈ 0.5 to (2 to 3)%, in the case of straggling from (4 to 5) to (15 to 20)%. The smallest error corresponds to the case of high-energy helium ions in the thickest targets, the greatest to that of low-energy protons in thin targets.

The experimental values of $\varepsilon \equiv dE/dt$ and $D = d\Omega^2/dt$ were calculated as follows:

$$\varepsilon_{\text{exp}}/_{E=\bar{E}} = \overline{\Delta E}_{\text{exp}}/t \quad \text{and} \quad D_{\text{exp}}/_{E=\bar{E}} = \Omega^2_{\text{exp}}/t \,, \tag{2}$$

where $E = E_0 - \Delta E_{\text{exp}}/2$. For the thickest targets, the additional error thus introduced, which is associated with the finite thickness of the target, was 1 to 2%.

3. Results and Discussion

Fig. 1 presents the results of measurements of the specific energy losses, ε_{exp}. As can be seen, the agreement with the experimental data [17 to 19] and with the results of semiempirical calculations [20] is on the whole reasonably good. The deviations from the latter exceed 5% only for helium ions in carbon at energies > 400 keV, reaching +9% in the region from 700 to 1000 keV. For helium ions in polystyrene, the values of ε_{exp} in the overlapping energy region (from 250 to 300 keV) turn out to be somewhat lower (by 7 to 5%) than those of [17], although for protons in polystyrene the agreement with the results obtained therein is very good (see curves 1 in Fig. 1).

The results of the measurements of the specific straggling of the energy losses D_{exp} are shown in Fig. 2. One can notice that for polystyrene $(C_8H_8)_n$ there is a good agreement with the results obtained previously [9] in the overlapping ion-energy regions. In the cases of carbon and aluminum, the dependences $D_{\text{exp}}(E)$ obtained for various targets made of the same material are similar, but the absolute values of D_{exp} can, at a given energy, be different by a factor of 1.5 to 2 (see, e.g., the data for different carbon targets presented in Fig. 2c). The lowest spread in the values of D_{exp} was observed for aluminum targets. This spread was in all appearance, due to different degrees of uniformity of the various targets in thickness, so that the lowest values of D_{exp} should correspond to the "smoothest" targets. Fig. 2b illustrates the depend-

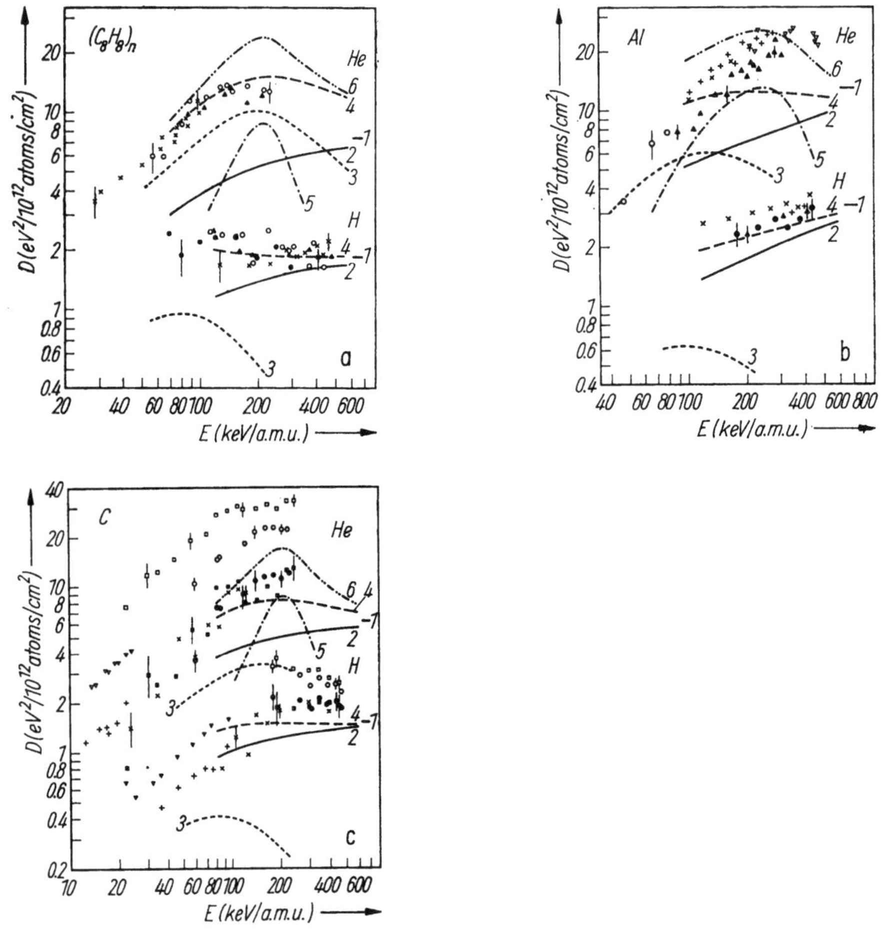

Fig. 2. Energy straggling of protons and helium ions a) in polystyrene, b) in aluminum, and c) in carbon. The results of measurements: a) ●, ○, ▲ with targets of thicknesses of 28.7, 31.5, and 42.8 μg/cm², respectively; × data of [9]. The calculated results: (1) D_B (by the Bohr formula [16]); (2) D_C (the results of Chu [13]); (3) D_S (the correlation correction of Sigmund); (4) $D_C + D_S$; (5) D_V (calculated according to the Vollmer theory); (6) $D_C + D_S + D_V$. b) ●, ○, ▲ with targets thickness of 129, 230, and 293 μg/cm², respectively; ×, + data of [9]; ▽ [5]. c) ○, □ with targets of thicknesses t of 71 and 27.5 μg/cm², respectively; ●, ■ the same results including the corrections calculated by formula (3) (τ = 2.80 and 2.65 μg/cm², respectively); ×, + data of [10]; ▽ data of [11]; ▼ data of [12]. The calculated results in b) and c) are as in Fig. 2a

ences $D_{exp}(E)$ for such targets made of aluminum. Their curves are lower than those obtained previously [9] by ≈ 20% for protons and ≈ 40% for helium ions. Within the limits of the spread of experimental values D_{exp} our results for helium ions in Al agree with the results of measurements obtained in the energy region between ≈ 800 and 1900 keV [5].

The $D_{exp}(E)$ dependences for the various carbon targets lie above the values obtained previously [10], using targets prepared at the Humboldt University of Berlin. The Berlin group has also reported recently [11, 12] values of energy-loss straggling in carbon overrated with regard to those of [10]. The results of these measurements can be considered to agree if it is assumed that the reason for the difference resides in the uncontrollable thickness nonuniformity of the targets. Assuming that the deviations of a local target thickness from the mean value t obey the Gaussian distribution with dispersion τ^2, one can readily show that the "true" straggling of energy losses Ω^2_{true} is given by

$$\Omega^2_{\text{true}} = \Omega^2_{\text{exp}} - (\bar{\varepsilon}\tau)^2\,, \tag{3}$$

where $\bar{\varepsilon} = \varepsilon(E)$. The value of τ^2 for each target was found from the condition that the data of [10] for protons with energies from 400 to 500 keV, in carbon were "true" $(D_{\text{true}} = \Omega^2_{\text{true}}/t = 1.9\ \text{eV}^2/(10^{12}\ \text{atoms/cm}^2))$. The data on the specific energy loss of protons in carbon were taken from [20], for helium ions we used the values of $\varepsilon_{\text{exp}}(E)$ obtained in the present study (see Fig. 1). The corrected values of energy-loss straggling are given in Fig. 2c as dark symbols. It can be seen that they agree farily well with the data of [10] within the overlapping energy regions. A significant spread of D_{true} values for helium ions results from the fact that the correction term in (3) turns out to be comparable in magnitude with Ω^2_{exp}. Accordingly, the results become highly sensitive to the errors in the values of Ω_{exp}, $\bar{\varepsilon}$, and τ, so that the errors in the determination of D_{true} increase to 15 or 20%.

Fig. 2 a to c compare the experimental values of D_{exp} with the results of calculations based on the currently adopted theory. Curves 1 and 2 correspond to the model of independent binary collisions of ions with free electrons [16] and with the electrons forming the Fermi liquid [13]. The correlative corrections (curves 3) were calculated by the formula [15]

$$D_{\text{S}} = \xi\varepsilon_{\text{e}}^2/4\pi Z_2 a_2^2\,, \tag{4}$$

where $a_2 = 0.885 a_0 Z_2^{-1/3}$ is the Thomas-Fermi screening radius ($a_0 = 0.529 \times \times 10^{-8}$ cm), ε_{e} the "electronic" stopping power of the material, ξ a numerical factor close to unity (in calculations it has been assumed that $\xi = 1$).

The contribution from the mechanism associated with the charge fluctuations of ions moving through the medium to the total energy-loss straggling was estimated by the formula [14]

$$D_{\text{V}} = \left[\frac{(q_{\text{a}}^2 - q_{\text{b}}^2)\,\omega(A)}{(q_{\text{a}}^2 - q_{\text{b}}^2)\,\bar{A} + q_{\text{b}}t}\right]^2 \varepsilon_{\text{e}}^2\,t\,, \tag{5}$$

in which q_{a} and q_{b} are the prevailing values of ionic charge when $E = \bar{E}$; $\bar{A}$ and $\omega(A)$ are the average total path travelled by an ion in a charge state "a" and the fluctuation of this quantity over a path segmente qual to the target thickness t. The values of A/t and ω/A were taken from [14].

Since there are no data on the cross-sections for loss and capture of electrons by fast ions in solids, we used in the calculations such data for light ions in gases, whose atomic numbers were closest to those of the target materials, viz. nitrogen for carbon and polystyrene targets and neon for aluminum targets. Curves 5 correspond to this mechanism.

4. Conclusions

The comparison between the results of the calculations and measurements enables one to draw the following conclusions. In the case of sufficiently uniform targets, the theory of "single-electron straggling" (curves 1 and 2 in Fig. 2) describes reasonably well the energy straggling of protons with energies ≥ 250 keV. At lower energies, the contribution of the correlation mechanism noticeably increases and, taking it into account, one considerably improves the agreement between theory and experiment (curves 4 in Fig. 2).

In the case of helium ions, the experimental values of straggling in the region of the maximum of the energy-loss curve exceed by a factor of two or three the corresponding values calculated using the one-electron theory. The estimates show that in this energy region the correlation and charge-exchange mechanisms may play a significant, and even dominant, role (curves 3 and 5 in Fig. 2). Taking them into account considerably improves the agreement between theory and experiment (curves 4 and 6 in the same figures). Formulae (4) and (5) which have been used to estimate the contribution from the above mechanisms, appear to overestimate somewhat the energy-loss straggling values, but the tentative nature of our calculations prevents us from concluding with certainty to which degree the calculated results and experimental data agree quantitatively.

References

[1] E. BONDERUP and P. HVELPLUND, Phys. Rev. A **4**, 562 (1971).
[2] P. HVELPLUND, Kong. Danske Vid. Selsk., mat.-fys. Medd. **38**, No. 4 (1971).
[3] E. LEMINEN and A. ANTTILA, Ann. Acad. Sci. Fenn. **AVI**, 370 (1971).
[4] J. M. HARRIS, W. K. CHU, M.-A. NICOLET, Thin Solid. Films **19**, 259 (1973).
[5] J. M. HARRIS, M.-A. NICOLET, Phys. Rev. B **11**, 1013 (1975).
[6] A. A. BEDNYAKOV, YU. V. BULGAKOV, V. S. NIKOLAEV, V. P. SOBAKIN, B. M. POPOV, Zh. eksper. teor. Fiz. **68**, 2067 (1975).
[7] A. A. BEDNYAKOV, YU. V. BULGAKOV, V. S. NIKOLAEV, V. P. SOBAKIN, and V. L. CHERNOV Proc. VII. All-Union Conf. Phys. Interactions of Charged Particles with Single Crystalls Izd. MGU, Moscow 1976 (p. 290) (in Russian).
[8] G. E. GOFFMAN and D. POWERS, Phys. Rev. A **13**, 2042 (1976).
[9] A. A. BEDNYAKOV, YU. V. BULGAKOV, V. S. NIKOLAEV, V. P. SOBAKIN, and V. L. CHERNOV, Phys. Letters A **62**, 183 (1977).
[10] A. A. BEDNYAKOV, F. BERNHARD, YU. V. BULGAKOV, V. L. CHERNOV, and U. MÜLLER-JAHREIS, phys. stat. sol. (b) **84**, K79 (1977).
[11] G. KREYSCH and U. MÜLLER-JAHREIS, Radiat. Eff. **31**, 101 (1977).
[12] D. KALZ, G. KREYSCH, and U. MÜLLER-JAHREIS, Radiat. Eff. **36**, 119 (1978).
[13] W. K. CHU, Phys. Rev. A **13**, 2057 (1976).
[14] O. VOLLMER, Nuclear Instrum. and Methods **121**, 373 (1974).
[15] P. SIGMUND, Phys. Rev. A **14**, 996 (1976); Prelim. Progr. and Abstr. of Papers VII. ICACS, Moscow 1977 (p. 232).
[16] N. BOHR, Kong. Danske Vid. Selsk., mat.-fys. Medd. **24**, No. 19 (1948).
[17] D. C. LORENTZ and E. J. ZIMMERMAN, Phys. Rev. **113**, 1199 (1959).
[18] S. GORODETSKY, A. CHEVALLIER, A. PAPE, J. U. SENS, A. M. BERGDOLT, M. BRES, and R. ARMBRUSTER Nuclear Phys. **A91**, 133 (1967).
[19] D. KAHN, Phys. Rev. **90**, 503 (1953).
[20] H. H. ANDERSON (Ed.), Stopping Powers and Ranges of Ions in Matter, Vol. 3 (Protons) and Vol. 4 (He Ions), Pergamon Press, New York 1977.

(Received February 17, 1981)

phys. stat. sol. (a) 68, 193 (1981)

Subject classification: 1.2; 21.4

Institute of High Pressure Physics, Academy of Sciences of the USSR, Troitsk[1])

Polymorphism of Cerium under Hydrostatic Pressure up to 8.0 GPa

By

L. G. KHVOSTANTSEV and N. A. NIKOLAEV

The phase diagram of cerium is proposed at hydrostatic pressure up to 8 GPa and temperature from 300 to 600 K. The high temperature β′-phase is found to exist above 420 K and the unknown γ′-phase should exist at more elevated temperatures.

Предложена фазовая диаграмма церия при гидростатическом давлении до 8 GPa и температуре от 300 до 600 K. Обнаружено, что при температуре выше 420 K существует β′-фаза и при более высоких температурах должна существовать неизвестная γ′-фаза.

1. Introduction

The data of the boundaries between the high pressure phases of cerium obtained under non-hydrostatic pressure conditions are contradictory [1 to 5].

The transition between α- and α′-phases was supposed to be connected with a constant increase of the electrical resistance R/R_0 by 6% at room temperature and 5.4 GPa and at more elevated temperatures up to 473 K and low pressures [1]. The pressure hysteresis of the $\alpha \leftrightarrows \alpha'$ transition is 1.3 GPa at room temperature. The slope of the equilibrium line between α- and α′-phases has negative sign. This straight line was supposed extend to the metling curve of cerium.

The sample of cerium contains a fraction of the α″-phase when the α-phase transforms to the α′-phase under non-hydrostatic conditions above 5.1 GPa and room temperature [2, 3]. The α″-phase disappears completely by pressure cycling. Contrary to [1] the reverse transition of the high pressure phase to the α-phase was fixed at 2.7 GPa. The authors supposed the high pressure phase to be the α′-phase.

The authors [4] have found a tetragonal phase at pressures above 12 GPa and proposed a phase diagram of cerium with two straight equilibrium lines between α, α′ and a tetragonal phases which remains up to the metling curve of cerium.

In [5] an unusual phase diagram of cerium with bent equilibrium line between α- and α′-phases is presented.

Our work reports new data on the phase diagram of cerium obtained by studying the behaviour of kinetic coefficients under hydrostatic pressure conditions.

2. Experiment

The method of measuring the kinetic coefficients was described earlier [6]. In the present case the ampoule was filled with a mixture of methyl and ethyl alcohol (4:1), solidifying approximately at 10 GPa [7]. The pressure was measured by a manganine gauge calibrated by bismuth upstroke and downstroke transitions at room temperature. The cerium sample was heated by cylindrical and/or two flat heaters placed on

[1]) 142092 Troitsk, Moscow region, USSR.

the sample ends. The electrical resistance of the cerium sample and manganine gauge was measured with a four-probe technique. Potential leads were welded on the side surface of the sample. All copper leads were 0.15 mm in diameter.

A modification of the toroid-type device for high pressure generation [8] creates hydrostatic pressures up to 8.0 GPa in an ampoule of 0.5 cm³ size. The cerium samples with purity of 99.93% had a length of 7 mm and a cross-section of 1.5×1.5 mm².

3. Results and Discussion

The new phase diagram of cerium is presented in Fig. 1. Dashed-dotted lines represent the data of [4, 5].

The equilibrium line between α- and α'-phases has a positive slope contrary to [1] and extends to the triple point ($P = 6.4$ GPa, 420 K) contrary to [5].

The lines of direct and reverse transitions between α- and α'-phases were obtained by distinctive changes of the relative electrical resistance R/R_0 on the isoterms $R/R_0(P)$ (Fig. 2, 3). The direct $\alpha \rightarrow \alpha'$ transition is accompanied by an R/R_0 increase of 7.9% and by a 2.5 time increase of the thermo-e.m.f. in the pressure range from 5.0 to 6.5 GPa (Fig. 2 and 4). These values exhibit the same changes at the reverse $\alpha' \rightarrow \alpha$ transition in the pressure range from 6.1 to 4.7 GPa. The R/R_0 and thermal-e.m.f. curve exhibit a hysteresis loop at the $\alpha \rightleftarrows \alpha'$ transition, the pressure of the onset of the reverse $\alpha' \rightarrow \alpha$ transition being higher than the pressure of the onset of the direct $\alpha \rightarrow \alpha'$ transition. The pressure cycling does not change the shape of the hysteresis loop (Fig. 5a, b). The hysteresis loop keeps its shape at high temperatures up to 445 K (Fig. 3). These data confirm conclusively that such character of direct and reverse transitions between α- and α'-phases is caused by the nature of the $\alpha \rightleftarrows \alpha'$ transition. The domain on the P–T diagram restricted by lines of direct and reverse $\alpha \rightleftarrows \alpha'$ transitions is the domain of mixture of α- and α'-phases.

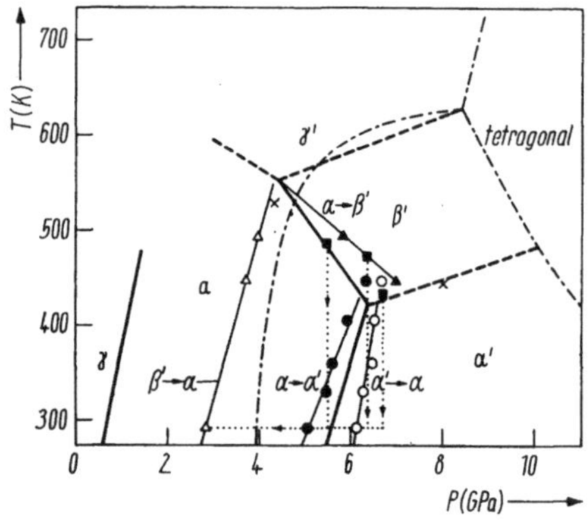

Fig. 1

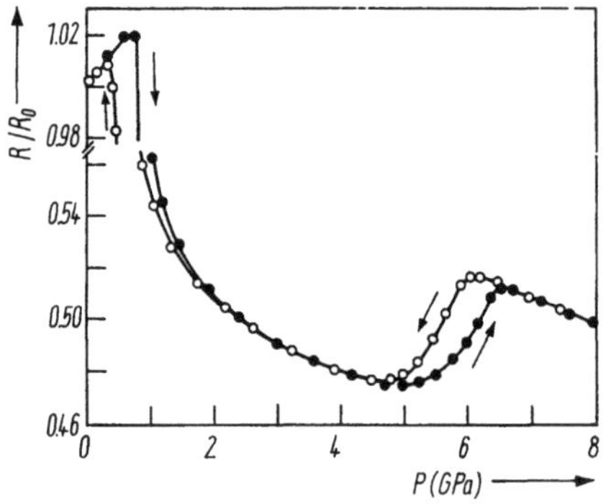

Fig. 2

Fig. 1. P–T phase diagram of cerium: ●, ○ locations of the onset of changes in the relative resistance in the isothermal dependences $R/R_0(P)$ at direct (●) and reverse (○) transitions between the α-phase and the α'-phase; ▲, △ direct and reverse transitions between the α-phase and the β'phase, respectively; ■ the β'-phase obtained at isobaric heating of cerium samples; × pure α-phase and pure α'-phase obtained at isobaric heating. Dash-dotted lines are the equilibrium lines in the P–T diagram proposed in [5]. Dash lines are boundary lines given approximately

Fig. 2. Hydrostatic pressure dependence of the relative resistance of cerium at increasing (●) and decreasing (○) pressure at room temperature

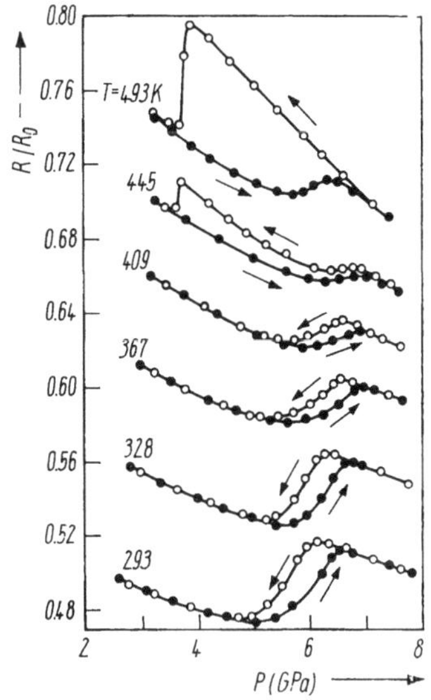

Fig. 3. Hydrostatic isothermal pressure dependences of the relative resistance of cerium at increasing (●) and decreasing (○) pressure

The pressure of the onset of direct and reverse transitions between α- and α′-phases of cerium increases with temperature up to 445 K (Fig. 3). The increase and decrease of R/R_0 diminish with temperature and they are equal to 0.45% at 445 K contrary to [1].

The shape of the hysteresis loop formed by the $R/R_0(P)$ curves at the $\alpha \rightleftarrows \beta'$ transition differs completely from the loop at the $\alpha \leftrightarrows \alpha'$ transition. The resistance increases by 1.2% at the direct the $\alpha \to \beta'$ transition at 5.8 GPa and 493 K and drops sharply by 7.5% at the reverse $\beta' \to \alpha$ transition at 3.9 GPa and 493 K (Fig. 3).

The isotherm of R/R_0 at 445 K is close to the triple point and the cerium sample consists of a mixture of α′- and β′-phases above 6.4 GPa. Therefore R/R_0 exhibits a first drop at the reverse $\alpha' \to \alpha$ transition at 5.8 GPa and a second sharp drop at the reverse $\beta' \to \alpha$ transition at 3.7 GPa. The $\alpha \to \beta'$ transition was obtained also at 472 K by isobaric heating of the sample at 5.4 GPa, being accompanied by a 4.1% increase of R/R_0 (Fig. 6).

The isobar of R/R_0 obtained at cooling of the sample from 490 to 293 K at 5.4 GPa does not exhibit a reverse drop and R/R_0 has a higher value at 293 K in comparison with the initial value of the α-phase at 5.4 GPa and 293 K. Hence the β′-phase appears to be metastable at 293 K and transforms to the α-phase only at 2.8 GPa (see dotted lines in Fig. 1), the drop of R/R_0 being by 28% (see $R/R_0(P)$ curve, symbols □ in Fig. 7). The availability of a metastabile state of the β′-phase at 293 K and pressure down to 2.8 GPa can be used to obtain the line of direct $\alpha' \to \beta'$ transition.

A special experiment with complete and incomplete transition of α′-phase to β′-phase in the region of the triple point was performed by isobaric heating of the α′-phase of

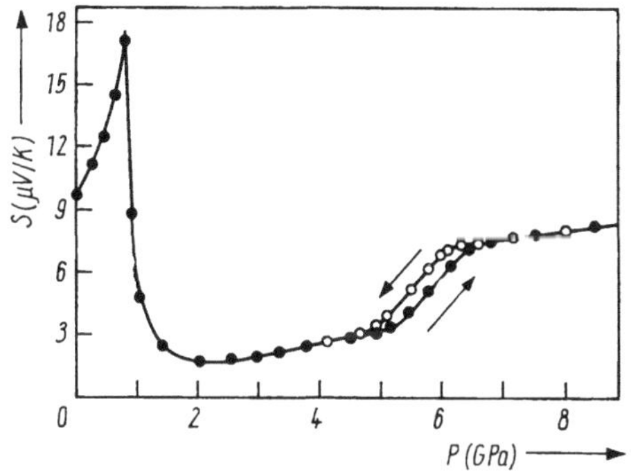

Fig. 4. Hydrostatic pressure dependence of the thermo-e.m.f. of cerium at increasing (●) and decreasing (○) pressure at room temperature

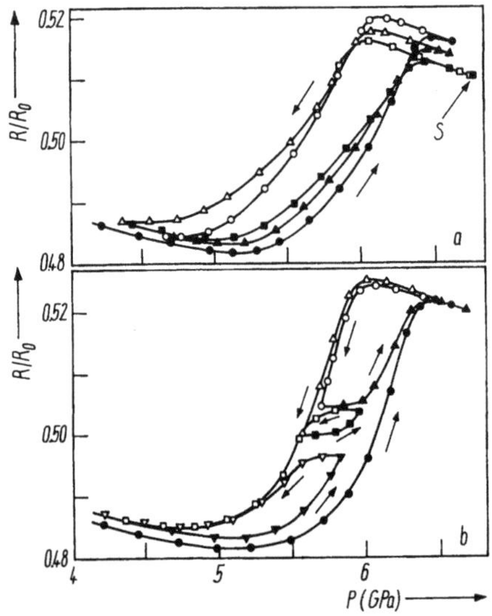

Fig. 5. Hydrostatic pressure dependences of the relative resistance of cerium at a) three and b) four cycles of increasing and decreasing pressure at room temperature: ●, ○ first cycle of increasing and decreasing pressure, respectively, ▲, △ second cycle, ■, □ third cycle, ▲, △ fourth cycle, point S constant pressure for 12 h

cerium (Fig. 7). The downstroke pressure curve $R/R_0(P)$ (symbol ○) was obtained after isobaric heating of the sample up to 432 K and cooling down to 293 K at 6.7 GPa. The first drop of R/R_0 at 5.4 GPa is due to the transition of the α'-phase to the α-phase and the sharp drop of R/R_0 at 2.8 GPa is due to the transition of the β'-phase to the α-phase. It means that the point ($P = 6.7$ GPa, 432 K) (symbol ■) (Fig. 1) is in the vicinity of the line of direct transition of the α'-phase to the β'-phase and the cerium sample consists of a mixture of α'- and β'-phases. The $R/R_0(P)$ curve (symbol □) obtained after isobaric heating up to 475 K at 6.3 GPa and cooling down to 293 K (point H_2, Fig. 7) has a sharp drop of R/R_0 by 28% at 2.8 GPa caused by complete transition of the β'-phase to the α-phase.

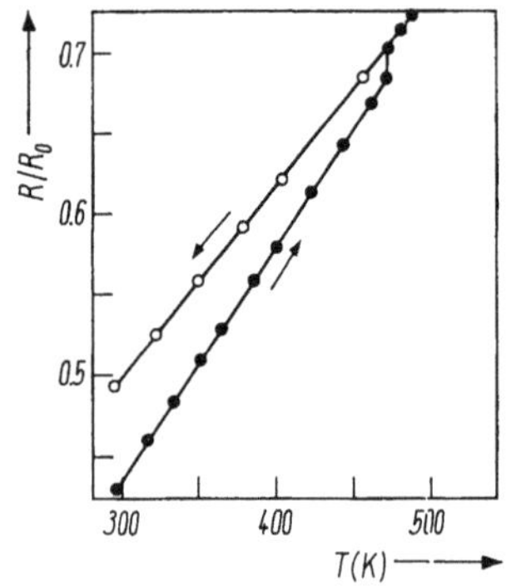

Fig. 6

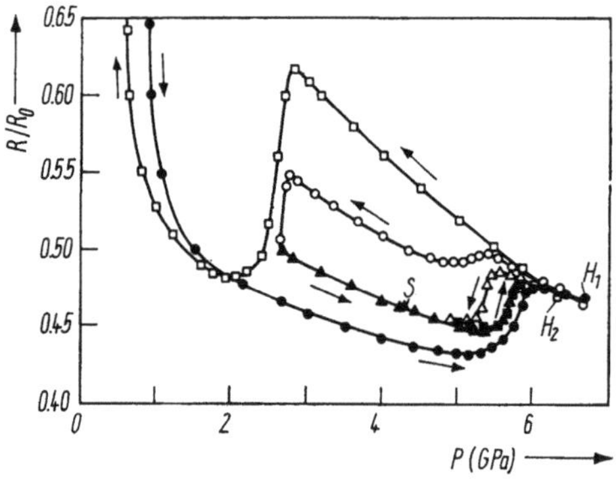

Fig. 7

Fig. 6. Temperature dependence of the relative resistance of cerium at increasing (●) and decreasing (○) temperature at 5.4 GPa

Fig. 7. Pressure dependences of the relative resistance of cerium at three cycles of increasing and decreasing pressure at room temperature: ●, ○ first cycle of increasing and decreasing pressure respectively, ▲, △ second cycle, ■, □ third cycle; point H_1 the sample was heated up to 432 K and cooled down to room temperature after increasing the pressure up to 6.7 GPa in the first pressure cycle; point H_2 the sample was heated up to 475 K and cooled down to room temperature after increasing the pressure up to 6.3 GPa in the third pressure cycle; point S constant pressure for 12 h

The line of direct transition of the α-phase to the β'-phase intersects the line of reverse transition of the β'-phase to the α-phase in the point ($P = 4.4$ GPa, 550 K). This point can be obtained also by intersection of the equilibrium line of the $\alpha \rightleftarrows \beta'$ transition with the high temperature part of the bent line obtained in [5]. Thus, the high temperature line [5] restricts the region of stability of the β'-phase at high temperatures and the point ($P = 4.4$ GPa, 550 K) has to be a triple point. This point has to be the origin of a third equilibrium line separating the α-phase and the hitherto unknown γ'-phase. Also, the high temperature part of the phase line obtained in [5] is the $\beta' \rightleftarrows \gamma'$ transition line.

The authors of [5] have reported no anomalies in the electrical and thermal properties of cerium in the 620 to 820 K temperature range and pressures up to 8.5 GPa which would evidence a first-order phase transition. It means that the properties of α- and γ'-phases are very close or the $\alpha \rightleftarrows \gamma'$ phase line ends in the critical point ($P \approx 3.0$ GPa, ≈ 600 K).

The authors of [2, 3] reported X-ray data on the structure of the α'-phase. They have found that a mixture of α'- and α''-phases exists after the transition. Pressure cycling removes the α''- phase, the structure of the retained α'-phase being orthorhombic α–U type.

The reverse $\alpha' \rightarrow \alpha$ transition was reported to occur at 2.7 GPa at room temperature contrary to [1]. The present data show that just the β'-phase is metastable down to 2.8 GPa at room temperature and that the structure of the β'-phase (labeled α' in [2, 3]) orthorhombic. The α''-phase [2, 3] was determined as distorted f.c.c. We believe that this α''-phase corresponds to the phase labeled α' in our phase diagram (Fig. 1).

The direct and reverse transition between α- and α'-phases is a transition between f.c.c. and distorted f.c.c. structures and may be of the martensitic type and therefore occurs in a wide pressure range. Therefore the pressure of the onset of the reverse $\alpha' \rightarrow \alpha$ transition is higher than the pressure of the direct $\alpha \rightarrow \alpha'$ transition.

The high temperature β'-phase can be obtained by heating to temperatures higher than 420 K under hydrostatic pressure or under shear stress at room temperature [2, 3]. Fig. 3 in [5] shows that it can also be obtained at 373 K and in the presence of non-hydrostatic stress in the sample [5]. The reverse $\beta' \rightarrow \alpha$ transition in the latter case occurs at 3.3 GPa and the drop of R/R_0 is 18%.

In view of such a strong role of the shear stresses [9] a study of cerium at high pressure should be performed under hydrostatic conditions.

Acknowledgement

We are grateful to A. M. Buckhartsev for useful assistance.

References

[1] E. King, J. A. Lee, I. R. Harris, and T. F. Smith, Phys. Rev. B **1**, 1380 (1970).
[2] F. H. Ellinger and W. H. Zachariasen, Phys. Rev. Letters **32**, 773 (1974).
[3] W. H. Zachariasen and F. H. Ellinger, Acta cryst. **A33**, 155 (1977).
[4] S. Endo, U. Sasaki, and T. Mitsui, J. Phys. Soc. Japan **42**, 882 (1977).
[5] T. E. Antonova, J. T. Belash, and E. G. Ponyatovskii, Fiz. Metallov i Metallovedenie **5**, 132 (1981).
[6] L. G. Khvostantsev and N. A. Nikolaev, phys. stat. sol. (a) **51**, K57 (1979).
[7] G. I. Piermariny, R. A. Forman, and S. Block, Rev. sci. Instrum. **49**, 1061 (1978).
[8] L. G. Khvostantsev, L. F. Vereshchagin, and A. P. Novikov, High Temp-High Press **9**, 637 (1977).
[9] J. C. Jamieson and A. W. Lawson, J. appl. Phys. **33**, 776 (1962).

(Received May 22, 1981)

phys. stat. sol. (a) 68, 199 (1981)

Subject classification: 13.4 and 20.3; 22.4.2

Institute of Physics, N. Copernicus University, Toruń[1])

Thermoluminescence in Conductive n-Type ZnSe Crystals

By

H. L. Oczkowski

A study is made of the mechanism of storing up minority non-equilibrium carriers in n-type ZnSe ($N_d = 2.2 \times 10^{16}\ cm^{-3}$, $N_a = 3 \times 10^{15}\ cm^{-3}$, $\varepsilon_d = 0.041$ eV) for explaining the thermoluminescence and thermally stimulated current effects. A hypothesis which assumes the existence of metastable donor (0.3 to 0.5 eV) - acceptor ($\approx$ 0.6 eV) pairs allows to propose a model which consistently describes the results of investigations.

Es wird der Mechanismus der Speicherung von Nichtgleichgewichtsträgern in n-ZnSe ($N_d = 2{,}2 \times 10^{16}\ cm^{-3}$, $N_a = 3 \times 10^{15}\ cm^{-3}$, $\varepsilon_d = 0{,}041$ eV) untersucht, um die Thermolumineszenz und den thermostimulierten Strom zu interpretieren. Unter der Annahme eines metastabilen Donator (0,3 bis 0,5 eV) - Akzeptor ($\approx$0,6 eV)-Paares wird ein Modell entwickelt, das die erhaltenen Ergebnisse befriedigend beschreibt.

1. Introduction

As early as 1663, Sir Robert Boyle reported to the Royal Society in London about "Experiments and Considerations upon Colours with Observations on Diamond That Shines in the Dark". Today we know that this report contains the primary fundamental observation on thermoluminescence phenomenon [1]. Two centuries later, despite of various investigations a lot of questions connected with the thermoluminescence mechanism has not been satisfactorily answered.

The purpose of this paper is to examine the mechanism, which allows to keep minority, non-equilibrium carriers in the metastable state in spite of the high concentration of majority carriers and high probability of the radiative recombination in ZnSe crystals.

2. Crystal Growth

Zinc selenide single crystals were grown by a vertical Bridgman method from 6N, electronic grade ZnSe powder obtained from Koch-Light. During the growth argon atmosphere (80 atm) surrounding the crystal was used. The ZnSe crystal was cut into 1 mm thick, {111} oriented slices, then polished and etched [2]. Next, these samples were heated in molten 6N zinc at 650 °C for 75 h. At the end of the heat treatment the tube was rapidly cooled. Crystal plates were then chemically polished with 50% HCl and subsequently with hot 10% NaOH after mechanical lapping to a mirror-like surface. Crystals were not intentionally doped during growth and heat treatment.

Four contacts to a 59-F ZnSe sample were made with the fused-in In. The fusing-in was carried out at 360 °C in the atmosphere of a dry mixture of nitrogen and hydrogen.

3. Electrical Transport Measurements

The temperature dependence of the free carrier concentration was determined from 64 to about 350 K with the help of the van der Pauw method [3]. For these measurements a maximum magnetic field of 0.75 T and a stabilized current through the sample

[1]) 87-100 Toruń, Poland.

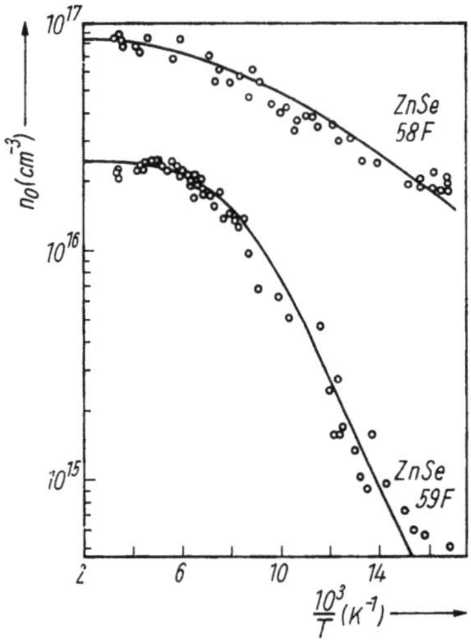

Fig. 1. Temperature dependence of the electron concentration for ZnSe crystals annealed in liquid Zn. The solid line represents the electron concentration calculated from equation (1) with $N_d - N_a = 2.5 \times 10^{16}$ cm^{-3}, $N_a = 3.3 \times 10^{15}$ cm^{-3}, $\varepsilon_d = 0.041$ eV for the 59-F ZnSe crystals

of 0.1 mA corresponding to at most 0.3 V was used. After heating up to 360 K the sample was slowly cooled in dark to 64 K.

Almost the same temperature range was then traversed starting from the lowest temperature. A very low natural heating rate ($\approx 10^{-2}$ K s^{-1}) justifies the supposition that the carrier density determined here is close to the equilibrium value n_0. Additional measurements performed with the sample illuminated at 64 K give the same result probably because of the low heating rate.

The sign of the Hall voltage shows that electrons predominate in electrical transport. The variation of ln n_0 with $1/T$ is shown in Fig. 1.

The temperature dependence of the free electron concentration can be described by the formula specific for non-degenerate statistics and strong compensation [4],

$$n_0 = (N_d - N_a)\left(1 + \frac{2N_a}{N_c} \exp \frac{\varepsilon_d}{kT}\right)^{-1}, \tag{1}$$

where N_c is the density of states in the conduction band given by $2(2\pi m^* kTh^{-2})^{3/2}$, m^* is the density of states effective mass of electrons ($m^* = 0.17m_0$) [5 to 7]. The quantities N_d and N_a are the donor and compensating acceptor concentrations, ε_d and k are the ionization energy and the Boltzmann constant, respectively. A value of 1/2 was used here for the degeneracy factor. Equation (1) is reasonable provided the following relations are fulfilled:

$$n_0 \ll N_c, \tag{2}$$

$$N_d - N_a \lessapprox \frac{(N_c' + N_a)^2}{4N_c'}, \tag{3}$$

where $N_c' = (N_c/2) \exp(-\varepsilon_d/kT)$. It is worthwhile to note that inequalities (2) and (3) are satisfied in the range of temperatures used in the present experiment. Similar interpretation has been done in other papers concerning ZnSe [6, 8 to 10].

The solid line in Fig. 1 represents the best square fit of equation (1) and data (60 points) for the following parameters: $N_d = 2.2 \times 10^{16}$ cm^{-3}, $N_a = (3 \pm 1) \times 10^{15}$ cm^{-3}, $\varepsilon_d = (0.041 \pm 0.003)$ eV. The correlation factor is equal to $r = 0.9$.

4. Photoluminescence and Thermal Quenching of Emission

Photoluminescence (PL) was excited using the ultraviolet radiation from an Hg arc lamp in conjunction with an interference absorption filter — Hg 365 nm (3.4 eV) or 330 nm (3.72 eV).

At 83 K the PL consists of several bands corresponding to the energies 1.92 eV (red), 2.22 eV (green), and 2.6 eV (blue). Widths at half-height of these broad emission peaks are about 0.19 eV for red and green bands and 0.1 eV for the blue band.

It is important to note here that in the vicinity of 170 K the temperature dependence of luminescence intensity reveals a very "strange" behaviour. The red band intensity has a local minimum while the green one changes activation energy for thermal quenching and decreases rapidly at higher temperatures. In the temperature range 170 to 290 K the intensity of the red band increases with temperature and for temperatures higher than 290 K rapidly diminishes.

Fig. 2 represents the temperature dependence of the intensity of the red and green bands plotted as lg I versus $1/T$. If it is assumed that the intensity can be described by the simple model of thermal quenching, this plot should yield a straight line [4]. In the low-temperature region the slopes for both bands are similar and give 0.015 and 0.012 eV for the red and green band, respectively. In the high-temperature region, where the strong thermal quenching is observed, the activation energy of 0.26 eV for the green band was found. For the red band a value greater than 0.6 eV was estimated.

To be clearer we should emphasize that the "strange" temperature dependence of the red band, near 170 K, is exactly the same regardless the temperature is increasing or decreasing.

The ZnSe crystal under investigation was not intentionally doped and therefore the identification of luminescent centres is a rather complicated problem. Earlier papers [11 to 15] suggest that the observed PL spectrum is connected with the presence of Cu, Ag dopants or so-called self-activated centres. This latter is responsible for the PL spectrum in the region of 1.9 eV. In particular the green (2.22 eV) and the blue bands (2.6 eV) have been explicitly related to the Ag impurity [11].

From subsequent papers [16 to 20] it follows generally that the recombination centres in ZnSe are not connected with only one impurity defect but their form is composed. These suggestions include a complex defect, compensation mechanism, and pair recombination.

In the present paper special attention will be paid to the red band. In agreement with the cited papers several assumptions can be made. First, the red emission is a result of donor–acceptor recombination. Here the acceptor can be Cu^{2+} [16] or an ionized complex of the type $(V_{Zn}^{-2}Al^{+})$, [5, 18, 19]. Second, the luminescence can be

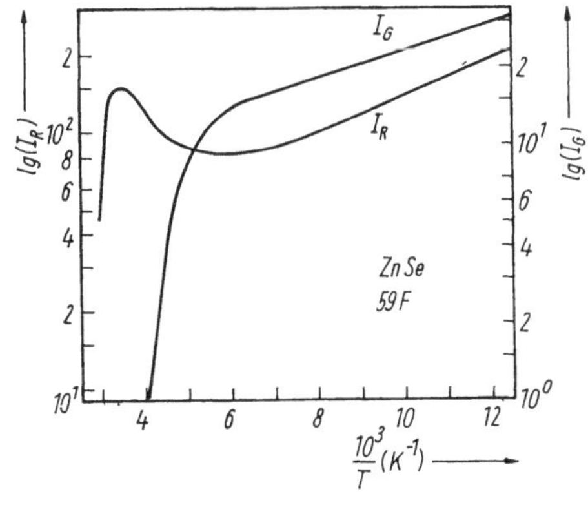

Fig. 2. Peak relative emission intensity of steady-state photoluminescence (I) as a function of temperature on a lg I vs. $1/T$ plot. I_R red band. I_G green band

explained by electronic transitions within a neutral complex defect of the type ($V_{Zn}^{-}Al^{+}$) [18], and finally the Schön-Klasens model for the Cu^{2+} acceptor cannot be excluded [16].

The exact nature of the recombination mechanism for the green and blue emission is not well understood. Neglecting at this moment silver as a dopant it has been noted [18 to 20] that this emission is due to a defect involving Al or Ga and an alkali impurity Na or Li.

5. Thermally Stimulated Current

Thermally stimulated current (TSC) measurements were made with the accuracy 10^{-8} A and four significant digits using a differential preamplifier, selective nanovoltmeter, and a digital voltmeter. The applied electric field is sinusoidal (10 mV, 480 Hz). In Fig. 3 the curve I_0 shows the dark current of a sample which before cooling was heated up to 360 K.

The specimen has been irradiated at 80 K with light of 365 nm (3.4 eV) from a high pressure mercury lamp for about 15 min. After a few minutes in the dark, on raising the temperature with a constant heating rate the current of the excited sample, I, is measured. Fig. 3 shows the dependence of the current difference $I - I_0$ upon temperature.

If we suppose that the observed difference between two currents corresponds to the density of excess free electrons only, the following simple relation is obtained:

$$\Delta n = \frac{\Delta I}{I_0} n_0 , \tag{4}$$

which allows one to calculate Δn with the help of Fig. 3 and equation (1).

Equation (4) is valid under the condition that the electron mobility of the excited sample is the same as in the equilibrium state. Unfortunately, in our case it was stated that the Hall measurement for the heating rate used for TSC cannot be performed because of thermoelectric effects, but this experiment has been realized by Stringfellow and Bube [14]. They have found that TSC in self-activated ZnSe results mainly from the increasing free electron density.

The data presented here also suggest that thermally stimulated carriers are electrons. In the opposite case (holes), the following equation should be fulfilled:

$$\Delta p = \frac{\mu_n \Delta I}{\mu_p I_0} n_0 , \tag{5}$$

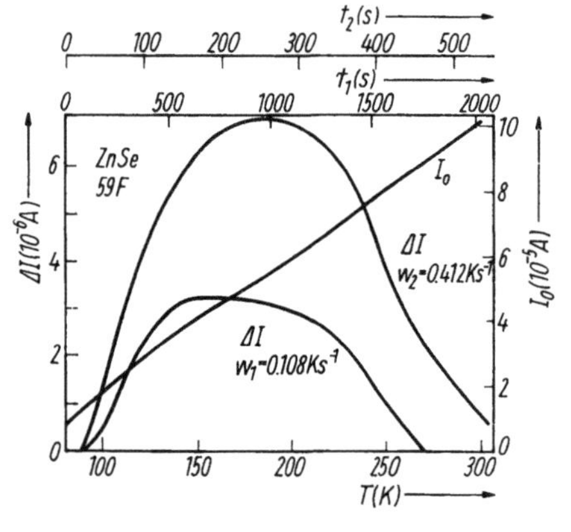

Fig. 3. Thermally stimulated current (ΔI) and dark current (I_0) vs. temperature

where μ_n and μ_p is the mobility of electrons and holes, respectively. Holes are here minority carriers, therefore, in the n-type, wide-gap semiconductor $\mu_n \gg \mu_p$. In this experiment we have $\Delta I/I_0 \approx 0.1$, $n_0 \approx 10^{16}\,cm^{-3}$, thus (5) gives $\Delta p \gg 10^{15}\,cm^{-3}$. This estimation gives a very high density of excess free minority carriers, which leads this assumption to absurdity because of the very short lifetime of free holes in ZnSe of the order 2×10^{-9} s [21].

It seems that the TSC curves shown in Fig. 3 are composed and perhaps consist of two peaks near 150 and 210 K. Similar results which were reported for ZnSe (SA) [14, 22] and ZnSe:Al [23] support this perception.

6. Thermoluminescence

The ZnSe crystals under investigation reveal thermoluminescence (TL). In the range 80 to 360 K two peaks of TL positioned near 100 and 180 K are observed. Special attention was paid to the second TL band (180 K) which is stronger than the first one.

TL phenomenon was excited at 77 K with 365 nm (3.4 eV) radiation from a mercury lamp provided with a suitable filter. Special electronic arrangement enables linear and saw-tooth programming of temperature. Our measuring apparatus allows detection and recording of T, $1/T$, the intensity of TL emission, I, $\log I$, and the light sum $L = \int I\,dt$ emitted during a defined temperature cycle.

The spectrum of TL was measured by means of a low-resolution monochromator (about 0.05 eV) just at the moment when the glow curve reaches maximum. The result is represented in Fig. 4 TL together with the photoluminescence spectrum (PL) which was measured at 171 K. As it is seen from Fig. 4 the TL emission contains mainly the 1.92 eV band.

In agreement with the basic theory for TL [24 to 27] the maximum of the glow curve shifts toward higher temperatures with increasing heating rate. In the simplest case this dependence is described by the well-known formula of Hoogenstraaten [24],

$$\ln\left(\frac{w}{T_*^2}\right) = \ln\left(\frac{ks}{\varepsilon}\right) - \frac{\varepsilon}{kT_*}, \tag{6}$$

where T_* is the temperature of the glow curve maximum, w the heating rate, s the so-called frequency factor, and ε the trap depth. The TL glow curves for different heating rates are given in Fig. 5. Despite of the wide range of heating rates the shape of the main band is simple and does not suggest any composed form.

According to (6) a plot of $\ln(T_*^2/w)$ versus $1/T$ should yield a straight line with a slope ε/k. The parameters determined from the fit of the data are given by $\varepsilon = (0.37 \pm \pm 0.07)$ eV, $s = 1 \times 10^8\,s^{-1}$ $(1 \times 10^6\,s^{-1} \leqq s \leqq 8 \times 10^9\,s^{-1})$ and the correlation factor $r = 0.95$.

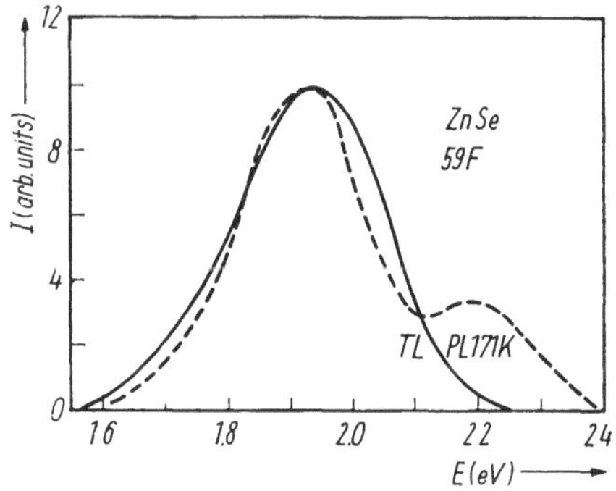

Fig. 4. Comparative luminescence spectra. The curves marked TL and PL show thermoluminescence and photoluminescence spectrum, respectively

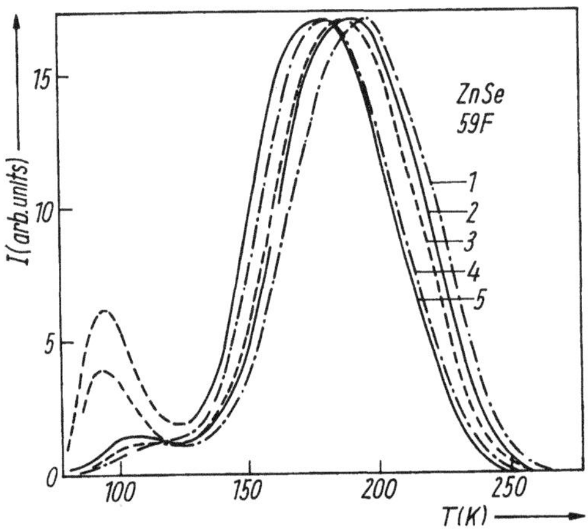

Fig. 5. Thermoluminescence glow curves for selected heating rates. (5) $w = 0.0282$, (4) 0.0379, (3) 0.108, (2) 0.173, (1) 0.404 K s^{-1}

These parameters were used for the mono- and bimolecular glow curve calculations (see Appendix A). The calculated glow curve is shown in Fig. 6, curve 1. In Fig. 6, curve 3 the solid curve represents the measured glow curve. It is easy to see that the calculated curve is about three times narrower than the observed one. This result suggest that the investigated TL band is composed. For this reason we applied the technique of trap spectroscopy using the Gobrecht-Hofmann (GH) method [28, 29].

To determine the trap spectrum five independent measurements using the GH method were realized. Each measurement had a different number of temperature oscillations from 80 to 150 cycles in the range of temperature 85 to 250 K. Fig. 7 shows the averaged values of the energetic trap density, $\overline{H}_\varepsilon$. It was found that the trap distribution in the range 0.1 to 0.2 eV corresponds to the glow curve maximum near 100 K, whereas the main band of TL is connected with the trap distribution from 0.3 to 0.53 eV. In this region the trap spectrum reveals two maxima at about 0.4 and 0.48 eV.

The energy distribution of traps was tested for the calculation of the shape of the main glow curve. It was assumed here that each fraction of traps from the energy inter-

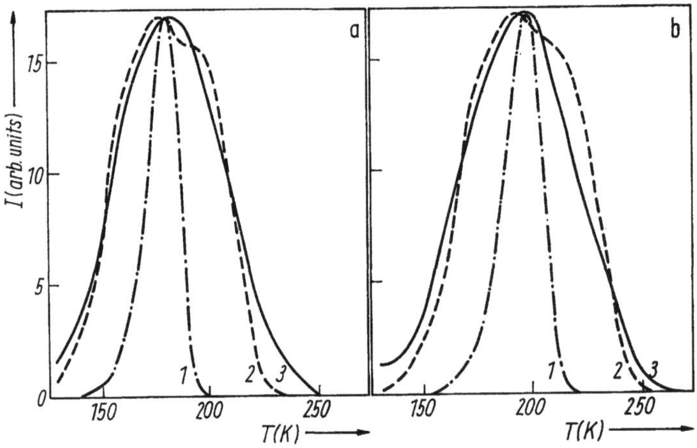

Fig. 6. The shape of the thermoluminescence glow curves near 180 K for two different heating rates a) $w = 0.0379$ K s^{-1} and b) $w = 0.404$ K s^{-1}; (1) calculated from (A1) with parameters: $\varepsilon = 0.368$ eV, $s = 1.07 \times 10^8$ s^{-1}; (2) calculated with regard to trap distribution (see text); (3) experimentally determined

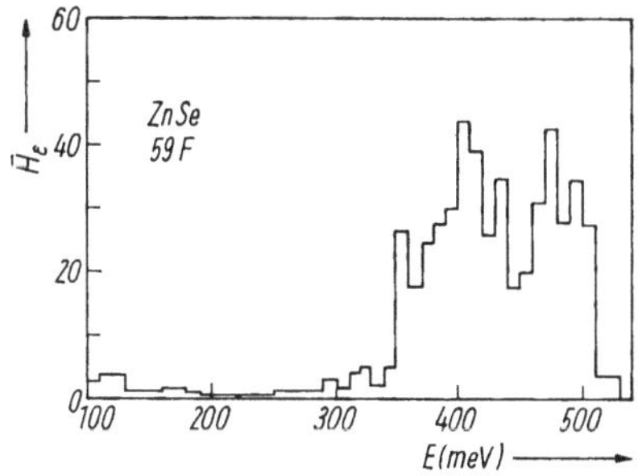

Fig. 7. Trap distribution determined by the fractional glow technique

val (ε, $\varepsilon + \Delta\varepsilon$) emits an independent band described by (A1) in Appendix A, where c_0 was replaced by the suitable $\overline{H}_\varepsilon\ \Delta\varepsilon$. The intensity of the whole TL band was obtained by summing all the constituent bands.

In order to attain the dependence $s(\varepsilon)$ a rough estimation was performed — two wide fractions of traps from intervals $0.31\ \text{eV} < \varepsilon_1 \leqq 0.45\ \text{eV}$ and $0.45\ \text{eV} < \varepsilon_2 \leqq 0.53\ \text{eV}$ were described by two different values of $s_1 = 3.5 \times 10^9\ \text{s}^{-1}$ and $s_2 = 5 \times 10^9\ \text{s}^{-1}$ (see Appendix B). To cover the main TL band the following other parameters were chosen:

$$\Delta\varepsilon = 0.02\ \text{eV}\,, \qquad 0.31\ \text{eV} < \varepsilon \leqq 0.53\ \text{eV}\,,$$

$$\Delta T = 5\ \text{K}\,, \qquad 130\ \text{K} < T \leqq 270\ \text{K}\,.$$

The results of these calculations are shown in Fig. 6, curve 2.

It is easy to see that these rough calculations reproduce fairly well the position and non-typical width of the investigated band. The inflexion near the maximum can be removed by a modification of s_1 and s_2.

7. Discussion

The investigations discussed above lead to serious difficulties when one tries to develop a model which should consistently describe the results presented in this paper. Firstly, it is necessary to propose a mechanism which is responsible for metastable trapping of excess holes in a n-type semiconductor with a high density of free electrons. It appears also that holes are trapped immediately at recombination centres. The second difficulty arises from the correlation of thermally stimulated current and thermoluminescence, which explicitly suggests that the electron transition from trap to conduction band is followed by relaxation and radiative recombination with resultant 1.92 eV photon emission.

It is interesting to note that a hypothesis which assumes the existence of intra-pair metastable states [30] allows to propose a consistent model. As a metastable donor–acceptor pair one ought to understand here electron and hole bound to deep donor (0.3 to 0.5 eV) and acceptor (≈ 0.6 eV) with no probability for recombination with a trapped hole. The ionization of a deep donor varies the state of the centre and makes possible radiative hole annihilation. This radiative recombination corresponds to transitions of the type free electron–acceptor or shallow donor–acceptor, which gives the red emission (1.92 eV).

In the case of a metastable pair the deep donor plays the role of an electron trap and also when it is emptied enables the radiative recombination of trapped holes. If we assume additionally that the free hole recombines with the electron trapped on the deep donor level, then a consistent model is obtained, which explains all the observations, in particular, the temperature correlation between the "strange" minimum of red

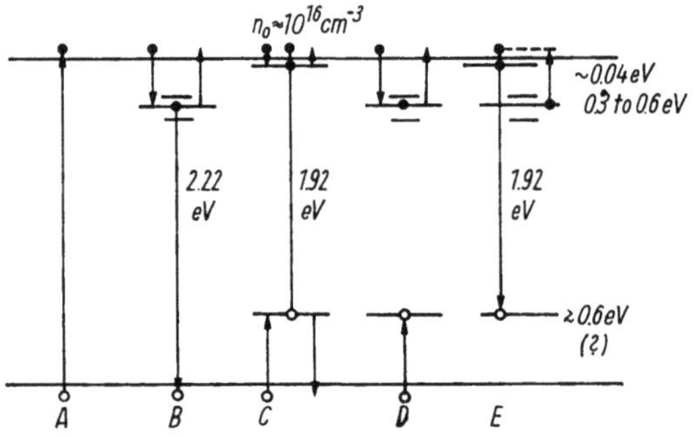

Fig. 8. Proposed energy diagram illustrating the transitions leading to: A excess carrier generation, B green steady-state emission (thermal quenching energy is near 0.3 eV), C red emission (shallow donor–acceptor or free electron–acceptor recombination), D metastable donor–acceptor pair (the recombination with the trapped hole is forbidden), and E thermoluminescence; the thermal ionization of the deep donor in the metastable pair enables the radiative recombination of the trapped hole giving the red emission

photoluminescence efficiency, green band quenching, thermally stimulated current, and luminescence spectrum. It is also clear that only the red band is emitted during the thermoluminescence (Fig. 8).

In conclusion, the possibility mentioned above can explain the observed correlations between results of different experiments. It seems to be the simplest way for ordering observations but, of course, this idea will be valuable only if the observed correlations are not accidental. It is possible, that this doubt will be removed by following investigations of somewhat different ZnSe crystals.

Acknowledgements

The author would like to thank H. Łożykowski for his continuous support of this work and to F. Firszt for preparing of ZnSe crystals and competent assistance during experiments.

This work was supported in part by a grant from the Istituto di Fisica, Universita degli Studi di Parma. The cooperative help of N. Romeo and L. Tarricone in making the Hall measurements are gratefully acknowledged.

Appendix A

For the calculations of monomolecular curve the relation [24]

$$I_M = sc_0 \exp\left[-x - \lambda\varphi(x)\right] \tag{A1}$$

was used, and in the bimolecular case

$$I_B = sc_0[1 + \lambda\varphi(x)]^{-2} \exp(-x) \tag{A2}$$

was applied, where

$$x = \frac{\varepsilon}{kT}, \tag{A3}$$

$$\lambda = \frac{\varepsilon s}{wk}, \tag{A4}$$

c_0 is the concentration of trapped electrons at depth ε. In the region of the maximum of the glow curve we have $x > 10$. Hence the following approximation for $\varphi(x)$ is justified:

$$\varphi(x) = \int_x^\infty z^{-2} \exp(-z) \, dz \approx \exp(-x) \sum_{n=1} \frac{n!}{(-x)^{n+1}}. \tag{A5}$$

Appendix B

In the calculation the following formula equivalent to (6) was used:

$$s = \frac{w\varepsilon}{kT_*^2} \exp\left(\frac{\varepsilon}{kT_*}\right). \tag{B1}$$

Taking into account the glow curve (Fig. 5) and the trap spectrum (Fig. 7) it was assumed that the appropriate values (ε_1, T_{*1}) and (ε_2, T_{*2}) can be approximated by the following relations:

$$T_{*1,2} = T(I_{max}/2)\,,$$

where T_* was taken from Fig. 5 from the ascending $-T_{*1}$ and the descending $-T_{*2}$ branches of the glow curve, and

$$\varepsilon_{1,2} = \varepsilon(\bar{H}_{max}/2),$$

which in agreement with Fig. 7 gives $\varepsilon_1 = 0.35$ eV and $\varepsilon_2 = 0.51$ eV. After putting ε and T_*, suitable for five different values $-$ w fairly coherent values for s_1 and s_2 were obtained. The averaged values of s are $s_1 = 3.5 \times 10^9$ s^{-1} and $s_2 = 5 \times 10^9$ s^{-1}.

References

[1] T. Cairns, Analyt. Chem. **48**, 266A (1976).
[2] H. L. Oczkowski and Z. Popławski, J. Crystal Growth **23**, 154 (1974).
[3] L. J. van der Pauw, Philips Res. Rep. **13**, 1 (1958).
[4] J. S. Blakemore, Semiconductor Statistics, Pergamon Press, Oxford 1962.
[5] M. Aven and J. S. Prener (Ed.), Physics and Chemistry of II–VI Compounds, North-Holland Publ. Co., Amsterdam 1967.
[6] M. Aven and B. Segall, Phys. Rev. **130**, 81 (1963).
[7] D. T. F. Marple, J. appl. Phys. **35**, 1879 (1964).
[8] M. Aven, J. appl. Phys. **42**, 1204 (1971).
[9] H. H. Woodbury and M. Aven, Phys. Rev. B **9**, 5195 (1974).
[10] K. Igaki and S. Satoh, Japan. J. appl. Phys. **18**, 1965 (1979).
[11] R. E. Halsted, M. Aven, and H. D. Coghill, J. Electrochem. Soc. **112**, 177 (1965).
[12] W. Lehmann, J. Electrochem. Soc. **113**, 449 (1966).
[13] W. Lehmann, J. Electrochem. Soc. **113**, 788 (1966).
[14] G. B. Stringfellow and R. H. Bube, Phys. Rev. **171**, 903 (1968).
[15] G. B. Stringfellow and R. H. Bube, J. appl. Phys. **39**, 3657 (1968).
[16] G. Jones and J. Woods, J. Lum. **9**, 389 (1974).
[17] Ph. Ged, J. Phys. Chem. Solids **40**, 439 (1979).
[18] J. C. Bouley, P. Blanconnier, A. Herman, Ph. Ged, P. Henoc, and J. P. Noblanc, J. appl. Phys. **46**, 3549 (1975).
[19] F. Firszt, Thesis, N. Copernicus University, Toruń 1980 (in Polish, unpublished).
[20] P. K Chattarjee, A. J. Rosa, and B. G. Streetman, J. Lum. **8**, 176 (1973).
[21] S. Mora, N. Romeo, and L. Tarricone, private communication, to be published.
[22] S. Satoh and K. Igaki, Japan. J. appl. Phys. **19**, 485 (1980).
[23] J. P. Fillard, J. Gasiot, and J. C. Manifacier, Phys. Rev. B **18**, 4497 (1978).
[24] W. Hoogenstraaten, Philips Res. Rep. **13**, 515 (1958).
[25] P. Bräunlich, Thermoluminescence of Geological Materials, Ed. D. J. McDougall, Academic Press, London 1968 (p. 61).
[26] P. Kivits and H. J. L. Hagebeuk, J. Lum. **15**, 1 (1977).
[27] P. Kivits, J. Lum. **16**, 119 (1978).
[28] H. Gobrecht and D. Hofmann, J. Phys. Chem. Solids **27**, 509 (1966).
[29] H. L. Oczkowski, J. Lum. **17**, 113 (1978).
[30] F. Williams, phys. stat. sol. **25**, 493 (1968).

(Received February 17, 1981)

phys. stat. sol. (a) 68, 209 (1981)

Subject classification: 9; 22.9

Groupe de Dynamique des Phases condensées (L.A. 233) (a), Laboratoire de Chimie Macromoléculaire (b), and Laboratoire de Microscopie Electronique et Microanalyse (c), Université des Sciences et Techniques du Languedoc, Montpellier[1])

Diffusion Study of Iodine and Antimony Pentafluoride in Polyacetylene

By

C. BENOIT (a), M. ROLLAND (a), M. ALDISSI (b), A. ROSSI (c), M. CADENE (a), and P. BERNIER (a)

Polyacetylene $(CH)_x$ is a simple conjugated polymeric system, the electrical behaviour of which is considerably modified by doping processes: from an insulating behaviour it successively becomes semi-conducting and even metallic. In this work first the dopant profiles through the sample is measured and the macroscopic diffusion coefficients of iodine and SbF_5 is deduced. In the second part of this work a model is proposed which simultaneously take into account the diffusion and a chemical fixation of dopant molecules along the polyacetylene fibrils. Comparison between experimental and computed results shows that this model allows a better comprehension of the observed phenomena than the simple diffusion laws.

Le polyacétylène $(CH)_x$ est un polymère dont la conductivité électrique est considérablement modifiée par dopage: d'isolant il devient successivement semiconducteur puis métallique. Dans ce travail nous avons tout d'abord déterminé quantitativement les profils de concentration en dopant au travers du film et déduit les coefficients de diffusion de l'Iode et du SbF_5 dans de matériau. La deuxième partie de cet article est consacrée à l'interprétation théorique des résultats expérimentaux, tout d'abord au moyen de la théorie de la diffusion; les résultats n'étant pas satisfaisants nous proposons un modéle basé sur deux phénomènes; diffusion du dopant associée à une fixation chimique de ce dernier sur les fibrilles du polymère. Les résultats obtenus montrent que ce modèle permet de rendre compte des résultats expérimentaux.

1. Introduction

Polyacetylene is a largely studied metal since ITO et al. [1] obtained flexible films, also available in a large thickness range. This simple conjugated organic polymeric system available in two isomeric structures, cis and trans, possesses very interesting physical properties in various fields: mechanical stretching, optical values of its band gap and absorption coefficient [2, 3] etc. Furthermore its transport properties lead to think that it may be a powerful, and low cost, material for electronic device improvement: with a high electrical resistivity (10^9 Ω cm for the cis isomer, 10^5 Ω cm for the trans one) it can be doped with AsF_5, halogens, etc.) [4, 5]. At low doping level, a wide range of semiconducting material is obtained and then a semiconductor to metal transition occurs when the doped product reaches the formula $(CH\ A^-_{0.01})_x$.

In the metallic behaviour a conductivity as high as $10^3\ \Omega^{-1}\ cm^{-1}$ is reached when $(CH)_x$ is AsF_5 doped.

At this time, many works have been performed about the undoped or doped polyacetylene both in pure and applied physics. In order to explain the observed transport properties and chiefly the electrical conductivity two theories are in proposal: a classical semiconducting behaviour with large band states in the gap [6] and a percolation

[1]) Place Eugène Bataillon, 34060 Montpellier Cédex, France.

process between high conducting metallic clusters [7, 8], these two hypothesis being associated with a solution deformation along the $(CH)_x$ chains; however, a few studies have been turned in the field of dopant localization in the sample.

The aim of this work is to study the doping process and dopant localization through the film thickness in order to determine the dopant diffusion coefficient in the film; such a knowledge is essential for preparing semiconducting polymers in order to realize electronic devices.

2. Experimental Techniques and Results

Cis-rich $(CH)_x$ films were grown onto the vertical surfaces of the reactor at −78 °C using Shirakawa's technique. The sample were doped with iodine and SbF_5 in vapour phase at room temperature. The mean value y of the doping level was obtained, measuring the weight uptake of the samples. About 200 μm thick films were then included in an epoxy resin and carefully polished. Then by means of Castaing microprobe analysis we have performed a quantitative analysis of all the essential com-

Table 1

Castaing microprobe analysis of various elements through a SbF_5 doped $(CH)_x$ film

		Sb	Ti	Al	F	O	C	H (3)
gas side	(1)	0.091	0.007	0.004	0.135	—	0.450	0.037
	(2)	0.102	0.007	0.004	0.135	—	0.446	0.037
middle	(1)	0.033	0.008	0.005	0.042	0.0003	0.516	0.046
	(2)	0.030	0.008	0.004	0.042	0.0003	0.513	0.046
glass side	(1)	0.070	0.008	0.006	0.110	0.0002	0.300	0.02
	(2)	0.086	0.010	0.007	0.112	0.0002	0.293	0.2

(1) one sample, (2) other sample, (3) deduced values from carbon measurements (assuming the same number of carbon and hydrogen atoms because H, too light, cannot be measured). The concentration is given in mass by comparison with the pure element, the concentration of which corresponds to unity

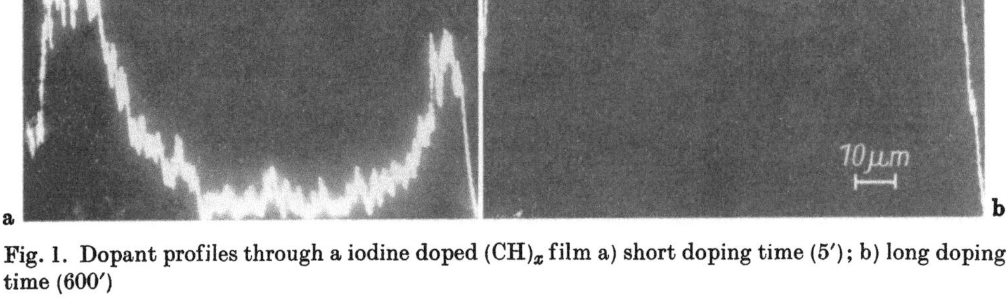

Fig. 1. Dopant profiles through a iodine doped $(CH)_x$ film a) short doping time (5′); b) long doping time (600′)

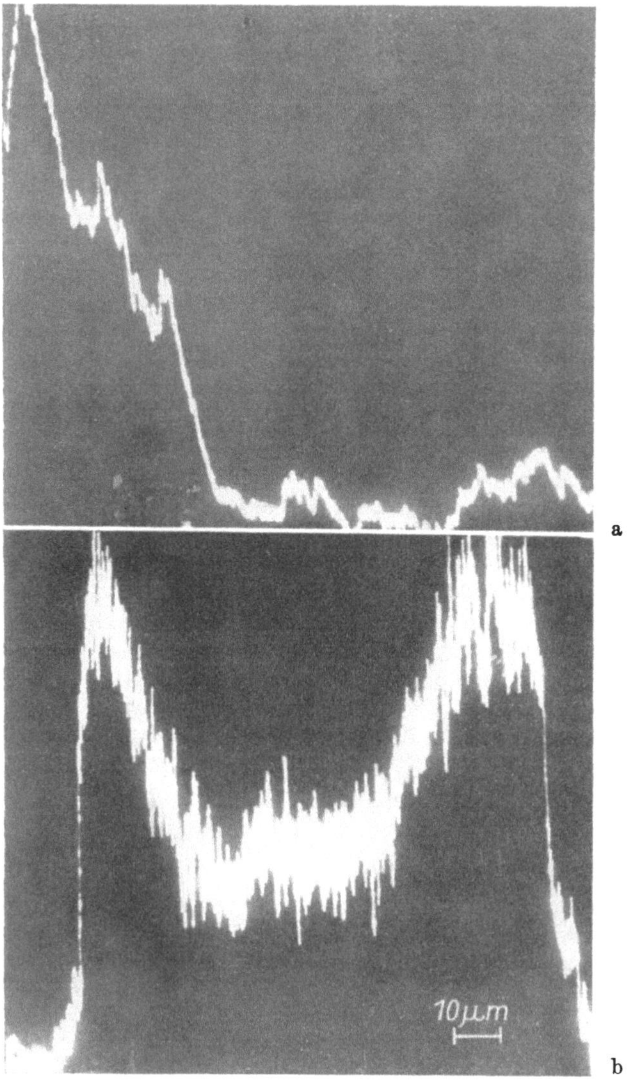

Fig. 2. Dopant profiles through a SbF_5 doped $(CH)_x$ film a) short doping time (5′); b) long doping time (600′)

ponents of the doped samples through their thickness. The experimental procedure was described elsewhere [9]. In these experiments the electron beam diameter was 1 μm and the ionization sphere diameter 5 μm, we must notice that they are very small compared to the sample thickness (≈ 200 μm). First we have reported the quantitative analysis of a SbF_5 doped $(CH)_x$ film performed near each side of the sample and in its middle. In order to estimate the reproducibility of the results, two profiles are reported, which were measured at different sides of the films. The results are summarized in Table 1. On the other hand, in order to determine the dopant diffusion we have measured, by a same way, the antimony and iodine profiles of SbF_5 and I_2 doped polyacetylene films. Typical dopant profiles are given on Fig. 1 (iodine doped)

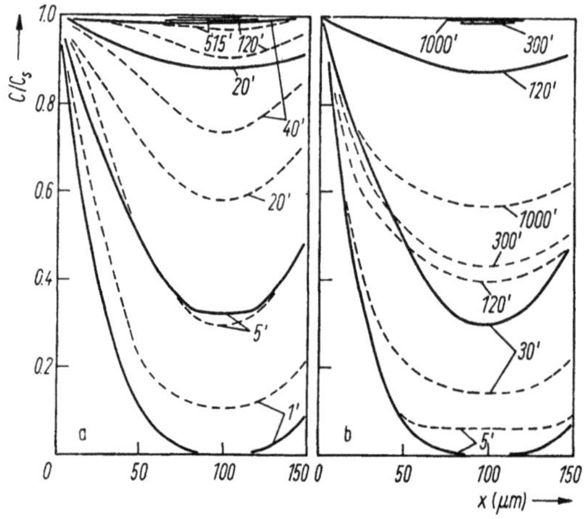

Fig. 3. Comparison between experimental and computed profiles (——— theory, – – – experimental). a) iodine ($\alpha = 0$, $\lambda = 0$, $D = 7.89 \times 10^{-8}$); b) SbF_5 ($\alpha = 0$, $\lambda = 0$, $D = 1.31 \times 10^{-8}$)

and Fig. 2 (SbF_5) for various values of the parameter y in the formula $[CH\ A_y]_x$ corresponding to different doping times. The results obtained with iodine and SbF_5 are quite similar; at low doping time only the film surfaces are covered by the dopant as proved by the schemes Fig. 1a and 2a. Then, with increasing the doping time, the dopant diffuses inside the film as observed on Fig. 1b and 2b. At last the dopant concentration becomes constant throughout the sample (Fig. 1b), we have also drawn the relative dopant concentration $C(x)/C_s$ through the sample, where C_s is the surface concentration and $C(x)$ the concentration at a depth x, for a large range of doping times. The curves obtained for iodine and SbF_5 are drawn on Fig. 3 a and b (dashed lines).

3. Discussion

3.1 Fick's law diffusion

The previous results look like the diffusion curves in solid state, so, we first tried to explain them using the well-known Fick's laws which allow to determine the concentration $C(x, t)$ at a depth x and time t by solving the equation

$$\frac{\delta C(x, t)}{\delta t} = \frac{\delta}{\delta x}\left[D(C)\,\frac{\delta C(x, t)}{\delta x}\right]. \tag{1}$$

Generally the diffusion coefficient D is taken as a constant and (1) can be written as

$$\frac{\delta C(x, t)}{\delta t} = D\,\frac{\delta^2 C(x, t)}{\delta x^2}, \tag{2}$$

in our case the boundary conditions are at $t = 0$: $C = 0$ for $0 < x < l$ and $C = C_0$ for $x \leq 0$ and $x \geq l$. At $t > 0$ where C_0 is the gas concentration of the dopant outside the film (thickness l), the well-known solution $C(x, t)$ in this case is

$$C(x, t) = C_0\left[1 - \frac{4}{\pi}\sum_{k=0}^{\infty}\frac{1}{2k+1}\sin\frac{(2k+1)\,\pi x}{l}\,\mathrm{e}^{-\left(\frac{2k+1}{l}\pi\right)^2 Dt}\right]; \tag{3}$$

if the diffusion time is long enough, this equation reduces to

$$C(x, t) \approx C_0\left[1 - \frac{4}{\pi}\sin\left(\frac{\pi x}{l}\right)\mathrm{e}^{-\left(\frac{\pi}{l}\right)^2 Dt}\right]. \tag{4}$$

The relation between C_0 and the mean value $\bar{C}$ of the dopant in the film (measured by weight uptake) is

$$\bar{C} = C_0 \left[1 - \frac{8}{\pi^2} \sum_{k=0}^{\infty} \frac{1}{(2k+1)} e^{-\left(\frac{2k+1}{l}\pi\right)^2 Dt}\right]. \quad (5)$$

The limits of $\bar{C}$ deduced elsewhere [10] are

$$\text{for a small time } t: \quad \bar{C} \approx 4 \frac{C_0}{l} \sqrt{\frac{Dt}{\pi}}, \quad (6)$$

$$\text{for a long time } t: \quad \bar{C} \approx C_0 \left[1 - \frac{8}{\pi^2} e^{-\frac{Dt}{l^2}}\right]. \quad (7)$$

3.2 Comparison with experimental results

Some remarks can be made from the sample analysis of Table 1: the two profiles obtained in different places are quite similar, so the sample appears homogeneous enough.

The amounts of oxygen and catalytic residuals (Ti, Al) are weak and of the same order of the value measured by chemical analysis.

Such results give a confirmation of the fact that a quantitative analysis can be performed even on a non-compact material such as $(CH)_x$. The Sb or F profiles indicate that an inhomogeneous concentration of the dopant is observed in the film: high concentration on the surfaces and light in the middle; elsewhere we observe (Fig. 1a) that the dopant concentration is higher on one surface than on the other: this effect can be related to the various density of the material in its two sides already observed [10].

On the other hand, we must notice that the obtained concentration of carbon (≈ 0.4) in the film, which was determined by comparison to graphite and diamond, is in good agreement with the known film density (≈ 0.4) (the true density of the material obtained by flotation technique is ≈ 1.1). We also observed a quasi-constant ratio Al/Ti through the $(CH)_x$, which is different from that of the catalyst solution. At last from the experimental curves of Fig. 3, using the above relations (5) to (7), we have estimated the diffusion coefficients of I_2 and SbF_5 in $(CH)_x$; they lie near 10^{-7} cm^2 s^{-1} for iodine and 10^{-8} cm^2 s^{-1} for SbF_5, respectively. In the case of iodine

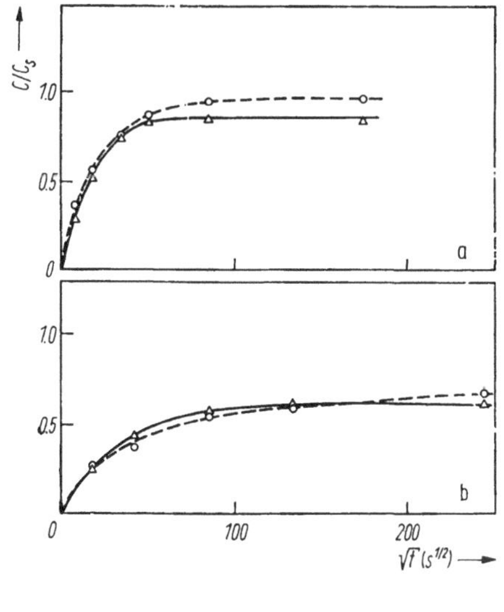

Fig. 4. Relative mean value $\bar{C}/C_s$ of dopant contained in the $(CH)_x$ films versus the square root of doping time (——— theory, – – – experimental) a) iodine ($\alpha, \lambda \neq 0$); b) SbF_5

doped $(CH)_x$ we observed that this result is near that of Beniere [10] (10^{-8} cm^2 s^{-1}), however, the discrepancy may be explained by the different doping process (vapour and liquid phase). Elsewhere, relations (6) and (7) are well verified as proved by Fig. 4 a, b.

3.3 Discussion

The classical theory of diffusion may be a good explanation of the observed phenomena of dopant fixation inside polyacetylene films with macroscopic diffusion coefficients lying in the range 10^{-7} to 10^{-9} cm^2 s^{-1} for the various species studied; however, we must notice several remarks concerning discrepancies between this theory and experiments:

(i) the surface concentration of the film, C_s is much higher than the gaseous concentration of the dopant C_0; we have observed that Beniere [10] also obtained such a result on $(CH)_x$ doped from a liquid solution of iodine. On the other hand, C_s does not increase largely with the doping time.

(ii) if the doping process is stopped after a time t_1, we immediately identify a dopant concentration profile $C(x, t_1)$ which is characteristic of t_1, but, even if the sample is kept during a very long time $t_2 \gg t_1$ in an inert atmosphere N_2 or under vacuum, this profile $C(x, t_1)$ keeps almost unchanged: however, if the diffusion was the unique phenomena, the profile $C(x, t_1)$ must change reaching first an x independant value $C(t)$ and even decreasing down to zero if the sample is kept under dynamic vacuum,

(iii) if the diffusion coefficient is calculated from a short doping time curve, it is higher than that obtained from a high level doped sample and it seems that the diffusion coefficient is largely dependent on the dopant concentration in the sample, decreasing when the concentration increases. These three different observations drive us to think that another phenomenon is supplied in addition to the diffusion and lead us to propose another model to interpret the experimental results.

4. Model of Dopant Diffusion

When the sample is plunged in the dopant atmosphere, an adsorption phenomenon takes place and the dopant concentration on its surface immediately increases to a higher value than that of the dopant atmosphere. We then suppose that this concentration does not depend on the one inside the slide. Such an effect is well enough verified but in order to simplify this study we shall neglect this effect and only consider the relative concentrations $C(x)/C_s$ normalized to unity. Such a hypothesis does not change anything in the theoretical calculations as the surface concentration is a free parameter. The remarks pointed out in the discussion and the observed adsorption phenomena drive us to think that another effect is supplied in addition to the diffusion. We have reported on Fig. 5 the schematic diagram of a possible process

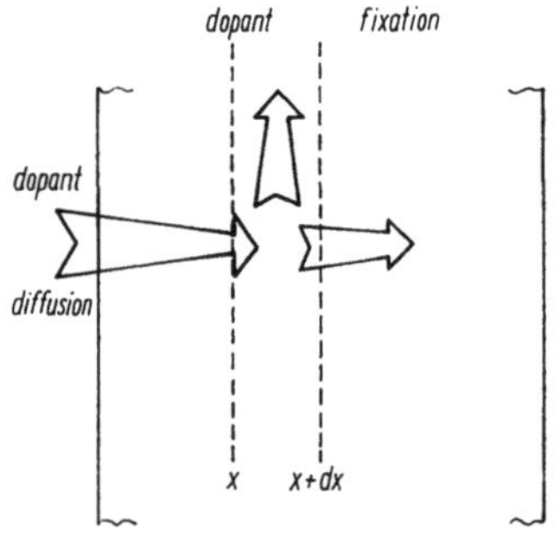

Fig. 5. Schematic model dopant diffusion and dopant fixation in the polyacetylene film

including simultaneously:

(i) the diffusion,

(ii) a chemical binding energy between the dopant and the $(CH)_x$ chains or fibrils, this latter effect taking into account both the fixation of dopant molecules around the fibril and the solid state dopant diffusion inside the fibrils. If none of these processes is very fast in respect to the other the observed profiles are a superposition of them.

We must take into account that the measured dopant concentration by means of microprobe analysis corresponds to all the molecules of the dopant lying outside and inside the fibrils.

From Fig. 5, calling $J(\Sigma)$ the dopant flow through the surface Σ, $C(x, t)$ its concentration at time t and abcissa x, the total number of dopant molecules varies in a volume $\mathrm{d}V = S\,\mathrm{d}x$ by a quantity $\mathrm{d}N = \mathrm{d}C\,\mathrm{d}V$ during the time $\mathrm{d}t$ so we can write

$$\mathrm{d}C\,\mathrm{d}V = [J(\Sigma_2) - J(\Sigma_1)]\,S\,\mathrm{d}t - \alpha C\,\mathrm{d}V\,\mathrm{d}t\,, \tag{8}$$

where the first term of the right-hand side corresponds to the diffusion (Fick's law) and the term $-\alpha C\,\mathrm{d}V\,\mathrm{d}t$ takes into account the dopant *fixation* on/in the fibrils.

We obtain

$$\frac{\delta C(x, t)}{\delta t} = \frac{\delta J}{\delta x} - \alpha C(x, t) \tag{9}$$

and using equation (2)

$$\frac{\delta C(x, t)}{\delta t} = D\,\frac{\delta^2 C(x, t)}{\delta x^2} - \alpha C(x, t) \tag{10}$$

with the same boundaries conditions as for the above calculations changing

$$C(x, t) = \overline{C}(x, t)\,\mathrm{e}^{-\alpha t} \tag{11}$$

we obtain

$$\frac{\delta \overline{C}(x, t)}{\delta t} = D\,\frac{\delta^2 \overline{C}(x, t)}{\delta x^2} \tag{12}$$

and the total (experimental) dopant concentration measured is given by

$$C_t(x, t) = C(x, t) + \alpha \int_0^t C(x, t')\,\mathrm{d}t'\,, \tag{13}$$

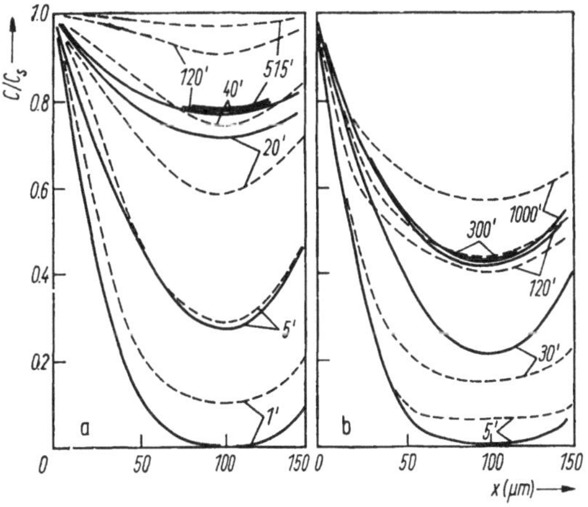

Fig. 6. Comparison between experimental and computed profiles (——— theory – – – experimental). a) iodine ($\alpha = 0.455 \times 10^{-3}$, $D_0 = 7.89 \times 10^{-8}$, $\lambda = 0$); b) SbF_5 ($\alpha = 0.3 \times 10^{-3}$, $D_0 = 1.31 \times 10^{-8}$, $\lambda = 0$)

where $C(x, t)$ is obtained from (3) and (11). Results computed from (13) are reported on Fig. 3 a, b and 6 a, b: Fig. 3 a and b correspond to $\alpha = 0$ (no chemical dopant fixation): we always note a bad fitting of experimental results. On Fig. 6 a and b where α is not zero, we observe a better fitting of experimental curves corresponding to the intermediate doping times (20′ and 40′ for iodine 30′ and 120′ for SbF_5).

However, these hypothesis must be modified in order to take account of the fact that the dopant amount absorbed inside the fibrils depends on the quantity already absorbed (α is certainly depending on concentration inside the fibril). This leads to a more complicated second term of (10) so it will be not valid for very long doping time, when fibrils are dopant saturated. At last, experiment shows that with doping, the fibrils get on size; this produces a modification of the diffusion coefficient during the doping process. Taking this into account, we have chosen a diffusion coefficient depending on the intra-fibrilar doping, $C'(x, t)$,

$$C'(x, t) = \alpha \int_0^t C(x, t)\, dt\,. \tag{14}$$

We have made a first order development of D by

$$D = D_0[1 + \lambda C'(x, t)]; \tag{15}$$

equation (9) using (1) and (15) leads to

$$\frac{\delta C}{\delta t} = \frac{\delta}{\delta x}\left(D\,\frac{\delta C}{\delta x}\right) + \alpha C$$

and finally

$$\frac{\delta C}{\delta t} = \lambda\,\frac{\delta C'}{\delta x}\,\frac{\delta C}{\delta x} + D_0\,\frac{\delta^2 C}{\delta x^2} + \lambda D_0 C'(x, t)\,\frac{\delta^2 C}{\delta x^2} + \alpha C\,.$$

In order to linearize this equation we take

$$C(x, t) = C_0(x, t) + \lambda C_1(x, t) \tag{16}$$

which leads to

$$\frac{\delta C_0}{\delta t} + \lambda\,\frac{\delta C_1}{\delta t} = \lambda\,\frac{\delta C'}{\delta x}\,\frac{\delta C_0}{\delta x} + \lambda^2\,\frac{\delta C'}{\delta x}\,\frac{\delta C_1}{\delta x} + D_0\,\frac{\delta^2 C_0}{\delta x^2} + D_0\lambda\,\frac{\delta^2 C_1}{\delta x^2} +$$

$$+ \lambda D_0 C'\,\frac{\delta^2 C_0}{\delta x^2} + \lambda^2 D_0 C'\,\frac{\delta^2 C_1}{\delta x^2} - \alpha C_0 - \alpha\lambda C_1\,. \tag{17}$$

We shall assume that the terms in λ^2 are negligible with respect to the linear term in λ and those small in λ compared to the terms in C_0 (this does not assume λ small). We check at the end of the calculus that these hypothesis is well verified, and obtain the two equations

$$\frac{\delta C_0}{\delta t} = D_0\,\frac{\delta^2 C_0}{\delta x^2} + \alpha C_0\,, \tag{18}$$

$$\frac{\delta C_1}{\delta t} - D_0\,\frac{\delta^2 C_1}{\delta x^2} - \alpha C_1 = D_0\,\frac{\delta C'}{\delta x}\,\frac{\delta C_0}{\delta x} + D_0 C'\,\frac{\delta^2 C_0}{\delta x^2}. \tag{19}$$

The boundary conditions of (19) are

$$C_1(x, 0) = 0 \quad \text{for even } x \quad \text{and} \quad C_1(0, t) = C_1(l, t) = 0 \quad \text{at each time } t\,, \tag{20}$$

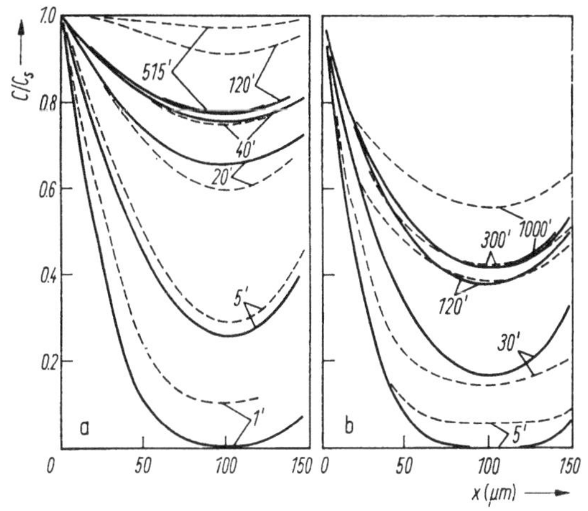

Fig. 7. Comparison between experimental and computed profiles with $D = D(C)$ (——— təqory, – – – experimental) a) iodine ($\alpha = 0.455 \times 10^{-3}$, $D_0 = 7.85 \times 10^{-8}$, $\lambda = -2.5$); b) SbF_5 ($\alpha = 0.3 \times 10^{-3}$, $D_0 = 1.31 \times 10^{-8}$, $\lambda = -2.8$)

the second term of (19) takes the role of a source. It may be solved using Green's functions with Dirichlet's conditions given by (20) (see Appendix) and we obtain

$$C_1(x, t) = \frac{2}{l} D_0 \int_0^{t'} \int_0^{l} \sum_{k'} e^{-D_0 \frac{(2k+1)^2 \pi^2 (t-t')}{l^2}} \sin \left[\frac{(2k+1)\pi x}{l}\right] \times$$

$$\times \sin \left[\frac{(2k+1)\pi x'}{l}\right] \left(\frac{\delta C'}{\delta x} \frac{\delta C_0}{\delta x} + C' \frac{\delta^2 C_0}{\delta x^2}\right) e^{-\alpha(t-t')} \, dx' \, dt \,, \qquad (21)$$

where $C_0(x, t)$ is given by (19) and C' by (14). Results obtained from (16) and (21) are plotted on Fig. 7 a and b. We can observe that those effects are not negligible and that the fitting of experimental curves is much improved.

We must notice that the interfibrilar amount of dopant is not necessary the amount of ionized dopant: indeed, the experiment shows that the doping process is partially reversible; this involve the fact that even inside the fibrils, an amount of dopant is in the molecular state I_2 or SbF_5. At last we have reported on Fig. 4 a and b, experimental and theoretical plots of the mean concentration. We can observe that the fit is good except for very long doping time, but we have already seen that our model is not valid in this case because the high intrafibrilar dopant concentration modifies the possible amount of absorption by the fibrils.

5. Conclusion

We reported here the essential part of our work concerning the doped polyacetylene films and its related properties. First we have analysed the concentration of all the components of these films exhibiting agreement with chemical analysis and essentially point out a considerable gradient of dopant concentration in lightly doped samples; such a result has a considerable impact about the previous measurements made on such a material and many of them may be reviewed, for example the electrical conductivity, optical transmission, etc. Such a result may be an explanation of the discrepancy observed in doping effect versus the employed doping process. Elsewhere far-infrared measurements of Montaner et al. [11] lead to the conclusion that there exist dopant inhomogeneities through the samples yielding variable optical refractive index n. On the other hand, we have measured the macroscopic diffusion coefficients of iodine and antimony pentafluoride in $(CH)_x$. The values reported are, in the I_2

case, in good agreement with the previous results of Beniere [10]. Concerning SbF_5, we have observed that its diffusion coefficient, one order of magnitude lower than the iodine one, is in agreement with our previous work concerning the relative stability of these dopants in $(CH)_x$ [12].

At last we have shown that it is necessary to include a chemical reaction in addition to diffusion, to interpret the measured profiles: a quantitative model taking into account these both processes was elaborated.

Appendix

Equation (19) may be written as

$$\frac{\delta C_1}{\delta t} - D_0 \frac{\delta^2 C_1}{\delta x^2} + \alpha C_1 = f(x, t) , \tag{A1}$$

where $f(x, t) = D_0((\delta C')/(\delta x)\ (\delta C_0)/(\delta x) + D_0 C'(\delta^2 C_0)/(\delta x^2))$ is a known function of space and time coordinates.

The boundary condition is the homogeneous Dirichlet one ($C_1 = 0$ for $x = 0$, $x = l$) and $C_1(x, t = 0) = 0$, for each x, we introduce the function $g(x, t)$ such that

$$C_1(x, t) = g(x, t)\ e^{-\alpha t} .$$

Hence (A1) becomes

$$\frac{\delta g}{\delta t} - D_0 \frac{\delta^2 g}{\delta x^2} = f(x, t)\ e^{\alpha t} . \tag{A2}$$

The Green's function for the diffusion equation satisfies the equation

$$\frac{\delta G(x, x', t - t')}{\delta t} - D_0 \frac{\delta^2 G}{\delta x^2} = \delta(x - x')\ \delta(t - t') \tag{A3}$$

and the solution of (A2) is

$$g(x, t) = \int_0^t \int_0^l G(x, x', t - t')\ f(x', t')\ e^{\alpha'}\ dx'\ dt' . \tag{A4}$$

We may expand G in terms of a function $\psi_n(x)$ satisfying the homogeneous boundary conditions and being a solution of the Helmholtz equation,

$$\left.\begin{aligned} &\nabla^2 \psi_n + k_n^2 \psi_n = 0 , \\ &\int_0^l \psi_n \psi_m\ dx = \delta_{nm} N_m . \end{aligned}\right\} \tag{A5}$$

We set

$$G(x, x', t - t') = \sum_n C_n(x', t - t')\ \psi_n(x) . \tag{A6}$$

Introducing this expression into (A3), one obtains

$$\sum_n \frac{\delta C_n}{\delta t} \psi_n(x) - D_0 \sum_n C_n(x', t - t')\ \psi_n''(x) = \delta(x - x')\ \delta(t - t') , \tag{A7}$$

$$\sum_n \frac{\delta C_n}{\delta t} \psi_n(x) + D_0 \sum_n C_n(x', t - t')\ k_n^2 \psi_n(x) = \sum_m \frac{\psi_m(x)\ \psi_m(x')}{N_m}\ \delta(t - t') .$$

With the help of the "closure" relation with

$$C_n(\tau) = \int C_n(\omega)\ e^{-i\omega\tau}\ d\omega$$

one obtains

$$\sum_n \int -i\omega\, e^{-i\omega\tau}\, d\omega\, C_n(\omega)\, d\omega\, \psi_n(x) + D_0 \sum_n k_n^2 \int C_n(\omega, x')\, e^{-i\omega\tau}\, \psi_n(x)\, d\omega =$$

$$= \frac{1}{2\pi} \sum_n \frac{\psi_m(x)\,\psi_m(x')}{N_m} \int e^{-i\omega\tau}\, d\omega\,.$$

We multiply by $e^{i\omega'\tau}\,\psi_l(x)$ and integrate over τ and x and obtain

$$C_l(x', \omega') = \frac{\psi_l(x')}{2\pi N_l(D_0 k_l^2 - i\omega')}.$$

By integration in the complex plane, one obtains as usual

$$C_l(x', \tau) = e^{-D_0 k_l \tau}\, \frac{\psi_l(x)\,\psi_l(x')}{N_l} \quad (\tau > 0)\,, \tag{A8}$$

$$C_l(x', \tau) = 0 \qquad (\tau < 0)\,.$$

Solutions of (A8) are

$$\psi_n(x) = \sin\left(\frac{n\pi}{l}\, x\right), \qquad N_n = \frac{l}{2}, \quad \text{and} \quad k_n = \frac{n\pi}{l}.$$

We now obtain for G with (A5) and (A6)

$$G(x, x', t - t') = \frac{2}{l} \sum_n e^{-D_0 k_n^2 (t-t')} \sin\left(\frac{n\pi}{l}\, x\right) \sin\left(\frac{n\pi}{l}\, x'\right) \qquad (\tau > 0)$$

$$= 0 \qquad (\tau < 0)\,. \tag{A9}$$

On the other side, from symmetry, only $n = 2k + 1$ will give a non-zero contribution. With the help of (A4) and (A9) one obtains

$$C_1(x, t) = \frac{2}{l} \int_0^t \int_0^l \sum_{k=0}^{\infty} e^{-D_0 \frac{(2k+1)^2 \pi^2 (t-t')}{l^2}} \sin\left(\frac{2k+1}{l}\, \pi x\right) \times$$

$$\times \sin\left(\frac{2k+1}{l}\, \pi x'\right) f(x', t')\, e^{-\alpha(t-t')}\, dx'\, dt'\,.$$

References

[1] T. Ito, M. Shirakawa, and S. Ikeda, J. Polymer. Sci. **12**, 11 (1974).

[2] C. R. Fincher, M. Ozaki, R. J. Heeger, and A. G. MacDiarmid, Phys. Rev. **19**, 4140 (1979).

[3] C. R. Fincher, M. Ozaki, M. Tawaka, D. Peebles, L. Lauchlan, and A. J. Heeger, Phys. Rev. **20**, 1589 (1979).

[4] C. K. Chiang, Y. W. Park, R. J. Heeger, M. Shirakawa, and E. J. Louis, and A. G. Mac Diarmid, J. chem. Phys. **69**, 5098 (1978).

[5] C. K. Chiang, S. C. Gau, C. R. Fincher, Y. W. Park, A. G. MacDiarmid, and A. J. Heeger, Appl. Phys. Letters **33**, 1, 18 (1978).

[6] Y. W. Park, M. A. Druy, C. F. Chiang, A. G. MacDiarmid, and A. J. Heeger, M. Shirakawa, S. Ikeda, J. Polymer. Sci. Polym. Letters **17**, 195 (1979).

[7] M. J. Rice and E. J. Mele, Solid State Commun. **35**, 487 (1980).

[8] B. Horowitz, Phys. Rev. B 22, 1101 (1980).
[9] M. Rolland, M. Cadene, J. F. Bresse, A. Rossi, D. Riviere, M. Aldissi, C. Benoit, and P. Bernier, Mater. Res. Bull., to be published.
[10] F. Beniere, S. Haridoss, J. P. Louboutin, M. Aldissi, and J. M. Fabre, J. Phys. Chem. Solids, in the press.
[11] A. Montaner, M. Galtier, C. Benoit, and M. Aldissi, in press.
[12] M. Rolland, S. Lefrant, M. Aldissi, P. Bernier, E. Rzepka, and F. Schue, J. Electr. Mater., in the press.

(Received June 5, 1981)

phys. stat. sol. (a) 68, 221 (1981)

Subject classification: 18.2; 2; 21

Department of Electrical Engineering, Imperial College of Science and Technology, London[1])

Magnetostriction Measurements on Amorphous Ribbons by the Becker-Kersten Method

By

T. H. O'DELL

The Becker-Kersten method of measuring magnetostriction, which involves observing the (M, B) loop of a long straight sample when it is in tension, is applied to two kinds of amorphous ribbon material both having very small negative magnetostriction.

Die Becker-Kersten Methode zur Messung der Magnetostriktion, die auf der Beobachtung der (M, B)-Schleife einer langen geraden Probe unter Spannung beruht, wird auf zwei Arten von amorphen Bändern angewandt, die beide eine sehr geringe negative Magnetostriktion haben.

1. Introduction

The importance of magnetostriction measurements on amorphous ribbons has been discussed by Narita et al. [1] who described a novel method involving small angle magnetisation rotation in the ribbon plane. This method is as sensitive as the methods involving strain gauges [2], capacitance transducers [3] or interferometers [4] and avoids the very difficult sample preparation problem which these direct methods require.

The saturation magnetostriction in amorphous ribbons may be very small indeed, values as low as 10^{-7} have been reported [5], and the sensitivity of the small angle magnetisation rotation method is 2×10^{-7}, [1]. To increase this sensitivity an increase in the drive field frequency would be called for and the errors introduced by skin effect, which are discussed by Narita et al. [1], would be increased. For this reason it is interesting to consider a very old method of measuring magnetostriction which was first described by Becker and Kersten [6] and involves an observation of the very low frequency, or static, (M, B) loop of the material when the sample is put under tension. This method can only be used when the sample has a negative magnetostriction, or a very small positive magnetiostriction, but it can be very sensitive indeed when applied to the amorphous ribbons because of the very small applied fields which are needed to saturate these materials. The method also has the advantage of working at very low frequency so that the finite conductivity of the material is no problem.

2. Theory

The Becker-Kersten method of measuring magnetostriction depends upon observing the (M, B) loop of the material when it is put in tension [6]. For an amorphous ribbon we may assume a simple uniaxial system because, ideally, the material itself may be assumed isotropic: the tension, σ_x, and the applied field, B_x, being applied along the sample length. We use (x, y, z) such that x lies along the length of ribbon and z is normal to the ribbon plane. The magnetisation, $\boldsymbol{M}$, is assumed to lie in the plane of the ribbon and the saturation magnetisation is M_s.

---

[1]) Exhibition Road, London SW7-2AZ, Great Britain.

A magneto-elastic energy density, $K_{\text{m-el}}$, may then be associated with the material,

$$K_{\text{m-el}} = |\lambda_{\sigma,\text{s}}\,\sigma_x| \,, \tag{1}$$

where $\lambda_{\sigma,\text{s}}$ is the saturation magnetostriction observed when stress σ_x is applied. As discussed in the standard texts [7],

$$\lambda_{\sigma,\text{s}} = 3/2\lambda_{0,\text{s}} \,, \tag{2}$$

where $\lambda_{0,\text{s}}$ is the saturation magnetostriction observed under conditions of zero applied stress and is the standard measure of the simplest kind of magnetostriction, often referred to as λ_{s}.

The effect of the magneto-elastic energy density can be represented by writing

$$Q_\lambda = 2\,|\lambda_{\sigma,\text{s}}\sigma_x|/\mu_0 M_{\text{s}}^2 \,, \tag{3}$$

a symbolism adopted from magnetic bubble domain theory [8], and representing the effect of magnetostriction by introducing an effective magnetic field,

$$B_\lambda = \mu_0 Q_\lambda M_y \tag{4}$$

directed along y. This applies for negative magnetostriction. Positive magnetostriction would be represented in just the same way but the effective field would be directed along x.

To understand the changes which take place in the (M, B) loop of the ribbon when it is in tension we must consider how the state of magnetisation changes as the applied field is increased from just above the coercive field, typically a few μT in these amorphous ribbons, to a value around 100 μT where the sample may be effectively magnetised to saturation.

The state of magnetisation of the ribbon in near-zero field may be deduced by considering the ratio Δ_λ/a where the ribbon thickness is $2a$ and

$$\Delta_\lambda = (A/K_{\text{m-el}})^{1/2} \,. \tag{5}$$

In (5), A is the exchange constant which has been deduced from the spin wave spectra of an amorphous alloy film with very low magnetostriction [9] to be around 4×10^{-12} J/m. Taking this value of A as typical, a very small value of $\lambda_{\sigma,\text{s}}$ at -10^{-7} and a typical stress of 10^8 Pa, which is well within the elastic limit of these materials, (5) gives a value of $\Delta_\lambda = 0.63$ μm. Because a typical ribbon will have $a \approx 10$ μm it follows that $\Delta_\lambda/a \ll 1$ and we could expect to find the ribbon magnetised, in near-zero applied field, in the manner sketched in Fig. 1.

In Fig. 1 there are two domains, magnetised along $\pm y$ because of the large B_λ field given by (4). This arrangement gives the lowest magnetostatic energy, the net magnetisation along y being zero, and also the lowest domain wall energy because there is only one wall. This wall, separating the two domains, may be assumed to have the simple Landau-Lifshitz structure [10] with a wall width parameter given by (5). We assume that $\boldsymbol{M}$ lies along $+x$ at the centre of the wall because Fig. 1 represents a remanent state obtained after a large positive B_x field has been applied. Under this assumption the remanent magnetisation, M_{rem}, predicted by this very simple model may be found

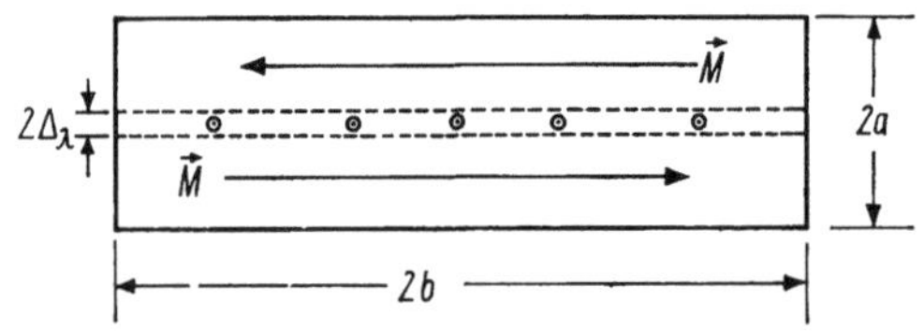

Fig. 1. When no field is applied along the ribbon axis, x, the tension, σ_x, should cause the magnetisation $\boldsymbol{M}$ to lie along y in the two domains shown

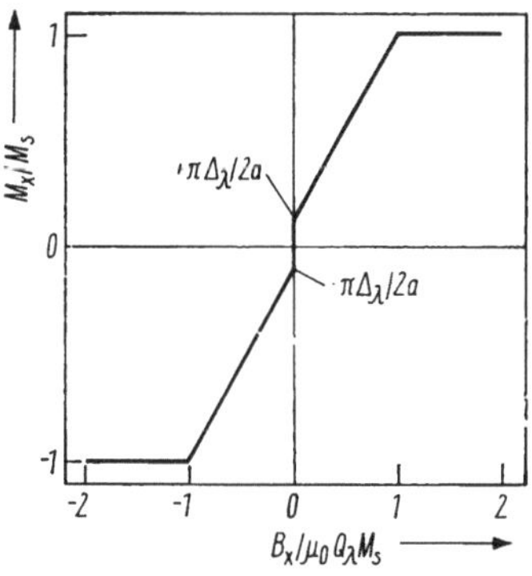

Fig. 2. The magnetisation characteristic of a ribbon under tension, predicted by the model of Fig. 1

by recalling that

$$M_x = M_s \operatorname{sech}(z/\Delta_\lambda) \tag{6}$$

in the Landau-Lifshitz wall [8] so that integration of (6) between the limits $-a$ and $+a$ and then dividing by $2a$ gives

$$(M_{\text{rem}}/M_s) = (\pi\Delta_\lambda/2a)\,, \tag{7}$$

where we assume $\Delta_\lambda \ll a$, as before.

We now consider what happens when a magnetic field, B_x, is applied along the length of the ribbon. If we continue to assume $\Delta_\lambda \ll a$, $\boldsymbol{M}$ does not vary with z over the major part of the ribbon thickness and exchange may thus be neglected. We may also assume that the magnetostatic field is simply $\mu_0\boldsymbol{M}$, when $\boldsymbol{M}$ lies in the ribbon plane because we shall have $a \ll b$ in practice and some kind of simple closure of the $\boldsymbol{M}$-field will occur at the edges of the ribbon and reduce div $\boldsymbol{M}$ to zero. It follows that only the applied field, B_x, and the effective field, B_λ, given by (4), need be considered in solving the micromagnetic equation [11]

$$(\boldsymbol{M} \times \boldsymbol{B}) = 0. \tag{8}$$

Only the z component of (8) need be considered and this can be solved at one to give

$$(M_x/M_s) = B_x/\mu_0 Q_\lambda M_s\,, \tag{9}$$

when $|B_x| < \mu_0 Q_\lambda M_s$. The magnetisation given by (9) must be added to the remanent magnetisation given by (7), because (9) was obtained by considering only the domains in Fig. 1 while (7) was obtained by considering only the wall region. The result is a magnetisation characteristic (M_x, B_x) of the form sketched in Fig. 2.

If the experimental situation can be made to tend towards the ideal situation, summarised by Fig. 1 and 2, it is clear that $\lambda_{0,s}$ may be determined by measuring the field, $B_x = \mu_0 Q_\lambda M_s$, at which the sample saturates. Alternatively, the slope of the magnetisation characteristic, (9) will give a measure of $(\mu_0 Q_\lambda)^{-1}$.

3. Experimental

Fig. 3 shows the low frequency (M, B) loop, under zero tension, of a 180 mm length of $Co_{56}Ni_{10}Fe_5Si_{11}B_{10}$ amorphous ribbon, 750 × 20 μm² cross-section, taken using a loop tracer which has been described elsewhere [12]. The value of $\mu_0 M_s$ for this ribbon is 0.6 T so that Fig. 3 shows a fairly high remanence, at 0.42 T, and a very low coercivity around 1.5 μT.

Fig. 4 shows the effect of loading the sample and, under the larger tension, it is possible to approach the kind of magnetisation characteristic shown in Fig. 2. There

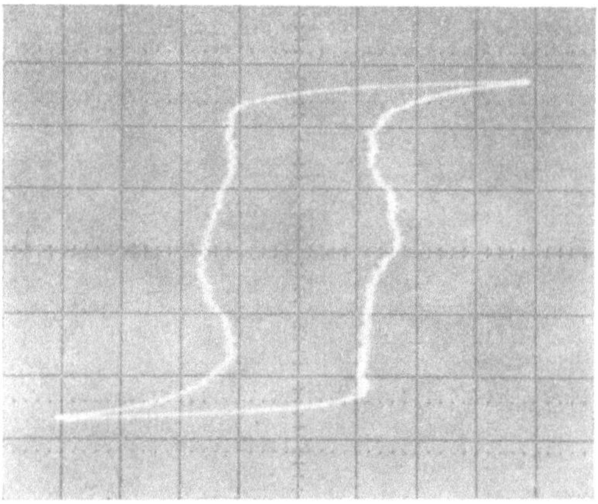

Fig. 3. The (M, B) loop taken at 9 Hz of a 180 mm length of $750 \times 20\ \mu m^2$ ribbon having a composition $Co_{58}Ni_{19}Fe_5Si_{11}B_{16}$. No tension is applied and the scales are: $\mu_0 M_s$ (vertical), 0.167 T/div; B_x (horizontal) 1.05 μT/div

is an experimental difficulty, however, in that the apparatus cannot produce an applied field large enough to saturate the sample when maximum tension is applied, Fig. 4b. Taking the slope, $d(\mu_0 M_x)/dB_x$, of the (M, B) loop shown in Fig. 4b at the points where it cuts $B_x = 0$ and using (9) to write

$$d(\mu_0 M_x)/dB_x = (Q_\lambda)^{-1} \tag{10}$$

gives a value of $\lambda_{0,s} = -3.6 \times 10^{-7}$ when (2) and (3) are used with $\mu_0 M_s = 0.6$ T and $\sigma_x = 1.38 \times 10^8$ Pa. This value of $\lambda_{0,s}$, used in (1), (2), (5), and (7) predicts a value of $\mu_0 M_{rem} \approx 0.03$ T which is in fair agreement with the value ≈ 0.04 T observed in Fig. 4b. This supports the very simple model, shown in Fig. 1, which is being used.

When tension on the sample is relaxed, Fig. 4a, the (M, B) characteristic no longer resembles the idealisation of Fig. 2 but becomes curved. The remanent magnetisation also increases more rapidly, as σ_x is reduced, than (7) predicts. Both these facts indicate that the sample, which has not been annealed, has some frozen-in anisotropy causing the magnetisation to develop an alternating component normal to the ribbon plane. Such a magnetisation structure has been observed in alloys with high magnetostriction by means of Bitter decoration technique [13], but an attempt to decorate the alloys used here was not successful. The frozen-in anisotropy should remain, superimposed upon any externally applied stress, however, so that a measure of how the applied

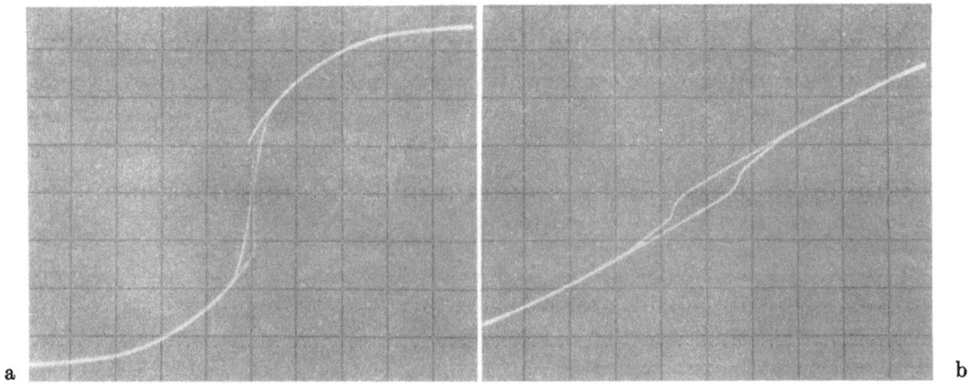

Fig. 4. Showing the effect of tension on the (M, B) loop originally shown in Fig. 3. The B_x scale is now 50.25 μT/div and the tension is a) 0.41×10^8 Pa, b) 1.38×10^8 Pa

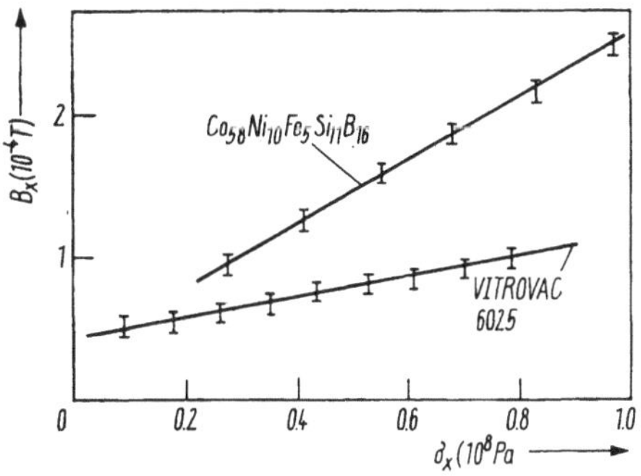

Fig. 5. Showing how the applied field, B_x, required to give $M_x/M_s = 0.9$, increased with applied tension, σ_x, in the sample used to obtain Fig. 3 and 4 and in a sample of VITROVAC 6025

field required for saturation changes with applied tension should give a result for $\lambda_{0,s}$ which may then be compared with the value deduced above from the slope of the (M, B) loop.

As Fig. 4a shows, the applied field required for saturation, which is at 3.6 vertical divisions, is not well defined. For this reason the applied field required for 90% saturation was measured as a function of applied tension. The results are shown in Fig. 5 where it can be seen that B_x increases linearly with tension and if we attribute this increase to the magnetostriction we may write

$$dB_x/d\sigma_x = 2.7 \lambda_{0,s}/M_s \tag{11}$$

from (9), (2), and (3), where B_x is now the applied field required to make $M_x/M_s = 0.9$.

Applying (11) to the results shown in Fig. 5 for $Co_{58}Ni_{10}Fe_5Si_{11}B_{16}$ gives a value $\lambda_{0,s} = -3.9 \times 10^{-7}$ which compares well with the value $\lambda_{0,s} = -3.6 \times 10^{-7}$ obtained from the slope of the (M, B) loop taken under maximum tension.

Fig. 5 also shows some results taken on unannealed VITROVAC 6025, a commercially available amorphous ribbon which is known to have very low magnetostriction. This ribbon was 3.75 mm $\times$ 30 μm cross-section and the magnetostriction was, in fact, so small that it was not possible to apply a tension great enough to make the sample show an (M, B) loop of the kind shown in Fig. 4b. The tension required would have been well above the yield stress. For this reason it was only possible to measure the magnetostriction by the second of the two methods discussed above: the slope of the (B_x, σ_x) characteristic shown in Fig. 5. The slope of the line shown in Fig. 5 for VITROVAC 6025 gives a value of $\lambda_{0,s} = -1.2 \times 10^{-7}$ when the value $\mu_0 M_s = 0.55$, given by the manufacturers, is used in (11). This result is well within the specification for this material: $|\lambda_{0,s}| < 5 \times 10^{-7}$.

4. Conclusions

The method of Becker and Kersten [6] has been shown to be useful for making a reasonably accurate measurement of the very small magnetostriction which can be found in some amorphous ribbon samples. It has been shown for one material, $Co_{58}Ni_{10}Fe_5Si_{11}B_{16}$, that close argeement is obtained between a value of $\lambda_{0,s}$ obtained from the slope of the (M, B) loop, when the sample is under very high tension, and a value obtained from the slope of the (B_x, σ_x) characteristic, where B_x is the applied field required to give $M_x/M_s = 0.9$ and σ_x is the tension.

While the method has been demonstrated only for materials with negative magnetostriction, it should be possible to use it in cases where the magnetostriction is very small and positive. Increasing tension should then bring about a small drop in the applied field required to make $M_x/M_s = 0.9$.

Acknowledgements

The author would like to thank Prof. H. Kronmüller, Max-Planck-Institut für Metallforschung, Stuttgart, and Dr. W. Kunz, Vacuumschmelze GmbH, Hanau, for the samples of amorphous ribbon used in this work.

References

[1] K. NARITA, J. YAMASAKI, and H. FUKUNAGA, IEEE Trans. Magnetics **16**, 435 (1980).
[2] R. C. O'HANDLEY, L. I. MENDELSOHN, and E. A. NESBITT, IEEE Trans. Magnetics **12**, 942 (1976).
[3] N. TSUYA, K. I. ARAI, and T. OHSAKA, IEEE Trans. Magnetics **14**, 946 (1978).
[4] P. J. M. VAN DER STRATEN, TH. KWAAITAAL, and W. M. M. M. EIJNDEN, J. Phys. Chem. Solids **41**, 925 (1980).
[5] R. HASEGAWA and R. C. O'HANDLEY, J. appl. Phys. **50**, 1551 (1979).
[6] R. BECKER and M. KERSTEN, Z. Phys. **64**, 660 (1930).
[7] L. F. BATES, Modern Magnetism, Cambridge University Press, Cambridge 1961 (p. 410).
[8] T. H. O'DELL, phys. stat. sol. (a) **48**, 59 (1978).
[9] G. SURAN and R. J. GAMBINO, J. appl. Phys. **50**, 7671 (1979).
[10] L. D. LANDAU and E. M. LIFSHITZ, Electrodynamics of Continuous Media, Pergamon Press, Oxford 1960 (p. 160).
[11] H. KRONMÜLLER, Z. angew. Phys. **32**, 49 (1971).
[12] T. H. O'DELL, INTERMAG Conf. Grenoble 1981 (paper 38.2, accepted for publication in IEEE Trans. Magnetics).
[13] R. KRISHNAN, S. PRASAD, and K. BRANSKA, J. appl. Phys. **50**, 7639 (1979).

(Received June 16, 1981)

phys. stat. sol. (a) 68, 227 (1981)

Subject classification: 14.3.1; 22

Department of Physics and Astrophysics, University of Delhi[1])

Electrical Transport Properties of Ag-Doped p-Type PbTe Epitaxial Thin Films

By

A. L. Dawar[2]), S. K. Paradkar[3]), P. Kumar, O. P. Taneja[4]) and P. C. Mathur

Hall coefficient and dc conductivity measurements are made on p-type PbTe films doped with silver. The analysis of Hall coefficient R_H and mobility μ_H ($R_H\sigma$) data shows that at higher concentrations of silver ($>$ 0.5 at%), silver provides neutral impurity centres.

Es werden Messungen des Hall-Koeffizienten und der Gleichstromleitfähigkeit an p-leitenden silberdotierten PbTe-Schichten durchgeführt. Die Analyse des Hallkoeffizienten R_H und der Beweglichkeit μ_H ($R_H\sigma$) zeigt, daß bei höheren Silberkonzentrationen ($>$ 0,5 At%) Silber neutrale Störstellen liefert.

1. Introduction

Lead telluride is a promising material because of its various device applications [1 to 3]. The extrinsic carrier concentration in this material, like other IV–VI compounds, is attributed to its native defects, the density of which depends on the deviations from stoichiometry [4, 5]. Due to the defective nature of the material the electrical properties of PbTe can be significantly changed by doping with donor or acceptor impurities. The influence of various impurities on the charge carrier concentration, however, depends to what extent the Fermi level is stabilized. In the case of complete stabilization of the Fermi level the carrier density may become even independent of impurity concentration [6, 7]. In such a case most of the impurity atoms are incorporated in the crystal lattice as electrically neutral impurity which, however, may act as scattering centres, thereby reducing the mobility of the charge carriers. Besides this, the introduction of certain materials, e.g. In in PbTe [8] makes the position of Fermi level insensitive to other donor and acceptor impurities and the location of the Fermi level thus depends upon the nature of crystal matrix and ambient temperature.

The effect of thallium impurities on the properties of bulk PbTe has been investigated by many authors [8 to 12]. It has been shown that the experimental data can be understood in terms of the existence of an impurity band introduced inside the valence band. In an earlier work [13], the authors have shown that in epitaxial thin PbTe films, thallium acts as an acceptor and the carrier concentration increases monotonically with the addition of thallium.

The present work deals with the elucidation of the influence of silver on the electrical properties of PbTe films. It has been found that silver acts as an acceptor. However, at higher concentration of silver ($>$ 0.5 at%) the carrier concentration becomes prac-

[1]) Delhi-110007, India.

[2]) Present address: Defence Science Laboratory, Metcalfe House, Delhi, India.

[3]) Present address: Regional College of Education, Ajmer (Raj.), India.

[4]) Present address: Government College, Karnal (Haryana), India.

tically independent of the silver content while the mobility decreases monotonically. The analysis of R_H and μ_H data suggest that at higher silver concentration ($>$ 0.5 at%), silver provides neutral impurity centres.

2. Experimental Details

The films (21×4 mm^2) rectangular in shape were grown from bulk PbTe (99.999% pure) by the vacuum evaporation technique under a vacuum of $\approx 5 \times 10^{-6}$ Torr on a freshly cleaved mica substrate preheated to 200 °C. The doping of silver was made by co-evaporating it from a separate heater. To obtain the required dopant concentration the rates of evaporation of the material was controlled using a quartz-crystal thickness monitor. The thickness of the films studied was in the range 0.9 to 1.1 μm. After the growth, the films were annealed in vacuum at 200 °C for 2 h and were cooled slowly in the vacuum chamber for about 4 h to stabilize the structure. The details of the measurements of parameters R_H and σ are already given in our earlier work [13]. The overall error in the Hall mobility ($R_H\sigma$) is estimated to about 5%. The sign of the Hall voltage showed that all the samples were p-type.

3. Results and Discussion

The measurements were taken on a large number of samples with different silver contents in thickness range 0.9 to 1.1 μm. No significant effect of thickness was observed. Typical data on five films with different silver contents is reported here. The variation of the Hall coefficient R_H with temperature (lg R_H versus $1/T$) for all the five films is shown in Fig. 1. The variation of free carrier concentration p ($R_H = 1/pe$) as a function of concentration of Ag at 100 K is shown in Fig. 2. It is observed from these figures that in the initial stages of doping, the free carrier concentration increases steadily with the increase of silver concentration until at about 0.5 at% of silver, saturation in p is reached. These results show that Ag acts as an acceptor impurity in PbTe, which is expected also because the univalent silver atoms substitute divalent lead atoms in PbTe, resulting in electron deficiency [15, 16]. The saturation in p for silver concentration exceeding 0.5 at% can be attributed to the impurity states of silver which stabilizes the Fermi level slightly below the valence band edge. Most of the Ag atoms in the saturation region are incorporated in the PbTe lattice as electrically neutral impurities without affecting significantly the Hall coefficient. It may be mentioned that similar saturation effects in the concentration of free electrons have been observed in In-doped PbTe and PbSnTe films [17]. The saturation effect in Ag_2Te doped p-type bulk PbTe samples has also been observed by Borisova and Dimitrova [16] for Ag_2Te concentrations exceeding about 0.5 mol%. It can be

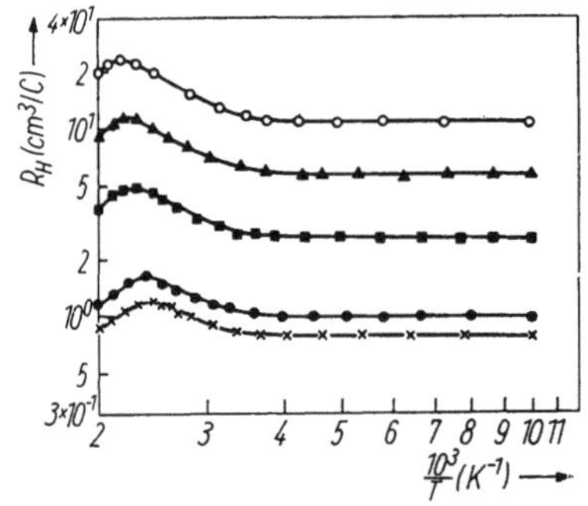

Fig. 1. Variation of Hall coefficient with temperature for various doping concentrations of Ag in films: ○ 0.1, ▲ 0.15, ■ 0.25, ● 0.6, × 1.0 at%

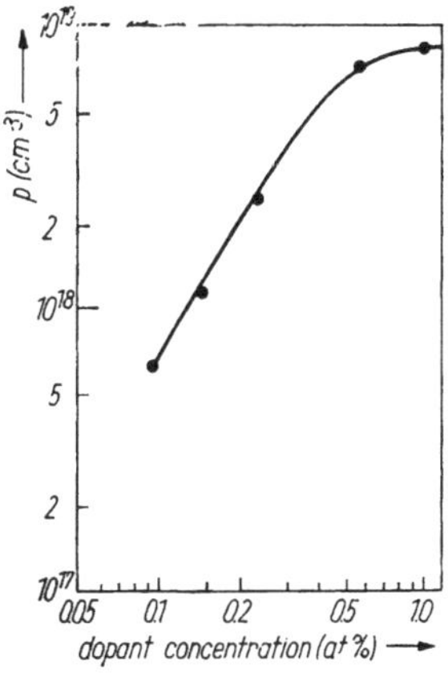

Fig. 2. Variation of carrier concentration p with dopant concentration (at%)

further observed from Fig. 1 that in the temperature range 100 to 275 K the value of R_H remains practically constant, which is characteristic of nearly extrinsic degenerate samples. The value of R_H starts increasing with a further increase of temperature beyond 275 K and reaches a maximum R_{HP} at about 460 K. The Hall coefficient again falls with the further increase of temperature. The increase in R_H above 275 K is due to the transfer of carriers from light hole to heavy hole valence band in the degenerate PbTe samples. The value of Hall coefficient above 275 K can be expressed as [18]

$$R_H = \frac{c}{ep_l} \frac{(1 + b^2 t)}{(1 + bt)^2}, \tag{1}$$

where $b = \mu_h/\mu_l$ and $t = p_h/p_l$, μ_h and μ_l being the mobilities in the heavy and light hole valence band and t is the carrier density ratio (population ratio) in the two valence bands. The temperature corresponding to the R_H peak signifies that at this temperature, the contribution of light and heavy holes to the conductivity are equal and the edges of light and heavy hole valence bands are at the same level [19]. The position of the Hall coefficient peak is observed to shift towards the lower temperature side with the increase of silver content. This may be due to either or both the following two factors:

a) reduction in the energy gap ΔE_V between light and heavy hole valence bands,
b) increase in the ratio of the mobilities of the heavy to light holes with the increase of Ag content.

Both of these possibilities have been examined by calculating the various band parameters in the following manner:

The Hall coefficient can be expressed in relation to its low temperature limit R_L by the relation [20]

$$\frac{R_T - R_L}{R_L} = (1 - b)^2 \left(\frac{m_2^*}{m_1^*}\right)^{3/2} \exp(-\Delta E_V/kT), \tag{2}$$

where R_T corresponds to the Hall coefficient at a temperature T. The variation of $(R_T - R_L)/R_L$ with temperature $(\lg (R_T - R_L)/R_L$ versus $1/T)$ for all the films is shown in Fig. 3. The value of R_L was obtained by extrapolating the low temperature Hall coefficient value from the region in which R_H remains constant [9]. The values of ΔE_V and m_2^*/m_1^* have been calculated for all the films and are shown in Table 1.

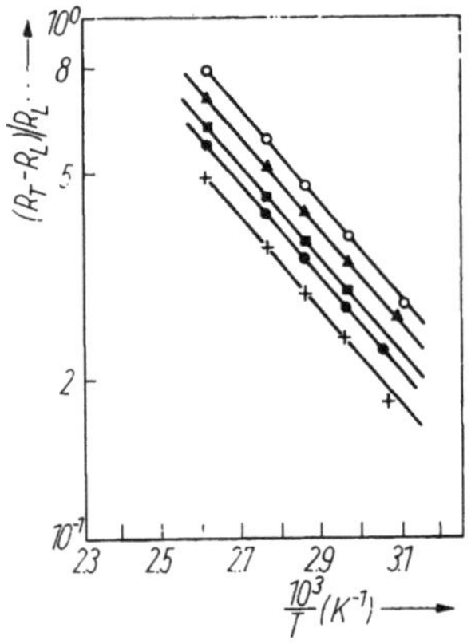

Fig. 3. Variation of Hall coefficient ratio with temperature for various doping concentrations of Ag in films: ○ 0.1, ▲ 0.15, ■ 0.25, ● 0.6, × 1.0 at%

The value of b used in these calculations was estimated from the maximum value of R_H using the relation

$$\frac{R_{HP} - R_L}{R_L} = \frac{(1-b)^2}{4b}. \tag{3}$$

These values of b are also shown in Table 1. It is observed from Table 1 that the value of ($b = \mu_h/\mu_l$) increases with the increase of carrier concentration. The possible cause of this deffect is that the increase of the carrier concentration decreases μ_l, while μ_h remains constant (since the heavy mass valence band is approximately parabolic by virtue of its distance in energy from the conduction band and large effective mass [2]). The values of b estimated in the present investigations are in good agreement with some of the reported values on bulk PbTe [9 to 11]. It can further be observed from Table 1 that the value of ΔE_V is practically independent of the concentration of silver. The value of ΔE_V determined in the present experiment corresponds to 0 K [22] if the gap is assumed to vary linearly with temperature. It may be mentioned that in our calculations for ΔE_V we have assumed the effective mass ratio m_2^*/m_1^* to be independent of temperature. If the allowance for the temperature dependence of the effective masses is taken into account, these observed values of ΔE_V will be slightly higher. It is also observed from Table 1 that the ratio m_2^*/m_1^* decreases slightly with the increases of carrier concentration, this may be due to the light increase in the light hole effective mass m_1^* with carrier concentration. Such an increase is expected in heavily doped materials on the basis of Kane's model [23].

Table 1

The doping concentration of silver, hole mobility ratio, energy separation between the two valence bands, and effective mass ratio of the heavy to light holes in various films doped with silver

sample No.	symbol	concentration of silver (at%)	ratio b of heavy to light hole mobilities	valence band separation ΔE_V (eV)	ratio m_2^*/m_1^* of heavy to light hole effective masses
1	○	0.10	0.15	0.175	39.82
2	▲	0.15	0.16	0.179	38.14
3	■	0.25	0.18	0.18	36.92
4	●	0.60	0.22	0.18	36.21
5	×	1.00	0.27	0.177	35.83

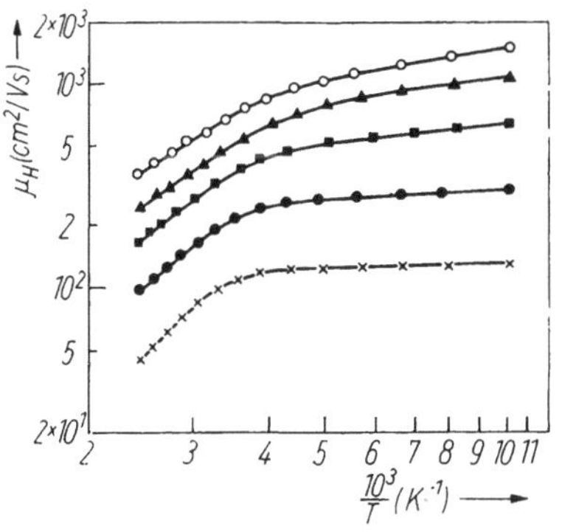

Fig. 4. Variation of Hall mobility μ_H with temperature T for various doping concentrations of Ag in films: ○ 0.1, ▲ 0.15, ■ 0.25, ● 0.6, × 1.0 at%

The variation of Hall mobility $\mu_H = (R_H\sigma)$ with temperature ($\lg \mu_H$ versus $1/T$) for all the films is shown in Fig. 4. It is found that in the entire temperature range the mobility of charge carriers decreases with the increase of silver content in the films. However, for all the films the rate of decrease of mobility for temperature exceeding about 275 K is much faster than that at low temperatures. The fast decrease of μ_H in the high temperature region is due to the onset of heavy hole conduction. The reduction in the mobility due to the increase of the silver content in the films is partly due to increase in the degeneracy of carrier distribution in the two valence bands and partly due to the predominance of neutral impurity centres provided by silver impurities. The low values of slopes of ($\lg \mu_H$ versus $1/T$) curves in the low temperature region, reduction in the value of μ_H with the increase of silver content, and saturation effect in R_H having higher silver contents suggest that silver atoms act as neutral impurity centres in PbTe.

References

[1] M. V. Fazio and W. L. Wilson, Jr., Infrared Phys. **19**, 6, 609 (1979).
[2] D. V. Eddalls, Infrared Phys. **16**, 47 (1976).
[3] Yu. I. Ravich, B. A. Efimova, and I. A. Smirnov, in: Semiconducting Lead Chalcogenides, Plenum Press, New York 1970 (p. 334).
[4] M. Fujimoto and Y. Sato, J. appl. Phys. **5**, 128 (1966).
[5] A. J. Strauss and R. F. Brebrick, J. Physique **29**, C4-21 (1968).
[6] A. A. Averkin, V. I. Kaidanov, and R. B. Melnik, Fiz. Tekh. Poluprov. **5**, 91 (1971); Soviet Phys. – Semicond. **5**, 75 (1971).
[7] I. A. Drabkin, Yu. Ya. Eliseeva, G. F. Zakharyugina, I. V. Nelson, and Yu. I. Ravich, Fiz. Tekh. Poluprov. **8**, 1947 (1974); Soviet Phys. – Semicond. **8**, 1261 (1975).
[8] B. F. Gruzinov, I. A. Drabkin, G. F. Zakharyugina, A. V. Matveenko, and I. V. Nelson, Soviet Phys. – Semicond. **13**, 190 (1979).
[9] A. J. Crocker and L. M. Rogers, Brit. J. appl. Phys. **18**, 563 (1967).
[10] R. S. Allgaier and B. B. Houston, J. appl. Phys. **37**, 302 (1966).
[11] B. F. Gruzinov, I. A. Drabkin, and Yu. I. Ravich, Soviet Phys. – Semicond. **13**, 315 (1979).
[12] B. F. Gruzinov, I. A. Drabkin, Yu. Ya. Eliseeva, E. Ya. Lev, and I. V. Nelson, Soviet Phys. – Semicond. **13** (7), 767 (1979).
[13] A. L. Dawar, O. P. Taneja, S. K. Paradkar, Partap Kumar, and P. C. Mathur, Thin Solid Films **78**, 153 (1981).
[14] Juh Tzeng Lue, J. Phys. E **10**, 161 (1977).
[15] L. D. Borisova, phys. stat. sol. (a) **53**, K19 (1979).
[16] L. D. Borisova and S. K. Dimitrova, phys. stat. sol. (a) **61**, K25 (1980).

[17] D. Eger, A. Zemel, Hadas Shtrikman, and N. Tamari, Mater. Res. Bull. **15**, 1338 (1980).
[18] T. O. Farinre and J. N. Zemel, Thin Solid Films **7**, 341 (1971).
[19] A. A. Andreev and V. N. Radionov, Soviet Phys. — Semicond. **1**, 145 (1967).
[20] L. W. Aukerman and R. K. Willardson, J. appl. Phys. **31**, 939 (1960).
[21] L. M. Rogers, Brit. J. appl. Phys. **18**, 1227 (1967).
[22] Yu. I. Ravich, B. A. Efimova, and I. A. Smirnov, in: Semiconducting Lead Chalcogenides, Plenum Press, New York 1970 (p. 119).
[23] E. O. Kane, J. Phys. Chem. Solids **1**, 249 (1967).

(Received July 10, 1981)

phys. stat. sol. (a) 68, 233 (1981)

Subject classification: 18; 1.2; 22.8

Zentralinstitut für Festkörperphysik und Werkstofforschung der Akademie der Wissenschaften der DDR, Dresden[1])

The Magnetic Properties of Cobalt-Doped Iron Sulphide $Fe_{1-y}Co_yS$ ($y \leqq 0.13$)

By

E. I. TERUKOV[2]), S. ROTH, G. KRABBES, and H. OPPERMANN

The temperature dependence of magnetic susceptibility of $Fe_{1-y}Co_yS$ ($0 \leqq y \leqq 0.13$) between 300 and 1070 K is investigated. There are three phase transitions within this temperature range: 1. a metal–semiconductor transition; 2. a spin-rotation transition; 3. a transition from an antiferromagnetic to a paramagnetic state. A phase diagram is drawn up based on the dependence of the phase transitions on the Co-concentration. The established dependence are qualitatively discussed on the basis of known models.

Es wird die Temperaturabhängigkeit der magnetischen Suszeptibilität von $Fe_{1-y}Co_yS$ ($0 \leqq y \leqq 0{,}13$) zwischen 300 und 1070 K untersucht. In diesem Temperaturbereich liegen drei Phasenübergänge: (1) ein Metall–Halbleiter-Übergang, (2) ein Spin-Umorientierungsübergang und (3) der Übergang vom antiferromagnetischen zum paramagnetischen Zustand. Aus der Abhängigkeit der Phasenübergänge von der Co-Konzentration wird ein Phasendiagramm konstruiert. Die gefundenen Abhängigkeiten werden an Hand bekannter Modelle qualitativ diskutiert.

1. Introduction

The structure of FeS corresponds to the NiAs structure. This is characterized by a compact hexagonal anion sublattice; the metal ions occupy the octahedron holes in the anion lattice. Deviations from the stoichiometry result in non-occupied sites in every second metal ion layer. They are always associated with a deviation from the ideal c/a ratio ($c/a < 1.633$) [1]. Stoichiometric FeS has three phase transitions with respect to its magnetic or structural behaviour. Above $T_N = 600$ K samples of stoichiometric composition are paramagnetic with the simple NiAs structure. Below this Néel temperature an antiferromagnetic order appears in the NiAs structure. At $T_M = 445$ K a spin rotation transition of Morin type occurs whilst the NiAs structure remains. Finally at $T_\alpha \approx 420$ K the simple NiAs-like cell changes into a superlattice cell with $c = 2C$ and $a = A\sqrt{3}$; C and A are the unit cell parameters of the simple NiAs cell. This $2C$ superstructure results from a cation shift normally to the c-axis within the base plane. This structural transition is associated with changing the electrical conductivity from a metallic above T_α to a semiconducting behaviour below T_α [2 to 4].

The physical properties of pure, stoichiometric and non-stoichiometric $Fe_{1-\delta}S$ ($0 \leqq \delta \leqq 0.14$) have been studied in detail in [2 to 5]. The influence of doping on the phase transitions in FeS are not known as well. In previous papers [6, 7] we reported studies on the influence of Co doping on the metal–semiconductor transition in FeS. This was established by conductivity measurements, structure investigations, and

[1]) Postfach, DDR-8027 Dresden, DDR.

[2]) From A. F. Ioffe Physico-Technical Institute, Academy of Sciences of the USSR, Leningrad.

DTA. The aim of this paper is the investigation of the magnetic properties of the system $Fe_{1-y}Co_yS$ ($0 \leqq y \leqq 0.13$) within the temperature range 300 to 1070 K which covers all three transitions.

2. Experiments

Susceptibility was measured by the Faraday method. The field strength at the point of the sample was about 0.7 T. HgCo $(CNS)_4$ was used as calibrating substance.

FeS and $Fe_{1-y}Co_yS$ were produced in the form of small single crystals by chemical transport [6]. A large number of such crystallites of no preferential orientation were introduced in quartz ampoules for susceptibility measurement. The ampoules were evacuated and sealed for measurement at higher temperatures.

Special care was taken to ensure that the employed samples did not have a detectable deviation from stoichiometry of the S:Me ratio because such deviations can conceal the influence of doping on the properties. To overcome this disturbing factor, a starting solid was employed for transport which contained the sulphide phase and the metallic (Co, Fe) alloy [6], or crystals with an initially higher S content were subjected to aftertreatment in temperature gradients without transport agent; stoichiometric composition was virtually reached [8] (see [6] for chemical analysis).

3. Measured Results

Fig. 1 shows the susceptibility and its inverse respectively, of $Fe_{1-y}Co_yS$ of different composition as a function of temperature. Typical anomalies corresponding to the phase transitions are revealed. Appropriate selection of the sample preparation can exclude nearly entirely any falsification resulting from uncontrolled deviations from stoichiometry. The metal–semiconductor transition at T_α and the spin rotation

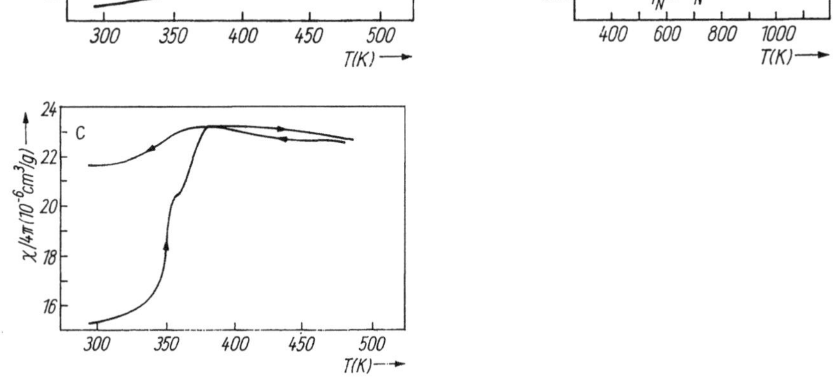

Fig. 1. Temperature dependence of a) magnetic susceptibility of $Fe_{0.97}Co_{0.03}S$ (with $T_{\alpha\uparrow}$ during heating, and $T_{\alpha\downarrow}$ during cooling); b) reciprocal magnetic susceptibility of (1) $Fe_{0.91}Co_{0.09}S$ and (2) $Fe_{87}Co_{13}S$, c) magnetic susceptibility of $Fe_{0.87}Co_{0.13}S$

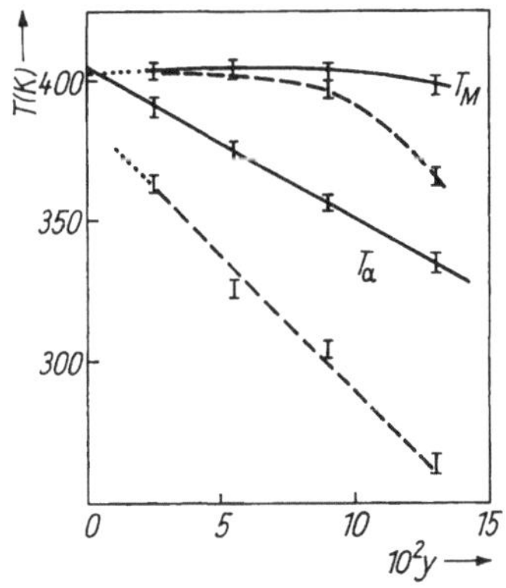

Fig. 2. Phase diagram of metal–semiconductor transition (T_α) and Morin transition (T_M) related to cobalt concentration. — — — values measured during cooling

transition (Morin transition at T_M) do not proceed suddenly, as is the case with pure FeS, but rather in a certain temperature interval. The temperatures, taken as T_α and T_M, are marked in Fig. 1a.

Corresponding to earlier results [7], doping of FeS with Co leads to a reduction of T_α, and to an increase of the hysteresis of the physical properties.

T_M of about 400 K was measured while heating up all samples of $Fe_{1-y}Co_yS$ ($0 \leqq y \leqq 0.13$). At small Co concentration ($y \leqq 0.09$) $\chi(T)$ does not show any hysteresis of T_M on cooling but hysteresis of T_M could be observed at higher concentration.

Fig. 1b shows that doping with Co leads to an increase of the Néel temperature (T_N). According to Horwood et al. [3], T_N in FeS is 600 K, and it is virtually independent of the stoichiometry. In $Fe_{1-y}Co_yS$ we measured for $y = 0.09$ and 0.13 Néel temperatures of $T_N = 620$ and 680 K, respectively.

On cooling, in $Fe_{0.87}Co_{0.13}S$ the structure transition takes place only at 263 K [7]. With this composition, therefore, it is possible to stabilize the metallic phase at room temperature. The cooling curve $\chi(T)$ (Fig. 1c) clearly contains the anomaly at the Morin transition. A change of the room temperature susceptibility $\chi(298)$ can be observed when this sample is stored at room temperature. Immediately after cooling from 480 to 298 K: $\chi(298)/4\pi = 22 \times 10^{-6}$ cm^3/g; 14 h later: $\chi(298)/4\pi = 19 \times 10^{-6}$ cm^3/g. The initial condition was $\chi(298)/4\pi = 15 \times 10^{-6}$ cm^3/g. The time dependence of magnetic susceptibility obviously correlates with the time dependence of the electrical properties of strongly doped $Fe_{1-y}Co_yS$ ($0.15 > y \geqq 0.09$) [7].

Fig. 2 presents the phase diagram, drawn on the basis of the magnetic measurements, for the metal–semiconductor transition and for the Morin transition.

4. Discussion of the Results

4.1 The influence of doping on metal–semiconductor transition

The influence of different interactions on the metal–semiconductor transition in FeS has been discussed in detail in the literature. According to Cocy et al. [9], the main reason for this transition in FeS lies in the electron–phonon interaction, and not in the electron–electron correlation. The change of the electron structure associated with this transition is caused by lattice distortion. In the one-dimensional case, the process is known as the Peierls transition [10]. This transition is also possible in the three-dimensional case if the electron spectrum fulfils, among others, the following condition:

$$\varepsilon(\boldsymbol{k} + \boldsymbol{Q}) = -\varepsilon(\boldsymbol{k}) , \tag{1}$$

where $2\boldsymbol{Q}$ corresponds to a vector of the reciprocal lattice in the undistorted state.

In many cases, condition (1) is fulfilled for uniaxial crystals due to the anisotropy of the wave functions of d-electrons; hence, in VO_2 where the metal–semiconductor transition is explained by the splitting of the quasi-one-dimensional d-band [11]. The precondition for the existence of such a band is fulfilled also in FeS. The wave function d_{z^2} from the t_{2g} triplet are aligned along the c-axis between the anions, and they can overlap each other. This can lead to the formation of a narrow d-band.

Due to the low anisotropy of the properties, a second band has to be considered in addition to this one-dimensional band. This second band originates from the other d-functions and from s–p hybridization. The metal–semiconductor transition with lattice distortion in such a two-band system was dealt with by Kuzmin and Ovchinnikov [12]. Accordingly, the width of the energy gap depends upon the deviation of the cations from the equilibrium state. Above the critical temperature, the system is metallic with undistorted lattice; at lower temperatures it is semiconducting. According to these concepts, distortion of the lattice (that means: a metal to semiconductor transition) is only possible, if in addition to (1), the dimensionless constant of electron–phonon interaction

$$\lambda = g^2/2\omega_1\omega_g \tag{2}$$

exceeds a critical limit λ_{crit}. In this context, g is the energy of the electron–phonon interaction, $2\omega_1$ the width of the one-dimensional d-band, and ω_g the frequency of the phonons with the wave vector $\boldsymbol{Q}$.

As it was shown by Ovchinnikov and coworkers [13, 14], doping increases the value of λ_{crit}. This results in narrowing the existence range of the distorted, semiconducting structure, i.e. increasing disorder by doping diminishes the phase transition temperature. Consequently, the experimentally established reduction of the temperature of the metal–semiconductor transition T_α of Co-doped FeS can qualitatively be well understood within the framework of this two-band model.

4.2 The influence of doping on spin flip transition (T_M)

This transition was observed at $T_M = 445$ K in stoichiometric FeS. The magnetic moments are arranged parallel to the c-axis below T_M, and normal above T_M. According to Horwood et al. [3], T_M declines markedly when the Fe deficiency in $Fe_{1-\delta}S$ is increased. A completely different behaviour can be observed when FeS is doped with Co. The substitution of $Fe^{2+}(3d^6)$ by $Co^{2+}(3d^7)$ with an additional electron does not change T_M (measured while heating up) within the investigated range of concentration. This points to the fact that magnetic anisotropy, which exerts a major influence on this transition, is virtually unchanged by substitution. According to [3], the following equation applies to T_M:

$$K = k_B T_M \ln \cosh \left(\frac{\lambda l_z S}{k_B T_M} \right), \tag{3}$$

where λ is the constant of spin–orbit interaction, l_z the z-component of the orbital angular momentum, K an anisotropy constant composed of the weighted anisotropy constants of the 1st and 4th order.

Analysis of the experimental results (reduction of T_M in $Fe_{1-\delta}S$, Mössbauer measurements), and (3), enabled Horwood et al. [3] to attribute the reduction of T_M to a reduction of the spin–orbit coupling as a result of intensified covalence at growing deviation from stoichiometry.

The independence of the Morin temperature T_M on doping with Co^{2+} ions sheds new light on the problem. According to X-ray examinations, both doping with cobalt and deviation from stoichiometry, lead to a reduction of c and c/a. According to [15], the reduction of the ratio c/a is an indication of an intensified covalence of the bond between the metal ions. Thus, if the observations of [3] were to be applied to Co-doped iron sulphide, a reduction of T_M must be expected, but this contradicts the experimental findings. That is why one has to assume that the dependence of T_M on the composition in $Fe_{1-\delta}S$ must be attributed to a change of the crystalline anisotropy. This assumption is confirmed by the hysteresis behaviour of T_M in strongly doped $Fe_{1-y}Co_yS$. In [7], the time dependence of the electrical properties observed with such samples is attributed to disturbances of the crystal lattice during the metal–semiconductor transition. The change of T_M in strongly doped samples during cooling indicates a pronounced dependence of the magnetic anisotropy on the disturbances in the crystal lattice. Apparently, the vacancies in non-stoichiometric FeS act in a similar manner. The vacancies, statistically distributed in the layers, influence primarily the interaction of neighbouring orbits and cause local disturbances in the crystal. These two effects, together with the spin–orbit coupling, are the cause of crystal anisotropy. It is, therefore, by no means a coincidence when the reduction of T_M was only observed in disturbed samples, i.e. in $Fe_{1-\delta}S$ ($\delta > 0$) and in $Fe_{1-y}Co_yS$ ($y > 0.1$), during cooling. In samples kept at room temperature for some days the independence of T_M with substitution of Fe^{2+} by Co^{2+} ions indicates the magnetic similarity of these ions, as well as a weak change of the spin–orbit coupling by lattice changes.

4.3 *The influence of doping on the Néel temperature*

The magnetic susceptibility $\chi(T)$ of $Fe_{1-y}Co_yS$ ($y = 0.09$, 0.13) in the paramagnetic range obeys the Curie-Weiss law (Fig. 1c). For these samples, the Néel temperature $T_N = 620$ and 680 K, respectively, is larger than $T_N = 600$ K for pure FeS. The effective magnetic moment per atom, and the paramagnetic Néel temperature θ, were determined (Table 1) on the basis of the Curie-Weiss law.

Table 1

μ_{eff}, T_N, and θ of stoichiometric FeS and $Fe_{1-y}Co_yS$

	$\mu_{eff}(\mu_B)$	T_N (K)	θ (K)
FeS [3]	5.5	600	−1160
$Fe_{0.91}Co_{0.09}S$	5.2	620	−1200
$Fe_{0.87}Co_{0.13}S$	4.9	680	−1045

The decrease of μ_{eff} with rising Co content can be understood within the framework of approximation of free ions. The small effective moment of Co^{2+} ions ($\mu_{eff}^{theor} = 3.87\mu_B$) leads to a reduction of the mean effective moment per atom when the Fe^{2+} ions ($\mu_{eff}^{theor} = 4.9\mu_B$) are substituted. This conforms with the experiment.

It is generally known that an antiferromagnetic order is caused by a negative exchange interaction between two sublattices ($A_{12} < 0$). Moreover, the exchange interaction within a sublattice, i.e. between next-but-one neighbours, has to be taken into account [16].

Evaluation of the experimental results within the framework of molecular field approximation shows that both A_{11} as well as A_{12} are negative. Yet A_{11} changes only slightly, whereas A_{12} increases continuously. The cause of this increase of A_{12} could be found in the reduction of the spacing between neighbouring metal ions along the c-axis. The fact that T_N remains constant in non-stoichiometric $Fe_{1-\delta}S$ argues against this, in spite of the fact that a reduction of c was observed here, too. It can be concluded from this that the exchange parameter A_{12} does not depend upon the spacing between the hexagonal planes. This statement agrees with the explanation given above concerning the behaviour of T_M with non-stoichiometry or doping. That is why we assume that the cause of intensified antiferromagnetic interaction is not due to a change of the lattice parameters, but rather to a change of the number of conduction electrons. The behaviour of T_M shows that the two kinds of ions are magnetically approximately equivalent, but the Co^{2+} ions have an additional valence electron. This additional electron, however, can have a strong influence on the exchange interaction, and this can be regarded as the cause of the increase of T_N.

References

[1] J. P. Suchet, Crystal Chemistry and Semiconduction in Transition Metal Binary Compounds, Academic Press, New York/London 1971.
[2] T. Hihara, J. Sci. Hiroshima Univ., Ser. A **24**, 31 (1960).
[3] J. L. Horwood, M. G. Townsend, and A. H. Webster, J. Solid State Chem. **17**, 35 (1976).
[4] H. Haraldsen, Z. anorg. allg. Chem. **246**, 169 (1941).
[5] W. Moldenhauer and W. Brückner, phys. stat. sol. (a) **34**, 565 (1976).
[6] G. Krabbes and H. Oppermann, Z. anorg. allg. Chem. **450**, 27 (1979).
[7] G. Krabbes, E. I. Terukov, and E. Wolf, Fiz. tverd. Tela, to be published.
[8] G. Krabbes, H. Oppermann, and E. Wolf, Z. anorg. allg. Chem. **423**, 212 (1976).
[9] I. M. P. Cocy, H. Roux-Buisson, and R. Brusetti, J. Physique **37**, C4–1 (1976).
[10] L. N. Bulaevskii, Uspekhi fiz. Nauk **115**, 263 (1975).
[11] J. B. Goodenough, J. Solid State Chem. **3**, 490 (1971).
[12] E. V. Kuzmin and S. G. Ovchinnikov, Fiz. tverd. Tela **18**, 750 (1976).
[13] S. G. Ovchinnikov and E. V. Kuzmin, Fiz. tverd. Tela **18**, 2074 (1976).
[14] G. V. Loseva, S. G. Ovchinnikov, V. V. Sokopovich, and E. P. Petuchov, Fiz. tverd. Tela **19**, 2922 (1977).
[15] J. B. Goodenough, in: Proceedings in Solid State Chemistry, Vol. 5, Ed. H. Reiss, Pergamon Press, Oxford 1971 (p. 145).
[16] S. V. Vonsovskii, Magnetism, Izd. Nauka, Moscow 1971 (in Russian).

(Received September 10, 1981)

phys. stat. sol. (a) 68, 239 (1981)

Subject classification: 1.1; 22.8

Department of Physics, University of Ottawa[1])

Crystal Structure of I III IV Se_4 Compounds

By

R. G. Goodchild, O. H. Hughes[2]), and J. C. Woolley

Polycrystalline samples are produced for 13 different compounds of the form I III IV □ Se_4 where I is Cu, Ag, III is Al, Ga, In, and IV is Si, Ge, Sn. Debye-Scherrer powder photographs are taken for each compound and all found to have tetragonal symmetry. The ordering lines are analyzed and it is found that all of the compounds have the same structure. Intensity ratios for the 103:112 and 211:220 lines are determined and compared with predicted values for various ordered arrangements. It is found that the compounds have the chalcopyrite structure with the lattice vacancy and the group I atom arranged at random on one cation sublattice and the group III and group IV atoms at random on the other, i.e. (I □) (III IV) Se_4.

Es werden polykristalline Proben für 13 verschiedene Verbindungen der Form I III IV □ Se_4 hergestellt, wobei I Cu, Ag, III Al, Ga und In und IV Si, Ge und Sn darstellen. Es werden Debye-Scherrer-Pulverdiagramme von allen Verbindungen gemacht und festgestellt, daß alle tetragonale Symmetrie besitzen. Die Ordnungslinien werden analysiert und gefunden, daß alle Verbindungen dieselbe Struktur haben. Intensitätsverhältnisse für die 103:112- und 211:220-Linien werden bestimmt und mit vorhergesagten Werten für unterschiedliche geordnete Anordnungen verglichen. Es wird gefunden, daß die Verbindungen Chalkopyrit-Struktur mit Gitterlücke aufweisen und das Atom der Gruppe I statistisch auf ein Kationenuntergitter und die Atome der Gruppe III und Gruppe IV statistisch auf das andere verteilt sind, d. h. (I □)(III IV) Se_4.

1. Introduction

Previous work [1 to 3] has established that many of the I III IV □ Se_4 compounds are adamantine with a tetragonal structure. It has been indicated that the c/a ratio for some of the compounds is approximately two and that the compounds probably have chalcopyrite structure. However, no detailed investigation of the structure has as yet been described. Here the X-ray diffraction results for 13 compounds of the form I III IV Se_4 have been considered in order to determine conclusively the c/a values and hence the detail of the atomic arrangement.

2. Experimental Methods

Except for the case of $CuGaSnSe_4$ [4], single crystal samples of these compounds were not available and hence the analysis has been made using the results of Debye-Scherrer powder photographs. The I III IV Se_4 compounds investigated were the twelve formed by combinations of I Cu, Ag, III Al, Ga, In, and IV Ge, Sn plus one silicon compound $AgGaSiSe_4$. These were prepared by the melt and anneal technique described previously [3] and for each compound, Debye-Scherrer powder photographs were taken using a Philips 114.6 mm camera.

---

[1]) Ottawa, Ontario K1N 6N5, Canada.

[2]) On leave from Physics Department, Nottingham University, Nottingham, England.

3. Analysis of Data

The I III IV □ Se_4 compounds are adamantine and can be considered as being developed from the cubic II–VI compounds with zincblende structure, the tetragonal distortion being due to ordering on the cation sublattice. Thus the lines observed in the powder photographs can be divided into two types: (i) those corresponding to the cubic lines, called, for convenience, 'structure lines' and (ii) those not occurring in the cubic case, caused by the ordering on the cation sublattice, i.e. 'ordering lines'. With the tetragonal structure, most of the cubic structure lines will be split, e.g. the cubic 220 and 202 lines will be separated and the separation of the structure lines gives a good indication of the c/a ratio.

For each compound, the positions of the structure lines have been measured and pseudo-cubic values obtained for the lattice parameters (a_{0c} and c_{0c}) using the Nelson-Riley extrapolation method [5] and imposing the condition in each case that the slopes of the lines of a and c against the function $1/2\,(\cos^2\theta/\sin\theta + \cos^2\theta/\theta)$ be the same.

The true tetragonal parameters can then be written as ma_{0c} and nc_{0c} where m and n are small integers. Values of m and n can then be determined from measurements of the ordering lines. For all compounds considered, the ordering line values indicated that $m = 1$ and $n = 2$, i.e. that $a_0 = a_{0c}$ and $c_0 = 2c_{0c}$. One effect of this is to double the value of the l parameter of the Miller indices from that for the cubic case, so that, for example, the 220 cubic structure line becomes the 220 and 204 lines in the tetragonal case. Ordering lines were considered only at low Bragg angles and were the same for all compounds considered, viz. 101, 103, 105, 211, 213, and 301. It was found difficult to try to detect ordering lines at higher Bragg angles, but the lines listed above were found sufficient for the analysis to be carried out.

Having determined the cell dimensions, the next problem to be considered is the ordering of the atoms on the lattice. Effectively, here there are four types of cation to be considered, viz. the I, III, and IV atoms and the lattice vacancy. The first step is to consider what arrangement of these will produce the combination of ordering lines, and hence absent spectra, listed above. The intensity of the line hkl can be written as [6]

$$I \sim \varphi(\theta)\, p A_p F^2\,, \tag{1}$$

where $\varphi(\theta) = (1 + \cos^2 2\theta)/\sin^2\theta\cos\theta$, p is the multiplicity factor for hkl, A_p is an absorption factor and

$$F^2 = \Big[\sum_n f_n \cos 2\pi(hx_n + ky_n + lz_n)\Big]^2 + \Big[\sum_n f_n \sin 2\pi(hx_n + ky_n + lz_n)\Big]^2 \tag{2}$$

the summation being over one unit cell, and f_n the atomic scattering factor of the atom at the coordinates $x_n y_n z_n$.

In the first part of the analysis, it is necessary only to determine whether I is zero or not and hence the variations of $\varphi(\theta)$, A_p and f_n with θ can be neglected. Thus f_n has been assumed proportional to the atomic number of the atom concerned and $\varphi(\theta)$ and A_p have been ignored in (1).

Since the same ordering lines are obtained with all of the compounds investigated, it is convenient to consider firstly the case of $CuGaGeSe_4$. In this case, the three cations can be taken as equivalent and the ordering lines will depend only upon the arrangement of the lattice vacancies. Initially, an analysis has been made to see what arrangement of lattice vacancies will give the observed ordering lines and absent spectra. To do this, eight different cation sites in the tetragonal lattice need to be considered — for convenience these will be labelled in terms of the cell coordinates

$x_n y_n z_n$ as follows:

A $0\,0\,0$, B $\frac{1}{2}\,\frac{1}{2}\,\frac{1}{2}$, C $0\,0\,\frac{1}{2}$, D $\frac{1}{2}\,\frac{1}{2}\,0$, E $\frac{1}{2}\,0\,\frac{1}{4}$,
F $0\,\frac{1}{2}\,\frac{1}{4}$, G $\frac{1}{2}\,0\,\frac{3}{4}$, H $0\,\frac{1}{2}\,\frac{3}{4}$.

Calculations of I from (1), (2) have been made with one vacancy kept on the site A and the second vacancy placed successively at each of the other sites. However, the combinations AC and AD were not considered since these appeared to be structurally impossible, so that five combinations, viz. AB, AE, AF, AG, and AH, were analysed. Table 1 shows the '*hkl*' values of the low-angle ordering lines for which non-zero values of I are obtained in each case. It is seen that no combination of A with a second site gives the observed combination of ordering lines and absent spectra. Thus the vacancies can not be completely ordered in these compounds.

Table 1

Ordering lines produced by the various indicated ordering of vacancies

AB	AE, AG	AF, AH
	001	001
002		
	003	003
	005	005
	010	100
101	101	101
	102	012
103	103	103
	014	104
105	105	105
110		
	111	111
	113	113
114		
	115	115
	201	201
202		
	203	203
	205	205
	210	120
211	211	211
	122	212
213	213	213
	214	124
215	215	215
	221	221
222		
	223	223
	225	225
	030	300
301	301	301
	302	032
303	303	303
305		

Another possibility is that the vacancies are arranged at random together with one of the cations on a cation sublattice. Thus the possible structures of stannite, thiogallate, and chalcopyrite have been investigated. For stannite, there are three cation sublattices, given by the combinations (i) AB, (ii) CD, and (iii) EFGH sites. The positioning of the vacancies on the AB and CD sublattices has already been considered in the above analysis but there is still the possibility of the vacancy and a cation at random on the EFGH sites. Calculations have been made for this combination again in the case of $CuGaGeSe_4$ and the resulting ordering lines are listed in Table 2. Again it is seen that this does not agree with the experimental values.

Table 2

Ordering lines produced by vacancies arranged on stannite, thiogallate, and chalcopyrite structures. Vacancies arranged at random on sites indicated

stannite EFGH	thiogallate CDFG	chalcopyrite ABEH
002	002	
	101	101
	103	103
	105	105
110	110	
114	114	
202	202	
	211	211
	213	213
	215	215
222	222	
	301	301
	303	303
	305	305

For thiogallate there are again effectively three cation sublattices viz. (i) AB, (ii) EH, and (iii) CDFG sites and again there is the possibility of random arrangement of vacancies on the CDFG sites. The ordering lines calculated for this case are shown in Table 2 and again they do not agree with the present experimental results. Finally, the chalcopyrite structure has two cation sublattices viz. (i) ABEH and (ii) CDFG sites, and the vacancies could be arranged at random on one of these sublattices. Calculations for this case give the ordering lines listed in Table 2 and it is seen that these are the same as those observed for the present quaternary compounds. Thus it is concluded that the I III IV □ Se_4 compounds have a defect chalcopyrite structure.

The final question which arises is the arrangement of the cations on the two cation sublattices. Three combinations are now possible, viz. (i) (I □) and (III IV), (ii) (III □) and (I IV), and (iii) (IV □) and (I III). To see which of these occurs, it is necessary to consider intensities of the ordering lines. It is convenient to consider these relative to the intensities of the structure lines, since the latter intensities will be the same for any of the three combinations indicated above. Thus in the powder photographs of the 13 compounds investigated, the 103 and 211 ordering lines have been used, the ratio of the intensities of the 103 to the 112 structure line and that of the 211 to the 220 line being determined. These ratios have been estimated by visual observation, two observers separately making the estimates. For each compound, the average of the four values thus obtained was determined and the resulting average values normal-

Table 3

Relative values for the ratio of the intensities of ordering to structure lines. Mean values for ratios 103:112 and 211:220

	exp.	(I □) (III IV)	(III □) (I IV)	(IV □) (I III)
$CuInSnSe_4$	10	10.0	2.1	2.1
$CuInGeSe_4$	7	6.2	0.4	6.2
$AgInSnSe_4$	7	4.7	4.7	4.7
$CuGaSnSe_4$	6	6.2	6.2	0.4
$CuAlSnSe_4$	3	3.0	12.9	0.05
$AgInGeSe_4$	3	2.1	2.1	10.0
$CuGaGeSe_4$	2	2.9	2.9	2.9
$AgGaSnSe_4$	2	2.1	10.0	2.1
$CuAlGeSe_4$	0.5	0.7	8.0	0.7
$AgAlSnSe_4$	0.5	0.5	18.0	0.5
$AgGaGeSe_4$	0.5	0.4	6.2	6.2
$AgAlGeSe_4$	<0.5	0.05	12.9	3.0
$AgGaSiSe_4$	<0.5	0.05	3.0	12.9

ized to a value of ten for the $CuInSnSe_4$ case, this having the maximum observed value. The normalized intensity values thus obtained are given in Table 3, the compounds being listed in order of the intensity values.

These values are to be compared with values calculated from (1), (2). Since each pair of lines, 103, 112 and 211, 220 are very close together in Bragg angle, again it was not necessary to allow for variation in $\varphi(\theta)$, A_p and f_n with Bragg angle in calculating the intensity ratios and so the calculations have been made as indicated above. In these calculations, the scattering factors of cations in the same row of the Periodic Table have been assumed equal, i.e. Al and Si taken as R_2; Cu, Ga, and Ge as R_3; and Ag, In, and Sn as R_4; R_2, R_3, and R_4 being taken as the mean of the appropriate atomic numbers. Calculations have then been made using (1), (2) of the intensity ratios for 103, 112 and 211, 220 lines for all possible combinations of cations obtained with one vacancy and any three of the R values. Again, the two intensity values were averaged and normalized to a value of ten for the case of (Cu □) $(InSn)Se_4$, i.e. $(R_3$ □$)$ $(R_4R_4)Se_4$.

Table 3 shows these calculated intensity ratios for each compound and for each of the three possible cation arrangements. Comparison of these values with those determined experimentally shows a very good agreement of the experimental values with those calculated for the case (I □) (III IV). The values for the compounds $CuInSnSe_4$, $CuInGeSe_4$, $AgGaSnSe_4$, $CuAlGeSe_4$, $AgAlSnSe_4$, and $AgAlGeSe_4$ rule out the possibility of (III □) (I IV) ordering, while those of $CuGaSnSe_4$, $CuAlSnSe_4$, $AgInGeSe_4$, $AgGaGeSe_4$, and $AgGaSiSe_4$ similarly eliminate the (IV □) (I III) case.

4. Conclusions

The results given above show that all 13 quaternary selenides investigated have a tetragonal structure with a c/a ratio of the order two. The same ordering lines occur in all cases and indicate that the structure is chalcopyrite with the vacancies arranged at random on one of the cation sublattices. The values in Table 3 indicate that the ordering is the same for all of the compounds with the vacancy mixed with the group I element on one sublattice and the III and IV elements mixed on the other, i.e. all compounds can be represented as (I □) (III IV)Se_4.

16*

One other effect which occurs with chalcopyrite ordering is the displacement of the anions, the selenium atoms in this case, from the ideal $\frac{1}{4}\frac{1}{4}\frac{1}{8}$ etc. positions. This displacement is small in the corresponding ternary chalcopyrite compounds and a similar result can be expected for the quaternary compounds. However, accurate intensity measurements on a large number of reflections, such as are obtained, for example, in a Weissenberg photograph, are needed in order for any calculation of the exact positions of the selenium atoms to be carried out and so that question has not been considered here.

References

[1] H. HAHN and G. STRICK, Naturwissenschaften **54**, 225 (1967).
[2] B. R. PAMPLIN, T. OHACHI, S. MAEDA, P. NEGRETE, T. P. ELWORTHY, R. SANDERSON, and J. H. WHITLOW, Ternary Compounds, Inst. Phys. Conf. Ser. **35**, 35 (1977).
[3] O. H. HUGHES, J. C. WOOLLEY, S. A. LOPEZ-RIVERA, and B. R. PAMPLIN, Solid State Commun. **35**, 573 (1980).
[4] J. A. LOPEZ-RIVERA and B. R. PAMPLIN, to be published.
[5] J. B. NELSON and D. P. RILEY, Proc. Phys. Soc. **57**, 160 (1945).
[6] N. F. M. HENRY, H. LIPSON, and W. A. WOOSTER, Interpretation of X-Ray Diffraction Photographs, MacMillan Co., Ltd., London 1960.

(Received June 25, 1981)

phys. stat. sol. (a) 68, 245 (1981)

Subject classification: 1.2 and 14.4.1; 12.2; 22.8

Sektion Physik der Martin-Luther-Universität Halle/Wittenberg, Halle (Saale)[1] *(a) and Institute of Crystallography, Academy of Sciences of the USSR, Moscow (b)*

Influence of the Mechanical Stress Component T_5 on the Elastic Behaviour of $KH_3(SeO_3)_2/KD_3(SeO_3)_2$ Crystals in the Vicinity of Their Ferroelastic Phase Transition

By

G. SORGE (a), A. ALMEIDA (a), L. A. SHUVALOV (b), and R. M. FEDOSYUK (b)

The influence of the mechanical shear stress component T_5 on the temperature dependence of the elastic stiffness coefficients c_{55} and c_{22} and of the normal stress components T_2 and T_3 on c_{11}, c_{22}, and c_{55} as well as on the corresponding attenuation coefficients is investigated experimentally. The results are in agreement with theoretical calculations based on a thermodynamical model. The shift of the minimum of the $c_{55}(T_5)$ curve is proportional to $T_5^{2/3}$.

Experimentell wird der Einfluß der mechanischen Scherspannungskomponente T_5 auf die Temperaturabhängigkeit der elastischen Koeffizienten c_{55} und c_{22} und der Normalkomponenten T_2 und T_3 auf c_{11}, c_{22} und c_{55} sowie auf die entsprechenden Dämpfungskoeffizienten untersucht. Die experimentell gewonnenen Ergebnisse stimmen mit auf der Grundlage eines thermodynamischen Modells berechneten Werten überein. Die Verschiebung des Minimums der $c_{55}(T_5)$-Kurve ist proportional zu $T_5^{2/3}$.

1. Introduction

Potassium trihydrogen selenite $KH_3(SeO_3)_2$ (KTS) and the deuterated analogous crystal $KD_3(SeO_3)_2$ (DKTS) belong to the pure fully pseudoproper ferroelastics [1 to 3]. The crystals undergo a second-order phase transition of the order–disorder type [4] with a symmetry change from the point group mmm to the point group 2/m. In the low-temperature phase a spontaneous shear component S_{5s} was observed [2]. This spontaneous shear component is the macroscopic characteristic of the phase transition [3, 5] and the order parameter η, a soft vibration mode couples linearly with the shear strain component S_{5s} (see, for example, [4]). Ferroelastic properties were found by elastic resonance measurements [2], ultrasound investigations [6, 7], and Brillouin scattering experiments [8, 9]. The generalized reciprocal susceptibility for the ferroelastic phase transition is the elastic stiffness coefficient c_{55} [6]. Further known properties of KTS and DKTS are given for example in [1 to 4].

In a recent paper we have published investigations on the influence of the mechanical stress components T_2 and T_5 on the temperature dependence of the elastic compliance coefficient s_{22} [10]. The aim of this paper is to complete the results given in [10] on the influence of the mechanical shear stress component T_5 on the ferroelastic phase transition in KTS/DKTS crystals, theoretically and experimentally. The shift of the minimum of c_{55} as well as of the anomaly of the c_{22} coefficients and the attenuation coefficients $\alpha_{z'z'}$ and α_{22}, respectively, are discussed.

---

[1]) Friedemann-Bach-Platz 6, DDR-4010 Halle (Saale), GDR.

2. Experimental

Investigations on the temperature dependence of elastic stiffness coefficients and corresponding attenuation coefficients were performed by measurements of ultrasound velocity and ultrasound attenuation with pulse technique [11] at 20 MHz. The samples will be rectangular parallelepipeds with lengths between 1.5 mm, if the velocity has an order of magnitude of about 500 m/s, and 8 mm and a quadratic side length of 5.5 mm. Surfaces for transducers were hand-lapped until they were parallel within 0.005 mm. Disk quartz transducers with a diameter of 5 mm were bonded with Nonaq for KTS and epoxy resin without activator for DKTS crystals. It was possible to detect variations of sound velocity greater than 0.3 m/s. The absolute error of the velocity is less than 5 m/s. The notation of axes is the same as used in [2]. The temperature control system is described elsewhere [12]. The accuracy of temperature control was 0.02 K and the absolute error of measurements 0.1 K.

The shear stress component T_5 ($0 \leq T_5 < 40 \times 10^5$ N m^{-2}) was produced by applying an external uniaxial stress T ($0 \leq T < 80 \times 10^5$ N m^{-2}) to the specimens as it is shown schematically in Fig. 1. Together with $T_5 = \frac{1}{2}|T|$ also T_1 and T_3 are produced. We have already noticed [10] that the normal stress component T_2 does not influence the behaviour of the velocity v_{yy} at the Curie point. The same is shown for example in Fig. 2 a and b for T_3 and T_2 experimentally and can also be shown theoretically. From our experimental values of T_n ($n = 1, 2, 3$) we could not observe any shift of θ_0 within our experimental accuracy of 0.1 K in the vicinity of the transition.

In order to investigate the shift of the minimum of the $c_{55}(\theta)$ curve as a function of T_5 theoretically we neglect terms with T_1 and T_3 in the theoretical treatment. For an experimental determination of c_{55} it is necessary to measure the velocity $\bar{v}_{z'z'}$ of a quasi-longitudinal mode. From the equation

$$\varrho \bar{v}^2_{z'z'} = \frac{1}{2} \{\frac{1}{2} (c_{11} + c_{33} + 2c_{55}) + [\frac{1}{4} (c_{11} - c_{33})^2 + (c_{13} + c_{55})^2]^{1/2}\}$$

the relation

$$\varrho \bar{v}^2_{z'z'} \approx c_{55} + \text{const}$$

follows in the vicinity of the transition temperature. One can find an explanation of this approximation for example in [6, 7].

Measurements of the quantities $\bar{v}_{z'z'}$, v_{yy}, $\alpha_{z'z'}$, and α_{yy} at the same specimens could be performed at the same temperature and at the same stress without a rearrangement of the sample and without a change of stress conditions (Fig. 1).

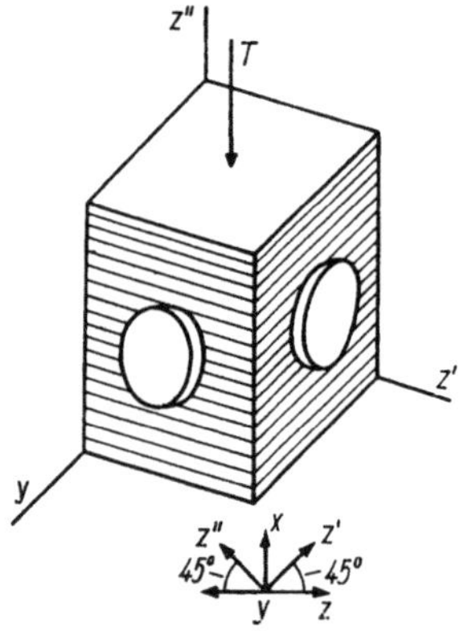

Fig. 1. Arrangement of quartz transducer on the specimen for determination of the velocities $\bar{v}_{z'z'}$ and v_{yy} and the stiffness coefficients c_{55} and c_{22}, schematically (for further explanation see the text)

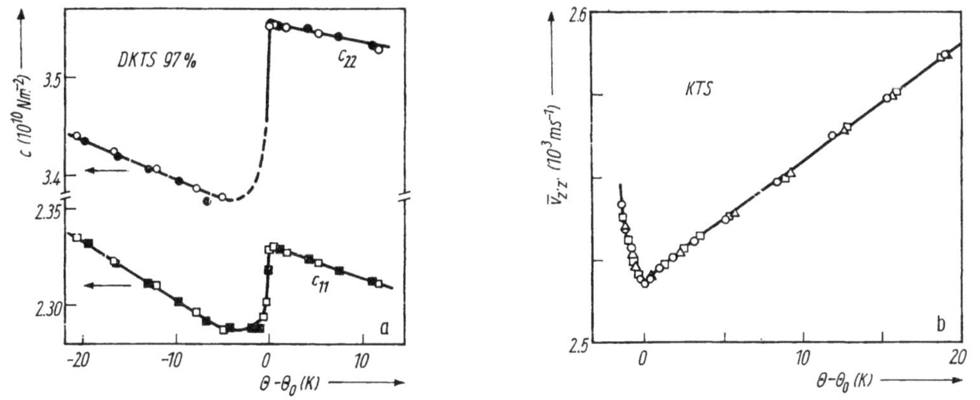

Fig. 2. a) Temperature dependence of the stiffness coefficients c_{22} and c_{11} ($f=20$ MHz) of $KD_3(SeO_3)_2$ 97% crystals in the vicinity of the phase transition with the normal stress component T_3 as the parameter. ○, □ $T_3 = 0$; ●, ■ 70.0×10^5 N m^{-2}. b) Temperature dependence of the stiffness coefficient c_{55} ($f = 20$ MHz) for KTS near the phase transition, with the normal stress component T_2 as the parameter. ○ $T_2 = 0$, △ 35.5×10^5, □ 70.0×10^5 N m^{-2}

3. Theory

We start from the same Gibbs function as reported in [10],

$$G(\theta, T_i, \eta_r) = G_0(\theta) - \tfrac{1}{2} s_{ij}^{\eta} T_i T_j + \tfrac{1}{2} A_{rr}\eta_r^2 + \tfrac{1}{4} B_{rrrr}\eta_r^4 - K_{ri}T_i\eta_r - \\ - M_{rij}T_iT_j\eta_r - N_{rri}T_i\eta_r^2 \,, \tag{1}$$

where θ is the temperature, s_{ij}^{η} the elastic compliance coefficients at constant order parameter, and K_{ri}, M_{rij}, N_{rri} are the coupling coefficients which describe the interaction between order parameter and mechanical stress components. Only the coefficient $A_{rr} = a_{rr}(\theta - \theta_0)$, where θ_0 represents the phase transition temperature for a mechanical free crystal, depends on temperature.

This Gibbs function (1) describes the temperature dependence of the elastic compliance coefficients s_{ij}. By a conversion of elastic stiffness coefficients into elastic compliance coefficients with the formula

$$s_{ij} = \frac{(-1)^{i+j} \Delta_{ij}^c}{\Delta^c},$$

where Δ^c is the determinant and Δ_{ij}^c a subdeterminant of the matrix of the elastic stiffness coefficients and analyzing experimental results on the temperature dependence

Table 1

Maximal error of conversion of elastic stiffness coefficients into elastic compliance coefficients for KTS/DKTS

coefficient	relation	error (%)
s_{11}, c_{11}	$s_{11} \approx 1/c_{11}$	20
s_{22}, c_{22}	$s_{22} \approx 1/c_{22}$	10
s_{33}, c_{33}	$s_{33} \approx 1/c_{33}$	30
s_{55}, c_{55}	$s_{55} = 1/c_{55}$	0

of elastic stiffness coefficients c_{ij} [6, 7], we can estimate a maximal error of the following relations (Table 1):
Taking into account these results it is possible to describe by (1) the temperature behaviour of c_{55} after conversion exactly and the temperature dependence of c_{22} with an accuracy of 10%.

If we assume that only $T_5 \neq 0$ and all other mechanical stress components are negligible (see Section 2) we must distinguish two cases in the vicinity of the phase transition:

(i) small stress component T_5,

$$T_5^2 \ll \frac{A_{55}^3}{3B_{55555}K_{55}^2}. \tag{2}$$

From the equilibrium conditions two temperatures θ_1 and θ_2 are distinguishable;

a) above θ_1

$$\theta > \theta_1 = \theta_0 - \frac{1}{a_{55}} \sqrt[3]{3B_{5555}K_{55}^2T_5^2},$$

$$\tilde{c}_{55} = \frac{A_{55}}{K_{55}^2} + 3B_{5555}\frac{T_5^2}{A_{55}^2}, \tag{3}$$

$$\tilde{c}_{22} = c_{22}^\eta - 4N_{552}^2 c_{22}^{\eta 2} \frac{K_{55}^2 T_5^2}{A_{55}^3}; \tag{4}$$

b) below θ_2

$$\theta < \theta_2 = \theta_0 - \frac{1}{a_{55}} \sqrt[3]{\frac{9}{4} B_{5555}K_{55}^2T_5^2},$$

$$\tilde{c}_{55} = \frac{1}{K_{55}^2}\left(-2A_{55} \pm \frac{3B_{5555}K_{55}T_5}{-A_{55}} \sqrt{\frac{-A_{55}}{B_{5555}}}\right), \tag{5}$$

$$\tilde{c}_{22} = c_{22}^\eta - \frac{2N_{552}^2 c_{22}^{\eta 2}}{B_{5555}} \pm \frac{N_{552}^2 c_{22}^{\eta 2} K_{55} T_5}{A_{55}^2 \left(1 + \frac{2N_{552}^2 c_{22}^\eta}{B_{5555}}\right)^2} \sqrt{\frac{-A_{55}}{B_{5555}}}. \tag{6}$$

If we assume a small stress T_5, the theory predicts a shift $\Delta\theta_a$ of the temperature θ_a of the minimum of the elastic stiffness coefficient c_{55} to higher temperatures with increasing stresses corresponding to the relation

$$\Delta\theta_a = \theta_a - \theta_0 = \frac{1}{a_{55}} \sqrt[3]{6B_{5555}K_{55}^2T_5^2} \sim T_5^{2/3}. \tag{7}$$

This shift is analogous to the known result of the shift of the maximum of the dielectric constant of proper ferroelectrics with a second-order phase transition;

(ii) very high stress component T_5,

$$T_5^2 \gg \frac{4A_{55}^3}{27B_{5555}K_{55}^2}. \tag{8}$$

In this case it follows from the equilibrium conditions that a phase is only stable for temperatures

$$\theta > \theta_0 - 3\sqrt[3]{B_{5555}K_{55}^2T_5^2}$$

and no transition takes place. We get

$$\tilde{c}_{55} = c_{55}^{\eta} = \text{const}\,, \tag{9}$$

$$\tilde{c}_{22} = c_{22}^{\eta} - \frac{4}{3}\,\frac{N_{552}^{2} c_{22}^{\eta 2}}{B_{5555}}\,. \tag{10}$$

We see later that this case could not be fulfilled experimentally, because of breaking of the specimens.

4. Experimental Results and Discussion

The temperature dependence of c_{55} (DKTS 97%) $\sim \bar{v}_{z'z'}^2$ and c_{22} (KTS and DKTS 97%) with T_5 as the parameter as well as of the corresponding attenuation coefficients are shown in Fig. 3 and 4 a and b. In Fig. 3 a shift of the minimum of c_{55} as well as of the anomaly in the attenuation coefficient $\alpha_{z'z'}$ for DTKS of 97% to higher temperatures with increasing stress T_5 is demonstrated. A shift is also observable by the anomaly (change of slope) of c_{22}. In more detail this is shown in Fig. 5 where the experimental data are plotted on an extended scale.

The straight lines in Fig. 6 where the shift of the minimum of c_{55} versus $T_5^{2/3}$ is plotted demonstrate the validity of (7) in KTS as well as in DKTS. It must be remarked that the straight lines do not cross the zero point. One reason for this behaviour seems to be the small internal mechanical stress which differs from specimen to specimen. For very small values of T_5 we could observe experimentally small shifts of the minimum of c_{55} to lower temperatures. An analogous behaviour is known in ferroelectrics if an external electric field is applied and internal fields are present in the specimen (see, for example, [13, 14]). But we must also take into account that we could not prove the stress state in the specimens. May be that the stress state is not homogeneous or further unknown mechanisms take place. The slope of the straight lines are nearly the same for KTS as well as for DKTS (Fig. 6). It is possible to calculate from the slope of the straight line in Fig. 6 the coefficient in (7),

$$\frac{1}{a_{55}} \sqrt[3]{6 B_{5555} K_{55}^2} \approx 2.17 \times 10^{-4}\, K N^{-2/3} m^{4/3}\,.$$

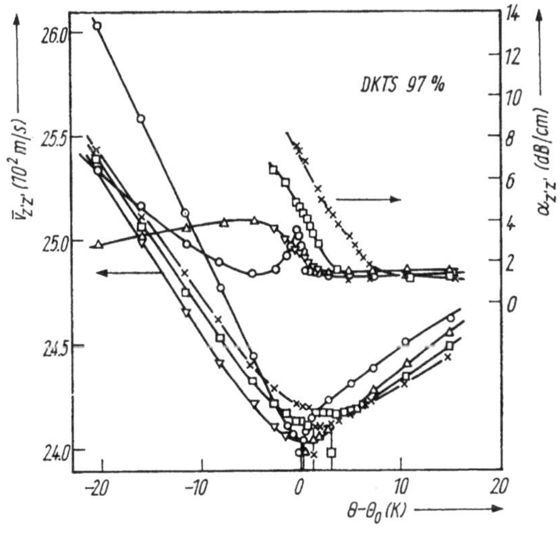

Fig. 3. Temperature dependence of the elastic stiffness coefficient $c_{55} \sim \bar{v}_{z'z'}$ and the corresponding attenuation coefficient $\alpha_{z'z'}$ in the vicinity of the phase transition with the shear stress component T_5 as the parameter. ○ $T_5 = 0$, △ 12.2×10^5, × 20.4×10^5, □ 37.6×10^5 N m^{-1}

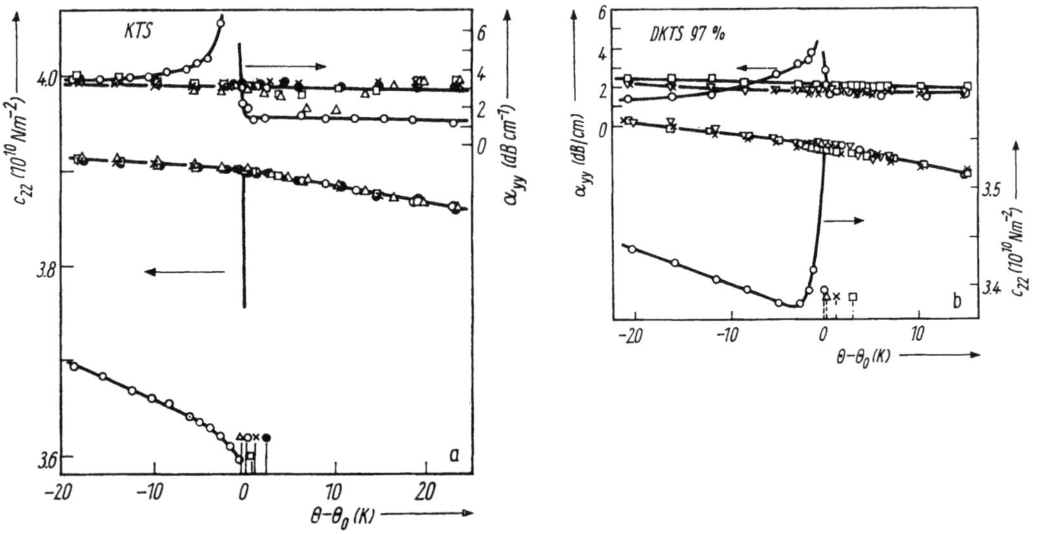

Fig. 4. Temperature dependence of the elastic coefficients c_{22} ($f = 20$ MHz) and the corresponding attenuation coefficients α_{yy} for KTS (a) and DKTS 97% (b) in the vininity of the phase transition with the shear stress component T_5 as the parameter. a) ○ $T_5 = 0$, △ $7.4 \times 10_5$, □ 15.0×10^5, × 17.6×10^5, ● 29.6×10^5 N m^{-2}; b) ○ $T_5 = 0$, △ 12.2×10^5, × 20.4×10^5, □ 37.6×10^{-5} N m^{-2}

From our measurements, together with known values for the elastic Curie constant $C^+ = 13.3 \times 10^{-9}$ N^{-1} m^2 K [3, 6] we can determine $a_{55} = 7.5 \times 10^7$ K_{55}^2 and $B_{5555} = 7.2 \times 10^{11}$ K_{55}^4. Since KTS/DKTS are pseudoproper ferroelastics it is impossible to separate the constants B_{5555}, a_{55}, and K_{55} by our experimental method.

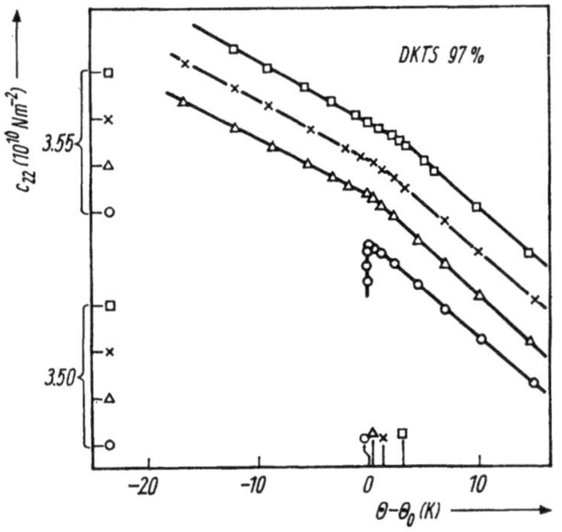

Fig. 5

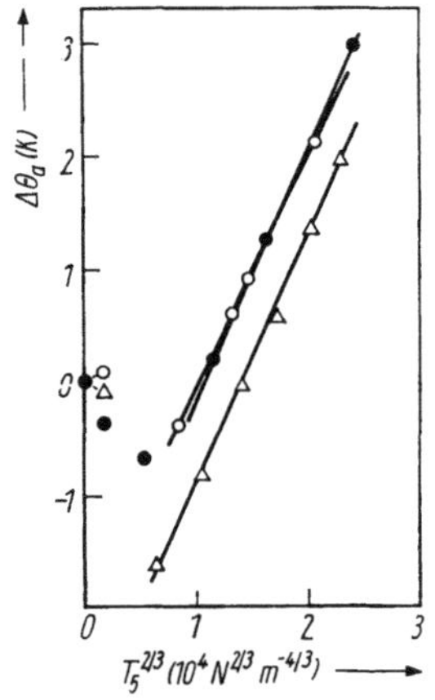

Fig. 6

Fig. 5. Temperature dependence of the stiffness coefficient c_{22} ($f = 20$ MHz) of DKTS 97% near the transition with the shear stress T_5 as the parameter (extended scale). ○ $T_5 = 0$, △ 12.2×10^5, × 20.4×10^5, □ 37.6×10^5 N m^{-2}

Fig. 6. The shift $\Delta\theta_a$ of the temperature θ_a of the minimum of the stiffness coefficient c_{55} of KTS and DKTS crystals versus the term $T_5^{2/3}$, where T_5 means the mechanical shear stress component. ○ KTS, △ DKTS 87%, ● DKTS 97%

The experimental results on c_{55} and c_{22} are conform with theoretical predictions. According to (3) we observe in Fig. 3 immediately above the temperature of the minimum (up to $\Delta\theta \approx 3$ K) that c_{55} increases with increasing stresses T_5 ($T_5 > 7 \times 10^5$ N m^{-2}). Below the minimum equation (5) describes the temperature behaviour of c_{55}. With increasing stress T_5 ($T_5 > 7 \times 10^5$ N m^{-2}) also c_{55} increases. From the shape of the c_{55}-curves in Fig. 3 we can conclude that from a theoretical point of view our stresses T_5 are small, because we always observe a minimum and not straight lines (equation (9)) up to the break stress. In agreement with (4) and (6) we can observe in Fig. 4 and 5 that the jump of c_{22} at the transition becomes smaller and smaller and than vanishes. The behaviour of c_{22} above θ_0 resembles equation (4) obviously (Fig. 6). But below θ_0 equation (6) disagrees with the experimental behaviour. We think that a main part of the jump is caused by domain effects [10] and only a very small part of the jump reflects the monodomain state or the theory. This means (equation (6)) $N_{552}^2 c_{22}^{\eta 2} \ll B_{5555}$. The theory which is valid for a monodomain crystal also predicts that the jump vanishes. It seems possible that deviations in the experimental behaviour of c_{22} below θ_0 are caused by effects of higher order or the error in the relation $c_{22} \approx 1/s_{22}$ of 10% is to high and an application of the potential (1) is impossible in this case.

The temperature dependence of the attenuation coefficient $\alpha_{z'z'}$ also demonstrates an influence of T_5. We observe a shift of the anomaly to higher temperatures with increasing stress T_5 according to the shift of the minimum of c_{55}. The different shapes of the curve for $T_5 = 0$ and $T_5 = 12.2 \times 10^5$ N m^{-2} or higher values give an indication on the domain state of the specimens.

The influence of T_5 on the domain state of the samples also is shown by the temperature dependence of the attenuation coefficient α_{yy} in Fig. 4. The anomaly in the α_{yy} curves disappears at stresses higher than the coercive stress of $\approx 7 \times 10^5$ N m^{-2} as already reported in [10, 15].

Acknowledgement

The authors would like to thank Prof. Dr. G. Schmidt for his continuous interest in this work, his helpful discussions and advices.

References

[1] L. A. SHUVALOV, N. R. IVANOV, and T. K. SITNIK, Kristallografiya **12**, 366 (1977).
[2] N. R. IVANOV, L. A. SHUVALOV, G. SCHMIDT, and E. STOLP, Izv. Akad. Nauk SSSR, Ser. fiz. **39**, 933 (1975).
[3] Y. MAKITA, F. SAKURAI, T. OSAKA, and I. TATSUZAKI, J. Phys. Soc. Japan **42**, 518 (1977).
[4] S. GRANDE, H.-D. MECKE, and L. A. SHUVALOV, phys. stat. sol. (a) **46**, 547 (1978).
[5] Y. MAKITA, Y. YAMAGUCHI, and S. SUZUKI, J. Phys. Soc. Japan **43**, 181 (1977).
[6] T. V. QUYNH, G. SCHMIDT, N. R. IVANOV, and L. A. SHUVALOV, phys. stat. sol. (a) **36**, K85 (1976).
[7] Y. MAKITA and F. SAKURAI, Phys. Letters A **55**, 435 (1976).
[8] Y. MAKITA, T. YAGI, and J. TATSUZAKI, Phys. Letters A **55**, 437 (1976).
[9] T. YAGI, H. TANAKA, and J. TATSUZAKI, J. Phys. Soc. Japan **41**, 717 (1976).
[10] G. SORGE, A. ALMEIDA, H. BEIGE, and L. A. SHUVALOV, phys. stat. sol. (a) **61**, 201 (1980).
[11] H. J. MC. SKIMIN, J. Acoust. Soc. Amer. **33**, 12 (1961).
[12] G. SORGE, G. SCHMIDT, W. FREIDANK, and U. KLAPPERSTÜCK, phys. stat. (a) **19**, K43 (1973).
[13] G. SORGE and W. FREIDANK, Kristall und Technik **9**, K1 (1974).
[14] G. SCHMIDT, phys. stat. sol. **8**, 41 (1965).
[15] P. CEVC, R. BLINC, R. SRINIVASAN, J. ZUPANČIČ, and L. A. SHUVALOV, phys. stat. sol. (a) **57**, 789 (1980).

(Received July 14, 1981)

phys. stat. sol. (a) 68, 253 (1981)

Subject classification: 10.2; 14.3.3; 22.1.2

Zentralinstitut für Elektronenphysik der Akademie der Wissenschaften der DDR, Abteilung Realstruktur, Berlin[1])

Nucleation of Oxygen Precipitations and Efficiency of Internal Gettering Centres in Czochralski Silicon

By

H. PEIBST and H. RAIDT

Nucleation and growth of $SiO_{1\,\mathrm{to}\,2}$ precipitates in Cz-silicon are studied by EBIC, HVTEM, X-ray and etching methods. Concentration, structure, and gettering efficiency of the microdefects are determined in dependence on material parameters and annealing conditions. By reason of the experimental results Cz-silicon wafers with a high concentration of gettering centres and a denuded surface layer can be obtained. The wafers prepared in this way can be successfully applied to an intrinsic gettering technique.

Es wird über Untersuchungen zur Keimbildung von $SiO_{1\,\mathrm{bis}\,2}$-Präzipitaten in Cz-Silizium berichtet. Mit EBIC-, HVTEM-, Röntgen- und Ätzmethoden werden Konzentration, Struktur und Getterwirksamkeit der Mikrodefekte in Abhängigkeit von Materialparametern und Temperbedingungen bestimmt. Im Ergebnis der Untersuchungen können Scheiben mit einer hohen Konzentration getterwirksamer Zentren und einer defektfreien Oberflächenschicht erhalten werden, die gut für eine "intrinsic"-Getterung geeignet sind.

1. Introduction

Since the first publication by Tan et al. [1] the fundamentals of an internal or intrinsic gettering technique were repeatedly investigated and as far as we know the technique itself is already well established in several device technologies. Nevertheless a great deal of problems are not yet solved. These problems are mainly due to the very sensitive influence of sometimes unknown material parameters on the pre-annealing procedure and on the specific behaviour of pre-annealed wafers in certain technologies of electronic devices. Lack of knowledge in fundamental parameters leads to more or less complicated annealing procedures and it is argued that the intrinsic gettering technique is not applicable in each case [2, 3].

The aim of this paper is to show some relations between the quality of wafers and the nucleation of gettering centres during a pre-annealing procedure, to give results about the influence of annealing ambients on the denudation of a surface layer, the gettering behaviour in pre-annealed wafers and the degradation of the centres in the following technological steps.

At any rate a good gettering technique requires

a high density of small defects in the bulk material acting as sinks for different harmful impurities,

a denuded surface layer which is free of crystal defects impairing the quality of electronic devices,

a high stability or regeneration of the gettering centres in the bulk to survive the different steps of device technology.

With the help of annealing procedures in different ambients, an extensive investigation of the crystallographic nature of defects and of their electric behaviour, some new aspects for an improved intrinsic gettering technique can be pointed out.

---

[1]) Mohrenstr. 40/41, DDR-1080 Berlin, GDR.

2. Experimental Procedure

B-doped ⟨111⟩ Cz-silicon wafers with 76 mm diameter, 380 μm thickness, and an oxygen content of (7 to 12) $\times 10^{17}$ atoms cm^{-3} were used in this investigation. Most of the samples were free of defects. Some of the wafers with high oxygen levels however contained grown-in microdefects ($\approx 2 \times 10^8\ cm^{-3}$). $1 \times 2\ cm^2$ pieces of the wafers were annealed by at least three different temperature–time programmes shown in Fig. 1. The annealing atmosphere was dry O_2 in most cases with an amount of 3% HCl in the first step of the three-step annealing procedure. Some experiments with an inert atmosphere (N_2, Ar) were done for comparison. For studying the reactions of the generated microdefects at higher temperatures the precipitate-forming annealing process was followed by a 10 to 20 h 1200 °C heat treatment with contamination of the sample.

After each of the annealing steps samples were taken to investigate the concentration, distribution, and crystallographic structure of the generated microdefects by etching, EBIC, X-ray, and HVTEM methods. By means of etching methods we got a general view on the distribution of the defects. X-ray topography could not resolve single defects because of their high concentration, but we got some informations from double crystal diffractometry.

EBIC micrographs proved to give the best results in determining the defect concentration in the range of 10^7 to $10^8\ cm^{-3}$. Therefore, Al Schottky diodes of $0.5\ cm^2$ were prepared and irradiated by 15 to 25 keV electrons from the Al side. For the determination of the depth distribution neighbouring samples were stepwise thinned (2 to 10 μm steps) by mechanical and/or chemical polishing to a depth of 50 μm.

Former investigations have shown that defects in silicon, which are small compared with the excited volume give only considerable contrast if they are decorated [4]. So the EBIC contrast can also be used for a qualitative estimation of the gettering efficiency of crystal defects. For a qualitative determination measurements of the diffusion length of minority carriers were done by means of EBIC-line scans on inclined {111} cleaving planes. The defect structure of some selected samples could be determined by HVTEM investigations.

None of the methods is able to detect the nuclei, but TEM investigations show the precipitates in a very early state (size 100 to 200 Å). The comparison of TEM and EBIC measurements shows that in the concentration range of 10^7 to $10^9\ cm^{-3}$ nearly all initial precipitates detectable with TEM grow during annealing to decorated precipitate–dislocation complexes. Therefore, we assume that the defect concentration measured by EBIC is nearly equal to the primary concentration of the nuclei

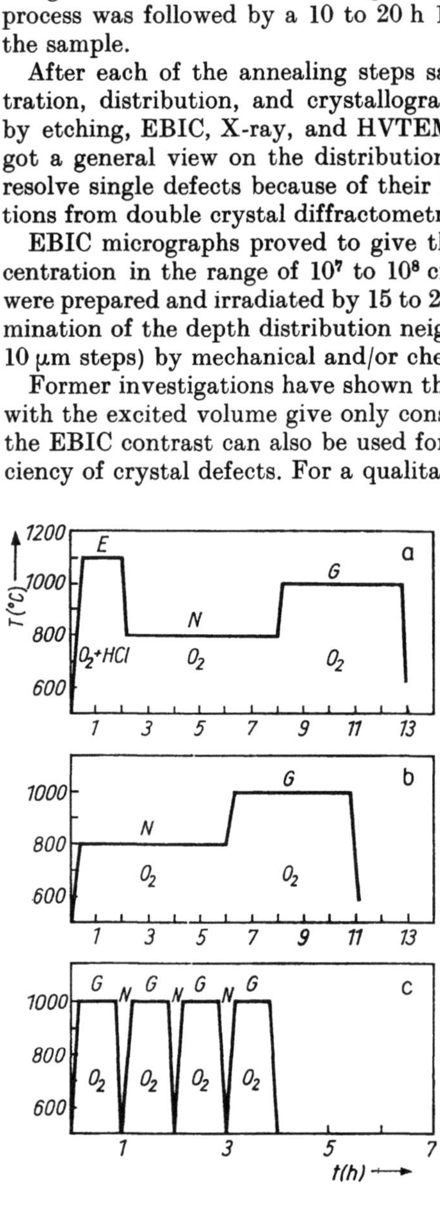

Fig. 1. Typical annealing programmes used in this study; a) three-step, b) two-step, c) oscillating; E extraction, N nucleation, G growth

3. Results and Discussion

3.1 Nucleation of $SiO_{1\ to\ 2}$ precipitates

In real crystals the nucleation of precipitates takes place mostly at lattice defects, but also in dislocation free very pure silicon the nucleation energy and therefore the nucleation rate is influenced by other impurities and even by the possible supersaturation of point defects. Therefore, to understand the different results reported up to now [5 to 9] the thermal history (e.g. growth conditions, donor killer annealing), the concentration of dopants and even an amount of impurities far from saturation have to be taken into account.

To realize the conditions for a successful gettering technique a high nucleation rate of precipitates is needed. On the other hand, if at a given supersaturation of oxygen only a small number of nuclei exists, then these nuclei will grow very fast and in their strain field secondary crystal defects are generated, which can extend into the defect free surface layer. For this reason the initial nucleation is the most important step in the getter centre formation processes.

3.2 Nucleation in wafers containing grown-in microdefects

To study the influence of grown-in microdefects on the $SiO_{1\ to\ 2}$ precipitation we compared two groups of wafers with nearly the same $[O_i]$ concentration but with and without grown-in microdefects. These defects are visible in EBIC micrographs (Fig. 2) and optically after etching but they are hardly detectable by TEM methods. Only very weak cloudy contrasts are visible, which may be due to clustering of point defects and/or impurities. Whether these clusters are identical with the so-called B- or C-swirl defects in Cz-silicon detected by de Kock et al. [10, 11] is not clear. The A- and B-defects should have a striated distribution which could not be observed in our case.

In wafers containing grown-in defects no influence of the annealing conditions and the $[O_i]$ content on the concentration of precipitates could be observed. Only their

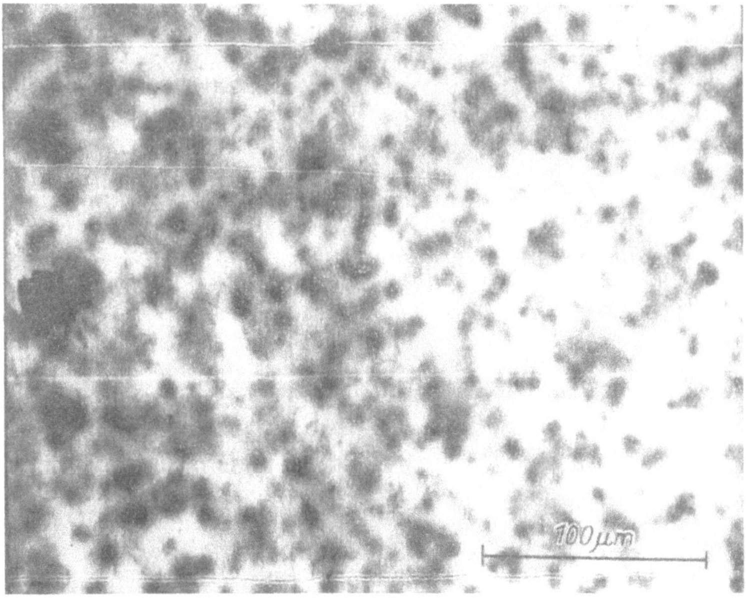

Fig. 2. EBIC micrograph (25 kV) of grown-in defects in Cz-silicon

crystallographic structure was altered. By annealing the wafers 2 to 5 h at 1000 °C the initial defects transform to heavily decorated aggregates of precipitates, loops, and stacking faults. During long time annealing up to 20 h at 1200 °C these complexes are destroyed and simple stacking faults are formed, which give only poor EBIC contrast (i.e. no gettering).

With crystals containing grown-in defects no defect free surface layer was obtained and no gettering effect during the following processes was detectable by measuring the diffusion length. Therefore, silicon wafers of this type are not applicable for an intrinsic gettering technique.

3.3 Relation between [O_i] concentration and nucleation rate

To investigate the effect of the initial [O_i] concentration, that means of the supersaturation, on the nucleation rate we used defect free wafers with different [O_i] concentrations varying from (7 to 12) $\times 10^{17}$ atoms cm^{-3}. Although this range is small the concentration of precipitates is changed by more than one order of magnitude as is shown in Table 1.

Table 1

[O_i] content and concentration of precipitates (some examples of experimental results)

[O_i] (10^{17} atoms cm^{-3})	800 °C, O_2 t(h)	1000 °C, O_2 t(h)	precipitate concentration (cm^{-3})	
			depth 20 to 24 μm	depth 2 to 6 μm
7.5	2	10	1×10^7	1×10^7
9.5	2	10	1×10^8	5×10^7
12.0	—	4.1	1.5×10^8	1×10^6

To get a sufficient precipitate concentration of about 10^8 cm^{-3} different annealing programmes are necessary for wafers with high and low oxygen level, respectively. Samples with [O_i] $= 1.2 \times 10^8$ atoms cm^{-3} show a high nucleation rate up to 1000 °C and nuclei in a sufficient number are formed already during heating with a rate of 1500 K h^{-1}. Periodic cooling and heating prevents the precipitates formed first to grow to an unwanted size (see annealing programme c in Fig. 1). Wafers with a low oxygen level (7×10^{17} atoms cm^{-3}) require a low temperature annealing step at the temperature of maximum nucleation rate (b in Fig. 1). As mentioned before this temperature depends not only on the oxygen content but also on the concentration of other impurities which in some cases, e.g. carbon, lower the nucleation energy. For the material used in this study we found an optimum nucleation rate at about 800 °C.

Using wafers with a low oxygen content the duration of the 800 °C anneal strongly determines the final precipitate concentration but at 1000 °C the supersaturation is too small, no additional nuclei are formed and only the size of the precipitates is controlled by this annealing time. Therefore, the interval anneal gives poor results and a 5 to 6 h 800 °C anneal is necessary to obtain a sufficient number of nuclei.

3.4 Suppression of nucleation in the surface layer

To avoid an unwanted near-surface nucleation [12, 13] there are two possible ways:

One can prepare a denuded zone by out-diffusion of oxygen so that the [O_i] supersaturation near the surface and therefore the nucleation rate tends to zero.

One can enhance the nucleation energy using special annealing ambients, e.g. O_2 or (O_2, N_2) mixtures with a few per cent HCl. Whether the lowering of nucleation rate in this case is due to an altered equilibrium state of point defects or to an extraction of fast diffusing metal atoms is not yet clear.

In wafers with a higher oxygen level (12×10^{17} atoms cm^{-3}) the nucleation in the bulk is already considerable at high temperatures, whereas an out-diffusion of oxygen suppresses the nucleation near the surface. In our experiments we obtained a defect free surface layer without a preceding high temperature step. However, at lower oxygen levels (7×10^{17} atoms cm^{-3}) we found a maximum of the nucleation rate at about 800 °C. The out-diffusion of oxygen at this temperature is unsatisfactory. A three-step anneal with a high temperature step at the beginning was necessary to prepare a denuded zone. Fig. 3 and 4 show a comparison of the depth distribution after the two-step and three-step anneal, respectively. The higher density of near-surface defects in the samples with two-step anneal may also be due to a contamination or to lattice distortions produced during lapping which are extracted or removed at higher temperatures.

We obtained the best results in denudation with a dry O_2 atmosphere during the nucleation step. An inert gas atmosphere (N_2, Ar) produces surface defects and pitting at the surface (see also [14]), but no evident change of the nucleation rate was found. However, an addition of 3% HCl to the O_2 used in the first high temperature step (1100 °C) reduces the nucleation rate. To get the same depth of the defect free layer one can shorten the annealing time to 1/4 in comparison with a pure O_2 atmosphere. This is in contradiction to the results of Craven [9] (results see Table 2). Using HCl mixtures in the 800 °C anneal no good results were obtained.

Table 2

Some experimental results of precipitation in O_2 + HCl atmosphere

1100 °C O_2 + HCl	800 °C O_2	1000 °C O_2	concentration of precipitates in different depths		
t (h)	t (h)	t (h)	2 μm	20 μm	40 μm
0	6	5	1.5×10^8	$(1 \text{ to } 2.5) \times 10^8$	$(6 \text{ to } 9) \times 10^7$
0	6 (+ HCl)	5	0*)	$(2.5 \text{ to } 5) \times 10^7$	$(1 \text{ to } 1.5) \times 10^8$
2	2	5	0	0*)	$(1 \text{ to } 3) \times 10^6$
4	6	5	0	0*)	0*)

*) Not detectable by SEM/EBIC.

The efficient suppression of nucleation in the surface layer can hardly be understood with simple models. Whether an extraction of heavy metal atoms, an enhanced out-diffusion of oxygen [15], an altered nucleation energy, or an influence of the annealing ambient on the equilibrium state of point defects is the dominant factor cannot be decided on the basis of our results.

That there is an extraction of impurities could clearly be seen by means of the very poor EBIC contrast of the precipitate complexes near the denuded zone. Using a 2 h first annealing step (1100 °C, O_2 + 3% HCl) the contrast measured after the complete annealing process is strongly reduced to a depth of 40 μm, however, the precipitate free layer is about 15 μm thick.

3.5 *Getter efficiency of the microdefects*

To get useful intrinsic gettering centres does not simply mean to generate SiO_2 precipitates and secondary defects, e.g. punching loops, but also to make sure that

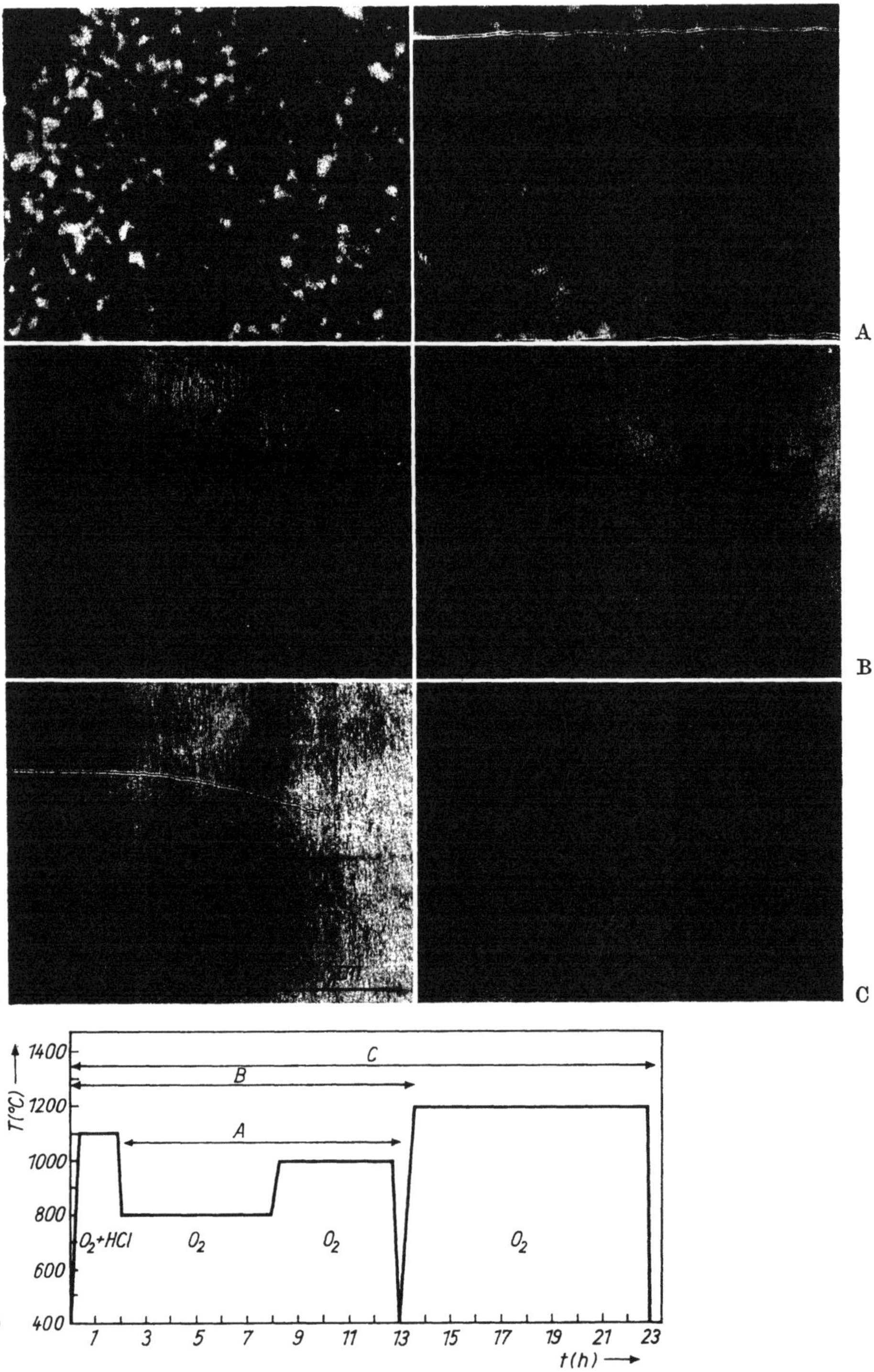
A
B
C
a
b
T(°C)
1400
1200
1000
800
600
400
C
B
A
O2+HCl
O2
O2
O2
1
3
5
7
9
11
13
15
17
19
21
23
t(h)

Fig. 3

Fig. 3. Result of different annealing processes on the formation and activation of gettering centres; a) EBIC micrographs (25 kV) at different distance from the surface: 2 μm (left-hand side), 20 μm (right-hand side) (the streaks are due to disturbances of the EBIC signal), b) corresponding annealing programmes

these microdefects are effective sinks for heavy metal impurities and keep this property during the following processes.

We tested this behaviour by submitting the sample to a high temperature anneal (1200 °C) up to 20 h with some contamination from the atmosphere. The gettering of the microdefects was observed by means of the EBIC method.

The precipitates nearly invisible in a surface region after HCl treatment showed a continually increasing contrast with annealing time. Contrary to the samples without precipitates the diffusion length of minority carriers was only slightly reduced after anneal.

Table 3

Decrease of the diffusion length (λ) after different heat treatments ($[O_i] = 9.4 \times 10^{17}$ atoms cm^{-3})

annealing conditions				λ (μm)	
1100 °C O_2 + HCl t (h)	800 °C O_2 t (h)	1000 °C O_2 t (h)	1200 °C O_2 t (h)	wafer I	wafer II*)
—	—	—	—	95	20 to 36
—	6	10 (+ HCl)		10	—
—	6	5	—	—	12
4	6	5	—	60	—
2	6	5	—		22 to 26
2	6	5	10	—	10 to 25

*) Wafer II with strong striations.

Table 3 shows examples of the decrease of the diffusion length after various heat treatments. It is remarkable that the getter efficiency of microdefects still exists after a high temperature treatment also in samples with low oxygen contents of $\approx 7.5 \times 10^{17}$ atoms cm^{-3}, which is only slightly above the saturation limit at 1200 °C.

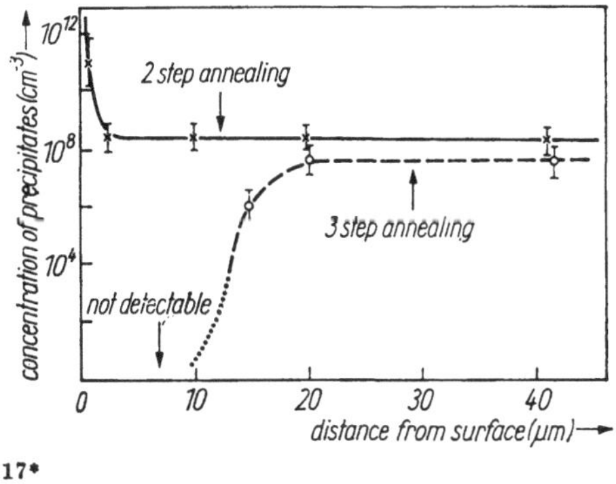

Fig. 4. Depth distribution of getter centres

4. Conclusions

For an improved gettering technique the presence of "grown-in" microdefects has to be taken into account. In this case the concentration of gettering centres is determined by the pre-existing defects and it is very difficult to suppress a heterogeneous nucleation of SiO_x precipitates in a near-surface layer.

In defect free wafers containing a high $[O_i]$ concentration ($\geqq 12 \times 10^{17}$ atoms cm^{-3}) at temperatures of about 1000 °C a sufficient nucleation rate is achievable. In consequence the nucleation of gettering centres and the formation of a denuded surface layer is possible in the same pre-annealing procedure.

Defect free wafers with a lower $[O_i]$ content ((7 to 10) $\times 10^{17}$ atoms cm^{-3}) can also be used for an intrinsic gettering technique but in this case at least a three-step annealing is necessary (1. out-diffusion of O and possibly extraction of impurities, 2. nucleation of $SiO_{1\ \text{to}\ 2}$ precipitates, 3. growth and stabilisation of the gettering centres). Each of the steps causes a different temperature. Commonly one or two steps are already present in some device technologies.

By annealing in O_2 with an admixture of a few per cent HCl the precipitate free surface layer is thicker or will be achieved in shorter time in comparison with a pure O_2 atmosphere. An inert gas atmosphere (N_2, Ar) sometimes produces erosion of the wafer surface.

Harmful secondary defects can be avoided by a sufficiently high density of small precipitates and slow heating rates. As it was shown by measurements of the diffusion length in the SEM an introduction of such annealing steps into the technology of electronic devices is very useful. The generated gettering centres were efficient even after high temperature steps.

Only if all parameters of Cz-silicon wafers are known (defect structure and impurity content) an optimum pre-annealing procedure for a given technology can be given.

A universal annealing programme for different material qualities and different technologies does not exist.

Acknowledgements

The authors wish to thank H. Fabig and D. Spaniol (ZIE) for performing the annealing experiments, W. Möhling and T. H. Jährling (ZIE) for X-ray investigations, and M. Pasemann (IFE, Halle) for HVTEM analysis.

References

[1] T. Y. Tan, E. E. Gardner, and W. K. Tice, Appl. Phys. Letters **30**, 175 (1977).
[2] W. J. Patrick, W. M. Hu, and W. A. Westdorp, J. appl. Phys. **50**, 1399 (1979).
[3] D. Huber, P. Stallhofer, and M. Blätte, in: Semiconductor Silicon, The Electrochem. Soc., Princeton (N.J.) 1981 (p. 756).
[4] H. Menniger, H. Raidt, and R. Gleichmann, phys. stat. sol. (a) **58**, 173 (1980).
[5] S. Kishino, Y. Matsushita, and M. Kanamori, Appl. Phys. Letters **35**, 213 (1979).
[6] F. Shimura, H. Tsuya, and T. Kawamura, Appl. Phys. Letters **37**, 483 (1980).
[7] H. Tsuya, K. Ogawa, and F. Shimura, Japan. J. appl. Phys. **20**, L31 (1981).
[8] N. Inoue, K. Wada, and J. Osaka, see [3] (p. 282).
[9] A. Craven, see [3] (p. 254).
[10] A. J. R. de Kock, W. T. Stacy, and W. M. van de Wijgert, Appl. Phys. Letters **34**, 611 (1979).
[11] A. J. R. de Kock and W. M. van de Wijgert, J. Crystal Growth **49**, 718 (1980).
[12] K. Kugimiya, S. Akiyama, and S. Nakamura, see [3] (p. 294).
[13] R. W. Series, K. G. Barraclough, and W. Bardsley, see [3] (p. 304).
[14] M. Futakami and M. Hamazaki, Japan J. appl. Phys. **17**, 1343 (1978).
[15] R. A. Rozgonyi and C. W. Pearce, Appl. Phys. Letters **31**, 343 (1977).

(Received August 24, 1981)

phys. stat. sol. (a) 68, 261 (1981)

Subject classification: 1.4 and 10.2; 13.4; 22.2.1

Institute of Inorganic Chemistry,
Academy of Sciences of the USSR, Siberian Branch, Novosibirsk (a) and
Institute of Solid State Physics and Electron Microscopy,
Academy of Sciences of the GDR, Halle (Saale)[1]) *(b)*

The Influence of the Residual Impurities (C, O) on the Crystal Structure of GaAs

By

L. A. Borisova (a) and M. Pasemann (b)

GaAs crystals undoped and doped with oxygen as well as with oxygen and carbon are investigated by means of TEM. In undoped crystals where the total concentration of residual impurities is about one order higher than the electrically active concentration small defects are observed which by comparison with oxygen doped crystals are supposed to be amorphous clusters due to a decomposition of the solid solution GaAs:O. In slowly cooled oxygen-doped GaAs crystals precipitations of As_2O_3 should be responsible for the decreasing of the electrical activity of oxygen by half an order.

Mittels TEM werden GaAs-Kristalle (undotiert oder dotiert mit Sauerstoff bzw. Sauerstoff und Kohlenstoff) untersucht. In undotierten Kristallen, in denen die Gesamtkonzentration an Verunreinigungen eine Größenordnung höher ist als die elektrisch aktive Konzentration, werden kleine Defekte beobachtet, die anhand des Vergleichs zu sauerstoffdotierten Kristallen als amorphe Cluster infolge einer Entmischung der festen Lösung GaAs:O interpretiert werden. In langsam abgekühlten sauerstoffdotierten GaAs-Kristallen sollten As_2O_3-Ausscheidungen verantwortlich für die um eine halbe Größenordnung verringerte elektrische Aktivität des Sauerstoffs sein.

1. Introduction

It is known that there is some disagreement between the amount of the residual impurities (C, O) in GaAs crystals and their electrical activity [1]. Therefore, it is the aim of this work to investigate the microstructure of single crystals of GaAs undoped or doped with oxygen and carbon and to reveal defects which prevent the electrical activity of the residual impurities.

2. Experimental Technique

Single crystals of GaAs were grown by direct crystallization in a horizontal rotating container [2] at a velocity of 1.5 cm/h. Oxygen doping was made by introducing As_2O_3 into the melt, carbon by addition of graphite powder enriched with the ^{14}C isotope.

Samples of three groups were investigated:
1. Undoped crystals, $[C] = 5 \times 10^{16}\ cm^{-3}$, $[O] = 10^{17}\ cm^{-3}$;
2. oxygen-doped crystals, $[C] = 5 \times 10^{16}\ cm^{-3}$, $[O] = 5 \times 10^{17}\ cm^{-3}$;
3. oxygen- and carbon-doped crystals $[C] = 2 \times 10^{17}\ cm^{-3}$, $[O] = 5 \times 10^{17}\ cm^{-3}$.

One part of the samples of each group was slowly cooled down during 3 to 5 h, another one was quenched in air.

[1]) Weinberg 2, DDR-4010 Halle (Saale), DDR.

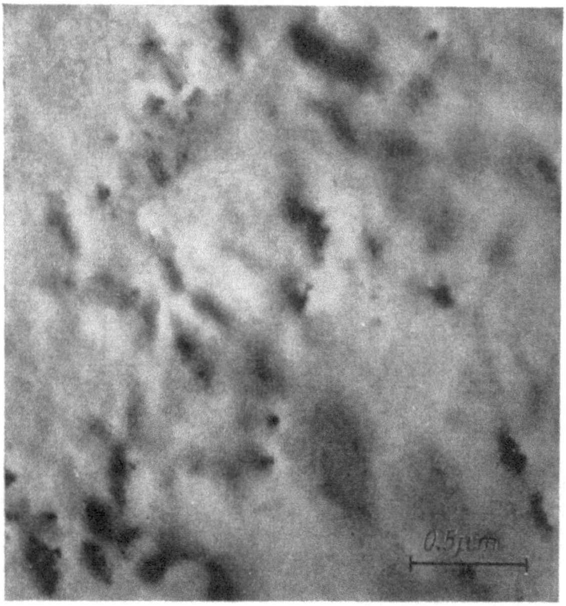

Fig. 1. Defect structure in an undoped slowly cooled GaAs crystal (HEM, bright-field image)

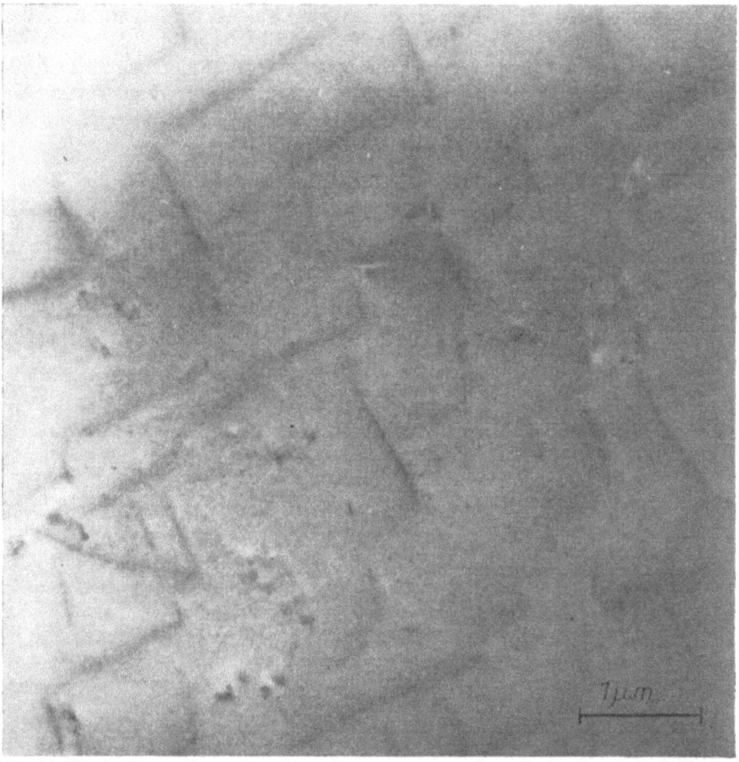

Fig. 2. Periodic surface structure in an oxygen-doped slowly cooled GaAs crystal (HEM, bright-field image)

Wafers were cut from the crystals perpendicular to the ⟨111⟩ growth axis and for the electron microscopic investigations chemically thinned in a solution of Br in methyl alcohol. The samples have been examined in a HVEM (1 MeV), and for the identification of very small precipitations the high-resolution microscope JEM-100C was used in the selected area diffraction mode as well as in the lattice plane imaging mode.

The total concentration of carbon was determined by a radioactive indicator method [3], that of oxygen by mass spectrometry.

The electrically active oxygen was equal to the concentration of the deep level donor $O'_{As}(E_c - 0.78\ eV)$, that of carbon to the shallow level acceptor $C'_{As}(E_v + 0.03\ eV)$. The former was determined by absorption spectroscopy [4], the latter by IR spectroscopy [5]. The total concentration of ionized impurities $\sum N_i$ was determined from the fundamental absorption edge [6].

2.1 Undoped crystals

In slowly as well as in quickly cooled crystals only clean dislocations and etch hillocks with diameters of 1 to 3 μm could be found. In the areas of these hillocks black dots exhibiting structure factor as well as diffraction contrast have been observed. In quickly cooled crystals these defects have diameters of about 10 nm and the defect density is very low (only a few defects being under one hillock). In slowly cooled crystals the defect density increases and colonies of black dots can be found with diameters of about 20 nm under one hillock (Fig. 1).

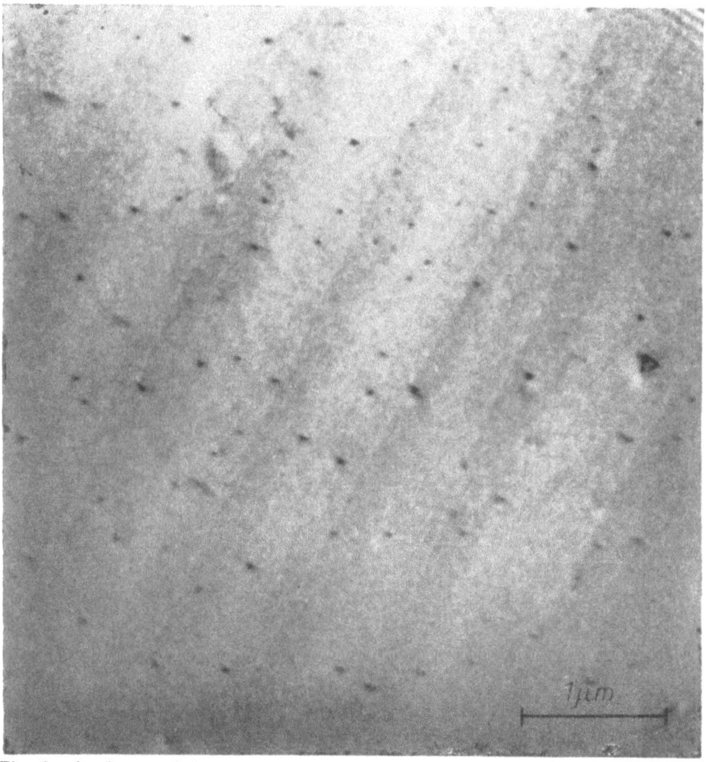

Fig. 3. As_2O_3 precipitations in an oxygen-doped slowly cooled GaAs crystal (HEM, bright-field image)

Since the specimens are very thick in the region of a hillock (5 to 7 μm) the selected area diffraction pattern does not exhibit extra reflections of a second phase.

It should be mentioned that the surface of a hillock has a rough structure compared with the normal specimen surface. So it can be concluded that during the preparation process selective etching of the second phase particles takes place not only on a macroscopic scale (formation of a hillock) but also on a microscopic one.

2.2 *Oxygen-doped crystals*

The defect structure of oxygen-doped crystals depends on the cooling mode.

In the quenched crystals stacking faults on (111) planes, small-angle grain boundaries, strain rosettes, dislocations, and twins as well as a number of black dots were usually found.

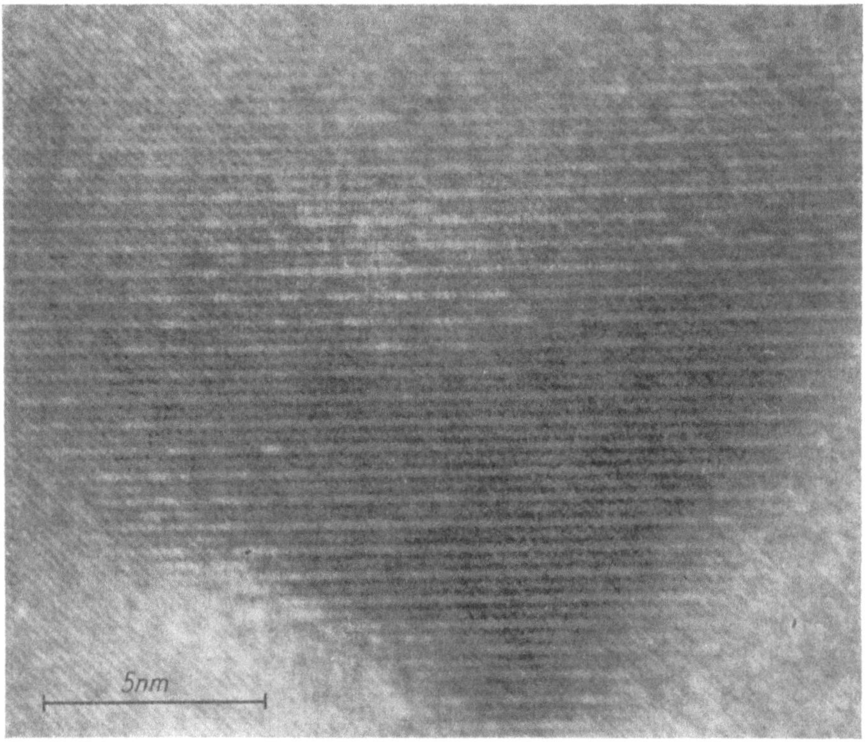

a

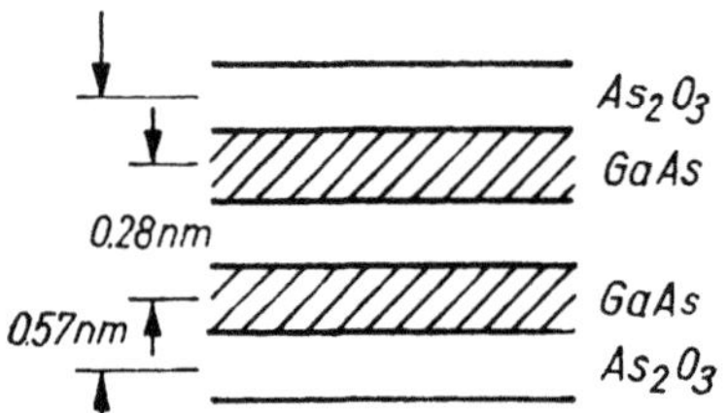

b

Fig. 4. a) Lattice plane image of a precipitation at the surface structure (Fig. 2) together with the (220) planes of the matrix; b) scheme of the layer structure (see a)) alternately consisting of two layers of GaAs and one layer of As_2O_3, respectively, in ⟨100⟩ direction

In slowly cooled crystals a defect structure can be observed which is shown in Fig. 2. The contrast of the periodic arrangements of lines lying nearly perpendicular to each other especially at the foil edge indicates a surface structure. But no further information could be received from selected area diffraction and from lattice plane imaging of the (220) planes of the ⟨111⟩ oriented matrix, respectively. We suppose that this surface structure is due to a periodic decomposition of the solid solution GaAs:O which alters the etching behaviour at the surface during chemical thinning and which can frequently be found in minerals.

In other specimen regions a high density of black spots could be found. In these crystals contrary to the more irregular shape in the undoped crystals they have a special triangular shape in most cases (Fig. 3) oriented in the same direction indicating a crystalline structure with a preferential orientation relationship to the matrix. From the selected area diffraction and from lattice plane imaging the following interplanar spacings could be estimated: 0.64, 0.32, 0.27, and 0.23 nm. These values correspond to the cubic As_2O_3 phase with $a_0 = 1.107$ nm.

From the size and density of the precipitations (compare Fig. 3) the number of oxygen atoms which take part in the crystallization of As_2O_3 can be estimated using the equation

$$n = \frac{NVZ}{a^3},$$

where N is the volume density of the black dots ($10^{11}/cm^3$), V their mean volume (diameter about 5 nm), Z the number of oxygen atoms in the elementary cell of As_2O_3 (8), and a the mean cell size (1.1 nm). The calculated value $n \approx 5 \times 10^{16}/cm^3$ is in good agreement with the observed decrease of electrically active oxygen in the slowly cooled crystals (see Table 1 and compare $[O'_{As}]$ in slowly cooled and quenched crystals with [O]). In some cases the black dots appeared directly together with and

Table 1

Total concentrations and electrically active defects connected with O and C in GaAs (in 10^{16} atoms/cm^3)

type of crystals and mode of cooling	$[O'_{As}]$	$[C'_{As}]$	$\sum N_i$	[O]	[C]	$\sum N_{tot}$
1. undoped						
slowly cooled	1	1	1 to 3	10	5	10
quenched	1	1	1 to 3	10	5	10
2. doped by O						
slowly cooled	5	1	3 to 6	50	4 to 6	50
quenched	10	1	10	50	4 to 6	50
3. doped by O and C						
slowly cooled	50	10	60	50	20	70
quenched	50	10	60	50	20	70

$[O'_{As}]$ concentration of the deep level donor O'_{As} ($E_c - 0.78$ eV).
$[C'_{As}]$ concentration of the shallow level acceptor C'_{As}($E_v + 0.03$ eV).
$\sum N_i$ total concentration of ionized impurities.
$\sum N_{tot}$ total concentration of residual impurities.

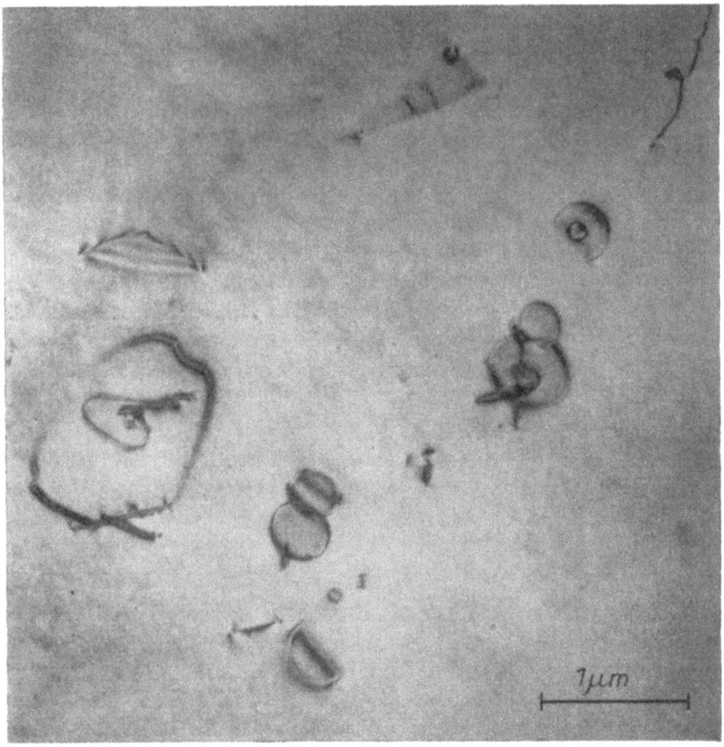

Fig. 5. Defect structure in oxygen- and carbon-doped slowly cooled crystals (HEM, bright-field image)

at the lines of the periodic structure (Fig. 2). Fig. 4a shows the lattice plane image of such a defect. It exhibits a typical double layer structure with periodicities of 0.57 and 0.28 nm, respectively. These defects can be interpreted as a special layer superstructure between GaAs and As_2O_3 in ⟨100⟩ direction which is given schematically in Fig. 4b. This explanation is supported by the orientation relationship between the (100) planes of the defect and the (220) planes of the matrix having an angle of 45°.

2.3 Oxygen- and carbon-doped crystals

In these crystals the defect structure depends also on the cooling mode.

In quenched samples dislocations and dislocation loops (15 nm diameter) have been observed. In slowly cooled crystals the formation of large amounts of different types of defects can be seen (Fig. 5): stacking faults of different size and shape, dislocation loops, and dislocations. In spite of the great number of defects the electrical activity is not changed and almost equal to the total concentration of impurities (see Table 1).

The surface structure mentioned above was found here, too, but with larger distances and only in one direction (Fig. 5). But no As_2O_3 precipitations have been observed.

3. Discussion

As follows from Table 1 the results received from the undoped crystals where the total concentration of the residual impurities $\sum N_{tot}$ is about one order higher than the concentration of ionized impurities $\sum N_i$ are the most interesting ones.

In these crystals only one type of defects was found and is therefore supposed to be responsible for the electrical inactivity of the impurities: the black dots. But such defects are always under the hillocks and hence due to the specimen thickness an analysis was not possible. From a comparison with the observation on oxygen-doped crystals it is assumed that they are amorphous clusters. The same type of defects was found before in silicon [7]. Some information about the nature of these defects can be received from the results on oxygen-doped crystals where the concentration of the black dots is higher. Here the defects began to crystallize forming precipitates of As_2O_3 or its layer superstructure with GaAs in the GaAs matrix. This fact proves that the black dots only occur in connection with oxygen.

Furthermore, with the oxygen-doped crystals good qualitative results were obtained to explain the electrical activity by the defect structure. Here the electrical activity depends on the mode of cooling. At quenched crystals the total amount of impurities is about the same as the amount of ionized impurities. But after a slow cooling two kinds of decomposition of the solid solution GaAs:O take place and decrease the electrical activity of oxygen by half an order. Firstly, there is a precipitation of As_2O_3 particles and secondly, there is a periodic arrangement of zones having an etching behaviour different from that of the matrix. This decomposition of the solid solution GaAs:O may be promoted by doping with carbon. At such a complex doping all the impurities are electrically active, both at quenching and slow cooling. The absence of one kind of decomposition, namely the formation of As_2O_3 precipitates, at double doping can be used as a proof for the well-known fact that the impurity solubility may be increased by the presence of another impurity (O'_{As} is a deep donor, whereas carbon only forms acceptor states C'_{As}).

Another conclusion concerns the effect of the other kind of decomposition, the regular surface structure. The fact that after double doping this structure is also present, but all the impurities are electrically active seems to show the low effect of such a structure but the high effect of the formation of As_2O_3 precipitations or their layer structure with the GaAs matrix.

Besides the results obtained on the electrical activity of oxygen nothing is found about the electrical inactivity of carbon. It is most probable that it only forms electrically inactive complexes as it was already assumed in [8].

References

[1] L. A. Borisova, in: Thesis of V. Vses. Simp. po sintesu and rosty poluprov. kristallov i plenok, Novosibirsk, Izd. Nauka, Moskva 1978 (p. 253).

[2] A. N. Kirgintsev and Yu. A. Ribin, Fiz. tverd. Tela **10**, 2361 (1968).

[3] L. A. Borisova, P. I. Artukhin, and Z. L. Akkerman, Izv. Akad. Nauk SSSR, Ser. neorg. Mater. **14**, 1790 (1978).

[4] M. F. Resnichenko and L. A. Borisova, see [1] (p. 56).

[5] Z. L. Akkerman and L. A. Borisova, in: Thesis of VI. Vses. Soveshch. po fiziko-khimicheskim osnovam legirovaniya poluprovodnikov, Moskva 1978 (p. 126).

[6] M. F. Resnichenko and V. N. Vertoprahov, Fiz. Techn. Poluprov. **11**, 1162 (1977).

[7] R. Köhler, W. Möhling, and M. Pasemann, phys. stat. sol. (a) **53**, 509 (1979).

[8] G. B. Stringfellow and H. Künrel, J. appl. Phys. **51**, 3254 (1980).

(Received August 20, 1981)

phys. stat. sol. (a) 68, 269 (1981)

Subject classification: 1.3; 10.1; 21; 21.1

Laboratoire d'Optique Electronique, Laboratoire propre du C.N.R.S., Associé à l' Université Paul-Sabatier, Toulouse[1])

Simulation du contraste de précipités en théorie dynamique de la diffraction des électrons

Application aux plaquettes θ' dans Al–4% Cu

Par

Y. Bouazra, C. Roucau et F. Reynaud

Nous avons étudié le contraste produit par une plaquette de précipité de composition, de structure et d'orientation quelconques par rapport à une matrice monocristalline, par simulation sur ordinateur en théorie dynamique à faisceaux multiples. Grâce à des hypothèses simplificatrices que nous exposons, nous avons pu rendre compte, avec un temps de calcul raisonnable, des contrastes d'orientation, de déplacement et de facteur de structure. Nous comparons les images simulées et les micrographies expérimentales dans le cas des plaquettes θ' dans un alliage Al–4% Cu en poids.

The contrast produced in transmission electron microscopy by a plate-like precipitate embedded in a single crystal matrix is computed by the dynamical theory of electron diffraction, whatever the composition, structure, and orientation relationships between the precipitate and the matrix. With the aid of certain clearly stated hypotheses, displacement, orientation, and structure factor contrasts can be accounted for satisfactorily in a reasonable computer time. Computed micrographs are compared with experimental ones in the case of θ' plates in Al–4 wt% Cu.

1. Introduction

Lorsqu'on étudie le contraste observé en microscopie électronique par transmission sur un précipité situé dans une matrice monocristalline, on distingue le contraste de la matrice de celui du précipité [1].

1.1 *Le contraste de la matrice*

Il provient de la déformation de la matrice provoquée par la présence du précipité. Il a été étudié en détail dans le cas d'un champ de déformation de symétrie sphérique par Ashby et Brown [2], dans le cadre de la théorie dynamique à deux faisceaux. Ce type de contraste est relativement bien connu [1] et récemment Lepski [3] en a présenté des images simulées en théorie dynamique à deux faisceaux, dans le cas de précipités ellipsoïdaux dans une matrice isotrope.

1.2 *Le contraste du précipité*

On distingue en général [1] les principaux types de contraste suivants:

i) *Le contraste de facteur de structure*, observé lorsque le vecteur de diffraction $\boldsymbol{g}$ excité est commun à la matrice et au précipité; les coefficients de Fourier du potentiel interne correspondants étant différents ($V_g^{\text{matrice}} \neq V_g^{\text{précipité}}$) par suite de la différence de structure ou (et) de composition entre la matrice et le précipité.

ii) *Le contraste d'orientation*, observé lorsque les réflexions de la matrice et du précipité sont distinctes; on peut alors former, outre le champ clair, les champs sombres en sélectionnant avec le diaphragme de contraste soit une réflexion de la

[1]) BP 4347, F-31055 Toulouse Cedex, France.

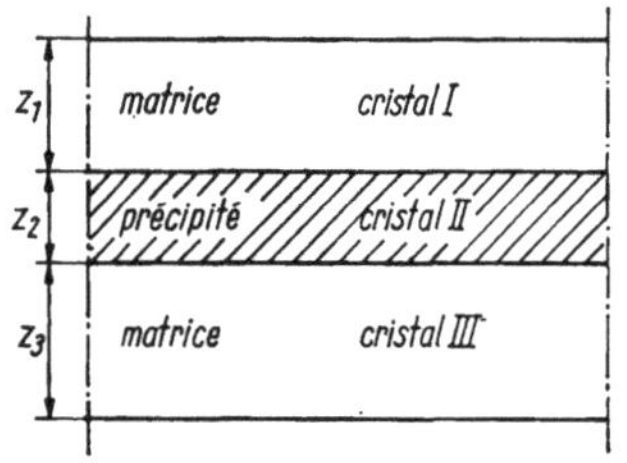

Fig. 1. Coupe d'une lame mince (cristaux I et III) contenant un précipité (cristal II) parallèle à la surface de la lame

matrice, soit une réflexion du précipité. C'est par exemple le contraste que l'on observe en utilisant une réflexion de surstructure de précipités ordonnés dans une matrice désordonnée [4].

iii) *Le contraste de déplacement*, observé dans le cas de certains précipités semi-cohérents (précipités θ' dans Al–Cu par exemple) et qui présente des analogies avec le contraste des défauts d'empilement (franges α).

iv) *Le contraste de moiré*, observé lorsque les vecteurs diffractants de la matrice et du précipité sont voisins, dont l'explication géométrique simple est connue [1].

Les deux premiers types de contraste de précipités ont été étudiés en théorie dynamique à deux faisceaux par Gleiter [5] et Guyot [6 à 8]; le contraste de moiré a fait l'objet d'une étude récente en théorie dynamique à faisceaux multiples par Pardo et al. [9].

Dans le cadre d'une étude des marches de croissance, nous présentons ici les premiers résultats qui concernent le contraste des précipités en microscopie électronique par transmission. Les calculs sont effectués en théorie dynamique à faisceaux multiples de la diffraction des électrons; les résultats sont présentés sous la forme d'images simulées analogues à celles de Head et al. [10] dans le cas d'une ou de deux dislocations éventuellement séparées par un défaut d'empilement. La composition, la structure et l'orientation des précipités par rapport à la matrice peuvent être quelconques et nous présentons l'exemple des précipités θ' dans les alliages Al–Cu. Nous rendons compte (à l'aide de certaines hypothèses simplificatrices) des trois premiers types de contraste, mais on peut envisager d'étudier ultérieurement le contraste de moiré [9].

2. Principe du calcul

2.1 Notations

Nous avons représenté sur la Fig. 1 la géométrie du problème étudié dans le cas d'un précipité en forme de plaquette parallèle à la surface de l'échantillon (nous considèrons ensuite (Section 2.5) une plaquette inclinée par rapport à cette surface). Le faisceau d'électrons incident traverse d'abord la matrice d'épaisseur z_1 (cristal I), puis le précipité d'épaisseur z_2 (cristal II) et enfin de nouveau la matrice d'épaisseur z_3 (cristal III). Dans certains cas, nous introduirons de plus un défaut d'empilement sur l'un des deux interfaces entre la matrice et le précipité.

Pour étudier le contraste, nous traitons successivement les deux interfaces matrice/précipité (I/II) et précipité/matrice (II/III). La structure et l'orientation du précipité pouvant être quelconques par rapport à la matrice (inclusion incohérente), nous ne pouvons pas utiliser le formalisme de Gratias [11] dans le cas général; en effet, ce formalisme qui semble plus élégant que celui que nous avons utilisé, ne s'applique que dans le cas particulier où il y a un réseau de coïncidence. Nous avons donc utilisé la méthode générale du traitement du contraste sur un interface développée par Gevers et al. [12 à 14] en théorie dynamique à deux faisceaux et étendue par Sheinin et Corbett [15], Roucau et Ayroles [16] et Sutton et Pond [17] à la théorie dynamique à faisceaux multiples.

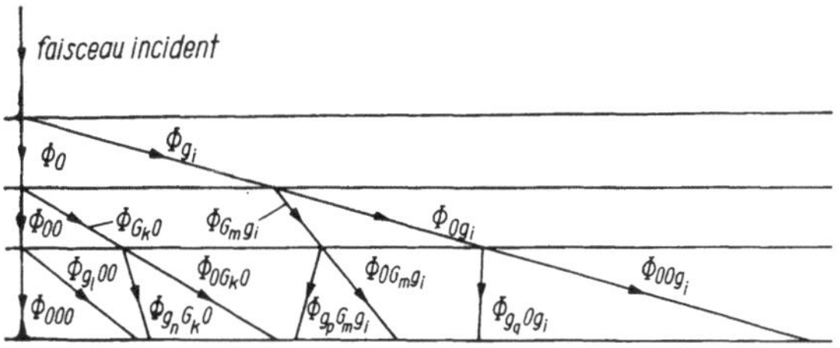

Fig. 2. Notation des amplitudes des différentes ondes excitées dans la matrice (g_i dans le cristal I, g_l, g_m, g_n, g_p, et g_q dans le cristal III) et dans le précipité (G_k et G_m dans le cristal II)

Nous allons tout d'abord définir les notations des différents faisceaux considérés (Fig. 2). Nous conviendrons de noter par φ et Φ les amplitudes des ondes:

i) *dans le cristal I*, le faisceau incident donne naissance à un faisceau transmis φ_0 et à des faisceaux diffractés $\{\varphi_{g_i}\}$;

ii) *dans le cristal II*, les faisceaux φ_0 et $\{\varphi_{g_i}\}$ sont considérés comme des faisceaux incidents qui donneraient naissance (si le module de leur amplitude était égal à un) à un ensemble de faisceaux transmis ou diffractés $\{\varphi_{G_m g_i}\}$ (G_m ou (et) g_i pouvant être égal à zéro);

iii) *dans le cristal III*, les faisceaux $\{\varphi_{G_m g_i}\}$ sont considérés comme des faisceaux incidents qui donneraient naissance (si le module de leur amplitude était égal à un) à un ensemble de faisceaux transmis ou diffractés $\{\varphi_{g_p G_m g_i}\}$.

A la sortie du cristal III, l'amplitude $\Phi_{g_p G_m g_i}$ du faisceau diffracté par g_i dans le cristal I, puis par G_m dans le cristal II et enfin par g_p dans le cristal III s'obtient en faisant le produit

$$\Phi_{g_p G_m g_i} = \varphi_{g_p G_m g_i} \varphi_{G_m g_i} \varphi_{g_i} . \tag{1}$$

Dans chaque tranche de cristal, les calculs sont effectués en théorie dynamique à faisceaux multiples sous sa forme matricielle. Pour rendre compte de l'absorption, nous utilisons les coefficients imaginaires du potentiel calculés par Radi [18]. Ces valeurs tiennent compte de l'interaction des électrons avec les phonons; elles ne rendent pas compte des autres causes d'absorption et elles sont généralement trop faibles. Nous avons donc été amenés à les multiplier par un coefficient de l'ordre de 2,5. On conçoit d'autre part que le nombre de faisceaux $\varphi_{g_p G_m g_i}$ puisse être grand; aussi, afin de réduire le temps de calcul, avons nous été amenés à introduire un certain nombre d'hypothèses simplificatrices.

2.2 Hypothèses simplificatrices

i) Dans la première partie de la matrice (cristal I), nous effectuons les calculs en théorie dynamique à faisceaux multiples en tenant compte généralement de huit ou neuf faisceaux; toutefois, nous ne considérons parmi les $\{\varphi_{g_i}\}$ et φ_0 comme incidents sur le cristal II que les deux ou trois faisceaux les plus intenses (φ_0 et φ_g lorsqu'une réflexion g est proche de la position de Bragg, φ_0, φ_g et φ_{-g} près de la position symétrique) et nous négligeons les autres.

ii) Bien que le formalisme utilisé permette de traiter les réflexions simultanées ou accidentelles, nous supposons ici qu'une seule famille de plans réticulaires est excitée dans la matrice et une seule dans le précipité, c'est à dire que les $\{g_i\}$ et les $\{g_l\}$ appartiennent à une rangée systématique de la matrice et les $\{G_k\}$ à une rangée systématique du précipité.

iii) Dans le cas du contraste d'orientation, nous supposons que les $\{g_i\}$ sont nettement distincts des $\{G_k\}$; plus précisemment on sélectionne avec le diaphragme de contraste soit une réflexion g de la matrice, soit une réflexion G du précipité sans les faire interférer (ce qui pourrait donner lieu à des franges de moiré). De plus, nous distinguerons deux cas principaux:

la matrice diffracte fortement et le précipité diffracte faiblement; nous négligeons les $\{\varphi_{G_k0}\}$ et les $\{\varphi_{G_mg_i}\}$ et nous ne considèrerons que φ_{00} et les $\{\varphi_{0g_i}\}$ comme incidents dans le cristal III. Comme dans le cas d'une cavité, le précipité n'interviendra que par son épaisseur.

le précipité diffracte fortement et la matrice diffracte faiblement; nous négligeons les $\{\varphi_{g_i}\}$ à la sortie du cristal I et nous ne considèrons que le faisceau φ_0 comme incident dans le cristal II. Nous effectuons le traitement à faisceaux multiples du précipité, mais nous ne considèrons comme faisceaux incidents dans III que φ_{00} et φ_{G0} (si G est au voisinage de la position de Bragg), ou φ_0, φ_{G0} et φ_{-G0} (près de la position symétrique).

2.3 *Vecteur d'onde des faisceaux considérés*

Pour traiter le problème des déphasages entre les divers faisceaux considérés en théorie dynamique à faisceaux multiples, il est important au préalable de préciser leur vecteur d'onde comme le propose Gevers [14]: c'est ainsi par exemple que le vecteur d'onde du faisceau diffracté d'amplitude φ_g à la sortie du cristal I (et donc à l'entrée du cristal II) s'écrit $\boldsymbol{k}_0 + \boldsymbol{g} + \boldsymbol{s}$ au lieu de $\boldsymbol{k}_0 + \boldsymbol{g}$.

Pour illustrer cette remarque, nous considérons les expressions utilisées par Gevers [14] en théorie à deux faisceaux. Les ondes transmise et diffractée à la sortie d'un cristal compris entre les cotes z et z_0 s'écrivent

$$\varphi_0 \exp(2i\pi\boldsymbol{k}_0 \cdot \boldsymbol{r}) = \big(\exp i\pi s(z - z_0)\big)\, T(z - z_0) \exp 2i\pi\boldsymbol{k}_0 \cdot \boldsymbol{r}\,, \tag{2}$$

$$\varphi_g \exp\big(2i\pi(\boldsymbol{k}_0 + \boldsymbol{g}) \cdot \boldsymbol{r}\big) = \big(\exp i\pi s(z - z_0)\big)\, S(z - z_0) \exp 2i\pi(\boldsymbol{k}_0 + \boldsymbol{g})\, \boldsymbol{r}\,. \tag{3}$$

Il est plus commode d'écrire (3) sous la forme équivalente

$$\exp(-2i\pi\boldsymbol{s} \cdot \boldsymbol{r})\, \big(\exp i\pi s(z - z_0)\big)\, S(z - z_0) \exp 2i\pi(\boldsymbol{k}_0 + \boldsymbol{g} + \boldsymbol{s}) \cdot \boldsymbol{r}\,. \tag{4}$$

En remarquant que l'erreur d'interférence $\boldsymbol{s}$ est dirigée le long de l'axe $0z$ (situé dans les plans diffractants [19], mais distinct de la normale à la face d'entrée ainsi qu'on l'écrit très souvent) on écrit $\boldsymbol{s} \cdot \boldsymbol{r} = sz$, soit

$$\exp(-2i\pi sz)\, (\exp i\pi s(z - z_0))\, S(z - z_0) \exp 2i\pi(\boldsymbol{k}_0 + \boldsymbol{g} + \boldsymbol{s}) \cdot \boldsymbol{r}\,. \tag{5}$$

Le traitement matriciel en théorie dynamique à deux faisceaux conduit aux amplitudes $A_0(z - z_0)$ et $A_g(z - z_0)$ des faisceaux transmis et diffracté telles que

$$A_0(z - z_0) = \exp i\pi s(z - z_0)\, T(z - z_0)\,, \tag{6}$$

$$A_g(z - z_0) = \exp i\pi s(z - z_0)\, S(z - z_0)\,. \tag{7}$$

Pour obtenir la fonction d'onde complète (3), on est donc amené à multiplier l'amplitude diffractée A_g obtenue par la méthode matricielle par un facteur de phase $\exp(-2i\pi sz)$. L'introduction de ce facteur se généralise en théorie matricielle à faisceaux multiples.

2.4 *Expression de la fonction d'onde des faisceaux à la sortie du cristal III*

Appelons $\Psi_{g_pG_mg_i}$ la fonction d'onde à la sortie du cristal III du faisceau ayant été diffracté par g_i dans le cristal I, par G_m dans le cristal II et par g_p dans le cristal III, et s_{g_i}, s_{G_m} et s_{g_p} les erreurs d'interférence correspondantes on a:

$$\begin{aligned} \Psi_{g_pG_mg_i} = {} & A_{g_i} \exp(-2i\pi s_{g_i} z_1)\, A_{G_mg_i} \exp\big(-2i\pi s_{G_m}(z_1 + z_2)\big) \times \\ & \times A_{g_pG_mg_i} \exp\big(-2i\pi s_{g_p}(z_1 + z_2 + z_3)\big) \times \\ & \times \exp 2i\pi(\boldsymbol{k}_0 + \boldsymbol{g}_i + \boldsymbol{s}_{g_i} + \boldsymbol{G}_m + \boldsymbol{s}_{G_m} + \boldsymbol{g}_p + \boldsymbol{s}_{g_p}) \cdot \boldsymbol{r} \end{aligned} \tag{8}$$

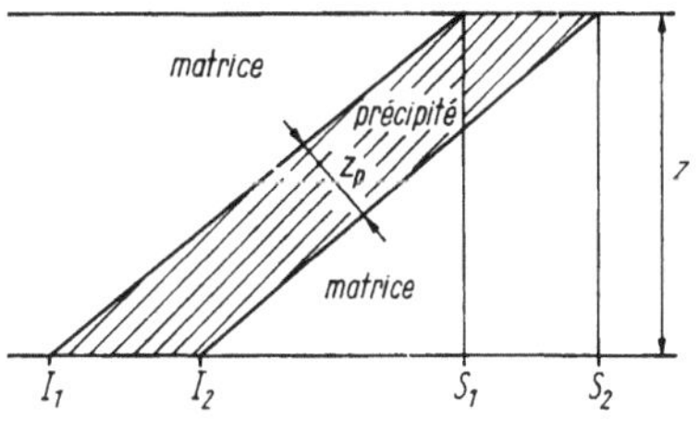

Fig. 3. Définition des repères tracés sur les images simulées: I_1, I_2, S_1 et S_2 sont les intersections des interfaces I/II et II/III avec les faces inférieure et supérieure de la lame mince

où la lettre A signifie une amplitude obtenue par la méthode matricielle. Enfin, on obtiendra la fonction d'onde Ψ_h des électrons aboutissant à une tache de diffraction h en effectuant la sommation

$$\Psi_h = \sum_{g_p + G_m + g_i = p} \Psi_{g_p G_m g_i} . \tag{9}$$

2.5 Précipité incliné par rapport à la lame

La méthode générale exposée précédemment s'applique également au cas d'une plaquette inclinée par rapport à la surface de l'échantillon (Fig. 3); les images simulées présentées dans cet article correspondent à cette configuration et la signification des rèpéres I_1, I_2, S_1 et S_2 est précisée sur la Fig. 3. Signalons enfin que l'échelle de onze teintes que nous avons utilisée est répartie entre l'intensité maximale et l'intensité minimale de chaque image simulée; nous obtenons ainsi un bon rendu des détails, mais les contrastes sont à considérer en valeur relative et non en valeur absolue.

3. Vérification du formalisme

Bien que l'étude des précipités en microscopie électronique ait commencé depuis assez longtemps, il n'y a pas de règles strictes concernant leur contraste. Afin de tester la validité de notre technique, nous avons été amenés à simuler des micrographies correspondant à des situations connues. Nous ne parlons pas ici de la plus simple d'entr'elles (une lame mince constituée de trois cristaux de même structure pour laquelle on doit obtenir une teinte plate pour toute valeur raisonnable de s), mais nous allons montrer que nous avons retrouvé les règles de contraste connues concernant les franges α et les cavités.

3.1 Contraste des franges α

Il suffit de prendre $Z_p = 0$; nous avons considéré par ailleurs une matrice d'aluminium pur. La Fig. 4 montre les images simulées d'un défaut d'empilement correspondant à $\alpha = \pm 2\pi/3$ et d'un autre défaut (imaginaire dans Al) qui produirait un déphasage $\alpha = \pi$. On vérifie aisément les règles de symétrie [14] des franges $\alpha = \pi$ et $\alpha \neq \pi$, en champ clair et en champ sombre.

3.2 Contraste des cavités

Dans le cas d'une cavité, nous avons également retrouvé par simulation les règles de contraste connues [8, 20].

4. Application à la simulation du contraste sur des plaquettes θ' dans un alliage Al–4% Cu (en poids)

Nous avons appliqué notre technique aux plaquettes de précipités θ' dans un alliage Al–4% Cu (en poids) mis en solution solide pendant 24 h à 540 °C, puis recuit pendant une heure et demi à 275 °C et trempé à l'eau. Nous allons tout d'abord discuter l'influence des principaux paramètres qui interviennent dans le contraste.

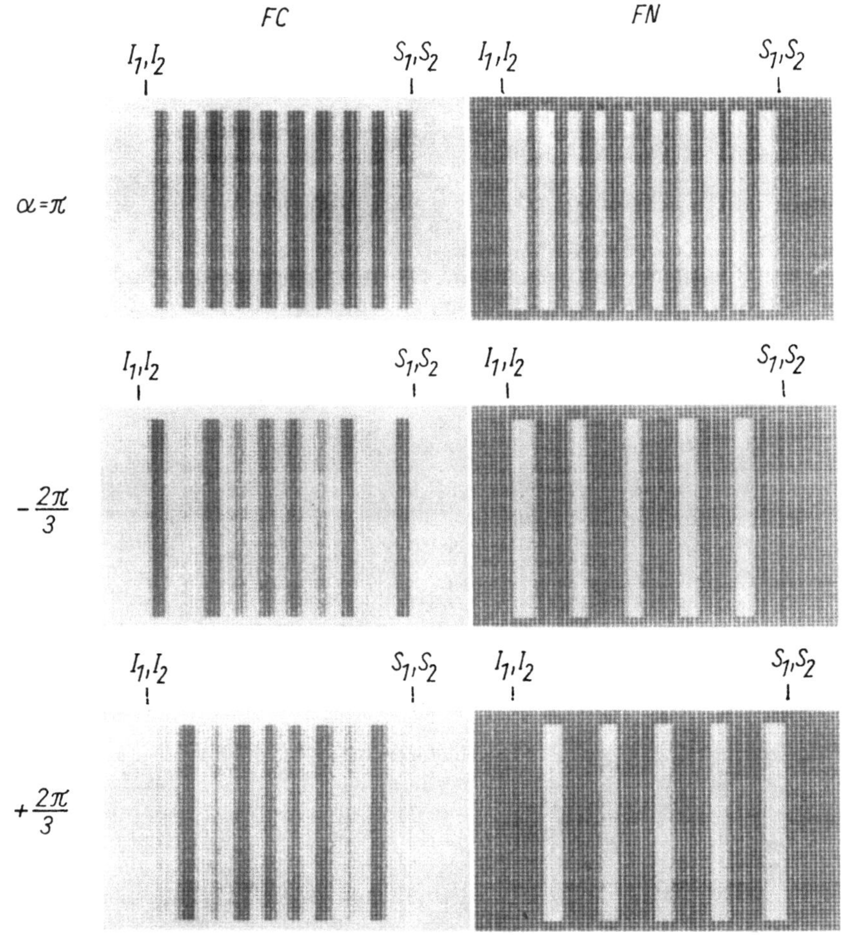

Fig. 4. Images simulées d'un défaut de translation (franges α) en fond clair (FC) et en fond noir (FN), pour différentes valeurs du déphasage α ($Z = 280$ nm, $g = 111$ et $s = 0$)

4.1 Influence des principaux paramètres

4.1.1 Epaisseur Z_p du précipité

L'épaisseur Z_p du précipité est un paramètre important. En effet nous ne pouvons généralement pas la mesurer expérimentalement, car nous ne pouvons pas mettre debout (parallèles au faisceau d'électrons) les plaquettes sur lesquelles nous travaillons (qui sont celles de la famille présentant la plus grande largeur apparente et donc proche du plan moyen de la lame). Nous présentons sur la Fig. 5 les images simulées en champ clair de précipités θ' d'épaisseur 15 et 30 nm; on remarque en particulier l'élargissement des franges externes avec l'augmentation de z_p, ainsi qu'une modification des franges pour les trois valeurs du déphasage α simulées.

4.1.2 Ecart par rapport à la position de Bragg

La Fig. 6 montre des images en champ clair simulées d'un précipité d'épaisseur 10 nm dans une lame d'épaisseur totale $Z = 280$ nm; nous avons considéré deux

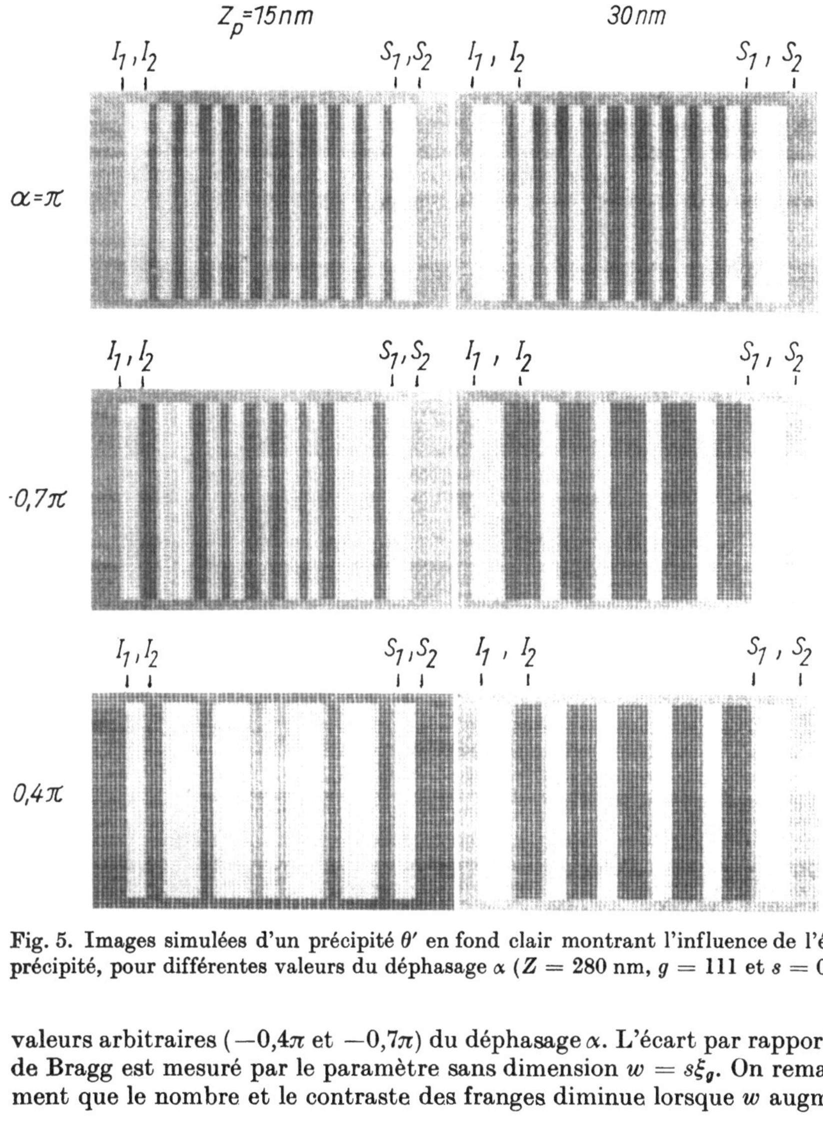

Fig. 5. Images simulées d'un précipité θ' en fond clair montrant l'influence de l'épaisseur Z_p du précipité, pour différentes valeurs du déphasage α ($Z = 280$ nm, $g = 111$ et $s = 0$)

valeurs arbitraires ($-0{,}4\pi$ et $-0{,}7\pi$) du déphasage α. L'écart par rapport à la position de Bragg est mesuré par le paramètre sans dimension $w = s\xi_g$. On remarque évidemment que le nombre et le contraste des franges diminue lorsque w augmente.

4.1.3 Valeur du déphasage α

La valeur du déphasage α peut varier d'un précipité à l'autre pour un même vecteur de diffraction [1]. C'est un paramètre important dont l'influence sur le contraste est mise en évidence sur les Fig. 5 et 6. On remarque en particulier que pour une épaisseur Z_p comprise entre 10 et 20 nm environ, le contraste des franges s'accentue et la périodicité en $\frac{1}{2}\xi_g$ s'impose d'autant plus que α croît vers la valeur π.

Par suite des effets de l'absorption, le signe du déphasage revêt une grande importance lorsque l'épaisseur traversée par les électrons est supérieure à plusieurs fois la distance d'extinction de la réflexion excitée; on peut en voir une illustration sur la Fig. 7 qui montre le changement de la nature des franges externes avec le changement du signe de α.

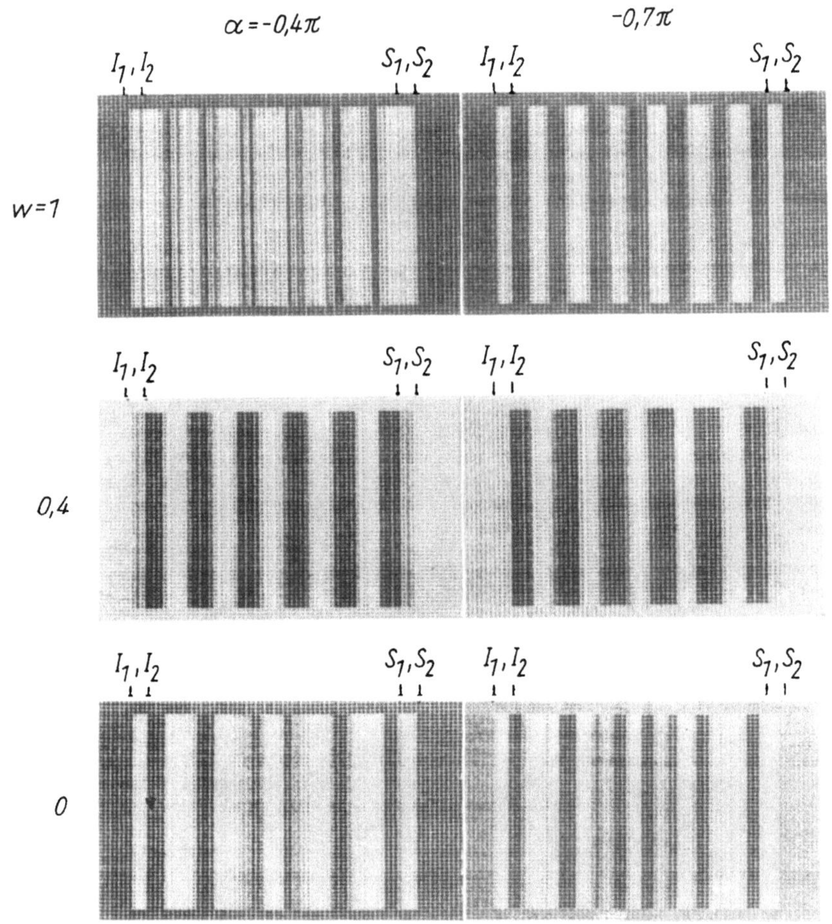

Fig. 6. Images simulées d'un précipité θ' en fond clair montrant l'influence de l'écart par rapport à la position de Bragg pour deux valeurs du déphasage $\alpha = -0{,}4\,\pi$ et $\alpha = -0{,}7\pi$ ($Z = 280$ nm, $Z_p = 10$ nm et $g = 111$)

4.2 Exemples de comparaison entre les images simulées et les micrographies expérimentales

Selon les vecteurs diffractants, on peut observer dans le cas des précipités θ' des contrastes d'orientation et de déplacement ou de facteur de structure.

4.2.1 Contrastes de déplacement et d'orientation

Nous avons déjà présenté sur la Fig. 7 un exemple de comparaison entre les images simulées et les micrographies expérimentales, dans le cas des contrastes de déplacement et d'orientation. L'épaisseur Z traversée par les électrons a été mesurée expérimentalement et l'épaisseur Z_p du précipité a été estimée à 14 nm en faisant la moyenne de la mesure de l'épaisseur des précipités voisins du précipité étudié, mais appartenant à des familles très inclinées par rapport au plan de la lame et que l'on peut donc mettre debout. L'écart par rapport à la position de Bragg est mesuré grâce aux lignes de Kikuchi. Des essais de différentes valeurs du déphasage α montrent qu'il est proche de $0{,}8\pi$.

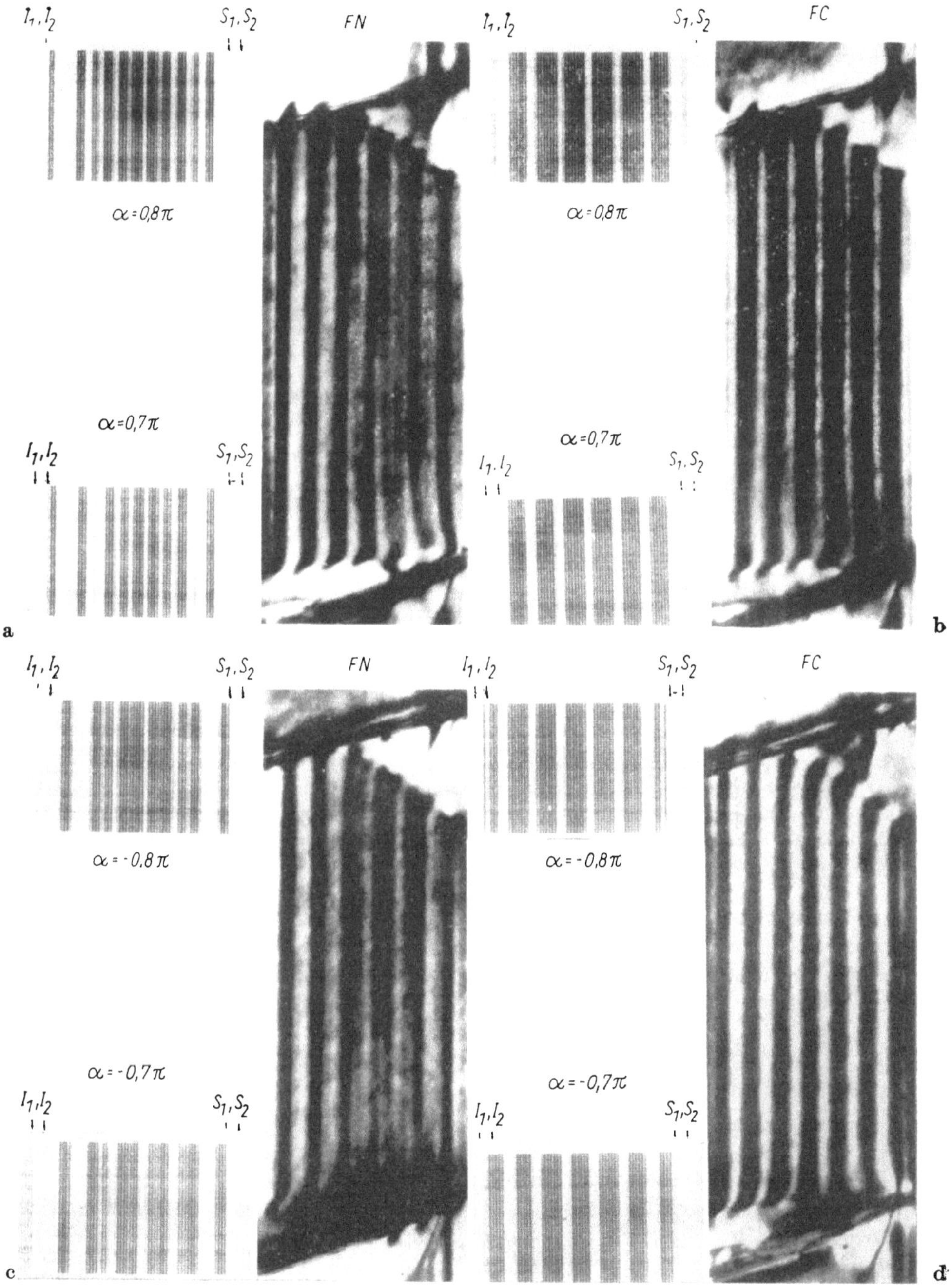

Fig. 7. Exemple de comparaison entre les micrographies expérimentales de précipités θ' (b, d) FC; a, c) FN) et les images simulées, dans le cas des contrastes de déplacement et d'orientation, montrant l'influence du signe de α ($Z = 355$ nm, $Z_p = 14$ nm, $g = \pm 11\bar{1}$, $\alpha = \pm 0{,}7\pi$ et $\alpha = \pm 0{,}8\pi$)

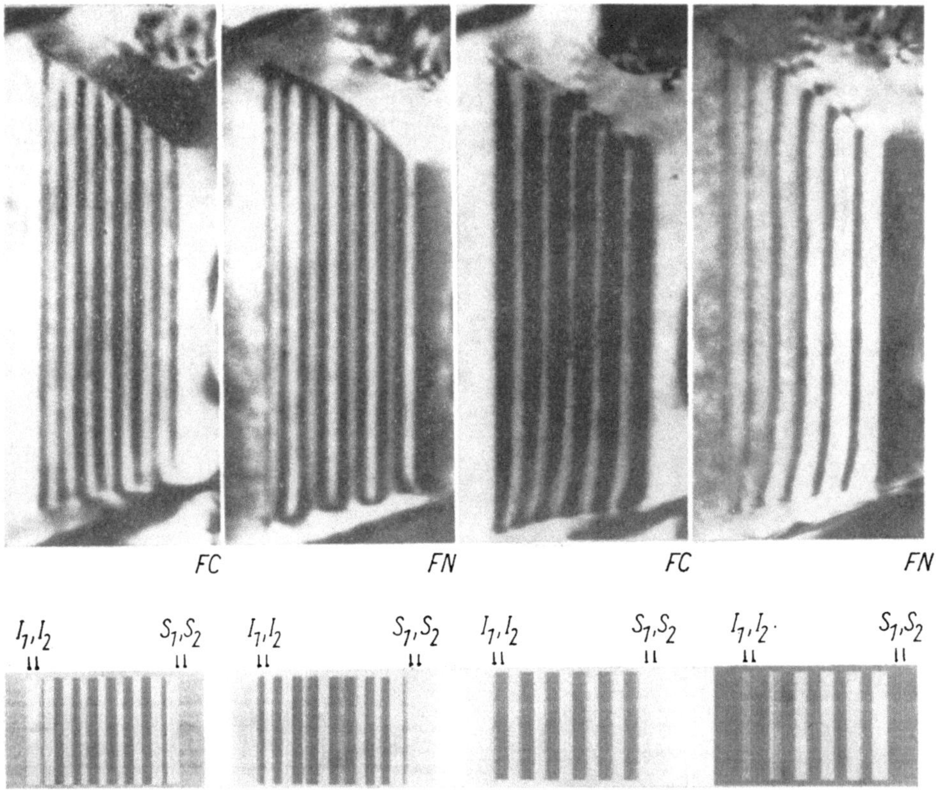

Fig. 8. Exemple de comparaison entre les micrographies expérimentales de précipités θ' (a, c FC; b, d FN et les images simulées correspondantes, dans le cas des contrastes de déplacement et d'orientation, montrant l'influence de l'écart par rapport à la position de Bragg ($Z = 225$ nm, $Z_p = 9{,}3$ nm, $\alpha = 0{,}9\pi$, $g = \bar{1}11$; a, b) $w = 0$, c, d) $w = 1$)

Nous présentons un second exemple de contraste de déplacement sur la Fig. 8; les simulations correspondent à $w = 0$ et $w = 1$, $Z = 225$ nm, $Z_p = 9{,}3$ nm et $\alpha \simeq 0{,}9\pi$. On observe un accord raisonable entre les micrographies expérimentales et les images simulées, en champ clair et en champ sombre.

4.2.2 Contraste de facteur de structure

Si le vecteur de diffraction $\boldsymbol{g}$ commun au précipité et à la matrice est en position de Bragg exacte, on n'observe pas de franges. Par contre, dans le cas de l'orientation symétrique ($+g$ et $-g$ également excités), on observe des franges (Fig. 9, $Z = 225$ nm, $Z_p = 10$ nm et $g = 200$). Le contraste observé est faible; il provient en effet de la différence entre les distances d'extinction des réflexions 200 du précipité et de la matrice, différence qui est de l'ordre de 5% seulement dans le cas des précipités θ' dans Al–4% Cu. Le déphasage α n'intervient pas (il n'y a pas de contraste de déplacement).

5. Conclusions

La simulation en théorie dynamique du contraste sur un précipité de structure, de composition et d'orientation quelconques par rapport à la matrice, est un problème

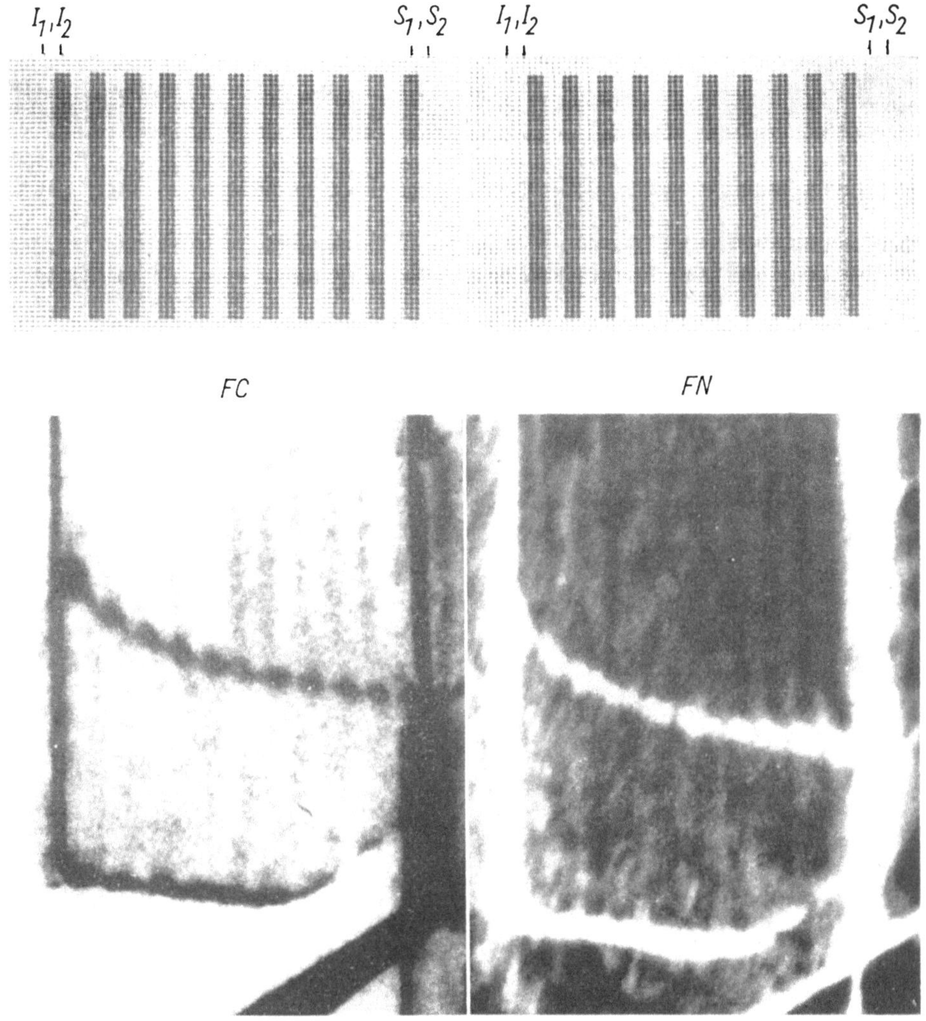

Fig. 9. Exemple de contraste de facteur de structure sur des précipités θ' dans Al–4% Cu ($Z =$ $= 225$ nm, $z_p = 10$ nm, $g = 200$ et position symétrique). On observe sur les micrographies expérimentales une marche de croissance qui n'est pas reproduite sur les images simulées

complexe. Toutefois, selon le type de contraste du précipité (contraste d'orientation, de déplacement ou de facteur de structure), on parvient à un temps de calcul raisonnable en faisant des hypothèses que nous avons précisées. La comparaison entre les images simulées et les micrographies expérimentales en champs clair et sombre dans le cas de plaquettes de précipités θ' dans un alliage Al–4% Cu en poids montre que, dans ce cas particulier au moins, ces hypothèses sont justifiées.

Dans un future proche, nous appliquerons la technique ainsi mise au point à l'étude des marches de croissance présentes sur les précipités θ'. On pourrait d'autre part traiter également le contraste de moiré en s'inspirant du traitement développé par Pardo et al [9].

Remerciements

L'un des auteurs (Y.B.) remercie M. le Doyen A. El Hili de la Faculté des Sciences de Tunis pour lui avoir permis de venir travailler en France de manière continue, M. le Directeur B. Jouffrey pour l'avoir accueilli au L. O. E. de Toulouse et lui avoir proposé ce travail et M. R. Ayroles pour ses conseils bienveillants.

Bibliographie

[1] P. B. Hirsch, A. Howie, R. B. Nicholson, D. W. Pashley et M. J. Whelan, Electron Microscopy of Thin Crystals, Butterworths, London 1965.
[2] M. F. Ashby et L. M. Brown, Phil. Mag. **8**, 1083, 1649 (1963).
[3] D. Lepski, phys. stat. sol. (a) **23**, 543 (1974).
[4] F. Reynaud, phys. stat. sol., to be published.
[5] H. Gleiter, Phil. Mag. **18**, 847 (1968).
[6] P. Guyot, phys. stat. sol. (a) **28**, 349 (1968).
[7] P. Guyot, phys. stat. sol. (a) **33**, 201 (1969).
[8] P. Guyot, dans: méthodes et techniques nouvelles d'observation en métallurgie physique, Ed. B. Jouffrey, S.F.M.E. 1972 (p. 281).
[9] B. Pardo, C. Pariset et D. Renard, phys. stat. sol. (a) **64**, 283 (1981).
[10] A. K. Head, P. Humble, L. M. Clarebrough, A. J. Morton et C. T. Forwood, Computed Electron Micrographs and Defect Identification, North-Holland Publ. Co., 1973.
[11] D. Gratias, Thèse de doctorat Es-Sciences, Université P. et M. Curie, Paris VI 1980.
[12] R. Gevers, P. Delavignette, H. Blank et S. Amelinckx, phys. stat. sol. **4**, 383 (1964).
[13] R. Gevers, P. Delavignette, H. Blank, J. van Landuyt et S. Amelinckx, phys. stat. sol. **5**, 595 (1964).
[14] R. Gevers, voir [8] (p. 145).
[15] S. S. Sheinin et J. M. Corbett, phys. stat. sol. (a) **38**, 675 (1976).
[16] C. Roucau et R. Ayroles, J. Microsc. Spectrosc. Electron. **3**, 197 (1978).
[17] A. P. Sutton et R. C. Pond, phys. stat. sol. (a) **45**, 149 (1978).
[18] G. Radi, Acta cryst. **A26**, 41 (1970).
[19] F. Reynaud, Radiat. Eff. **39**, 181 (1978).
[20] J. van Landuyt, R. Gevers et S. Amelinckx, phys. stat. sol. **10**, 319 (1965).

(Received July 27, 1981)

phys. stat. sol. (a) 68, 281 (1981)

Subject classification: 1.4

Sektion Mathematik der Bergakademie Freiberg[1]) *(a) and Zentralinstitut für Festkörperphysik und Werkstofforschung der Akademie der Wissenschaften der DDR, Dresden*[2]) *(b)*

Random Set Models in the Interpretation of Small-Angle Scattering Data

By

U. SONNTAG (a), D. STOYAN (a), and H. HERMANN (b)

Random set models describing two-phase structures are introduced in order to interpret small-angle scattering (SAS) data from random structural inhomogeneities with arbitrary volume fraction. First- and second-order characteristics of several stochastic models are calculated. For illustration, correlation functions obtained from experimental SAS data of different specimens are compared with theoretical ones.

Es werden Modelle zufälliger Mengen zur Beschreibung zweiphasiger Strukturen eingeführt, um Kleinwinkelstreudaten von stochastischen strukturellen Inhomogenitäten mit beliebigem Volumenanteil zu interpretieren. Hierfür werden Kenngrößen erster und zweiter Ordnung für unterschiedliche stochastische Modelle ermittelt. Zur Illustration werden Korrelationsfunktionen, die aus Messungen der Kleinwinkelstreuung verschiedener Proben bestimmt wurden, mit theoretischen Kurven verglichen.

1. Introduction

Small-angle scattering (SAS) is produced by structures the characteristic extent of which is about 1 to 100 nm. The basic methods for interpreting SAS data are explained in [1, 2]. Results of SAS experiments can be interpreted in terms of the reciprocal space and the real space, respectively. Recent applications of SAS to problems of physics and material science, e.g. spinodal decomposition and magnetism, are reviewed in [3] where the SAS intensity (that is reciprocal space) is discussed. In [4] the density–density correlation function containing the real-space information obtained from SAS experiments is examined for different types of single particles.

In the present paper, methods of stochastic geometry are used to calculate density–density correlation functions of two-phase structures which can be described by random set models. Such types of structures seem to appear in powder technology, ceramics, and in other fields where random inhomogeneities with arbitrary volume fraction occur.

Probably, the first who used the stochastic approach to SAS was Porod [5]; in the modern terminology, he has studied a "Boolean model" with non-random spherical grains (model 1 in Section 4).

Here we shall discuss some more general random set models, demonstrate the variability of the corresponding correlation functions, and use them for the interpretation of several experimental SAS data.

The density–density correlation function is defined by

$$\gamma(r) = (4\pi V)^{-1} \int_V \int_\Omega \mathrm{d}\Omega \, \mathrm{d}V_x \, [\varrho^e(\boldsymbol{x} + \boldsymbol{r}) - \varrho_0^e] \, [\varrho^e(\boldsymbol{x}) - \varrho_0^e] \,, \qquad r \geqq 0 \,, \qquad (1)$$

[1]) Bernh.-v.-Cotta-Str. 1, DDR-9200 Freiberg/Sa., DDR.

[2]) Helmholtzstr. 20, DDR-8027 Dresden, DDR.

$$\varrho_0^e = V^{-1} \int_V \varrho^e(\boldsymbol{x})\, dV_x = \varrho_0^c V_A/V = \varrho_0^c p\,,$$

$$\varrho^e(\boldsymbol{r}) = \begin{cases} \varrho_0^c & \text{for} \quad \boldsymbol{x} \in A\,, \\ 0 & \text{otherwise}\,. \end{cases}$$

The volume fraction of the scattering regions A is denoted by p, $\varrho^e(\boldsymbol{r})$ is the electron density distribution. Equation (1) is integrated over the volume V of the whole specimen and over all spatial directions Ω. (1) can be written in the form

$$\gamma(r) = (\varrho_0^c)^2[C_A(r) - p^2]\,,$$

$$C_A(r) = (4\pi V)^{-1} \int_V \int_\Omega d\Omega\, dV_x\, n(\boldsymbol{x} + \boldsymbol{r})\, n(\boldsymbol{x}) = P(\boldsymbol{0} \in A\,,\, \boldsymbol{r} \in A)\,,$$

$$n(\boldsymbol{r}) = \varrho^e(\boldsymbol{r})/\varrho_0^c\,,$$

where A is homogeneous and isotropic, $P(\boldsymbol{0} \in A, \boldsymbol{r} \in A)$ is the probability of finding an arbitrary point belonging to A at distance r from the origin which also belongs to A. Because absolute SAS experiments are difficult in many cases we shall restrict ourselves to discuss the correlation function k_A,

$$k_A(r) = \gamma(r)/\gamma(0) = [C_A(r) - p_A^2]/p_A(1 - p_A)\,, \qquad r \geqq 0\,. \tag{2}$$

The small-angle intensity $I(s)$ is related to $\gamma(r)$ by

$$\gamma(r) = (2\pi^2 r)^{-1} \int_0^\infty s I(s) \sin(sr)\, ds\,, \qquad r \geqq 0\,, \tag{3}$$

$$s = 4\pi\lambda_x^{-1} \sin\theta\,,$$

where 2θ is the scattering angle and λ_x the wavelength.

In Section 2 general properties of spatial random closed sets are stated. The Boolean model used for the construction of stochastic structures is explained in Section 3. The correlation functions of special model structures which may be useful for interpreting SAS data in some cases are given in Section 4 and, finally, experimental SAS results are discussed in Section 5.

2. Spatial Random Closed Sets

Spatial random closed sets (RACS's) are random variables the samples of which are closed sets of the three-dimensional space R^3 as, for example, collections of spheres, holes in Swiss cheese, non-metallic particles in iron, etc. ("closed" means that the boundary of the sets belongs to them). Mathematically satisfactory theories of RACS's have been established in the seventies by Kendall [6] and Matheron [7], basic references are the books of Matheron [7] and Serra [8].

Throughout this paper we suppose that the RACS's considered are motion-invariant (homogeneous and isotropic). A RACS A may be described by the first- and second-order characteristics

$$p_A = P(\boldsymbol{r} \in A) = P(\boldsymbol{0} \in A)\,, \tag{4}$$

$$C_A(r) = P(\boldsymbol{0} \in A\,,\, \boldsymbol{r} \in A)\,, \qquad r \geqq 0\,, \tag{5}$$

$$K_A(r) = C_A(r) - p_A^2\,, \tag{6}$$

$$k_A(r) = K_A(r)/p_A(1 - p_A)\,, \tag{7}$$

where $p_A = p$ is the volume fraction, $C_A(r)$ and $K_A(r)$ are the covariance and covariance function, respectively, $k_A(r)$ denotes the correlation function.

If the surface of A is sufficiently smooth, then the specific surface S_V (mean surface area per unit volume) is given by

$$S_V = -4C'(0)\,. \tag{8}$$

Note that from planar intersections $C(r)$ can be easily determined using electron microscopes. For this, it suffices to consider point pairs of distance r and to determine the fraction of them which are both in A (see Serra [8]).

Starting from given RACS's, the well-known set-theoretical operations yield further (potentially more complicated) RACS's. We consider the operations "complement" $A \to A^c$ (A^c: set of all points outside A, including the boundary of A) and "intersection" $A_1, A_2 \to A_1 \cap A_2$ ($A_1 \cap A_2$: set of all points which are both in A_1 and A_2). Clearly,

$$p_{A^c} = 1 - p_A\,, \tag{9}$$

$$C_{A^c}(r) = P(\boldsymbol{0} \notin A\,,\ \boldsymbol{r} \notin A) = C_A(r) - 1 + 2p_A\,, \tag{10}$$

$$k_{A^c}(r) = k_A(r)\,, \qquad r \geqq 0 \tag{11}$$

and, if A_1 and A_2 are independent,

$$p_{A_1 \cap A_2} = p_{A_1} p_{A_2}\,, \tag{12}$$

$$C_{A_1 \cap A_2}(r) = C_{A_1}(r)\, C_{A_2}(r)\,, \qquad r \geqq 0\,. \tag{13}$$

Expression (11) corresponds to "Babinet's theorem" [2].

3. The Boolean Model

A random set model very important and useful for practical applications is the Boolean model (see [7 to 9]). Here, the RACS is the union of randomly distributed particles where overlappings are possible. More precisely, there are a Poisson process (see, e.g. [9]) with points $\boldsymbol{x}_1, \boldsymbol{x}_2, \ldots$ ("germs") and a sequence of isotropic, independent, identically distributed, bounded RACS's $A_1, A_2, \ldots$ ("primary grains"). The A_n are shifted by the $\boldsymbol{x}_n$ and united,

$$A = \bigcup_{(n)} A_n + \boldsymbol{x}_n\,. \tag{14}$$

If, for example, A_n is a ball with radius r_n and midpoint in $\boldsymbol{0}$, then $A_n + \boldsymbol{x}_n$ is the ball with the same radius and midpoint $\boldsymbol{x}_n$. Let λ be the intensity (mean number of germs per unit volume) of the point process. Then

$$p = 1 - \exp\{-\lambda E V(A_n)\}\,, \tag{15}$$

$$C(r) = 2p - 1 + (1-p)^2 \exp\{\lambda \gamma^0(r)\} \tag{16}$$

with

$$\gamma^0(r) = EV\big(A_n \cap (A_n + \boldsymbol{r})\big)\,, \qquad r \geqq 0\,,$$

$$S_V = \lambda(1-p)\, ES(A_n)\,, \tag{17}$$

where E, V, S stand for "expectation", "volume", and "surface", respectively.

Important particular cases are spherical and Poisson-polyhedric grains. In the first case

$$\gamma^0(r) = (4\pi/3) \int_{r/2}^{\infty} x^3(1 - 3r/4x + r^3/16x^3)\, f(x)\, \mathrm{d}x\,, \qquad r \geqq 0 \tag{18}$$

applies, where $f(x)$ is the density of the radius distribution.

A Poisson polyhedron is an isotropic RACS which is a typical mosaic stone of a random tesselation of space by random planes (see [7] and [8]). Its main geometric characteristics are mean volume

$$EV(A_n) = 6/\pi^4\varrho^3 , \tag{19}$$

mean surface area

$$ES(A_n) = 24/\pi^3\varrho^2 , \tag{20}$$

and mean breadth

$$E\bar{b}(A_n) = 3/2\pi\varrho , \tag{21}$$

where ϱ is a parameter describing the density of random planes. (The breadth of a convex set in a given direction is the distance between the two parallel supporting planes that are perpendicular to the direction. Then $\bar{b}$ is the mean of all breadths, averaged over all directions. For a sphere, it is the diameter.) Furthermore,

$$\gamma^0(r) = 6 \exp(-\pi\varrho r)/\pi^4\varrho^3 , \qquad r \geqq 0 . \tag{22}$$

Note that $\gamma^0(r)$ is essentially the same as the "distance probability function" or "characteristic function" of single particles (cf. [4]). Thus, it is possible to use known formulae for $\gamma^0(r)$ for further grain types (e.g. ellipsoids, hollow spheres) in building corresponding Boolean models.

Of particular interest are RACS's with very small volume fraction p. Therefore, it is important that the limit of $k(r)$ for $\lambda \to 0$ and fixed A_n for a Boolean model is a proper function

$$k^0(r) = \gamma^0(r)/\gamma^0(0) , \qquad r \geqq 0 .$$

This is a very satisfactory fact, for it shows the relation between the one-particle correlation function and that of a stochastic structure built up by a random distribution of particles of the same form. Especially, in the case of Poisson polyhedra we obtain

$$k^0(r) = \exp(-\alpha r) , \qquad \alpha = \pi\varrho , \qquad r \geqq 0 , \tag{23}$$

a form of correlation function which has been already discussed by Debye and Bueche [10], but without a geometrical interpretation of α. Here, α is related to the size of primary grains, $\bar{b} = 3/2\alpha$.

4. Stochastic Model Structures

The models differ with respect to the choice of the "primary grains" and to the set-theoretical operations applied (cf. Section 3).

Model 1: Spheres of constant radius r_0 are used as primary grains. The correlation function is calculated from (7), (16), (18) where $f(x)$ in (18) is replaced by the delta-function $\delta(x - r_0)$,

$$k(r) = \begin{cases} \{\exp[\lambda\pi r_0^3(r^3/12r_0^3 - r/r_0)] - 1 + p\}/p & \text{if} \quad r \leqq 2r_0 , \\ 0 \quad \text{otherwise} , \end{cases} \tag{24}$$

$$\lambda = -3 \ln(1 - p)/4\pi r_0^3 .$$

Model 2: The primary grains are spheres the radii of which follow the exponential distribution $f(x) = \mu \exp(-\mu x)$, $x \geqq 0$. We get from (7), (16), (18)

$$\left.\begin{aligned} k(r) &= (1-p)\{\exp[\lambda\pi(2r/\mu^2 + 8/\mu^3)\exp(-\mu r/2)] - 1\}/p\,, \quad r \geqq 0\,, \\ \lambda &= -\mu^3 \ln(1-p)/8\pi\,. \end{aligned}\right\} \tag{25}$$

Model 3: Poisson polyhedra characterized by the parameter ϱ are used as primary grains. Inserting (22) into (7), (16) we get

$$\left.\begin{aligned} k(r) &= \{\exp[6\lambda\big(\exp(-\pi\varrho r) - 1\big)/\pi^4\varrho^3] - 1 + p\}/p\,, \quad r \geqq 0\,, \\ \lambda &= -\pi^4\varrho^3 \ln(1-p)/6\,. \end{aligned}\right\} \tag{26}$$

Model 4: Spheres are used as primary grains where the distribution of radii is supposed to be $f(x) = \alpha\mu\exp(-\mu x) + (1-\alpha)\,\delta(x - r_0)$. Then the correlation function is given by (6), (7) with the covariance

$$C(r) = \begin{cases} 2p - 1 + (1-p)^2 \exp\{\pi\lambda\,[\alpha(2r/\mu^2 + 8/\mu^3)\exp(-\mu r/2) + \ldots \\ \ldots + (1-\alpha)\,4r_0^3(1 - 3r/4r_0 + r^3/16r_0^3)/3]\}\,, \quad r \leqq 2r_0 \\ 2p - 1 + (1-p)^2 \exp\{\pi\lambda\alpha(2r/\mu^2 + 8/\mu^3)\exp(-\mu r/2)\}\,, \quad r > 2r_0 \end{cases}$$

$$\lambda = -\ln(1-p)/[8\pi\alpha/\mu^3 + 4\pi r_0^3\,(1-\alpha)/3]\,. \tag{27}$$

This mixture may be interpreted as an approximation of distributions such as log-normal or Weibull.

Generally, the covariance of models constructed by mixing two different types of primary grains is

$$C(r) = 2p - 1 + (1-p)^2 \exp\{\lambda[\alpha\gamma_1^0(r) + (1-\alpha)\,\gamma_2^0(r)]\}\,, \tag{28}$$

where $\gamma_1^0(r)$ and $\gamma_2^0(r)$ are the average density-density correlation functions of the primary grains (cf. (16)), and λ is determined from (15).

Model 5: We consider the intersection of two independent models A_1 and A_2 (where the resulting model A is given by all points belonging both to A_1 and A_2). If A_1 and A_2 are realized by two different sets of Poisson polyhedra we obtain $k(r)$ from (7), (12), (13) with

$$\left.\begin{aligned} C_i(r) &= 2p_i - 1 + \exp\{6\lambda_i[\exp(-\pi\varrho_i r) - 2]/\pi^4\varrho_i^3\}\,, \quad r \geqq 0\,, \\ \lambda_i &= -\pi^4\varrho_i^3 \ln(1-p_i)/6\,, \qquad i = 1, 2\,. \end{aligned}\right\} \tag{29}$$

Model 6: The model is constructed by intersecting A_1 and A_2^c where A_2^c is the complement of A_2. (A_1 and A_2 are the same as in model 5). The correlation function is

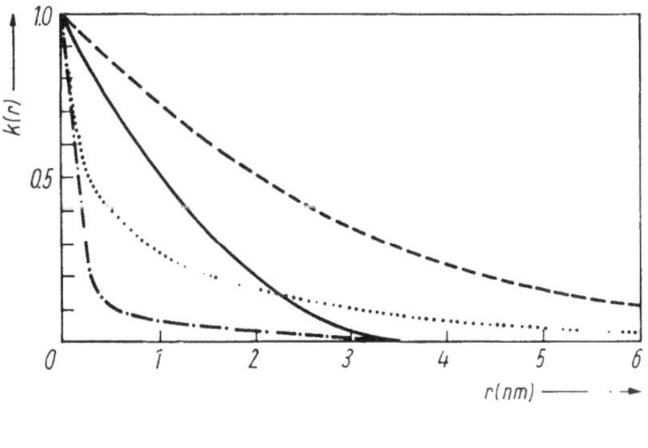

Fig. 1. Correlation functions for several Models. model 1 with $p = 0.5$, $r_0 = 2.0$ nm (———); Model 2 with $p = 0.5$, $\mu = 0.91\ \text{nm}^{-1}$ (– – – –); model 6 with $p_1 = 0.11$, $\varrho_1 = 0.12\ \text{nm}^{-1}$, $p_2 = 0.5$, $\varrho_2 = 1.0\ \text{nm}^{-1}$ (· · · · ·); the particles have the form of "sponges" with polyhedric contours and polyhedric holes; intersection of two sets such as model 1 with parameters $p_1 = 0.1$, $r_{0,1} = 1.1$ nm, $p_2 = 0.2$, $r_{0,2} = 0.2$ nm, (—·—); the particles have the form of clusters of small spheres

obtained by inserting (29) into

$$\left.\begin{aligned} k(r) &= \{C_1(r)\,[C_2(r) - 2p_2 + 1] - p^2\}/p(1-p)\,, \\ p &= p_1(1-p_2)\,. \end{aligned}\right\} \qquad (30)$$

In Fig. 1 correlation functions are plotted for several models.

5. Application to Experimental SAS Data

The models described in the previous section are used to interpret the SAS obtained from a WC powder sample [11] and a $Co_{75}Fe_5B_{20}$ alloy which was prepared by rapid quenching from the melt and annealed at 350 °C for 1 h [12]. In order to diminish the influence of the termination effect, the experimental SAS curves are extrapolated to $s = 0$ using the Guinier approximation and to $s \to \infty$ using the ansatz $I(s) \propto s^{-a}$. The statistical errors of the experimental data are removed by a spline analysis of the extrapolated $I(s)$ function. After this, the SAS intensity is desmeared for infinite slit length, and the correlation function $k(r)$ is calculated from (2), (3). The model functions are fitted to the experimental ones $k_{\exp}(r)$ using a Marquadt procedure which minimizes the quantity

$$\sigma^2 = N^{-1} \sum_{i=1}^{N} [k(r_i) - k_{\exp}(r_i)]^2\,.$$

(The r_i are equidistant r-values for which the experimental correlation function is calculated from the SAS data, $i = 1, \dots, N$.)

Table 1

Average deviation σ of the theoretical correlation function from the experimental one

model	1	2	3	4	5	6
WC powder	0.006	0.146	0.284	0.005	0.317	0.279
$Co_{75}Fe_5B_{20}$ (annealed)	0.219	0.020	0.009	0.004	0.041	0.027

We consider now the WC sample. The best fit (see Fig. 2) is reached for model 4 which is constructed by a mixture of spheres with constant radius and such with exponentially distributed radii. The σ-values (see Table 1) of all other models are substantially higher. In Table 2 the parameters resulting from the Marquadt procedure for models 1 and 4 are compared with particle diameter and volume fraction

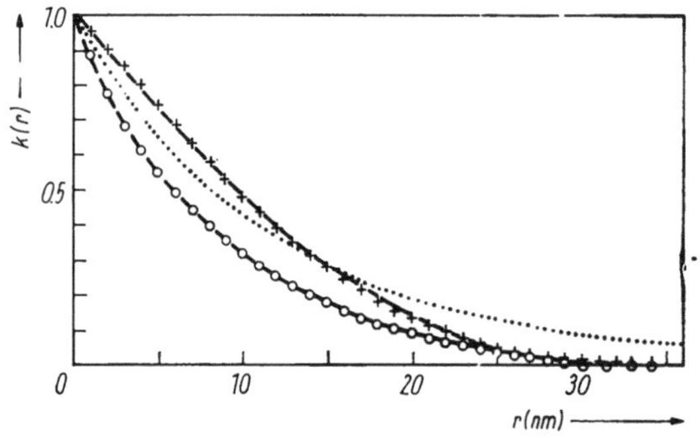

Fig. 2. Correlation functions for WC powder (theoretical curve (——) from model 4 and experimental values (+ + +)) and annealed $Co_{75}Fe_5B_{20}$ (theoretical curve (- - - -) from model 4 and experimental values (○)). The dotted curve represents the approximation to the WC correlation function obtained by model 3

which are estimated [11] by X-ray profile analysis and density measurements, respectively. The agreement between the parameters calculated from SAS data and those obtanined by other methods [11] is reasonable. Considering that the volume fraction p is about 0.2, the grains of the stochastic models can be identified approximately with the real WC particles.

Table 2

Model parameters for WC powder and annealed $Co_{75}Fe_5B_{20}$

	r_0 (nm)	μ^{-1} (nm)	p	α	$\bar{b}/2$ (nm)
WC powder					
model 1	14.9	—	0.22	—	—
model 4	14.1	7.9	0.18	0.07	—
ref. [11]	12.0	—	0.17	—	—
$Co_{75}Fe_5B_{20}$					
model 3	—	—	0.1	—	6.2
model 4	2.6	3.1	0.1	0.48	—

$2r_0$ is the particle diameter, μ^{-1} the average value of the exponential distribution of radii, p the volume fraction of scattering particles, α the weighting factor (cf. expression (28)), $\bar{b}$ the average diameter of the Poisson polyhedra.

Experimental results concerning the $Co_{75}Fe_5B_{20}$ sample show [12] that Co crystallites precipitate during annealing at 350 °C. The correlation function calculated from the SAS data [12] characterizes the Co precipitates within an amorphous matrix. The structure built up by the Co crystallites is approximated by models 1 to 6. As shown in Table 1, the Co precipitates cannot be described by a stochastic assembly of equally sized spheres. On the other hand, the σ-values of models 2 to 4 are of the same order of magnitude and the approximation by the very irregular models 5 and 6 is better than in the case of the WC sample by a factor of 10. The best fit (see Fig. 2) is obtained from model 4 constructed by a mixture of equal spheres and spheres with exponentially distributed radii and the simple Poisson polyhedron model (model 3), respectively. Because of the small volume fraction ($p = 0.1$), the grains of the stochastic models are related to the real Co precipitates. Therefore, the Co crystallites in the $Co_{75}Fe_5B_{20}$ alloy annealed for 1 h at 350 °C seem to be particles the shape of which deviates from that of a sphere and has some similarity with a Poisson polyhedron. The average diameters of the particles are distributed around an average value which is different for the models 3 and 4 by a factor of about 2. For a more detailed discussion of the shape of the Co precipitates additional information, e.g. from electron microscopy, must be required, and the margin of errors of the experimental correlation function must be discussed. The volume fraction $p = 0.1$ which is the same for both models can be understood physically (from X-ray phase analysis [12] follows that the volume fraction of the Co precipitates should lie in the region $0 < p \leqq 0.20$).

6. Conclusion

The methods of stochastic geometry have been shown to be useful for interpreting SAS data from random two-phase structures. The rules for establishing stochastic structure models have been outlined and a series of theoretical and experimental examples have been discussed.

Once a special model is assumed to be valid the model parameters such as volume fraction, average grain size, etc. can be determined by fitting the theoretical correlation function to the experimental one. Absolute SAS measurements are not necessary. In order to improve the reliability of the results, the experimental and numerical errors should be estimated and compared with the average deviation of the theoretical correlation function from the experimental one.

Acknowledgements

We are grateful to Dr. J. Henke of the Zentralinstitut für Festkörperphysik und Werkstofforschung der AdW der DDR, Dresden, for kindly allowing us the use of small-angle scattering data and for critical remarks concerning the experimental examples. Thanks are due to Dr. W. Kluge of the same institute for making available the computer program of the Marquadt procedure.

References

[1] A. Guinier and G. Fournet, Small-Angle Scattering of X-Rays, Wiley, New York 1955.
[2] R. Hosemann and S. N. Bagchi, Direct Analysis of Diffraction by Matter, North-Holland Publ. Co., Amsterdam 1962.
[3] V. Gerold and G. Kostorz, J. appl. Cryst. **11**, 376 (1978).
[4] O. Glatter, J. appl. Cryst. **12**, 166 (1979).
[5] G. Porod, Kolloid Z. **125**, 108 (1952).
[6] D. G. Kendall, in: Stochastic Geometry, Ed. E. F. Harding and D. G. Kendall, Wiley, New York 1974.
[7] G. Matheron, Random Sets and Integral Geometry, Wiley, New York 1975.
[8] J. Serra, Image Analysis and Mathematical Morphology, Academic Press, New York 1981.
[9] D. Stoyan, Biom. J. **21**, 693 (1979).
[10] P. Debye and A. M. Bueche, J. appl. Phys. **20**, 518 (1949).
[11] C. Wetzlaugk and J. Henke, unpublished.
[12] J. Henke, N. Mattern, and H. Hermann, 14. Metalltagung in der DDR "Amorphe metallische Werkstoffe", Dresden, April 6 to 8, 1981.

(Received July 6, 1981)

phys. stat. sol. (a) 68, 289 (1981)

Subject classification: 9; 18.2; 21.1.1

Institut für Metallforschung der Technischen Universität Berlin, Berlin (West)[1])

Diffusion of Zn in α-Fe Single Crystals

By

I. Richter and M. Feller-Kniepmeier

In the temperature range from 848 to 1169 K chemical diffusion coefficients of Zn in α-Fe are given for Zn concentrations between 0 and 12 at%. Above 1068 K the tracer diffusion coefficient ($n_{Zn} \to 0$) can be described by an Arrhenius law with $D_0 = 60\ cm^2\ s^{-1}$ and $Q = 262.6\ kJ\ mole^{-1}$. For lower temperatures the diffusion coefficients decrease below the extrapolated high temperature Arrhenius plot the deviations starting appr. 25 K above the respective Curie temperatures. Below 1068 K the experimental results can be fitted to the theory of Borg. Closely by the Curie temperature no decrease of the diffusion coefficients is found for $n_{Zn} \to 0$. This effect is discussed in terms of magnetic fluctuations causing dynamical processes in the atomic lattice.

Im Temperaturbereich zwischen 848 und 1169 K wird der chemische Diffusionskoeffizient (DK) von Zink in α-Eisen für Konzentrationen zwischen 0 und 12 at% Zink angegeben. Oberhalb 1068 K läßt sich der Tracer-DK ($n_{Zn} \to 0$) durch eine Arrhenius-Beziehung mit $D_0 = 60\ cm^2\ s^{-1}$ und $Q = 262{,}6\ kJ\ Mol^{-1}$ darstellen. Infolge der magnetischen Umwandlung kommt es etwa 25 K oberhalb der jeweiligen Curietemperaturen zu einer Absenkung des DK unter die extrapolierten Werte der Hochtemperatur-Arrheniusgeraden. Unter 1068 K können die Meßergebnisse mit dem theoretischen Modell von Borg beschrieben werden. In der unmittelbaren Umgebung des Curiepunktes bleibt die Absenkung des DK ($n_{Zn} \to 0$) aus. Dieser Effekt wird auf magnetische Fluktuationen und damit verbundene dynamische Reaktionen des Kristallgitters zurückgeführt.

1. Introduction

Concerning the influence of magnetic transformation on diffusion in iron three different types of behaviour are reported in the literature.

a) No influence has been found so far for the diffusion of tungsten [1], vanadium [2], silver [3], and zinc [4] in iron.

b) A deviation below the Curie temperature was postulated for the diffusion of copper [5], antimon [6], molybdenum [7], and manganese [8] in iron.

c) A decrease of the diffusion coefficients from a linear Arrhenius plot above the Curie temperature was measured for self-diffusion in iron [9] and the diffusion of cobalt [10, 11], nickel [11], gold [11], and tin [12] in iron.

However, for a given matrix element one should expect one uniform type of dependence on magnetic transformation. It seemed reasonable, therefore, to contribute an investigation on this topic [13]. The system Fe–Zn was chosen as so far very little work has been done on the diffusion of Zn in α-Fe [4, 14].

2. Experimental

Fe specimens of 2 mm length were cut by a diamond saw from high purity α-Fe single crystals [15] of 3 mm diameter. The residual resistivity ratio $R_{293\,K}/R_{4.2\,K}$ was better than 3000, the calculated content of carbon lower than 0.1 ppm. A surface layer of at least 50 μm was removed electrolytically. These crystals were sealed in quartz ampules under vacuum (4×10^{-4} Pa) and annealed for 100 h at 1173 K. For the

[1]) Straße des 17. Juni, 1000 Berlin (West) 12.

diffusion experiments a single crystal of α-Fe was sealed under vacuum in a quartz tube together with a piece of zinc (total impurity content less than 20 ppm), the two metals being separated by a necked-down portion. Diffusion annealings were carried through in an ac resistant heated furnace, except the diffusion couples at 1043 and 1039 K, which were annealed in a dc furnace. The iron crystals were always located at the temperature maximum of the furnace, while Zn pieces had temperatures from 25 to 50 K below that of the Fe crystals. Under these circumstances a constant vapour pressure was obtained throughout the diffusion experiment. The annealing temperatures were controlled with ± 0.2 K. Diffusion runs were made in the temperature range 848 to 1169 K. The diffusion times varied from 45 to 3792 h. A Cameca electron probe microanalyzer was used to measure the concentration gradients of Zn in Fe. Analysis of the corrected concentration profiles was achieved in three steps. First the finite diameter of the electron probe which deforms the concentration profile is taken into account. A correction of the concentration profile for the change of crystal parameters was not necessary, as an estimation showed that the error for a maximum concentration of 20 at% Zn and a penetration depth of 100 μm is 1% in profile length. Secondly a rough determination of the diffusion coefficients is carried through using the Matano method. This analysis, however, yields considerable oscillations in the $D = f(c)$ curves, as the slope of the profile directly enters the result. A measuring point deviating from its true position due to experimental error therefore evaluates two incorrect diffusion coefficients one being too low, the other being too high. In addition the Matano analysis yields extremely decreasing diffusion coefficients at the ends of the profiles. In order to avoid these difficulties the differential equation

$$\xi(\lambda) = \frac{\int\limits_0^{\lambda} \frac{1}{D(c)} \exp\left(-\frac{1}{2}\int\limits_0^{\bar{\lambda}} \frac{1}{D(c)}\,\mathrm{d}l\right)\mathrm{d}\bar{\lambda}}{\int\limits_0^{\infty} \frac{1}{D(c)} \exp\left(-\frac{1}{2}\int\limits_0^{\bar{\lambda}} \frac{1}{D(c)}\,\mathrm{d}l\right)\mathrm{d}\bar{\lambda}} \tag{1}$$

with $\lambda = x/\sqrt{t}$, $\xi = \mathrm{d}C/\mathrm{d}\lambda$, $C = \int D(c)\,\mathrm{d}c$, and l integration variable, which is the solution for the prevailing boundary conditions was solved by an iterative programme. In this programme an estimated $D(c)$ function from the Matano analysis is taken for the calculation of a $c(x)$ profile. By successively changing the $D(c)$ function, the calculated profile is fitted to the experimental one.

An estimation of error for the diffusion coefficients resulted in the values given in Table 1. Here $\Delta D/D$ (1) is due to the error in determining the profile length $\Delta x/x = \pm 2\%$ and $\Delta D/D$ (2) is the error due to uncertainties in concentration measurements.

Table 1
Errors in the determination of chemical diffusion coefficients

n_{Zn} (at%)	$T > 973$ K			$973 > T > 873$ K			$T < 873$ K
	$\frac{\Delta D}{D}$ (1) (%)	$\frac{\Delta D}{D}$ (2) (%)	Σ (%)	$\frac{\Delta D}{D}$ (1) (%)	$\frac{\Delta D}{D}$ (2) (%)	Σ (%)	Σ (%)
0	±5	0	± 5	±10	± 0	±10	±25
12	±5	±5	±10	±10	±10	±20	

Errors for the temperatures $\leqq$ 873 K are given separately because of their low profile length. The error in determining the activation energy was calculated to be $\Delta Q/Q = \pm 3.5\%$.

3. Results

3.1 Chemical diffusion coefficients

Chemical diffusion coefficients as function of temperature and Zn concentration are listed in Table 2 and plotted in Fig. 1. Except for the low concentrations at 1038,

Table 2
Chemical diffusion coefficients (in $cm^2\,s^{-1}$) in α-Fe/Zn

T (K)	n_{Zn} (at%) 0.0	4.0	8.0	12.0	A
1169	1.13 ± 5% E-10	1.62 ± 8% E-10	2.35 ± 9% E-10	3.36 ± 10% E-10	9.08
1148	6.56 ± 5% E-11	1.02 ± 8% E-10	1.58 ± 9% E-10	2.50 ± 10% E-10	11.23
1123	3.81 ± 5% E-11	5.84 ± 8% E-11	8.83 ± 9% E-11	1.37 ± 10% E-10	10.64
1098	2.06 ± 5% E-11	3.35 ± 8% E-11	5.39 ± 9% E-11	8.83 ± 10% E-11	12.11
1072	9.67 ± 5% E-12	1.62 ± 8% E-11	2.76 ± 9% E-11	4.56 ± 10% E-11	12.91
1048	3.20 ± 5% E-12	8.00 ± 8% E-12	1.44 ± 9% E-11	2.50 ± 10% E-11	(14.86)
1043	3.45 ± 5% E-12	6.20 ± 8% E-12	1.24 ± 9% E-11	2.26 ± 10% E-11	(15.31)
1039	3.50 ± 5% E-12	5.85 ± 8% E-12	1.02 ± 9% E-11	1.82 ± 10% E-11	(13.74)
1024	1.54 ± 5% E-12	3.57 ± 8% E-12	6.86 ± 9% E-12	1.33 ± 10% E-11	(16.22)
1018	1.07 ± 5% E-12	2.48 ± 8% E-12	5.24 ± 9% E-12	1.09 ± 10% E-11	(18.50)
1008	7.47 ± 5% E-13	1.63 ± 8% E-12	3.74 ± 9% E-12	8.19 ± 10% E-12	20.07
1000	4.51 ± 5% E-13	1.05 ± 8% E-12	2.48 ± 9% E-12	5.85 ± 10% E-12	21.48
975	1.50 ± 5% E-13	3.59 ± 8% E-13	8.80 ± 9% E-13	2.13 ± 10% E-12	22.11
948	4.70 ± 10% E-14	1.23 ± 14% E-13	3.45 ± 17% E-13	8.67 ± 20% E-13	24.29
926	1.64 ± 10% E-14	5.26 ± 14% E-14	1.69 ± 17% E-13	5.23 ± 20% E-13	28.87
873	1.7 ± 25% E-15	5.1 ± 25% E-15	—	—	26.5
848	4.0 ± 25% E-16	7.4 ± 25% E-16	—	—	28.1

1039 (cf. Section 4.2), and 1048 K which will be treated later, an experimental dependence

$$D^{ch}(n_{Zn}) = D(0) \exp (A n_{Zn})$$

was found. In conformity with Darkens equation

$$D^{ch} = (n_{Zn} D^*_{Fe} + n_{Fe} D^*_{Zn}) f_{th} R_L \tag{2}$$

the values of D^{ch} extrapolated for $n_{Zn} \to 0$ at% equal the tracer diffusion coefficients D^*_{Zn}.

In Fig. 2 the Arrhenius plots for various Zn concentrations between 0 and 12 at% are given. For a first view these curves may be divided into two regions, above 1068 K for all concentrations considered ln D^{ch} is a linear function of $1/T$, while below 1068 K deviations from that relation are apparent. We shall first consider the high temperature region which is in the paramagnetic range.

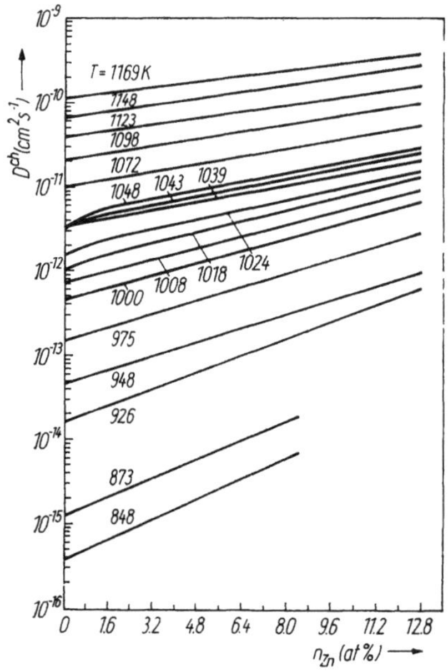

Fig. 1. Chemical diffusion coefficients in the α-Fe/Zn system as function of Zn concentration

3.2 Paramagnetic region above 1068 K

3.2.1 Tracer diffusion coefficient of Zn

Using results of Wriedt [16] on the partiale pressur of Zn in Fe–Zn alloys thermodynamic factors as listed in Table 3 were calculated. There is no influence of the magnetic transformation on f_{th}. In order to evaluate D^*_{Zn} from equation (2) in addition to f_{th} the diffusion coefficients $D^*_{Fe}(n_{Zn})$ in the alloys and the vacancy flow term R_L are needed. For an estimation of $D^*_{Fe}(n_{Zn})$ two limiting conditions can be assumed.

1. The lower limit of $D^*_{Fe}(n_{Zn})$ is set to be equal to $D^*_{Fe}(0)$.

2. From the fact that markers of Al_2O_3 brought to the surface of the α-Fe single crystals before diffusion were always found to be on the surface after diffusion, some minimum ratios of D^*_{Zn}/D^*_{Fe} for high concentrations of Zn could be calculated. These values give an upper limit of $D^*_{Fe}(n_{Zn})$.

Using these two limitations an estimation of $D^*_{Zn}(n_{Zn})$ can be given. The values are compiled in Table 4 and plotted in Fig. 3. The error given contains both, the uncertainty of $D^*_{Fe}(n_{Zn})$ and the error in $D^{ch}(n_{Zn})$. The vacancy flow term which is a function of the ratio of D^*_{Zn} and D^*_{Fe},

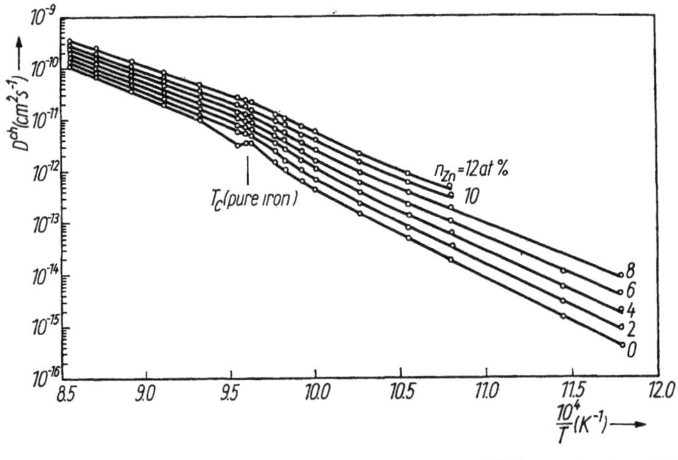

Fig. 2. Arrhenius plots for chemical diffusion coefficients in the α-Fe/Zn system

Table 3
Thermodynamic factors f_{th} in Fe/Zn

T (K)	n_{Zn} (at%)			
	2	5	10	15
1169	1.090	1.215	1.442	1.532
1148	1.095	1.220	1.450	1.547
1123	1.098	1.225	1.462	1.565
1098	1.102	1.232	1.475	1.582
1073	1.107	1.237	1.487	1.597
1043	1.112	1.245	1.502	1.618
1000	1.120	1.255	1.525	1.647
975	1.125	1.262	1.537	1.667

Table 4
Tracer diffusion coefficients (in cm² s⁻¹) of Zn in Fe/Zn together with errors of estimation

T (K)	n_{Zn} (at%)			
	0	5	10	15
1169	1.13 E-10	1.44 E-10	1.97 E-10	2.74 E-10
1148	6.56 E-11	9.26 E-11	1.38 E-10	2.07 E-10
1123	3.81 E-11	5.70 E-11	8.82 E-11	1.38 E-10
1098	2.06 E-11	3.06 E-11	4.83 E-11	7.62 E-11
1072	9.67 E-12	1.48 E-11	2.45 E-11	4.11 E-11
	±5%	±10%	±15%	±22%

$$R_L = 1 + \frac{2n_{Fe}n_{Zn}(D^*_{Zn} - D^*_{Fe})^2}{5.33\,(n_{Zn}D^*_{Fe} + n_{Fe}D^*_{Zn})\,(n_{Zn}D^*_{Zn} + n_{Fe}D^*_{Fe})}, \tag{3}$$

was taken to be equal to one in the first run and was calculated for a second run using the result of the first approximation.

3.2.2 Correlation factor

According to Le Claire [17] the correlation factor f_i can be calculated from the equation

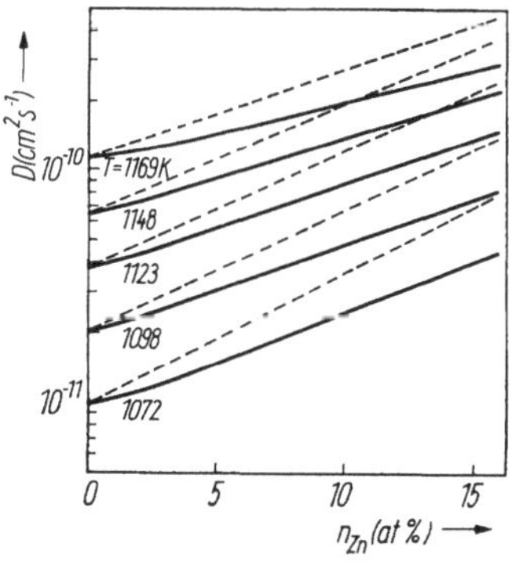

Fig. 3. Tracer diffusion coefficients of Zn in α-Fe/Zn as fun- tion of Zn concentration. - - - - D_{ch}, ——— D^*_{Zn}

$$f_i = 1 - \frac{D_i}{D_s(0)} \frac{6f_0}{b + 15 + \frac{1.53b + 9.18}{b + 9.06}}, \tag{4}$$

where b is a constant factor defined by the dependence of the matrix self-diffusion coefficient from the concentration of the second element

$$D^*_{Fe}(n_{Zn}) = D^*_{Fe}(0) (1 + bn_{Zn}) . \tag{5}$$

Using the limitations given above, b and $f_{Fe/Zn}$ can be estimated in the range from 0 to 15 at% Zn as given in Table 5 together with the ratio ω_2/ω_0 of the jump frequencies

Table 5

Values of quantity b, correlation factor f_i, and ratio of jump frequencies ω_2/ω_0 as function of temperature

T (K)	1169	1123	1072
f_i (max)	≈ 0.25	≈ 0.45	≈ 0.60
ω_2/ω_0	>8	>3	>2
b	10	16	23

(ω_2 frequency vacancy–Zn atom, ω_0 jump frequency for self-diffusion in the matrix).

The jump frequency ratios exhibit that there is a strong binding between vacancies and Zn atoms in the Fe matrix at least above 1123 K. Furthermore a possible temperature dependence of the correlation factor may have to be considered.

3.2.3 *Activation energy*

Activation energies were calculated using the seven diffusion experiments between 1072 and 1169 K. In Table 6 and Fig. 4 the values are given for chemical diffusion as a function of the Zn concentration. The least square line yields a linear relationship.

Table 6

Activation energies for chemical diffusion in α-Fe/Zn

n_{Zn} (at%)	D_{0i} (cm² s⁻¹)	Q^{ch} (kJ mol⁻¹)
0	60	262.6
2	38	256.2
4	19	247.5
6	11	239.9
8	6.2	232.9
10	3.9	226.4
12	2.0	218.1

For $n_{Zn} \to 0$ at% the activation energy of the tracer diffusion process is found to be $Q = 262.6$ kJ mol^{-1} with a pre-exponential factor of 60 cm^2 s^{-1}. These values do not regard a possible temperature dependence of $f_{Fe/Zn}$. The dependence of the pre-exponential factor D_{0i} from the activation energies is given by $\ln D_{0i} = a + mQ$, with $a = -16.1$ and $m = 7.6 \times 10^{-5}$.

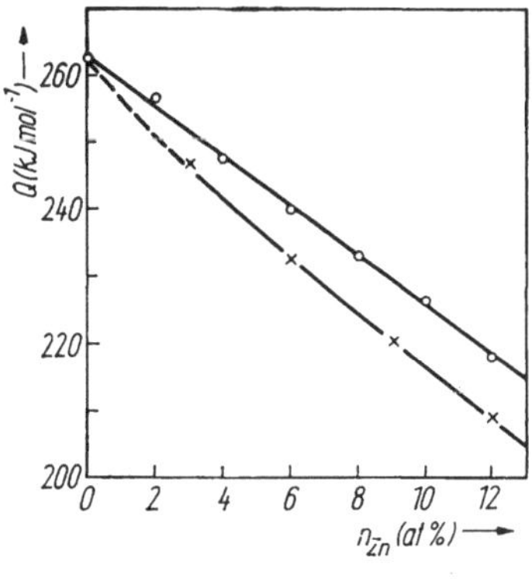

Fig. 4. Activation energies for chemical diffusion in the α-Fe/Zn system. —×— after [4], —○— this work

3.3 Transition and ferromagnetic region below 1068 K

In the temperature range below 1068 K the $\ln D$ over $1/T$ curve for $n_{Zn} \to 0$ at% differs in its qualitative appearance as compared with the curves for all other concentrations. While the latter show a continuous decrease from the extrapolated linear relationship the curve for $n_{Zn} \to 0$ at% is decreased at 1048 K, the amount of reduction being lower at 1043 K and not existing at 1039 K. Measuring points below that temperature again fit into the general behaviour. A discussion of the anomaly found at the two temperatures is postponed at the moment and the outline of results is carried on for decreasing diffusion coefficients due to magnetic transformation.

3.3.1 Temperature of decrease

The beginning of a decrease from the linear Arrhenius curves was evaluated as intersection between that curve and the extrapolated values from lower temperatures. The results are given in Fig. 5. The error for T_A quoted is ± 10 K. For pure iron the Curie temperature was taken to be 1042 K [18]. Curie temperatures for the Fe–Zn alloys were estimated taking data of Schramm [19] and Fallot [20], which found the Curie temperature for an alloy containing 19 at% Zn to be 943 K. Assuming a linear relationship between transformation temperature and concentration for non-magnetic foreign atoms the respective Curie temperatures were interpolated. Despite the lack of exact experimental data on transformation temperatures of the alloys there certainly is a decrease in diffusion coefficients in Fe–Zn for concentrations between 0 and 12 at% above the respective Curie temperatures. The temperature difference is approximately 25 K.

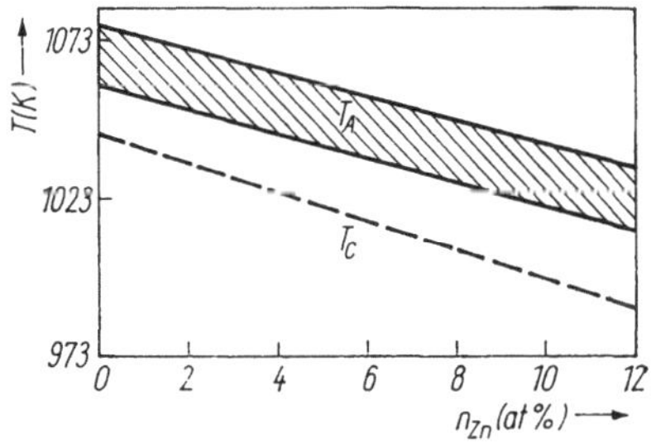

Fig. 5. Temperatures of decrease from Arrhenius line and Curie temperature in the Fe/Zn system

3.3.2 Comparison with theories

A comparison of our data with the theories of Ruch et al. [21] and Borg [22] is possible only for $D^*_{Zn}(n_{Zn} \to 0)$ because there are neither magnetic data nor elastic constants available for the alloys. In the theory of Ruch diffusion behaviour in a ferromagnetic crystal is connected solely to spontaneous magnetization

$$D = D_0^P \exp\left(-\frac{Q^P}{RT}\right) \exp\left(-\frac{\alpha_F M^2 + \alpha_M M^2}{RT}\right) \tag{6}$$

whit D_0^P pre-exponential factor in the paramagnetic region, Q^P activation energy of diffusion in the paramagnetic region, M spontaneous magnetization, $\alpha_F, \alpha_M = \text{const.}$, R gas constant, T temperature.

Borg's theory is based on ideas of Zener [23] regarding the movement of an atom across a saddlepoint. The energy necessary was postulated to be connected with a suitable elastic constant μ. Borg extended this idea to magnetic ordering, assuming that the change in free migration enthalpy due to magnetic transformation is proportional to the difference of two elastic constants. The change of the free vacancy formation enthalpy is as in Ruch's theory connected to the spontaneous magnetization. For the diffusion coefficient in a ferromagnetic crystal Borg's theory resulted in

$$D = D_0^P \exp\left(-\frac{Q^P}{RT}\right) \exp\left(\frac{\alpha_F M^2 + K\,\Delta\mu}{RT}\right). \tag{7}$$

The fit of equations (6) and (7) to the experimental data was done using the measured points at 926, 948, and 975 K. The constants $\alpha = (\alpha_F + \alpha_M)/Q^P$ and K are compiled in Table 7. Magnetization data were taken from Rocker et al. [18]. For the elastic

Table 7

Values of quantities α, $K(c')$, and $K(c_m)$ as function of temperature

T (K)	α	$K(c')$ (10^{-24} cm^3)	$K(c_m)$ (10^{-24} cm^3)
926	0.118	2.9	5.5
948	0.116	2.7	5.2
975	0.121	2.8	5.4
	$\bar{\alpha} = 0.118$	$\bar{K} = 2.7$	$\bar{K} = 5.4$

constant the shear modulus $c' = (c_{11} - c_{12})/2$ and the mean value of c' and c_{44} was used $c_m = (c' + c_{44})/2$. The result of the calculations is plotted in Fig. 6. At a first view there seems to be no unambiguous evidence for one of the two theories. However, setting apart the region between 1018 and 1043 K as will be given the reason for later, the decreased diffusion coefficient at 1048 K together with the decrease found above the Curie temperature for higher Zinc concentrations positively favours the fit due to Borg.

In principle a calculation of the change in activation energy of diffusion caused by magnetic transformation is possible using the values of α and K. Prerequisite is that D_{0i} is not a function of temperature down to the ferromagnetic region. A possible dependence of f_i has already been discussed, in addition a further dependence of f_i on magnetic ordering has to be taken into account.

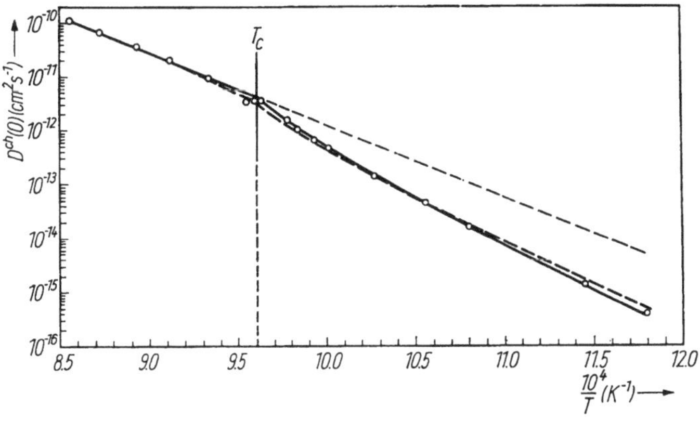

Fig. 6. Comparison of experimental values with theories of Ruch (———) and Borg (- - - -); o this work

For the diffusion system Fe–Zn a change in activation energy of 13 kJ is apparent at 923 K. For very low temperatures a limit of $\Delta Q^{ch} = 30$ kJ is calculated from the theory of Ruch.

4. Discussion

4.1 Comparison with the literature

During the last ten years only two investigations concerning diffusion of Zn in Fe in the reported temperature range have been undertaken. Budurov et al. [14] made diffusion runs between 1223 and 1424 K for $n_{Zn} = 11$ at%. Extrapolated into the temperature range we have chosen their values are about 20% below ours. Chemical diffusion coefficients were determined by Suzuki et al. [4] between 973 and 1253 K. Their data for temperatures above 1073 K are close to ours, whereas below that temperature they do not find an influence of magnetic transformation on diffusion.

Comparing the decrease of diffusion coefficients in iron due to magnetic transformation at a temperature of 923 K the greatest effect is found for self-diffusion whereas for the diffusion of foreign elements the ratio $D_{ex}(T)/D_f(T)$, where $D_{ex}(T)$ is the diffusion coefficient extrapolated from the paramagnetic region at the temperature T and $D_f(T)$ is the actual one, does not seem to be affected much by the type of element or its magnetic condition (Table 8). For self-diffusion in α-Fe the data of the literature as given by Hettich et al. [24] were used in order to make a fit to the

Table 8

Ratio D_{ex}/D_f of extrapolated and measured diffusion coefficients in the ferromagnetic region ($T = 923$ K)

Co	Borg et al. (1963)	5.8
Ni	Borg et al. (1963)	6.6
Ni	Hirano et al. (1961)	4.5
Cu	Salje et al. (1976)	4.7
Sb	Myers et al. (1977)	5.5
Zn	this work	4.3
Fe	Hettich et al. (1977)	25

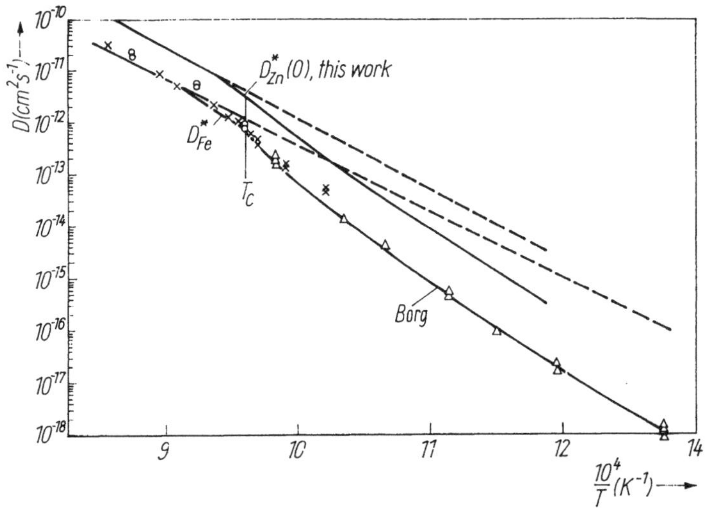

Fig. 7. Self-diffusion in α-Fe. Fit with Borg's theory. △ after [24], × after [9], ○ after [25]

theory of Borg. This calculated curve is plotted in Fig. 7. It is seen that there is excellent agreement with the experimental data of Hettich et al. [24], Borg and Lai [10], and Buffington et al. [25] and that the decrease starts about 50 K above the Curie temperature. In the work of James and Leak [26] and Kucera et al. [27], however, a decrease had not been detected. One of the systems which have been quoted to exhibit no modulation of diffusion coefficients above the Curie temperature is Fe–Cu [5]. Unfortunately in that system there is only one measuring point above the Curie temperature that could show a decrease from the linear Arrhenius line. Furthermore in that work the data obtained at 1163 and 1173 K were published but disregarded in the calculation of the fitting curve because they were thought to be off too much from the "true" line. Using all experimental data in the paramagnetic region for calculation an activation energy $Q = 275.03$ kJ mol^{-1} with $D_0 = 114.33$ cm^2 s^{-1} is obtained (Fig. 8). Deviations from that Arrhenius curve are given in Table 9. It is

Table 9

Deviations of experimental points from calculated Arrhenius curve for diffusion of Cu in α-Fe

T (K)	1045.16	1063.16	1080.16	1093.16	1111.16	
Δ (%)	−5.77	0.28	1.19	2.77	−0.18	
T (K)	1112.16	1130.16	1143.16	1153.16	1163.16	1173.16
Δ (%)	2.63	2.96	4.23	8.05	−7.70	−7.18

seen that for $T = 1045$ K there is a decrease of 5.8% but that this value is within the experimental error found at higher temperatures. So at the present time the question whether or not a deviation due to magnetic transformation starts above or at the Curie temperature cannot be answered for this system.

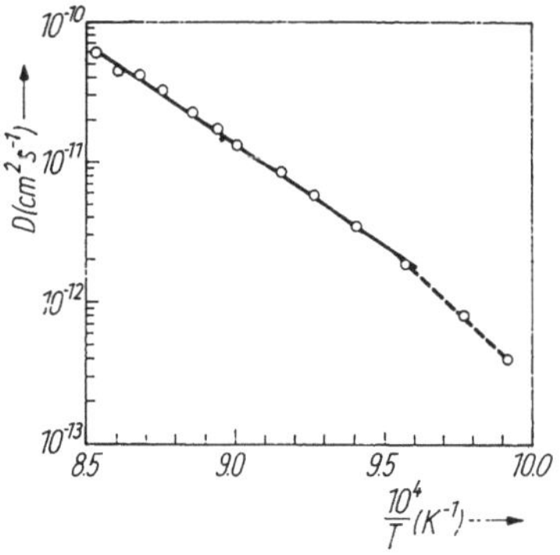

Fig. 8. Arrhenius plot for diffusion of Cu in α-Fe

4.2 Anomaly in the ferromagnetic environment for $n_{Zn} = 0$ at %

As there is a decrease of diffusion coefficients for all concentrations between 0 and 12 at% Zn above the Curie temperature, the lack of that decrease at 1039 and 1043 K for $n_{Zn} = 0$ at% has to be discussed. The results at these temperatures were obtained using a dc regulated furnace with a magnetic field of only 2000 A/m. In an additional experiment at 1038 K the diffusion specimen was annealed in an ac furnace the crystal lying in an alternating magnetic field of the same magnitude. For concentrations above 1.5 at% zinc the results of the experiments at 1038 and 1039 K are identical. Below that concentration, however, there is an increase of the diffusion coefficient for the specimen annealed in the alternating magnetic field.

$$D^*_{Zn}(0)\ 1039\ K_{H_-} = 0.35 \times 10^{-11}\ cm^2\ s^{-1}\,,$$
$$D^*_{Zn}(0)\ 1038\ K_{H_\sim} = 0.52 \times 10^{-11}\ cm^2\ s^{-1}\,.$$

In explaining these discrepancies the real structure of the ferromagnet has to be considered. Both the theories of Borg [22] and Ruch et al. [21] deal with diffusion in ordered magnetic domains only disregarding Bloch walls. An influence of a stationary Bloch wall on diffusion is a second-order effect and an increase of diffusion coefficients within the wall volume is considered possible utmost to values of the extrapolated Arrhenius line from the paramagnetic region. Considering the stimulation of Bloch wall motions by an alternating magnetic field an enlarged enhancement of diffusion in the total volume covered by the moving wall is expected. Via magnetostriction the change in spin direction of complete crystal volumes results in a change of crystal parameters. A periodic movement of Bloch walls therefore induces lattice oscillations and enhanced diffusion.

Following these arguments the increase of the diffusion coefficients measured at 1038 K in an alternating magnetic field is accounted for. The lack of a decrease at 1039 and 1043 K, however, still needs to be explained. This effect is discussed as a consequence of magnetic fluctuations above and thermal fluctuations below the Curie temperature. The latter causes magnetic fluctuations in accordance to the high gradient of the magnetization as a function of temperature. These magnetic fluctuations may result in an increase of diffusion by magnetostriction. Remembering the fact that a movement of a Bloch wall can be done without loss of energy only in ideal ferromagnetic systems it is supposed that magnetic fluctuations are very sensitive to non-magnetic atoms representing a disturbance of magnetic order.

Taking this picture there are diffusion supporting conditions only for low Zn concentration levels. The coincidence of low Zn concentration and areas of magnetic fluctuation exists only at the Curie temperature of pure iron. For lower temperatures the areas of possible magnetic fluctuations are connected with high Zn concentrations the latter suppressing the former.

The different types of diffusion behaviour described in the literature may now be better understood. In cases where no decrease of diffusion coefficients above the Curie temperature was found, the effect could have been overlooked, either because measuring points were not taken close enough to the Curie temperature or because of scattering. In cases where a decrease of diffusion at the Curie temperature was reported, the question arises if the iron matrix was clean enough to allow magnetic fluctuations. It should be reminded that the use of single crystals of high purity iron is possible just for a few years.

References

[1] J. KIESZNIEWSKI, Prace Instyt, Hutn 19, 253 (1967).
[2] J. KUCERA and B. MILLION, Z. Metallk. 61, 161 (1970).
[3] A. BONDY and V. LÉVY, C.R. Acad. Sci. (France) 272, C19 (1971).
[4] H. SUZUKI, Y. MASUDA, and K. NISHIDA, Nippon kinzoku gakhai shi 36, 509 (1972).
[5] G. SALJE and M. FELLER-KNIEPMEIER, J. appl. Phys. 48, 1833 (1977).
[6] S. M. MYERS and H. J. RACK, J. appl. Phys. 49, 3246 (1978).
[7] K. NOHARA and K. HIRANO, J. Japan. Inst. Metals 40, 1053 (1976).
[8] K. NOHARA and K. HIRANO, Proc. ICSTIS Suppl. Trans. ISIJ 11, 1267 (1971).
[9] R. S. BORG and C. E. BIRCHENALL, Trans. M S AIME 218, 980 (1960).
[10] R. S. BORG and D. Y. F. LAI, Acta metall. 11, 861 (1963).
[11] K. HIRANO and M. COHEN, Trans. Japan. Inst. Metals 13, 96 (1972).
[12] D. TREHEUX, D. MARCHIVE, J. DELAGRANGE, and P. GUIRALDENQ, C.R. Acad. Sci. (France) 274, C1260 (1972).
[13] I. RICHTER, Dissertation, TU Berlin, 1981.
[14] S. BUDUROV, P. KOVATCHEV, and Z. KAMENOVA, Z. Metallk. 64, 652 (1973).
[15] G. SALJE and M. FELLER-KNIEPMEIER, Z. Metallk. 68, 293 (1977).
[16] H. A. WRIEDT, Trans. MS AIME 239, 1120 (1967).
[17] A. D. LE CLAIRE, Phil. Mag. 21, 819 (1970).
[18] W. ROCKER, R. KOHLHAAS, and H. W. SCHÖPGENS, Z. angew. Phys. 32, 164 (1971).
[19] J. SCHRAMM, Z. Metallk. 28, 203 (1936).
[20] M. FALLOT, Ann. Phys. (France) 11, 420 (1937).
[21] L. RUCH, D. R. SAIN, H. L. YEH, and L. A. GIRIFALCO, J. Phys. Chem. Solids 37, 649 (1976).
[22] R. J. BORG, Trans. MS AIME 218, 980 (1960).
[23] C. ZENER, J. appl. Phys. 22, 372 (1951).
[24] G. HETTICH, H. MEHRER, and K. MAIER, Scripta metall. 11, 795 (1977).
[25] F. S. BUFFINGTON, K. HIRANO, and M. COHEN, Acta metall. 9, 434 (1961).
[26] D. W. JAMES and G. M. LEAK, Phil. Mag. 14, 701 (1966).
[27] J. KUCERA, B. MILLION, J. RUZICKOVA, V. FOLDYNA, and A. JAKOBORA, Acta metall. 22, 135 (1974).

(Received June 19, 1981)

phys. stat. sol. (a) 68, 301 (1981)

Subject classification: 8; 14.2; 21.1

Zentralinstitut für Festkörperphysik und Werkstofforschung der Akademie der Wissenschaften der DDR, Dresden[1])

Thermal Conductivity of V_3Si

By

A. GLADUN, C. GLADUN, M. KNORN, and H. VINZELBERG

The thermal conductivity of V_3Si single crystals is measured within the temperature range from 0.1 to 300 K. In the temperature range above 100 K the thermal conductivity increases with increasing temperature. This behaviour is related to the deviation from linearity of the electrical resistivity at high temperatures common to A15 superconductors. At the temperature of the lattice transformation a change of both the electron–phonon scattering coefficient and the electron–defect scattering coefficient is found. In the superconducting state, the electron conductivity steeples decreases with decreasing temperature. From a fit of the measuring points to theoretical estimations the energy gap can be determined. At the lowest investigated temperatures the lattice thermal conductivity is dominating. The different scattering mechanisms determining the temperature dependence of lattice thermal conductivity are discussed.

Die Wärmeleitfähigkeit von V_3Si-Einkristallen wird im Temperaturbereich 0,1 bis 300 K gemessen. Im Temperaturbereich oberhalb 100 K wächst die Wärmeleitfähigkeit mit steigender Temperatur an. Dieses Verhalten steht in Zusammenhang mit der für die A15-Supraleiter charakteristischen Abweichung vom linearen Ansteigen des elektrischen Widerstandes bei hohen Temperaturen. Bei der Temperatur der Martensitumwandlung wird eine Änderung der beiden Koeffizienten für die Elektron–Phonon- und die Elektron–Defekt-Streuung gefunden. Im supraleitenden Zustand fällt die Elektronenwärmeleitfähigkeit mit fallender Temperatur steil ab, aus der Anpassung der Meßwerte an theoretische Abschätzungen kann die Energielücke bestimmt werden. Bei den tiefsten untersuchten Temperaturen dominiert die Gitterleitfähigkeit. Die verschiedenen Streumechanismen, die die Temperaturabhängigkeit der Gitterleitfähigkeit bestimmen, werden diskutiert.

1. Introduction

The improvement of methods for preparation and characterization of A15 structure monocrystals has stimulated further investigation of their superconducting and normal-state properties.

In this paper we present our results of thermal conductivity measurements of V_3Si monocrystals within the temperature range 0.1 to 300 K. All the investigated crystals are prepared by the same method [1] and differ mainly by the degree of deviation from the ordered stoichiometric V_3Si compound.

Common to high T_c A15 superconductors is a deviation from linearity of the electrical resistivity at high temperatures, the origin of which is subject of lively discussions [2, 3]. We believe that the increase of thermal conductivity with rising temperature after passing a minimum near 100 K is to be related to this effect.

From our measurements we were able to determine the temperature of the martensitic phase transition in accordance with the values derived from measurements of other properties [3]. We observe a change of both the electron–phonon scattering coefficient and the electron-defect scattering coefficient at the transition.

[1]) Postfach, DDR-8027 Dresden, GDR.

2. Experimental

The investigated V_3Si single crystals were prepared by the method described by Jurisch et al. [1]. The crystals were rod-shaped. A characterization of the samples is given in Table 1. The residual resistivity ratio RRR and the specific resistivity at 0 K, ϱ_0, have been extrapolated from values measured above 17.5 K by means of the ideal electrical resistivity values given by Berthel and Pietrass [3].

Table 1

Dimensions, residual resistivity ratio RRR, residual resistivity at 0 K ϱ_0, critical temperature T_c, and energy gap at 0 K $\Delta(0)$ of the investigated V_3Si samples

sample	diameter (mm)	length (mm)	RRR	ϱ_0 (μΩ cm)	T_c (K)	$2\Delta(0)/kT_c$
1	4.14	52	65	1.11	16.7	3.2
2	4.34	58	43	1.73	16.8	3.5
3	3.60	52	29	2.51	16.7	3.8
4	4.50	31	25	2.99	16.7	3.8
5	3.96	61	6.2	14	16.9	3.8

According to Lange et al. [4] the residual resistivity of our V_3Si samples increases linearly with increasing deviation of the Si concentration from stoichiometry. The chemical composition of our samples varies with length as well as in the cross-section. This causes changes of the residual resistivity along the length of the samples. For example in sample 1 the residual resistivity varies from 0.8 to 1.2 μΩ cm. In Table 1 mean values of ϱ_0 and RRR are given.

Precipitations of second phases and dislocation density were determined with an optical metallographic microscope. Samples 1 to 4 are single-phased, in sample 5 α-V precipitates were found. The dislocation density decreases with rising deviation from stoichiometric composition from 2×10^5 cm^{-2} in sample 1 to 2×10^4 cm^{-2} in sample 4. This result agrees well with investigations carried out by Quyen [5] on similar samples.

The thermal conductivity was measured by the stationary method within the temperature range 0.07 to 290 K. Whereas the measuring principle was the same in the entire temperature range, the measurements were performed in two different cryostats [6, 7] and with three different pairs of thermometers. In the temperature range from 0.07 to 2 K specially prepared Speer resistors (Grade 1002, 0.5 W) [6], in the range from 2 to 70 K Allen-Bradley resistors (220 Ω, 1/8 W) [7], and in the range from 70 to 290 K constantan–copper thermocouples [8] were used. To avoid the injurious influence of radiation at temperatures above 70 K the sample was additionally covered by a radiation shield which has nearly the same temperature and temperature gradient as the sample.

3. Results

The thermal conductivity of the investigated crystals as a function of temperature is shown in Fig. 1 and 2. At temperatures below 1.7 K only samples 2 and 3 have been measured and at temperatures above 40 K only samples 1 and 5. At the lowest and highest investigated temperatures the conductivities tend to approach each other whereas between 2 and 50 K the varying purity of the samples causes different conductivities as well as different temperature dependences of conductivity. At temperatures near and above the critical temperature T_c electronic thermal conductivity dominates while at $T \ll T_c$ lattice conductivity is prevailing. According to the differ-

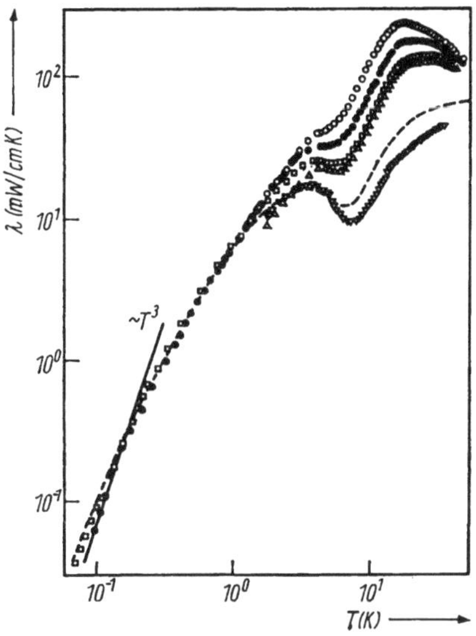

Fig. 1. Thermal conductivity of V_3Si in the temperature range 0.07 to 50 K. ○ sample 1, ● sample 2, □ sample 3, △ sample 4, ▽ sample 5. The dashed line shows experimental results obtained by Lou [25]. The solid line represents a T^3 proportionality

ent scattering mechanisms determining the heat conduction, the investigated temperature range may be divided into three sections: $T > T_c$, $6\text{ K} < T < T_c$, $T < 6\text{ K}$.

3.1 Temperature range $T > T_c$

In pure normal metals and intermetallic compounds thermal energy is carried by electrons. For the thermal resistivity at temperatures below $0.1\theta_D$ ($\theta_D \approx 420$ K [9], Debye temperature) in general the equation

$$\lambda_e^{-1} = W_e = W_{ed} + W_{eph} = AT^{-1} + BT^2 \quad (1)$$

is used. The first term describes the scattering of the conduction electrons by defects, the second term the scattering by phonons.

A graphical fit of (1) is shown in Fig. 3, where T/λ is plotted versus T^3 for samples 1 to 4. In Fig. 3 a kink in the fitting lines is observed at the temperature T_m of the cubic–tetragonal lattice transformation, which is typical for the A15 high T_c superconductors [9]. In the cubic and tetragonal phases the thermal conductivity fits to different values of A and B for each sample (Table 2). It should be noted that the kink appears in the fitting lines only. The constants A and B vary continuously from the value in the cubic phase to that in the tetragonal phase. This means W_{ed} and W_{eph} are steady at T_m.

Increasing the residual resistivity seems to decrease the lattice transformation temperature T_m as well as the magnitude of the kink in the lines T/λ at T_m. The same tendency has been observed by Berthel and Pietrass [3] in the electrical resistivity ϱ of V_3Si, where also at the temperature T_m a kink appears in the curves $\varrho(T)$ as a function of T^2. The authors are able to describe the observed behaviour of electrical resistivity by a simple two-band model assuming a logarithmic peak in the d-band density of states.

In sample 5 a lattice transformation is not observed, it is prevented by the α-V precipitates [5].

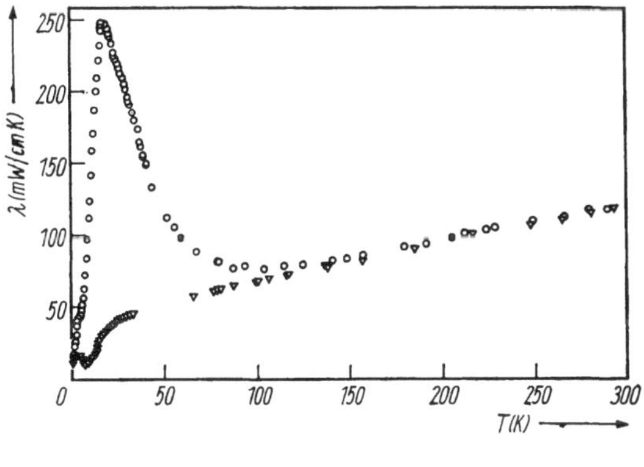

Fig. 2. Thermal conductivity of V_3Si as a function of temperature. Definition of symbols as in Fig. 1

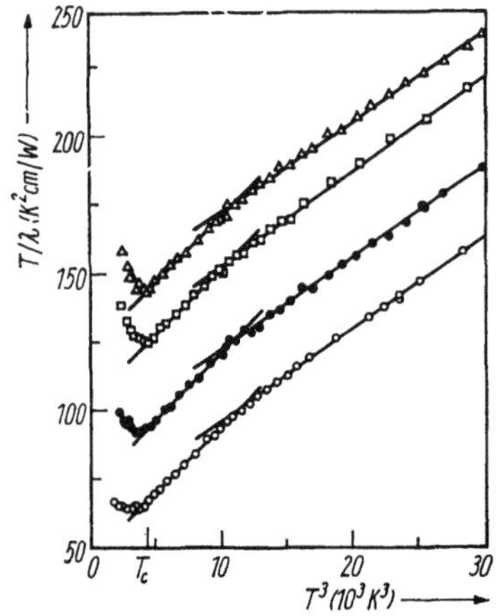

Fig. 3. Analysis of the measuring results in the temperature range T_c to 30 K according to equation (1). Definition of symbols as in Fig. 1. At the martensitic transition a kink in the fitting lines is seen

An anomalous temperature dependence of the thermal conductivity at low temperatures of the A-15 superconductor V_3Sn and related compounds has been observed by Morton et al. [10], which may be related to the lattice transformation. In these compounds the contribution of the lattice to the thermal conductivity may not be neglected already at temperatures above T_c. The observed anomaly is discussed as anomalous variation of the lattice component of the thermal conductivity due to a non-Debye-like shape of the phonon density of states.

Table 2

Temperature of the cubic–tetragonal lattice transformation T_m, thermal scattering coefficients A and B (see equation (1)) in the cubic and tetragonal phases, and Lorenz number L_0 of the investigated V_3Si samples

sample	T_m (K)	cubic A (cm K^2/W)	B (10^{-3} cm/WK)	tetragonal A (cm K^2/W)	B (10^{-3} cm/WK)	L_0 (10^{-8} V^2/K^2)
1	22.4	63	3.37	45	1.11	2.47
2	22.1	89	3.30	71	1.73	2.44
3	21.8	117	3.39	103	2.51	2.44
4	22.0	139	3.24	122	2.99	2.45
5	—	572	4.10	—	—	2.45

The fitted electron–phonon scattering coefficients B in the cubic and tetragonal phases of samples 1 to 4 scatter only by 2% around the mean values 3.33×10^{-3} cm/WK and 4.91×10^{-3} cm/WK, respectively (see Table 2). The B-value of sample 5 in the cubic phase deviates from the mean value, this means a deviation from Matthiesen's rule.

The fitted electron–defect scattering coefficients A of the tetragonal phase of the samples 1 to 4 and of the cubic phase of sample 5 are related to the residual resistivity ϱ_0 according to the Wiedemann-Franz law: $A = \varrho_0/L_0$ (L_0 Lorenz number). The agreement of L_0 with the theoretical Lorenz number is rather good (see Table 2).

High T_c superconductors and other highly resistive metals at room temperature and above have electrical resistivities rising less rapidly with T than the Bloch-Grüneisen theory predicts [2, 3, 11, 12]. Different theoretical models have been derived to explain this effect. An overall review is given by Allen [2]. At present it is not possible to prefer one of these models.

A saturation of the electrical resistivity means a constant mean free path l of electrons. From the simple relation $\lambda = \frac{1}{3} c v_F l$ for the thermal conductivity (heat capacity $c \sim T$, Fermi velocity $v_F = \text{const}$) a proportionality to T at high temperatures follows. In accordance to this our results in Fig. 2 show a linear increase of thermal conductivity in the upper temperature range, although the temperature is far below the temperature of saturation for electrical resistivity. In the temperature dependence of $\lambda(T)$ of the pure V_3Si monocrystal after a maximum at low temperatures a minimum must appear before the thermal conductivity increases at higher temperatures. We measured this minimum for sample 1 near 100 K.

A small minimum in the thermal conductivity with a depth of 3 to 5% has been observed in some pure metals, e.g. Al [13] and Na [14]. Up to high temperatures the electrical and thermal conductivities of these metals are well described by the Bloch-Grüneisen-Wilson theory. Therefore, the origin of this minimum may also be described according to the Wilson theory taking into account umklapp scattering [15].

3.2 Temperature range 6 K $< T < T_c$

In the range below the critical temperature of $T_c = 16.7$ K down to about $0.4T_c$ the heat conduction of pure superconducting intermetallic compounds is determined by thermally excited conduction electrons. The number of these quasiparticles is reduced continuously with decreasing temperature resulting in a steep decrease of thermal conductivity.

Based on the BCS theory Bardeen et al. [16] and Gejlikman et al. [17, 18] evaluated the relation of thermal conductivities in the normal and superconducting states. The electronic thermal conductivity λ_e^s in the superconducting state is given by

$$\frac{1}{\lambda_e^s} = \frac{A}{T f[t, \Delta(0)]} + \frac{BT^2}{g[t, \Delta(0)]}, \tag{2}$$

where A and B are defined by equation (1) and $f[t, \Delta(0)]$ as well as $g[t, \Delta(0)]$ are universal functions for the electron–defect and electron–phonon scattering, respectively, depending on the reduced temperature $t = T/T_c$ and the energy gap at zero temperature $\Delta(0)$ only.

The relations for BCS superconductors may be used for superconductors with strong coupling, too, if the temperature dependence of the energy gap $\Delta(T)$ is described by the following formula:

$$\Delta(T) = \Delta_{BCS}(T) \frac{\Delta(0)}{\Delta_{BCS}(0)}. \tag{3}$$

A comparison of our experimental values with (2) and (3) (Fig. 4) shows an increase of the energy gap of V_3Si at zero temperature with rising residual resistivity (see Table 1). By the fitting procedure the critical temperature was treated as a second free parameter and was obtained for all samples to $T_c = 16.4$ K. This value differs slightly from that determined from dc current measurements of the electrical resistivity, which is given in Table 1. The inhomogeneity of the samples causes a transition range of 0.2 to 0.3 K. The electrically measured T_c is defined as the highest kink in the transition curve [4]. The electrical resistivity disappears, if a small connected superconducting path within the sample is formed. The thermal conductivity changes only slightly at the transition to the superconducting state. Hence a change in the slope is observable only when most of the sample has become superconducting. Thus the difference in thermally and electrically determined T_c may be explained.

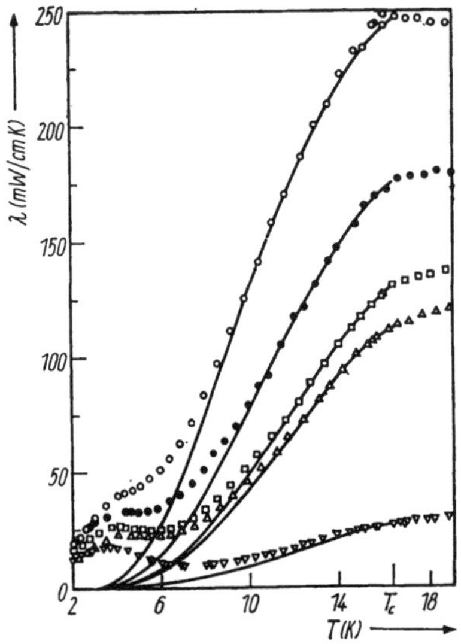

Fig. 4. Analysis of the measuring results in the temperature range below T_c according to equations (2) and (3) (solid lines). Definition of symbols as in Fig. 1. The deviation of the measuring points from the solid lines in the lower temperature range is caused by the appearance of lattice conductivity

Hegenbarth and Schmidt [19] derived $2\Delta(0) = 3.8kT_c$ from thermal conductivity measurements of a V_3Si single crystal with RRR = 24 in accordance with our results. From electron tunnelling experiments Schumann [20] obtained a value of $2\Delta(0) = (3.46 \pm 0.06)\, kT_c$ for several samples with RRR from 15 to 30. Muto et al. [9] determined $2\Delta(0) = 3.54kT_c$ for a sample with RRR = 17 from specific heat measurements.

From heat capacity and tunnelling investigations a value for $2\Delta(0)/kT_c$ near the predicted value of BCS theory is obtained, which seems to be independent of sample purity. The theory for strong-coupling superconductors leads to positive corrections to the BCS value and cannot explain the strong dependence of energy gap on stoichiometric composition found by us. So it may be concluded that the thermal conductivity formulae based on the BCS theory do not allow a precise determination of energy gap in strong-coupling superconductors.

3.3 Temperature range $T < 6$ K

Since the number of thermally excited conduction electrons decreases the heat is transferred by phonons at very low temperatures. The phonon contribution to the thermal conductivity λ^s_{ph} is seen in Fig. 4 as difference of the measured total λ^s and the electronic part λ^s_e calculated according to (2) and (3),

$$\lambda^s_{ph} = \lambda^s - \lambda^s_e .$$

The total conductivity of samples 2, 3, 4, and 5 passes a minimum at the temperatures 4.9, 5.4, 5.6, and 7.1 K, respectively, or an inflection point at 4.5 K in the case of sample 1 (see Fig. 1 and 4).

The lattice thermal conductivity is influenced by several scattering processes. Each of them causes a thermal resistance and the total lattice thermal resistance $W^s_{ph} = (\lambda^s_{ph})^{-1}$ is the sum of the individual resistances,

$$W^s_{ph} = W^s_{phe} + W_{php} + W_{phd} + W_{phb} + \dots \tag{4}$$

The first term in (4) describes the scattering of the phonons by conduction electrons, the second term the phonon–point defect scattering $W_{php} \sim T$, the third term the phonon–dislocation scattering $W_{phd} \sim T^{-2}$, the fourth term the phonon–boundary scattering $W_{phb} \sim T^3$, and further terms the scattering of phonons by additional lattice defects which may be present.

The different temperature dependences of the scattering processes cause a maximum of the lattice thermal conductivity. All investigated samples pass this maximum near 4 K (Fig. 5). The height of the maximum increases with decreasing deviation from sample stoichiometry.

Near 6 K the lattice thermal conductivity λ^s_{ph} is mainly determined by the scattering of phonons by conduction electrons ($\lambda^s_{ph} \approx \lambda_{phe}$) and increases steeply with decreasing

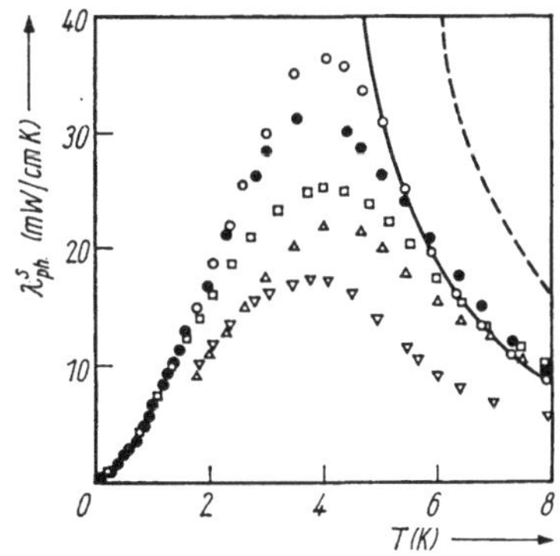

Fig. 5. Lattice thermal conductivity of V_3Si. Definition of symbols as in Fig. 1. The solid and dashed lines represent the theoretical curves of λ^s_{phe} for $2\Delta(0)/kT_c = 3.2$ and 3.8, respectively

temperature as the number of thermally excited conduction electrons decreases. The temperature dependence of the ratio $\lambda^s_{phe}/\lambda^n_{phe}$ for $2\Delta(0)/kT_c = 3.2$ and 3.8 has been calculated using the results of Wasserbäch [21]. For the lattice thermal conductivity in the normal state we use $\lambda^n_{phe} = aT^2$ with a as adjustable parameter. We found $a = 6.1 \times 10^{-6}$ W/cm K^3 by fitting the calculated curve λ^s_{phe} for $2\Delta(0)/kT_c = 3.2$ to the experimental results for sample 1. An estimation of lattice conductivity λ^n_{phe} by the approximation formula of Buttler and Williams [22] gives a value for a which is smaller by a factor 2. The calculated curves for λ^s_{phe} are plotted in Fig. 5.

Above 5 K a good description of the experimental lattice thermal conductivity is found using the same Δ-values as in Table 1. Especially the observed intersection of the lattice thermal conductivity may thus be explained.

At temperatures below 5 K the phonon–point defect scattering diminishes the steep increase of lattice conductivity. For samples 3, 4, and 5 its influence is already observable at higher temperatures.

At the lowest investigated temperatures the thermal conductivity is proportional to T^3 indicating dominating boundary scattering. Within the temperature range 0.07 to 4 K the lattice thermal resistivity of the investigated samples 2 and 3 can be fitted well to (4) with an additional term $W_{ad} \sim T^{-1}$ and without the term W^s_{phe}, which can be neglected below 4 K. The fitted curves are

$$W_{ph} = 0.006T^{-3} + 0.073T^{-2} + 0.057T^{-1} + 0.0036T$$

for sample 2 and

$$W_{ph} = 0.007T^{-3} + 0.038T^{-2} + 0.088T^{-1} + 0.0039T$$

for sample 3.

According to the theory of Casimir [23] the term for phonon boundary scattering can be estimated. Using the data for the elastic constants of Testardi et al. [24] $\lambda_{phb} = 0.23DT^3$ W/K^4 cm^2 is obtained, but using the data for specific heat and for mean sound velocity of Lou [25], $\lambda_{phb} = 0.955DT^3$ W/K^4 cm^2 is obtained. Our fit of the measured values gives for both samples $\lambda_{phb} = 0.4DT^3$ W/K^4 cm^2, i.e. within the limits of the above estimations.

The Rayleigh scattering of phonons by point defects such as stoichiometric deviations, vacancies and impurities dominates at the upper limit of the fitted temperature range, but is already masked by the phonon–conduction electron scattering. So the linear scattering rate in the two samples only roughly scale with their deviation from stoichiometry, i.e. with the ratio of their residual resistivities.

Dislocations have a pronounced effect on lattice thermal conductivity [26] and can cause a thermal resistivity proportional to T^{-2}. A comparison of the T^{-2} scattering terms of the two samples indicates that the dislocation density of sample 2 must be higher by a factor 1.9 than that in sample 3 being in good agreement with our results of dislocation density measurements. However, the dislocation concentration is too low to account for the observed magnitude of the T^{-2} scattering term [27, 28].

Lou [25] found a T^{-2} term for polycrystalline samples with V_5Si_3 inclusions, too, and supposed that this is caused by a scattering of phonons by electrons trapped in the normal inclusions. Because in our samples 2 and 3 inclusions were not detected, this explanation is not possible in our case. The experimental results obtained by Lou concerning a sample with cylindrically shaped grains having diameters ranging from 1 to 3 mm and a length of 30 mm is shown in Fig. 1. The good agreement with our results below 1 K is surprising. To fit the thermal conductivity curves Lou introduced an additional T^{-1} scattering term, too. Such scattering could arise from the scattering by planar imperfections, their existence in V_3Si is possible in view of the plate-like structures induced by martensitic lattice transformation.

Acknowledgements

The authors wish to thank Dr. M. Jurisch and Dr. G. Behr for the preparation of single crystals, Dr. D. Elefant for the performance of the measurements of residual resistivities and critical temperatures of the samples, Mister H. van der Waydbrink for his help during the measurements, and Dr. B. Pietrass for valuable discussions.

References

[1] M. JURISCH, K.-H. BERTHEL, and H.-J. ULLRICH, phys. stat. sol. (a) 44, 277 (1977).
[2] P. B. ALLEN, Superconductivity in d- and f-band Metals, Ed., H. SUHL and B. M. MAPLE, Academic Press, New York 1979.
[3] K. H. BERTHEL and B. PIETRASS, J. Physique suppl. 39, C6–383 (1978).
[4] F. LANGE, K.-H. BERTHEL, und P. PAUFLER, Reinstoffprobleme, Vol. 5, Akademie-Verlag, Berlin 1977.
[5] NGUYEN HONG QUYEN, Thesis, Dresden 1980.
[6] A. GLADUN, C. GLADUN, M. KNORN, and H. VINZELBERG, J. low-Temp. Phys. 27, 873 (1977).
[7] A. GLADUN, C. GLADUN, and H. VINZELBERG, phys. stat. sol. (a) 58, 409 (1980).
[8] H. VAN DER WAYDBRINK, Thesis, Dresden 1980.
[9] Y. MUTO, N. TOYOTA, K. NOTO, K. AKUTSU, M. ISINO, and T. FUKASE, J. low-Temp. Phys. 34, 617 (1979).
[10] N. MORTON, B. W. JAMES, G. H. WOSTENHOLM, and S. JONES, Cryogenics 21, 81 (1981).
[11] V. A. MARČENKO, Fiz. tverd. Tela 15, 1893 (1973).
[12] N. MORTON, B. W. JAMES, and G. H. WOSTENHOLM, Cryogenics 18, 131 (1978).
[13] R. W. POWELL, Contemp. Phys. 10, 579 (1969).
[14] J. G. COOK, M. P. VAN DER MEER, and M. J. LAUBITZ, Canad. J. Phys. 50, 1386 (1972).
[15] R. BERMAN, Thermal Conduction in Solids, Clarendon Press, Oxford 1976.
[16] J. BARDEEN, G. RICKAYZEN, and L. TEWORDT, Phys. Rev. 113, 982 (1959).
[17] B. T. GEJLIKMAN, Zk. eksper. teor. Fiz. 34, 1042 (1958).
[18] B. T. GEJLIKMAN, M. J. DUSENAT, and V. R. CECETKIN, Zk. eksper. teor. Fiz. 73, 2319 (1977).
[19] E. HEGENBARTH and B. SCHMIDT, phys. stat. sol. (b) 76, 307 (1976).
[20] J. SCHUMANN, phys. stat. sol. (b) 99, 79 (1980).
[21] W. WASSERBÄCH, phys. stat. sol. (b) 84, 205 (1977).
[22] W. H. BUTTLER and R. K. WILLIAMS, Phys. Rev. B 18, 6483 (1978).
[23] H. B. G. CASIMIR, Physica 5, 495 (1938).
[24] L. R. TESTARDI, W. A. REED, T. B. BATEMAN, and V. G. CHIRBA, Phys. Rev. Letters 15, 250 (1965).
[25] L. F. LOU, Phys. Rev. B 14, 3914 (1976).
[26] W. WASSERBÄCH, Phil. Mag. A38, 401 (1978).
[27] P. G. KLEMENS, Proc. Phys. Soc. A68, 1113 (1955).
[28] M. W. ACKERMANN and P. G. KLEMENS, Phys. Rev. B 3, 2375 (1971).

(Received June 16, 1981)

phys. stat. sol. (a) 68, 309 (1981)

Subject classification: 10.2 and 20.3; 22.8

Laboratoire de Physique Expérimentale (a) and
Laboratoire de Géologie et Géochimie (b), Université de Nice[1])

Chloropatite, $Ca_5(PO_4)_3Cl$: Thermoluminescent Properties

By

D. LAPRAZ (a) and A. BAUMER (b)

The thermoluminescence (TL) properties of chloroapatite $Ca_5(PO_4)_3Cl$, — the samples being either natural or obtained by hydrothermal method — are studied between 77 and 650 K after X-irradiation. TL peak temperatures are specific of crystal lattice and not of main activators (Mn^{2+}, Ce^{3+}). The wide emission band at 3.1 to 3.6 eV, especially observed at low temperatures in undoped samples, and the intense diminution of the TL by thermal annealing are interpreted as resulting from PO_4 molecule ions.

Les propriétés thermoluminescentes (TL) de cristaux de chloroapatite $Ca_5(PO_4)_3Cl$, soit obtenus par synthèse hydrothermale, soit naturels, ont été étudiées entre 77 et 650 K après irradiation X. Les températures des pics de TL sont spécifiques du réseau et non des principaux activateurs (Mn^{2+}, Ce^{3+}). La large émission à 3,1–3,6 eV, surtout observée à basse température dans les échantillons non dopés, ainsi que le fort affaiblissement de la TL après recuit thermique sont interprétés comme provenant principalement de groupements PO_4.

1. Introduction

The thermoluminescence (TL) of calcium chloroapatite $Ca_5(PO_4)_3Cl$ (here after referred to as ClAP) has not, to our knowledge been mentioned previously, except in some studies [1] indicating after UV excitation ($\geqq$ 185 nm) a very weak TL in commercial halophosphates, which are in fact mixtures of chloro- and fluoro-apatites ($Ca_{5-x}\cdot Sr_x(PO_4)_3F_{1-y}Cl_y:Mn^{2+}, Sb^{3+}$). This paper is a continuation of one we published on hydroxyapatite [2, 3], itself part of our research on the thermoluminescent properties of calcium apatite [4].

2. Experimental Techniques

2.1 Preparation and characterization of the crystals

We selected the hydrothermal method because it is closest to the natural process of endogenous rocks. The crystals have been obtained according to one of the following reactions:

a) $Ca_{9.4}H_{0.6}(PO_4)_6(OH)_{1.4} \cdot 4\,H_2O + CaCl_2$
$\rightarrow 2\,Ca_5(PO_4)_3Cl + 0.4\,Ca(OH)_2 + 4.6\,H_2O$

type A,

b) $6\,H_3PO_4 + 10\,CaCl_2 \cdot 2\,H_2O + 123\,H_2O$
$\rightarrow 2\,Ca_5(PO_4)_3Cl + 18\,HCl + 143\,H_2O$

type C.

The first reaction was obtained from Prolabo tricalcium phosphate between 673 and 973 K, with pressure ranging from (0.35 to 1.5) $\times\ 10^8$ N/m². As for type C apatite

[1]) Parc Valrose, 06034 Nice Cedex, France.

crystals, they were obtained from liquid mixture, usually at 873 K, 10^8 N/m² during 6 h; these crystals grow rapidly with few fluid inclusions [5].

Doping with Mn or Ce — the two main activators of apatites — is carried out by adding such an amount of $MnCl_2 \cdot 4\,H_2O$, CeO_2 or $CeCl_3 \cdot 6\,H_2O$ as impurity/calcium (mol) = 0.2% (type A) or 2% (type C) for Mn and 0.2% for Ce in the reactive mixture. A study with an electronic microprobe on our Mn(Ce)-doped type C crystals indicates that only 1/50 of Mn^{2+} (1/10 for Ce^{3+}) ions added to the reactive mixture is introduced into the sample [6].

Each sample is washed in distilled water, then dried at 325 K for several hours.

Type A chloroapatites are crystalline powders (< 10 μm), while type C samples are small monocrystals (0.1 to 1 mm) which, according to the "pressure–temperature" region, are in a compact or acicular form [7]. All the undoped or doped crystals are uncoloured. Thanks to X-diffractometry, we see that the undoped samples crystallize in the monoclinic system ($P2_1/b$), in accordance with Elliot and Mackie's results [8], while the doped samples crystallize in the hexagonal system ($P6_3/m$): for the hexagonal cell, $a = (9.635 \pm 0.006)$ Å and $c = (6.762 \pm 0.004)$ Å. The refraction index is $n = 1.663 \pm 0.001$, with a very weak birefringence $\delta < 0.001$ [9] in accordance with Mackie et al. [10]. Note that no trace of OH^- ions is observed by IR spectrometry in our crystals.

2.2 *Measurements of thermoluminescence*

After a slight crushing (< 5 μm), 2 mg of a sample are uniformly spread in a 10 mm diameter aluminium cupel, by sedimentation in acetone. TL is measured by fixing the cupel in a cell allowing the variation of its temperature between 77 and 650 K at a vacuum of 2×10^{-3} N/m². The sample is previously submitted at 77 K to an X-ray (W target, 45 kV, 2 mA) dose of 10 μC/kg, then it is heated in situ at 0.5 K s^{-1}.

The TL emission is detected by means of an S13 or S20 R photomultiplier cooled to −20 °C. The spectral analysis is realized between 250 and 800 nm by 10 to 15 nm bandwidth interference filters; all the TL spectral analyses presented are corrected according to the photomultiplier response and the spectral bandwidths of the filters and optical windows.

3. Experimental Results

3.1 *Influence of impurities*

3.1.1 *Undoped samples*

The undoped type A or C crystals are characterized by an intense TL peak at (118 ± 4) K, accompanied by some ten weak peaks (Fig. 1). Table 1 gives some parameters E (activation energy), s (frequency factor), and l (kinetic order) of the main TL peaks, determined by Chen [11] or the initial rise [12] method.

The emission band of the 118 K peak (Fig. 2) is in general centred at 3.3 eV (type A) or 3.1 eV (type C), with a very important bandwidth at half-maximum from 1.5 to 1.6 eV. Note that two components seem to exist at 3.5 and 3.1 eV, whatever the sample type. For type C, the main maximum can move from 3.1 to 3.5 eV according to synthesis conditions.

3.1.2 *Mn-doped samples*

As with the undoped samples, we observed the same TL peak temperatures (Fig. 1): the peaks at (118 ± 4) K (types A and C), 410 K (type C), and 445 K (type A) are the most intense ones. Their parameters E, s, and l are given in Table 1; the initial rise method shows that the large 410 K peak of type C samples is in fact formed by two

Table 1

Some values of the parameters E (eV), s (Hz), and l, calculated by various methods, for TL peaks in synthetic chloroapatite samples

sample	method	118 K			205–250*) K			310 K			370*)–410 K			445 K			500 K		
		E	s	l	E	s	l	E	s	l	E	s	l	E	s	l	E	s	l
A undoped	[11]	0.09	10^2	1.1										0.72	10^6	1.2			
Mn	[11]	0.10	10^3	1.6										0.60	10^5	1.7			
C undoped	[11]	0.14	10^4	0.9										0.56	10^4	1.4	1.35	10^{11}	2.1
Mn	[11]	0.16	10^5	0.9	0.50	10^{11}	1.6	0.80	10^{12}	1.8	0.47	10^4	1.1						
	[12]										0.87	10^9		1.08	10^{10}		1.20	10^{11}	
Ce	[11]	0.16	10^5	0.9	0.52	10^9	1.2				0.91	10^{11}	>2						

*) For Ce.

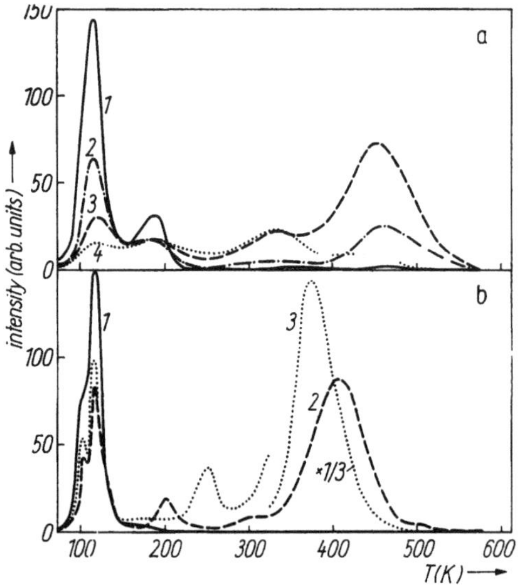

Fig. 1. Glow curves of $Ca_5(PO_4)_3Cl$ after X-irradiation at 77 K: a) type A: (1) undoped, (2) 0.02% Mn-, (3) 0.2% Mn-, and (4) 0.2% Mn-doped samples; b) type C: (1) undoped, (2) 2% Mn-, and (3) 0.2% Ce-doped samples

peaks centred at around (405 ± 15) K (also observed in the type A) and (445 ± 15) K, which explains the weak values of E and s calculated by Chen's method.

Two TL emission bands exist (Fig. 3):

a) *a yellow emission* which is predominant in most peaks: from 118 to 500 K, its maximum shifts from 2.10 to 2.15 eV, and its width increases from 0.25 to 0.34 eV. Its intensity increases in function of the Mn concentration,

b) the wide *UV emission*, already observed in undoped crystals, which is detected at low temperatures. As with OHAP [3], its intensity decreases when the Mn concentration increases (in the case of type A ClAP by a factor ≈ 7 from Mn/Ca = 0 to 0.2%).

3.1.3 Ce-doped samples

We find the same peak temperatures as with undoped and Mn-doped ClAP. No new peak due to Ce presence is detectable, only the relative intensities change. After X-irradiation at 77 K, the predominant peak is at 325 K (type A) and 370 K (type C), but the peak at 118 K is also intense (Fig. 1). The parameters E, l, s for these TL peaks are shown in Table 1.

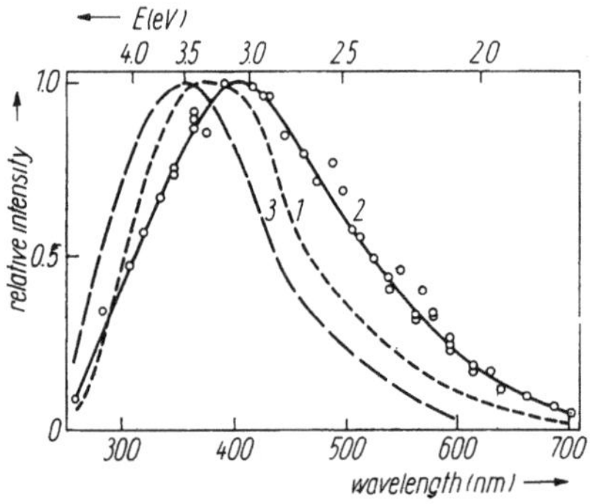

Fig. 2. Emission spectra of the 118 K TL peak of undoped ClAP: (1) type A (2) type C (synthesis: 10^8 N/m², 873 K) and (3) type C (1.5 × 10^8 N/m², 873 K) samples

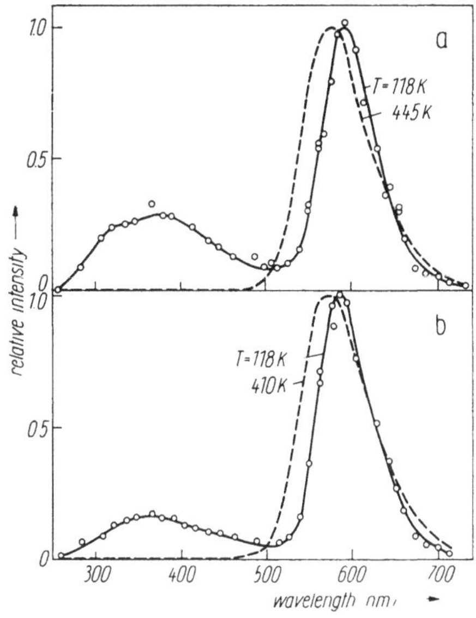

Fig. 3

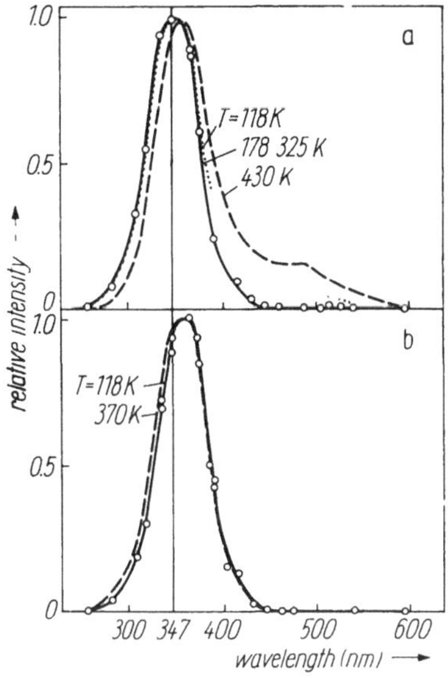

Fig. 4

Fig. 3. TL emission spectra of Mn-doped ClAP: a) type A samples; b) type C samples

Fig.4 . TL emission spectra of Ce-doped ClAP: a) type A samples; b) type C samples

Only one emission band characterizes the TL of ClAP:Ce samples (Fig. 4): at 118 K it is centred at (3,54 $\pm$ 0.02) eV with (0.66 $\pm$ 0.02) eV width for type A, while it is at 3.43 eV with 0.60 eV width for type C.

Note that all undoped or doped crystals, after a stay in oxygen atmosphere and without any ionizing irradiation, present a TL peak ("adsorption" TL) around (630 $\pm$ 10) K characterized by a wide emission band at 2.4 eV, as with OHAP samples [3].

3.2 Influence of synthesis conditions

3.2.1 Stoichiometry of the reactive mixture

When we modify Ca/P from 10/7 to 12/6 in the reactive mixture, the TL intensities of undoped or Mn-doped ClAP increase up to 10/6 (stoichiometric mixture), then, contrary to OHAP:Mn [3], they remain constant. For Ca/P $<$ 10/6, three new peaks at 273, 332, and 376 K are due to $Ca_2P_2O_7$ traces.

3.2.2 Temperature, pressure, and time of the synthesis

The TL interactions increase when the synthesis temperature varies from 673 to 973 K (10^8 N/m^2, 6 h) but the kinetic is different according to the peak: for ClAP (type A): Mn; by a factor of ≈ 2 at 118 K, but ≈ 10 at 445 K.

No significant variation is observed when synthesis time (4 to 24 h at 873 K and 10^8 N/m^2) and pressure ((0.35 to 1.5) $\times$ 10^8 N/m^2 at 873 K and 6 h) vary.

3.3 Influence of thermal annealing

As with hydroxyapatites [3], annealing at atmospheric pressure (argon or air) at $T \geqq 900$ K, before X-irradiation, considerably decreases all the TL peaks of undoped or doped samples (Fig. 5a).

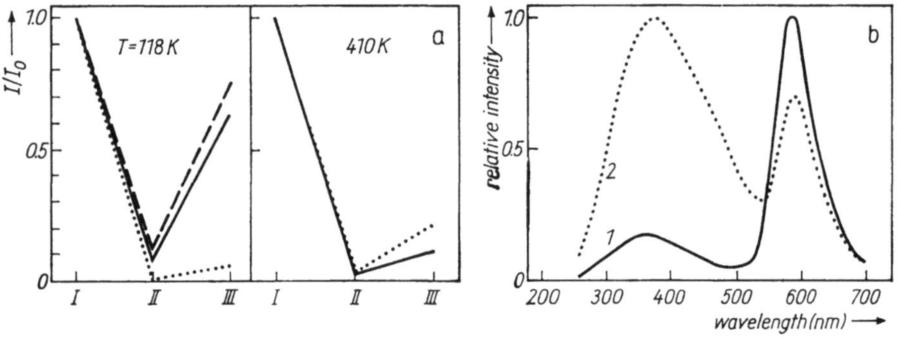

Fig. 5. a) Ratio I/I_0 of TL main intensities of Mn-doped ClAP (type C): (I) before ($I = I_0$), (II) after annealing at 1073 K, and (III) after the previous annealing and the "re-synthesis" (without ——— and with 347 nm – – – or 579 nm filter · · · ·). b) Emission spectra of the 118 K TL peak: (1) before annealing and (2) after annealing and "re-synthesis" of Mn-doped ClAP

If, after annealing, the sample is once again submitted to the same synthesis conditions, the TL re-appears, mainly for low TL temperatures (Fig. 5a).

As for the TL spectral emission of ClAP:Mn, Fig. 5 shows that

(i) the intensities of two emission bands decrease simultaneously at annealing;

(ii) the increase of the peak at 118 K under "re-synthesis" conditions is mainly due to that of the wide UV emission; the Mn emission centre or its excitation mechanism seems more perturbed in the course of these treatments than the emission centre responsible for the wide UV band.

N.B.: a) A light rosy coloration appears in certain ClAP (type C):Mn crystals after annealing: the analysis of such crystals by electronic microprobe (done by Dr. A. Peckett of Durham University (GB)) shows that the distribution of Mn^{2+} ions remained uniform in the crystal and that no trace of Mn oxydes could be detected after annealing.

Table 2

Origin and characterization of natural chloroapatites

denotation	geographic origin	provenance	colour	TLN	TLX	impurities per 100 at. Ca	ref.
ClAP BAMBLE	Brevik, Telemark, Norway	British Museum (GB) No. 51745	whitish	weak	very weak	Mn^{2+}: 0.04 Ce^{3+}: 0.02 Dy^{3+}: 0.04 Sm^{3+}: 0.03 Nd^{3+}: 0.17 Fe^{2+}: 0.39	[29]
ClAP SNARUM	Norway	Museum of Mineralogy, Oslo University (W. L. Griffin)	whitish	weak	very weak		

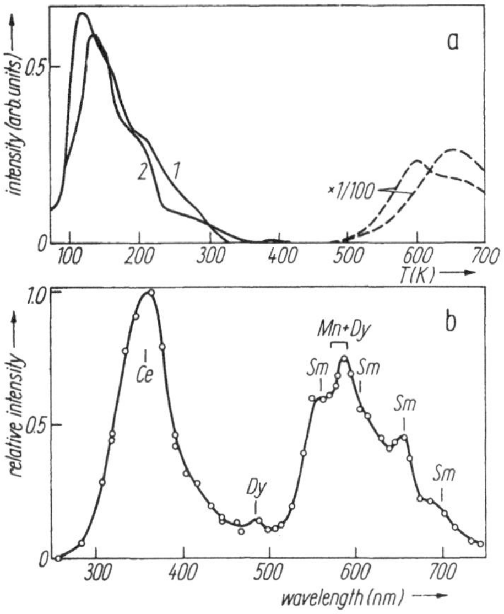

Fig. 6. Natural chloroapatites. a) Glow curve before (TLN: - - - -) and after X-irradiation at 77 K (TLX: ———) (1) ClAP BAMBLE and (2) ClAP SNARUM; b) emission spectrum of the 393 K TL peak of ClAP SNARUM, after intense X-irradiation at 290 K

b) If the "re-synthesis" of ClAP:Mn takes place at ambient temperature, TL remains unchanged when compared with annealed samples.

3.4 Comparison with some natural crystals

Table 2 indicates some characteristics of both the natural apatites studied: geographic analysis, origin, aspect, main impurities. All the crystals are ground, then put (2 mg) into cupels and studied under the same conditions as those used for synthetic apatites.

In TLN (natural TL), weak peaks appear at 603 and 668 K for BAMBLE ClAP and at the 653 K for SNARUM ClAP (Fig. 6a).

After X-irradiation at 77 K, the TL of these crystals is very weak: the main peak is at 120 K for one and at 120 to 155 K for the other (Fig. 6a). The temperature position of the TLX (TL after X-irradiation) peaks remains close to that of our synthetic ClAP crystals.

In the emission of the SNARUM ClAP peak at 393 K analyzed after an intense X-irradiation at 290 K, we observe the presence of two main bands (Fig. 6b):

(i) one centred at 3.46 eV, 0.66 eV wide, probably due to Ce^{3+},

(ii) the other at 2.14 eV, attributed to Mn^{2+}, with a long tail in the red and some weak maxima due to the presence of rare earths (Dy^{3+}, Sm^{3+}, etc.).

The emission spectra of the other peaks (at $T > 300$ K) are comparable to that of peak 393 K; however, the intensity of the various emission bands are slightly different.

4. Discussion and Conclusion

4.1 Comparison of the TL of synthetic chloroapatites with natural ones

All the crystals studied showed detectable thermoluminescent properties only after ionizing (X) irradiation.

4.1.1 Temperatures and intensities of the TL peaks

The comparison of all the TL curves of natural and synthetic, doped or undoped, chloroapatites reveals a fairly good coincidence in the temperature position of most

TL peaks, whatever the type or the concentration of the activator. After X-irradiation at 77 K we observe 14 peaks between 77 and 600 K. Most of them coincide with those found for OHAP [3] and FAP [4]. Beyond 600 K only a peak at (650 ± 25) K appears in the TLN of the natural crystal: as they are not generally observed in TLX, they can be ascribed to the action of more energetic, ionizing radiations (α, β) originating with natural radioactivity. Other peaks appear, but are most of them due to the presence of phases other than that of apatite (cf. Section 3.2).

The intensities of the TLX peaks, however, vary considerably from one sample to another: they depend not only on the concentration of the activator type, but also on synthesis conditions or on the influence of the thermal annealing. In general

(i) the most intense TL peaks are situated, after X-irradiation at 77 K, below 300 K, chloroapatite being characterized, whatever the doping element, by a main peak at (118 ± 4) K;

(ii) an annealing beyond 1000 K weakens considerably all the main TL peaks, whether the samples are doped or not.

4.1.2 Spectral emissions of the TL peaks

Two types of emission can be observed in the emission spectra of the TL peaks:

(i) an emission that we shall call "lattice emission", characterized by a wide band centred at 3.1 to 3.6 eV and a bandwidth at half-maximum of 1.5 to 1.6 eV; this emission is mainly observed at low temperature in undoped synthetic samples and is considerably weakened when the impurity rate (activators or not) increases;

(ii) a characteristic emission of the activator (mainly Mn and Ce); the band maximum shifts towards high energies as temperature rises, whereas the opposite takes place with the lattice emission.

4.2 Luminescence centres

4.2.1 Lattice emission

We were able, thanks to systematic control, to put aside the intervention

(i) of impurities, such as Ce^{3+}, Pb^{2+}, Y^{3+} ..., the presence of which might have explained the wide and sometimes structured band;

(ii) of phases found with apatite;

(iii) of fluid inclusions, easily eliminated through grinding and drying processes;

(iv) and of adsorbed elements on our ground single crystals, such as O_2, H_2O, acetone, or alcohol.

Three hypotheses have been kept in order to explain this emission: OH^- ion, V_K centre, and PO_4^{3-} group; we think that the last one is the most plausible one [4]. Despite all attempts, Knotterus [13, 14] was unable to identify any V_K centres in chloroapatites, concluding that the coloration mechanism of apatites is different from that Pooley [15] applied to alkali halides, and that PO_4^{3-} groups might well be concerned by the process. The sames goes for the emission resulting from a recombination of O^-/H^0 or OH^0/electron, formed by the ionizing irradiation of OH^-, because we do not explain why the intensity is weaker for ClAP than for OHAP when no trace of OH^- ions is seen in ClAP crystals by IR spectrometry.

This wide blue-UV emission evokes for us the one observed in self-activated compounds with oxygen groups such as WO_4^{2-}, MoO_4^{2-}, VO_4^{3-}, SiO_4^{4-}, BO_3^{3-}, ... [16, 17]. Self-activated phosphates (YPO_4, $LaPO_4$, ...) are scarce and their fluorescence intensity is generally very weak. Although Blasse thinks that there is no emission of the PO_4 group, several authors show that PO_4 groups introduced into lattices, such as sulfates or silicates, give a fluorescence under an excitation at 4.89 or 3.40 eV [18], or that undoped YPO_4 has a weak blue fluorescence (an emission band at 2.9 eV, with an 1.0 to 1.6 eV bandwidth at 300 K), with an excitation maximum at 4.6 eV [19 to 21]. Tale et al. [22] indicate two emission bands in the TL 80 K spectrum of undoped $Ba_3(PO_4)_2$: one, centred at about 3.8 to 4.1 eV with 1.1 eV halfwidth and observed at $T < 135$ K, is attributed to PO_4^{2-}/electron recombination; the other, called "intrinsic" and centred at 3.0 eV with 1.4 eV width, is predominant between 185 and 230 K. With the presence of Eu^{3+} ions, they show that the emission of that activator takes the place of the two previous emissions.

Finally, our ClAP and OHAP samples have a very weak photoluminescence with two main bands at 3.1 and 3.7 eV at 77 K temperature; the excitation maxima are weakly marked at 5.39, 4.77, and 4.35 eV [4].

As the P–O bonding is 39% ionic and the charge is $+0.85e$ on the P atom of PO_4^{3-} group [23], we think that Leverenz's hypothesis [16], which explains the emission of SiO_4^{4-} groups, is applicable to PO_4^{3-}, isoelectronic with SiO_4^{4-}. The excitation of PO_4^{3-}, which may be direct or consecutive to a PO_4^{2-}/electron recombination for example would cause the transfer of an electron 2p owing to one of four oxygens to a vacant orbital 3 s of the central P atom. The resulting hole may then be exchanged among the tetrahedrally arranged O atoms, until the excited electron on the P atom makes a radiative return to one of the ligand O atoms. In this hypothesis, the main emitting atom would be an oxygen, although the entire PO_4 group should be involved, explaining the wide emission band.

Our "lattice emission" is a wide band with at least two components: therefore, at least two types of emitting O atoms are involved [4]. So, the variation of the "lattice emission" position would be bound to O atoms owing the PO_4 tetrahedra, more or less disturbed by the crystal lattice. Furthermore, the great intensity of the "lattice emission" in ClAP samples, compared with that of OHAP [3] may be due to an "intensifying effect" of very electronegative Cl^- ions in the PO_4 emission, as in the case of the activators which intensify the SiO_4 emission [16].

4.2.2 *Characteristic emission of two main activators*

Manganese: The emission spectrum of the TL peaks of ClAP:Mn is attributed to Mn^{2+} ion in octahedral coordination; it is a yellow band, wide and asymmetric:

(i) its maximum E^* increases linearly in relation to the TL peak temperature T^*,

$$E^* = 2.095 + 14 \cdot 10^{-5}(T^* \text{ (K)} - 77) \pm 0.010 \text{ eV}; \tag{1}$$

(ii) its halfwidth $\Delta E_{1/2}$ varies according to the relation available with the localized centre,

$$\Delta E_{1/2}(T^*) = E_{1/2}(0) \left[\coth \frac{\hbar\omega_e}{2kT^*}\right]^{1/2}, \tag{2}$$

with a good correlation between the experimental and calculated values for

$$\Delta E_{1/2}(0) = (0.245 \pm 0.010) \text{ eV}$$

and $\hbar\omega_e = (0.050 \pm 0.005)$ eV, $\hbar\omega_e$ being the quantum vibrational energy in the excited state.

(iii) the Mn emission is to be attributed to the transition $^4G(^4T_1) \to {}^6S(^6A_1)$ of the Mn^{2+} ion which would localize on the site Ca(I) if we refer to fluorescence properties of $Ca_5(PO_4)_3$ (Cl, F):Mn^{2+} [24].

Cerium: Cerium, the other main activator of apatites, is responsible for an emission band at about (3.5 ± 0.1) eV, (0.65 ± 0.05) eV wide. It very slightly shifts towards shorter wavelengths and its halfwidth increases when TL peak temperature increases.

This emission is attributed to the transitions $5d(^2D_{3/2}) \rightarrow 4f(^2F_{5/2}, ^2F_{7/2})$ of the Ce^{3+} ion which probably substitutes itself to calcium in site I because there is no difference between the emission characteristics of various types of Ce-doped apatites (FAP, ClAP, or OHAP) [4].

4.3 Trapping centres

The position of the TL peaks of ClAP and the corresponding values of the E, s, and l parameters are independent of the impurities or of the synthesis type used.

E is varying linearly with T^* between 100 and 650 K as

$$E\ (\text{eV}) = \frac{T^*\ (\text{K}) - 55}{355} \tag{3}$$

with a correlation coefficient of 98% and a relative error of ≈ 15%.

When there are important differences between the E-value calculated by (3) and that resulting from Chen's method [11], it is generally due to a jumble of TL peaks (example: peaks at 405 and 445 K of ClAP (type C):Mn).

The form of relation (3) is the same as that proposed by Curie and Curie [25], $E = (T^* - T_0)/A$, for the most simplified TL model and for $l = 1$; but the values of coefficients T_0 and A may not be adjusted.

Whatever the TL peak between 100 and 550 K, in spite of the very important fluctuations of l and s, the mean value of l (kinetic order) is about 1.3 ± 0.4, and that of s (frequency factor) $10^{7\pm3}$ Hz.

So, we think that the main trapping centres are specific of the crystal lattice, as in the case of OHAP [3] or FAP [4]. No TL peak is connected with the thermal bleaching of the main colour or paramagnetic centres which are observed in apatites and due essentially to vacancies and O^{2-} defects in halide sites [26, 13, 14]. Therefore, the trapping centres seem linked with PO_4^{3-} groups. From Leverenz's hypothesis for SiO_4 groups, we can derive that an electron, which comes from an oxygen of the PO_4 group, can be trapped

(i) either by rotating around a charge $+e$ (P^{5+} central ion);

(ii) or by being in the field of two charges $+e$, the P central ion and the trapped hole in the oxygen; this hole can be exchanged with other oxygens belonging to further PO_4 groups.

We recognize here the two "hydrogenoid atom and molecule" trap models of Bethe [27] and Curie [28] which accurately describe ZnS traps. But their application turns out to be unsuccessful to explain the main TL peaks of ClAP. That is why another trapping mechanism must be considered, where defects in the immediate vicinity of PO_4 groups would occur which defects would also explain

(i) the "intensifying effect" of PO_4 radiative emission;

(ii) and the great decrease of ClAP thermoluminescence after thermal annealing at 900 to 1000 K.

We think these defects originate in the crystal lattice, thus explaining that annealing (which re-arranges the crystalline structure) reduces the PO_4 perturbations and so, the "lattice" emission.

Acknowledgements

The authors wish to express their thanks to Prof. G. Turco, director of the Laboratoire de Pétrologie-Minéralogie of Nice University, where the crystals have been synthetized, to Prof. P. Keller for his assistance in this work, and to Dr. J. Seriot for the photoluminescence measurements made at the Laboratoire de Physico-Chimie des Matériaux Luminescents of Lyon I University.

References

[1] L. Suchow, J. Electrochem. Soc. **108**, 847 (1961).
[2] D. Lapraz, A. Baumer, P. Keller, and G. Turco, C.R. Acad. Sci. (France) **285B**, 291 (1977).
[3] D. Lapraz, A. Baumer, and P. Iacconi, phys. stat. sol. (a) **54**, 605 (1979).
[4] D. Lapraz, Thesis, University of Nice, 1980.
[5] R. Argiolas, Thesis of Speciality, University of Nice, 1978.
[6] A. Baumer, A. Peckett, and R. Argiolas, Schweiz, mineral. petrogr. Mitt. **59**, 239 (1979).
[7] R. Argiolas and A. Baumer, Canad. Mineral **16**, 285 (1978).
[8] J. C. Elliott and P. E. Mackie, Coll. Internat. C.N.R.S. **230**, 69 (1973).
[9] R. Argiolas, A. Baumer, and D. Lapraz, Canad. Mineral. **3**, 573 (1979).
[10] P. E. Mackie, J. C. Elliott, and R. A. Young, Acta cryst. **B28**, 1840 (1972).
[11] R. Chen, J. Electrochem. Soc. **116**, 1254 (1969).
[12] G. F. Garlick and A. F. Gibson, Proc. Roy. Soc. **60**, 574 (1948).
[13] D. I. M. Knottnerus, Thesis, University of Groningen, 1976.
[14] D. I. M. Knottnerus, A. Lanjouw, and H. W. den Hartog, phys. stat. sol. (a) **41**, 95 (1977).
[15] D. Pooley, Brit. J. appl. Phys. **17**, 855 (1966).
[16] H. W. Leverenz, An Introduction to Luminescence of Solids, John Wiley & Sons, New York 1950.
[17] G. Blasse and A. Bril, Techniques Philips **2**, 13 (1971).
[18] Y. Kotera, M. Yonemura, and T. Sekine, J. Electrochem. Soc. **108**, 540 (1961).
[19] M. A. Aia, J. Electrochem. Soc. **114**, 367 (1967).
[20] W. L. Wanmaker and D. Radielovic, Bull. Soc. Chim. Franç. 1785 (1968).
[21] J. Grafmeyer, Thesis, University of Lyon I, 1979.
[22] I. Tale, P. Kulis, and V. Kronghauz, J. Lum. **20**, **343** (1979).
[23] L. Pauling, The Nature of the Chemical Bond and the Structure of Molecules and Crystals 3rd edition, Cornell University Press, New York 1960.
[24] F. M. Ryan and F. M. Vodoklys, J. Electrochem. Soc. **118**, 1814 (1971).
[25] G. Curie and D. Curie, J. Phys. Radium **16**, 199 (1955).
[26] W. W. Piper, L. C. Kravitz, and R. K. Swank, Phys. Rev. **138**, 1802 (1965).
[27] H. A. Bethe, Radiation Lab. Rep., M.I.T., No. 43 – 12 (1942).
[28] D. Curie, J. Phys. Radium **18**, 214 (1957).
[29] F. K. Taborszky, Neues Jahrb. Mineral. **5**, 79 (1972).

(Received June 23, 1981)

phys. stat. sol. (a) 68, 321 (1981)

Subject classification: 4; 22.2.4

Sektion Physik der Karl-Marx-Universität Leipzig,
Arbeitsgemeinschaft $A^{III}B^{V}$-Halbleiter[1])

Bond Charge Determination in Ternary Chalcopyrite Semiconductors by Means of X-Ray Structure Amplitude Investigations

By

U. PIETSCH

The bond charges (BC) of $A^{II}B^{IV}C_2^{V}$ chalcopyrite semiconductors are experimentally determined for the first time. X-ray structure amplitudes depending on the bond charges and influenced only by the anion displacement as well as the C atom scattering power are measured. The BC are shown to be dependent almost only on their next neighbours. The BC amount, position, and extension can be deduced from the sum of the effective charges of the atoms and the electronegativity difference of the atoms contributing to the respective bond. The B–C bond containing a higher number of electrons has a stronger covalent part with respect to the A–C bond. Because the A atoms transfer electrons as to B as to C atoms, the ionic part of the A–C bond increases and of the B–C decreases with respect to a bond formed by neutral atoms.

Erstmals wird die Bindungsladung in $A^{II}B^{IV}C_2^{V}$-Chalkopyrithalbleitern experimentell untersucht. Dazu werden Strukturamplituden von solchen Röntgenreflexen vermessen, die neben der Bindungsladung nur noch von der Anionenverrückung und der Streufähigkeit der C-Atome abhängen. Es wird nachgewiesen, daß in mehrkomponentigen Verbindungen jede Bindungsladung fast nur von ihren nächsten Nachbarn beeinflußt ist. Betrag, Position und Ausdehnung bestimmen sich aus der Summe der effektiven Ladungen und der Elektronegativitätsdifferenz der an der Bindung beteiligten Atome. Weil die B–C-Bindung mehr Elektronen enthält als die A–C-Bindung, ist der kovalente Anteil zur ersteren größer. Infolge des Transfers von Elektronen von A zu B und C-atomen wird der ionare Anteil zur A–C-Bindung erhöht und der zur B–C-Bindung vermindert bezüglich der aus neutralen Atomen bestehenden Bindung.

1. Introduction

The bond charge of ternary solid solutions composed of $A^{III}B^{V}$ semiconductors can be calculated with nearly linear interpolation between the end component parameters [1]. Because of the cubic symmetry we cannot distinguish both different bonds of a ternary compound.

Either each bond depends on the next atoms contributing to this bond or all bonds depend on the macroscopically measurable dielectric constant of this system.

The $A^{II}B^{IV}C_2^{V}$ chalcopyrites are the ternary relatives of the $A^{III}B^{V}$ semiconductors. The A- and B-atoms ocupy fixed places of a cation sublattice, the C-atoms those of the anion sublattice. The latter ones are shifted from their ideal positions. The chalcopyrite structure can be composed by doubling of the zincblende lattice in one direction. This tetragonal axis is varied in comparison with the doubled zincblende lattice constant. The shift of anions and the tetragonal distortion can be converted into one another [2, 3]. Some authors have measured two different lengths for both the A–C and B–C atom distances [2, 4, 5]. The bond angles C–B–C are nearly those

[1]) Linnéstr. 5, DDR-7010 Leipzig, DDR.

of a tetrahedron, but the C–A–C angles deviate from the ideal values. The B–C bond should be more covalent, the A–C bond more ionic. From this expectation the existence of two different bond charges (BC's) in our system follows.

In this work the author wants to show that each bond has a typical BC. For that reason some X-ray structure amplitudes influenced by BC's were measured integrally and were analysed by means of a BC model, already used for the determination of BC parameters of monoatomic and binary semiconductors and ternary solid solutions [1, 6, 7].

The BC is assumed to be an additional scattering particle exceeding the scattering power of the spherical atoms. The particle should be positioned at the maximum of the additional electron density between the next neighbours.

In the following the structure factor formalism of the chalcopyrite structure will be derived and the expected BC parameters are established by means of the extended dielectric theory. The experimental arrangement and parameters are described elsewhere [1].

2. The BC-Structures Factor Formalism

We describe the anion shift by a parameter a. The scattering power of the spherical matrix atoms f_i determines the structure amplitudes of the chalcopyrite elementary cell.

$$\left.\begin{array}{lll} \text{a.} & 4f_A + 4f_B + F_1(a)\,8f_C & \text{for } 220, 228, 400 \ldots, \\ \text{b.} & 4f_A + 4f_B \pm iF_2(a)\,8f_C & \text{for } 112, 332, \ldots, \\ \text{c.} & 4f_A + 4f_B - F_3(a)\,8f_C & \text{for } 200, 004, 600, \ldots, \\ \text{d.} & (2 \pm 2i)\,f_A + (2 \pm 2i)\,f_B - F_4(a)\,(1 + i)\,f_C & \text{for } 033, 035, 051, \ldots, \\ \text{e.} & F_5(a)\,f_C & \text{for } 150, 130, 402, \ldots . \end{array}\right\} \tag{1}$$

The values $F_1(a)$, $F_2(a)$, and $F_3(a)$ are smaller than unity, the factors $F_4(a)$ and $F_5(a)$ are somewhat greater than zero. The factor F_5 has either a real part F^R or an imaginary part F^I. For the group (1e) the scattering power of the matrix atom depends only on the anion shift a and the temperature factor T_C. Taking into consideration two BC's the structure amplitudes of the (1e) reflections are determined as follows:

$$\begin{aligned} F_{obs} = &\{F_5^R(f_C + \Delta f'_C)\,T_C - F_5^I\,\Delta f''_C\,T_C + V^R(q, r_{BC})\,f_{BV}(q) + U^R(q, r_{BC})\,f_{BU}(q)\} + \\ &+ i\,\{F_5^R\,\Delta f''_C\,T_C + F_5^I(f_C + \Delta f'_C)\,T_C + V^I(q, r_{BC})\,f_{BV}(q) + U^I(q, r_{BC})\,f_{BU}(q)\}\,. \end{aligned} \tag{2}$$

V^R, V^I, U^R, and U^I indicate the structure factors of the BC. They can be calculated as a function on the scattering length $q = 4\pi \sin\theta/\lambda$ and the BC position (3)

$$\left.\begin{aligned} U^R + iU^I &= \sum_{i=1}^{16} \exp\left(-2\pi i(hx_U + ky_U + lz_U)\right), \\ V^R + iV^I &= \sum_{i=1}^{16} \exp\left(-2\pi i(hx_V + ky_V + lz_V)\right). \end{aligned}\right\} \tag{3}$$

The indices U and V correspond with the A–C and the B–C bond, respectively.

The scattering factor function of BC is approximated by a Gaussian,

$$f_B(q) = f_B(0) \exp\left(-q^2/4k_F^2\right). \tag{4}$$

The distribution of BC is correlated with the Fermi wave number k_F as pointed out by Phillips [8]. The $\Delta f'$ and $\Delta f''$ mark the dispersion correction as given by Cromer and Libermann [10]. The f_C are listed in [9]. The temperature factor T_C is calculated

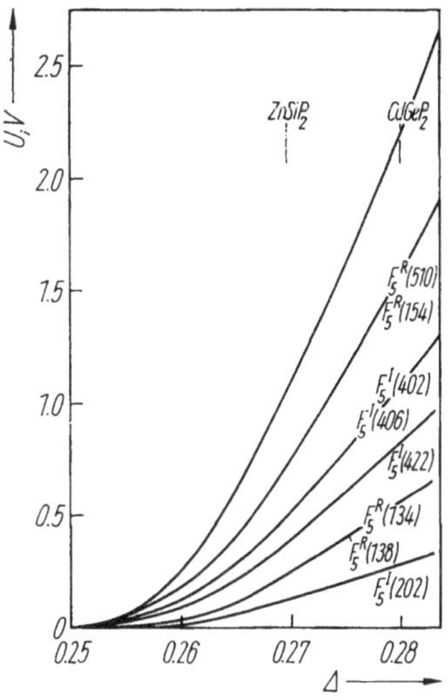

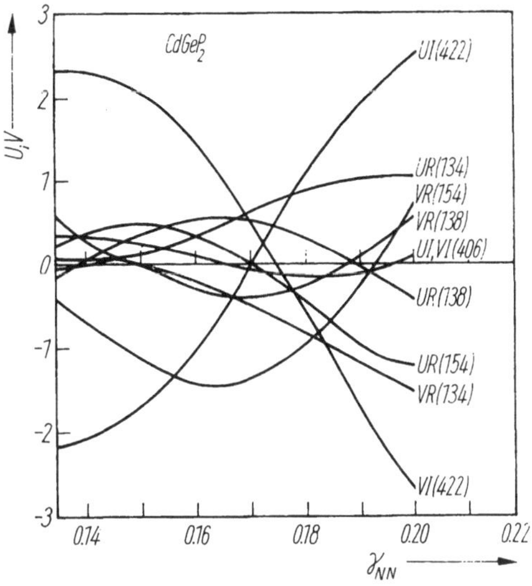

Fig. 1 Fig. 2

Fig. 1 Anion structure factors V, U for $A^{II}B^{IV}C_2^V$ compounds versus the anion displacement

Fig. 2 BC structure factors V, U versus the relative BL position

using a mean square displacement $\langle u_C^2 \rangle$

$$T_C = \exp\left(-\tfrac{1}{2} \langle u_C^2 \rangle q^2\right). \tag{5}$$

Fig. 1 shows the factor $F_5(a)$ versus the anion shift for the reflection group (1e). The scattering part of the C-atoms is smaller than unity for the most reflections. In Fig. 2 the structure factors of BC are represented as a function of the BC positions for $CdGeP_2$, for instance. The structure factor contributions are small, too. The sums $V^R + V^I$ and $U^R + U^I$ vanish for nearly all reflections in the case that both BC have the same position. But, really the sums are somewhat greater than zero, because of the different bond lengths. If both BC have extremely different positions the sums $V^R + V^I$ and $U^R + U^I$ of some reflections are nearly doubled. The correction of the scattering power of spherical atoms is small for identical BC positions and has a maximum for extremely different BC positions. Also the value of the correction depends on the anion shift.

3. Theoretical Estimation of the BC Parameters for $A^{II}B^{IV}C_2^V$ Compounds

The dielectric theory by Phillips and van Vechten [8, 11] is originally deviced for crystals with two electrons per bond. For the chalcopyrite semiconductors this theory has to be generalized. Simply, the BC can be described by

$$f_b(0) = \frac{ne}{\varepsilon_0} \tag{6}$$

with

$$n = \frac{\text{number of electrons in a given bonding type}}{\text{number of these in the elementary cell}}. \tag{7}$$

The BC amount depends on the static dielectric constant ε_0, which can be calculated from the Penn energy gap

$$\varepsilon_0 - 1 \sim \frac{n}{2(E_h^2 + E_C^2)^{1/2}} . \quad (8)$$

The Penn energy gap involves the covalent part E_h and the ionic part E_C. The position of BC depends on the electronegativity difference of the atoms taking part in the bond. Using the formula by Unger and Neumann [20] this position r_{BC} $\left(0 \leqq r_{BC} \leqq \frac{\sqrt{3}}{4} a\right)$ can be calculated as follows:

$$r_{BC} = \frac{1}{2}\left(1 + \frac{2}{\pi} \arcsin \frac{E_C}{E_h^2 + E_C^2}\right) \frac{\sqrt{3}}{4} a_0 . \quad (9)$$

Chemla [12] determined the mean energy gap and the bond ionicities for compounds containing Zn–P, Cd–P, Si–P, and Ge–P bonds. From the mean energy gap we can conclude on ε_0.

The calculation of the BC amount depends on the knowledge of the effective number of valence electrons in a bond. We can consider two cases. First, each bond contains the same number of valence electrons ($n = 2$) and second, the electron numbers correspond to the sum of valence electrons of the atoms building the bond ($n = 1.75$ and 2.25). For both cases Table 1 shows the estimated BC amounts. The presented BC positions were calculated from the ionicities taken by Chemla by means of (9).

Corresponding to the mean ionicities of ternary compounds taken from the binary relations Hübner [21] has calculated the effective charges of ions. It is shown that A- and B-atoms transfer electrons to the C-atom. But A-atom give more electrons than B-atoms. Taking into consideration the calculated effective charges of constituents, the BC amounts of both bonds differ slightly. The mean ionicity of the compound given by Hübner is by 5% greater than that found by Chemla.

The ionicity of a single bond depends on the effective charge of the ions. Because of the electronegativity difference the Cd–P bond is more ionic than the Zn–P. The same is true for Ge–P and Si–P bonds, respectively. The bond ionicities calculated from ionic atoms are greater than those from neutral atoms.

Table 1

Theoretical and experimental BC parameter of the single bonds for $ZnSiP_2$ and $CdGeP_2$

	r_{BC}	$f_b(0)$	r_{BC}	$f_b(0)$	r_{BC}	$f_b(0)$
Zn–P	0.190	0.317	0.173	0.316	0.190	0.30
Cd–P	0.195	0.286	0.178	0.279	0.195	0.36
Si–P	0.155	0.213	0.178	0.215	0.135	0.20
Ge–P	0.160	0.187	0.178	0.196	0.135	0.28
	[12] $n = 2$		[12] $n = 1.75$ or 2.25		experimental (this work)	

Phillips [3] has presented a theoretical relation in order to interpret the tetragonal distortions in terms of the electronegativities calculated by the dielectric theory. In this case A-atoms only transfer electrons to the B- and C-atoms. In this way the electronegativity difference of the B–C bond is smaller and the A–C bond greater than that for neutral atoms.

Chelikowsky and Phillips [13] have explained the tetragonal distortion by means of the structural theory by Simons and Bloch [14]. In this work the tetragonal distortion is caused by the different influence of the d-orbitals of the A-, B-, and C-atoms. The calculated bond electronegativity difference agrees qualitatively with that of Hübner, Boudriot and Oettel [15] have calculated new covalent radii for atoms in $A^{II}B^{IV}C_2^{V}$ compounds. The calculations are based on the measurement of the distances of A–C and B–C bonds different from those for neutral atoms. The radii of the fourth-group elements does not differ, but those of the second- and fifth-group elements are greater and smaller, respectively, than the covalent radii given by van Vechten and Phillips [16]. Corresponding to a relation indicated by St. Johns and Bloch [17], the valence radius is indirectly proportional to the effective ion charge.

In this concept the A-atoms transfer electrons to the C-atom only. The charge of the Si and Ge atoms are constant. The ionicity of the B–C bond alone is varied in comparison with neutral atoms.

4. Experimental Procedure and Results

Crystals having a great anion shift were chosen. Five to six reflections of the group (1e) from $ZnSiP_2$ and $CdGeP_2$ were measured in the asymmetric Bragg case.

For all reflections the extinction length was greater than the absorption length. The kinematic X-ray scattering theory was used for the analysis. Table 2 shows some material parameters. The structure amplitudes were obtained by the measurement of the absolute integrated intensity of the X-ray reflection using CuK_α radiation. The incident beam intensity was measured directly by means of an absorption method optimized by the author [18]. In addition, all results were normalized to the (600) reflection of GaAs precisely described by Bilderback [19].

Table 2

Material parameters for $ZnSiP_2$ and $CdGeP_2$

parameter	$ZnSiP_2$	$CdGeP_2$
lattice constants (Å)	5.399	5.74
	10.435	10.777
anion shift a	0.2691	0.2819
volume of the elementary cell (Å)	304.172	355.076
density ϱ_x (g cm^{-3})	5.9333	4.619
absorption coefficient (cm^{-1})	222.431	492.96
dispersion corr. $\Delta f'$	0.283	0.283
for CuK_α $\Delta f''$	0.434	0.434

Firstly the BC amounts were assumed to be the same for both kinds of bonds. The anion shift parameter a was taken from Abrahams and Bernstein [2] and from Grigorovici [5] for $ZnSiP_2$ and $CdGeP_2$, respectively. The temperature factor of C-atoms and the anion shift parameter were fitted. The results are indicated in Table 3. The variation of the anion shift found in this manner is smaller than the experimental error. The temperature factor in both cases is the same and corresponds to that determined for $ZnSiP_2$ by Abrahams and Bernstein [2]. However, we would like to point out that we can describe the scattering power of the spherical C-atoms by different parameter sets a and $\langle u_P^2 \rangle$. In each case, the necessary BC correction is the same.

In the case of $ZnSiP_2$ the BC effect was determined by means of the (134), (422), and (150) reflections. The (402) and (606) reflections were independent of the BC correction. For $CdGeP_2$ the (134) and (422) reflections were influenced strongly by the BC, too. The quality of the fit may be tested by the R-factor,

$$R = \frac{\sum |F_{obs} - F_{calc}|}{\sum F_{obs}}. \quad (10)$$

Table 3

Structure amplitudes for $A^{II}B^{IV}C_2^{V}$ compounds

$ZnSiP_2$

hkl	F_{exp}	F_{sph}	F_e		
			A	B	C
202	1.14 ± 0.02	1.16	1.16	1.25	1.29
134	1.65 ± 0.02	1.73	1.74	1.66	1.79
402	3.63 ± 0.02	3.67	3.63	3.65	3.66
422	2.50 ± 0.02	2.64	2.63	2.47	2.79
150	4.07 ± 0.02	4.23	4.20	4.12	4.65
606	4.51 ± 0.11	4.65	4.65	4.66	4.65
R-factor ($\times 10^{-2}$)		3.32	2.86	2.11	5.60

parameters: $a = 0.2702$, $\langle u_P^2 \rangle = (0.0200 \pm 0.0005)$ Å^2.
A: $v = u = 0.160$, $f_b(0) = 0.25$.
B: $v = 0.135$, $u = 0.190$, $f_b(0) = 0.25$.
C: $v = 0.190$, $u = 0.135$, $f_b(0) = 0.25$.

$CdGeP_2$

hkl	F_{exp}	F_{sph}	A	B	C
134	4.20 ± 0.06	4.40	4.29	4.52	4.20
422	4.94 ± 0.04	5.41	5.11	5.66	4.94
406	6.47 ± 0.04	6.48	6.50	6.49	6.51
138	3.24 ± 0.06	3.20	3.23	3.20	3.24
154	8.53 ± 0.04	8.55	8.57	8.54	8.57
R-factor ($\times 10^{-2}$)		2.70	1.24	4.05	0.29

parameters: $a = 0.2816$, $\langle u_P^2 \rangle = (0.0200 \pm 0.0005)$ Å^2.
A: $v = 0.135$, $u = 0.200$, $f_b(0) = 0.25$.
B: $v = 0.200$, $u = 0.135$, $f_b(0) = 0.25$.
C: $v = 0.135$, $u = 0.200$, $f_b(0) = 0.30$.

5. Discussion

The structure amplitudes were determined experimentally with an accuracy better than 2%. The correction by the BC is greater than the experimental error. The mean BC amount assumed to be $f_b(0) = 0.25$ electrons was somewhat too great for $ZnSiP_2$ and too small for $CdGeP_2$.

The structure amplitudes for BC influenced reflections (Table 3) calculated from spherical scattering factors differ from the experimental ones. For equal BC positions the BC correction is small. Different BC positions for the B–C and A–C bonds change the results qualitatively. In the case of the ionic A–C bond the fit becomes better, for an ionic B–C bond worse. For $ZnSiP_2$ the fitted BC positions are the same as the results predicted by Chemla.

For a changed BC amount the results remain the same, qualitatively. With respect to $CdGeP_2$, the BC correction was not sufficient for the reconstruction of the experimental results including extremely different BC positions. A BC amount of $f_b(0) = 0.30$ electrons yields the best fit disagreeing with the result found by Chemla.

The obtained BC positions are relative ones. Because of the same B–C distance for both substances we can conclude from the same BC positions on the same ionicities of the covalent bonds. The A–C bond length is different for both measured substances. The A–C bond of $CdGeP_2$ is more ionic than that of $ZnSiP_2$. The BC parameters deduced from the experiment are marked in Table 1. The difference between the experimental and calculated BC positions is caused mostly by the incorrect calculation from formula (6) rather than by experimental error.

From the BC positions and amounts we can reconstruct the effective charges of the ions. The obtained results correspond to the hypothesis by Phillips and Hübner. The A-atoms transfer electrons to B- as well as to C-atoms. The ionic part of the A–C bond increases and of the B–C bond decreases with respect to a bond formed by neutral atoms.

Acknowledgements

The author is greatly indebted to Prof. Dr. sc. K. Unger for stimulating discussions. The author would like to thank the Section Metallurgie and Werkstoffwissenschaften of the Bergakademie Freiberg for supplying the $ZnSiP_2$ and $CdGeP_2$ samples.

References

[1] U. Pietsch, phys. stat. sol. (b) **107**, 185 (1981).
[2] S. C. Abrahams and J. L. Bernstein, J. chem. Phys. **55**, 765 (1971).
[3] J. C. Phillips, J. Phys. Chem. Solids **35**, 1205 (1974).
[4] M. D. Lind and R. W. Grant, J. chem. Phys. **58**, 357 (1973).
[5] R. Grigorovici, Acta cryst. **B24**, 535 (1968).
[6] U. Pietsch, phys. stat. sol. (b) **102**, 127 (1980).
[7] U. Pietsch, phys. stat. sol. (b) **103**, 93 (1981).
[8] J. C. Phillips, Phys. Rev. **166**, 832 (1968).
[9] International Tables of X-Ray Crystals, Kynoch Press, Birmingham 1974.
[10] D. T. Cromer and D. Libermann, J. chem. Phys. **53**, 1891 (1970).
[11] J. A. Van Vechten, Phys. Rev. **182**, 891 (1969).
[12] D. S. Chemla, Phys. Rev. Letters **26**, 1441 (1971).
[13] J. R. Chelikowsky and J. C. Phillips, Phys. Rev. B **17**, 2453 (1978).
[14] G. Simons and A. Bloch, Phys. Rev. B **7**, 2754 (1973).
[15] H. Boudriot and H. Oettel, Thesis, Bergakademie Freiberg, 1979.
[16] J. A. Van Vechten and J. C. Phillips, Phys. Rev. B **2**, 2160 (1970).
[17] J. St. Johns and A. Bloch, Phys. Rev. Letters **33**, 1095 (1974).
[18] U. Pietsch, Exper. Tech. Phys. **28**, 75 (1980).
[19] D. H. Bilderbach, Thesis, Purdue University, West Lafayette 1975.
[20] K. Unger and H. Neumann, phys. stat. sol. (b) **64**, 117 (1974).
[21] K. Hübner, phys. stat. sol. (b) **52**, K33 (1972).

(Received June 22, 1981)

See Short Note by L. M. DEDUKH et al.

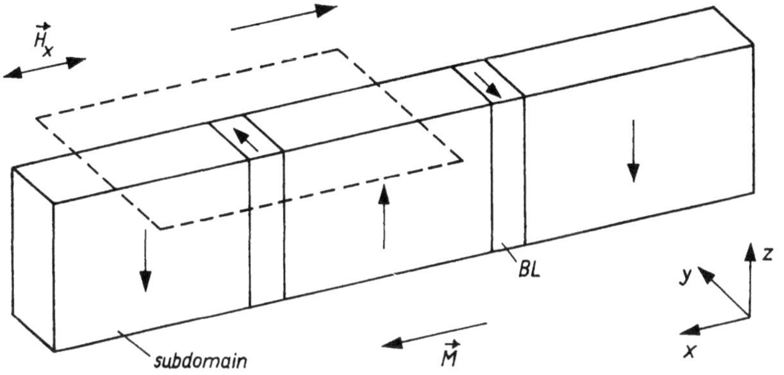

a

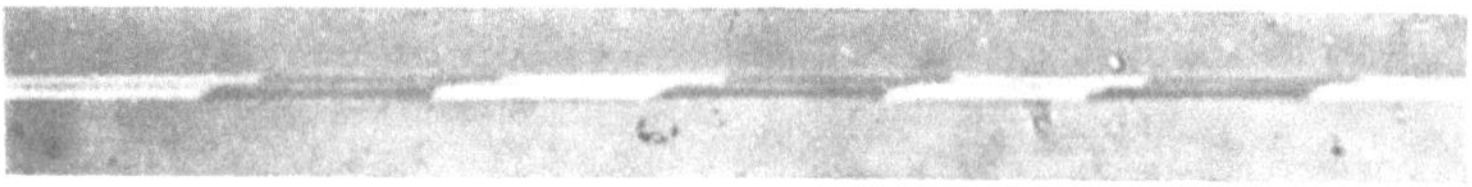

b

Fig. 1. a) Schematic view of 180° DW in a sample of YIG. b) Optical micrograph of an individual 180° DW vibrating in uniform sinusoidal field H_x of $\nu = 5$ kHz and of amplitude $H_x^0 = 0.6$ Oe

See Short Note by Bhanwar Singh and K. L. Chopra

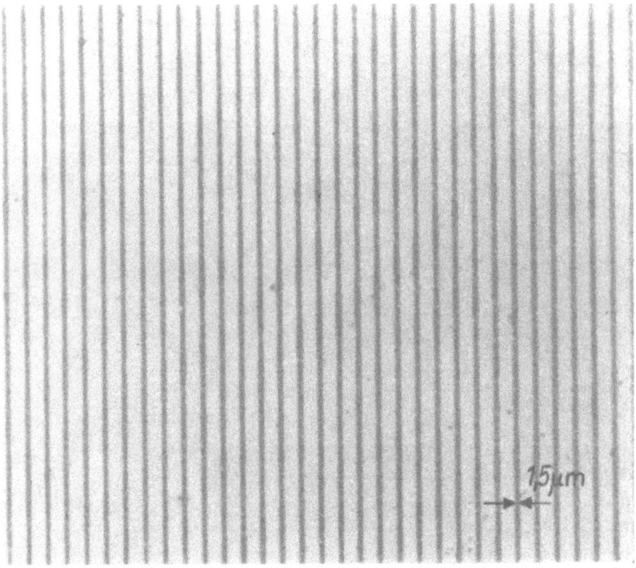

Fig. 3. Optical microphotograph of line patterns generated in 80°-incidence deposited a-$Se_{75}Ge_{25}$ film

See Short Note by W. T. McKenna and L. Ward

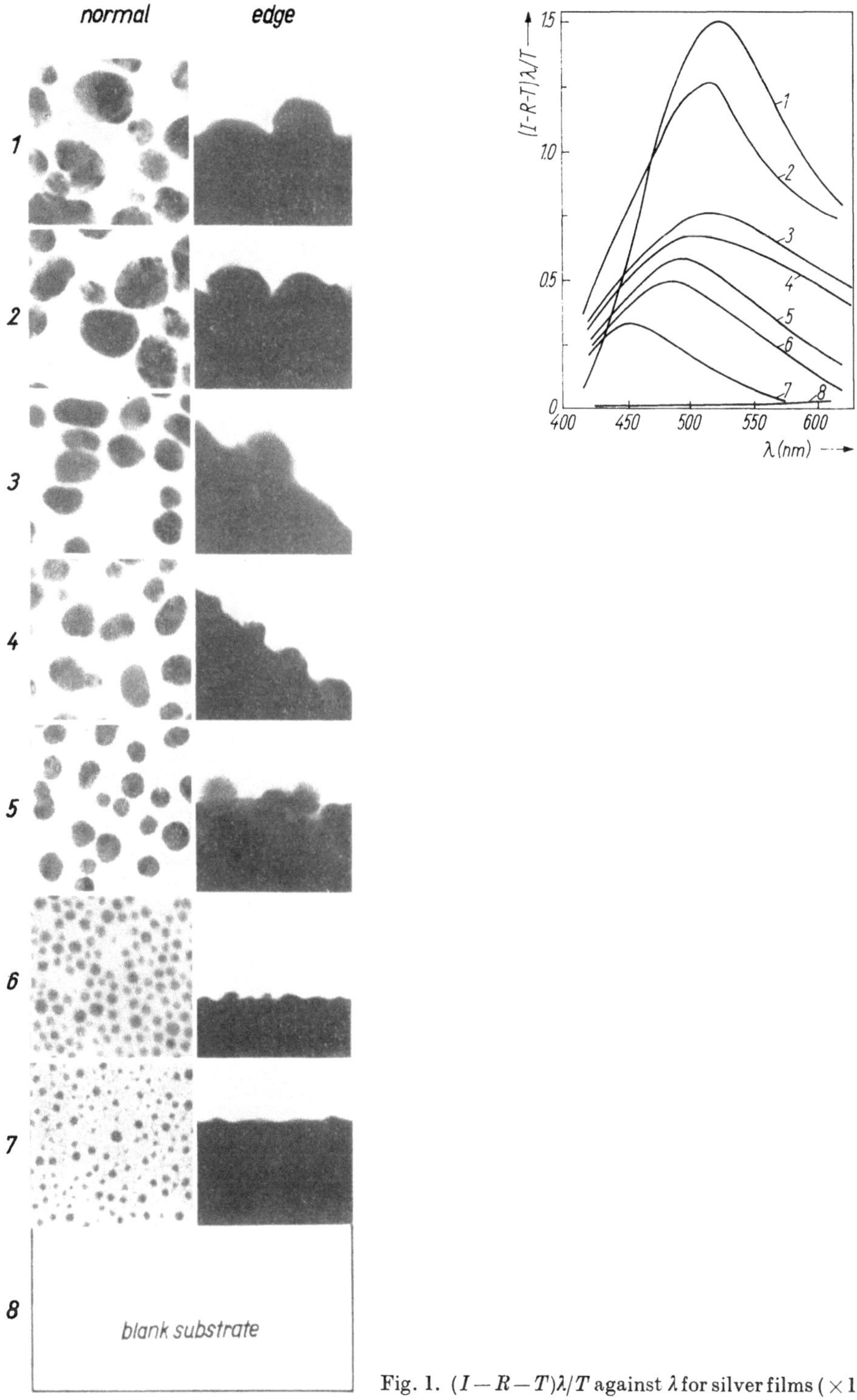

Fig. 1. $(I-R-T)\lambda/T$ against λ for silver films ($\times 188000$)

a b

Fig. 2. The simulated section topographs of the screw dislocations in the bent crystal obtained with use of the method [9]. a) $\boldsymbol{h} \cdot \boldsymbol{b} = 6\pi$, b) $\boldsymbol{h} \cdot \boldsymbol{b} = -6\pi$

See Short Note by H. H. Mende and H.-J. Kaiser

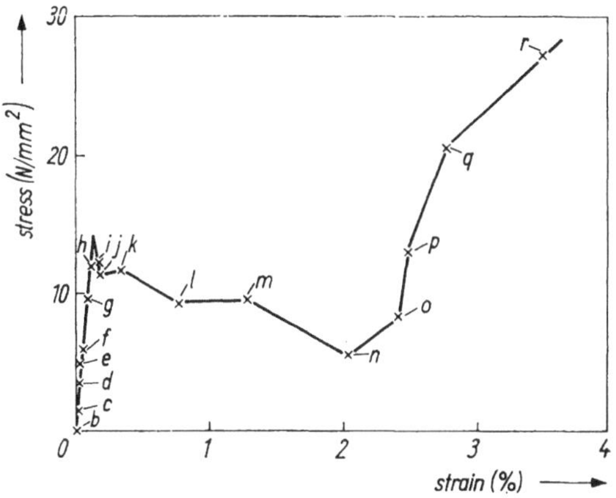

Fig. 2. X-ray topograms of well defined stress–strain states of a ⟨111⟩ copper whiskers transverse dimension 34 µm. a) Stress–strain curve with designated states. b) X-ray topograms from the designated states in Fig. 2a, ($\bar{2}$20) reflection

See Short Note by H. H. Mende and H.-J. Kaiser

Fig. 2b. First part

See Short Note by H. H. Mende and H.-J. Kaiser

Fig. 2b. Second part

phys. stat. sol. (a) 68, K1 (1981)

Subject classification: 18.3; 22.8.2

Institute of Solid State Physics,
Academy of Sciences of the USSR, Chernogolovka[1)]

Motion of a Bloch Line in YIG under the Action of a Magnetic Field Normal to It

By

L.M. DEDUKH, V.S. GORNAKOV, YU.P. KABANOV, and V.I. NIKITENKO

Recently, it has been experimentally shown /1/ that in YIG single crystals the essential dissipation of the energy by moving domain walls (DWs) is associated with motion of Bloch lines (BLs). Hence the study of the BL dynamics is necessary to describe the displacement processes of DWs in these crystals. The possibility of experimental investigation of the BL dynamics in YIG was found in /2/. In these experiments a magnetic field ($\vec{H}_z$) applied normal to the magnetization ($\vec{M}$) in 180° domains (see Fig. 1a on the photo pages before the Short Notes part) gave rise to magnetizing DWs due to the displacement of BLs. On the contrary, DWs itself remained fixed.

Apparently special interest concentrates on the study of the BL behaviour under conditions most like those at which DW dynamics is investigated namely when an applied field ($\vec{H}_x$) is oriented along $\vec{M}$ in domains. As it is seen from Fig. 1a, where the arrows indicate the direction of M at domains and at the centre of subdomains and of BLs in DW, $\vec{H}_x$ provides the torque $\vec{T} = \vec{M} \times \vec{H}_x$ on the magnetic moments $\vec{M}$ in BL, causing the BL to move in the direction depending on the type of $\vec{M}$ rotation in the BL transition region. If the crystal is infinite along the x direction, $\vec{H}_x$ would give rise to translational motion of all subdomains in the same direction (at single mode of rotation for $\vec{M}$ in all BLs). In real samples having finite dimensions such BL displacements will lead to a change in the size of subdomains. The associated increase in stray demagnetizing field arising near the DW transition region will counteract the

1) Chernogolovka, 142432 Moscow District, USSR.

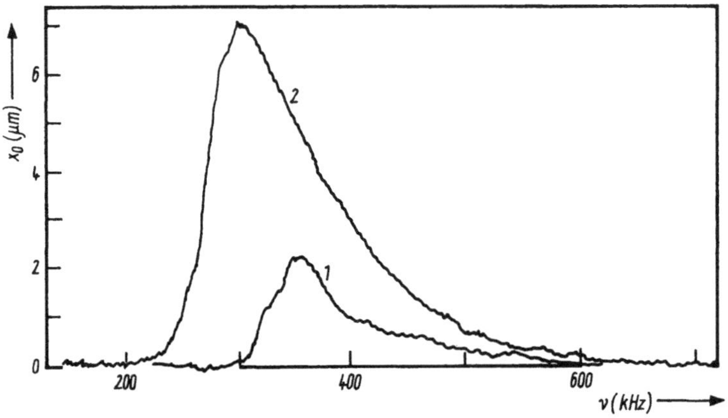

Fig. 2. Amplitude x_0 of the forced oscillation of a BL as a function of frequency ν of a sinusoidal field $\vec{H}_x$ generated by the parallel wires; (1) amplitude $H_x^0 = 3.8$; (2) $H_x^0 = 7.6$ mOe

BL motion. In the present work we succeeded in measuring the BL displacement under the action of the field $\vec{H}_x$ under different conditions in YIG single crystals.

The samples were cut in the form of rectangular (112) platelets of $0.05 \times 0.5 \times 5$ mm^3 size elongated along $[11\bar{1}]$ and having a single straight 180° DW that separated two domains magnetized along $[11\bar{1}]$ and $[\bar{1}\bar{1}1]$. The apparatus used is described in /2/. The intensity of polarized light (due to Faraday rotation) transmitted through a rectangular crystal region (shown by the dashed lines in Fig. 1a) including a single BL was measured directly. When the polarizers were slightly uncrossed the increment of the average intensity was linearly proportional to the dynamic BL displacement and was not dependent on the DW displacement.

Fig. 1b (see on the photo pages before the Short Notes part) shows the BL motion in DW oscillating in a uniform sinusoidal field $\vec{H}_x$ (produced by the Helmholtz coils) acting along $[11\bar{1}]$. Two images of DW correspond to two DW stops where the direction of the DW motion is reversed. It is seen that, during the DW motion from one final position to another, all subdomains (black and white segments of DW) are shifted in the same direction. This offers evidence that $\vec{M}$ in all BLs shown in Fig. 1b rotates in the same direction and that the

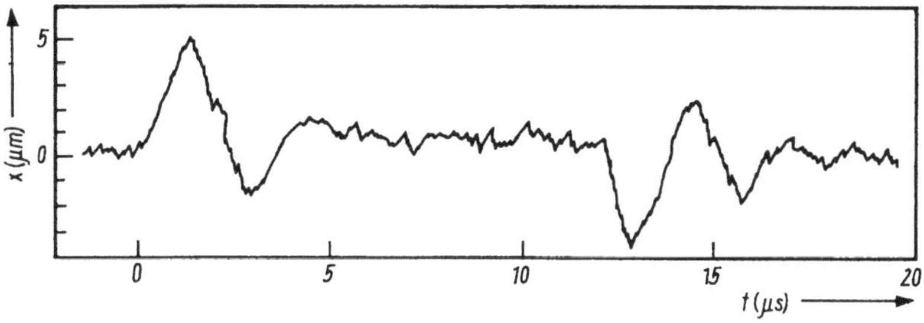

Fig. 3. Typical free oscillations of BL in pulse field $\vec{H}_x$ of duration 12 μs and of amplitude H_x^o = 0.04 Oe generated by the parallel wires

BLs do not contain so-called Bloch points /3/. In the case of its existence the latter BL would not move translationally because its regions with opposite polarity of $\vec{M}$ would displace in opposite directions. Unmoved BLs and BLs moving in opposite direction with respect to a common BL flow in field $\vec{H}_x$, those are likely due to both the internal structure of BLs and crystal defects, are also observed in different crystal regions.

Fig. 2 shows the resonance oscillations of BL in the sinusoidal field $\vec{H}_x$ acting on one BL only. This $\vec{H}_x$ was generated by a system of four parallel wires spaced such as to secure optimum homogeneity of $\vec{H}_x$. They were disposed perpendicularly to DW so that one BL was between them. The distances between the wires were smaller than the sizes of subdomains. The $x_o(\nu)$ curves in Fig. 2 are not symmetric with respect to the resonance frequency. Forced oscillations of BL were non-harmonic but periodic with the exciting frequency. With increasing $\vec{H}_x$ amplitude the peak in $x_o(\nu)$ shifted to lower frequencies. The value of the resonance frequency measured at minimum $\vec{H}_x$ amplitude roughly coincides with the natural frequency of BL found by applying the field $\vec{H}_z$. It also coincides with the frequency of free BL oscillations excited by a fast rising rectangular pulse field $\vec{H}_x^{\bullet}$ which was produced by the above wire system (Fig. 3).

The essential result of Fig. 3 is that the BL motion at the moments when a step field $\vec{H}_x$ was switched on (t = 0) and switched off (t = 12 μs) begins in opposite directions and that BL always remains at the initial equilibrium position after the cessation of the damped oscillations. This means that the BL displacement is not necessary for magnetizing a crystal in the x direction. However, the BL motion is excited by the field $\vec{H}_x$, especially at frequencies close to the natural frequency of BL. In particular, it turned out that in pulse fields $\vec{H}_x$ the exponential DW motion /4/ is accompanied by free vibrations of BLs.

References

/1/ L.M. DEDUKH, V.I. NIKITENKO, A.A. POLYANSKII, and L.S. USPENSKAYA, Zh. eksper. teor. Fiz., Pisma 26, 452 (1977).

/2/ V.I. NIKITENKO, L.M. DEDUKH, V.S. GORNAKOV, and YU.P. KABANOV, Zh. eksper. teor. Fiz., Pisma 32, 152 (1980); phys. stat. sol. (a) 63, K63 (1981).

/3/ J.C. SLONCZEWSKI, AIP Conf. Proc. 24, 613 (1975).

/4/ L.M. DEDUKH, V.I. NIKITENKO, and A.A. POLYANSKII, Zh. eksper. teor. Fiz. 79, 605 (1980).

(Received August 6, 1981)

phys. stat. sol. (a) 68, K5 (1981)
Subject classification: 2 and 20; 22.1.1; 22.1.3

Department of Physics, Indian Institute of Technology, Delhi[1)]

A Novel Bi-Layer System for High Resolution Optical Imaging

By
BHANWAR SINGH[2)] and K. L. CHOPRA

Amorphous chalcogenide films have shown potential as inorganic resists for high resolution lithography /1 to 3/. The fundamental functions of this resist are based on the silver doping phenomenon by the action of light /1/, electrons /2/, X-rays /3/, or ions /4/. Silver doping produces a drastic change in the chemical etch rate of chalcogenide films and this effect is used to define submicron patterns.

Yoshikawa et al. /5/ reported a chemical process for silver sentization of $Se_{1-x}Ge_x$ films by immersing them in aqueous solutions containing silver. It was estimated that a thin layer of silver (< 100 Å) was formed on the surface of $Se_{1-x}Ge_x$ films. For optical imaging applications, this thickness of silver is too small to give an optical contrast between silver deposited and virgin regions. We /6/ and Chen and Tai /7/ carried out extensive structural studies to understand the mechanism of the silver sensitization process. Transmission electron microscopy studies revealed the formation of polycrystalline domains of Ag_2Se and its oxide in silver sensitized $Se_{1-x}Ge_x$ films. By electrochemical studies, we established that the silver deposition (from aqueous solution) is due to the oxidation /8/ of germanium atoms present in amorphous $Se_{1-x}Ge_x$ films.

Currently, there is considerable technological interest in obliquely deposited amorphous Se-Ge films. Oblique deposition of chalcogenide films yields a columnar microstructure /9, 10/ with a high density of surface dangling bonds resulting in an increase in chemical reactivity. The film sensitivity as an ion /11/ and electron beam resist /12/ was found to increase considerably with deposition angle of these films. We found /6/ that the amount of silver deposited also increases with the deposition angle of Se-Ge films: in 80°-incidence films it was more than 15 times that for normal incidence films. This amount of silver de-

---

1) New Delhi 110016, India.

2) Present address: Department of Electronics and Electrical Engineering, Glasgow University, Glasgow, G12 8QQ, Scotland, Great Britain.

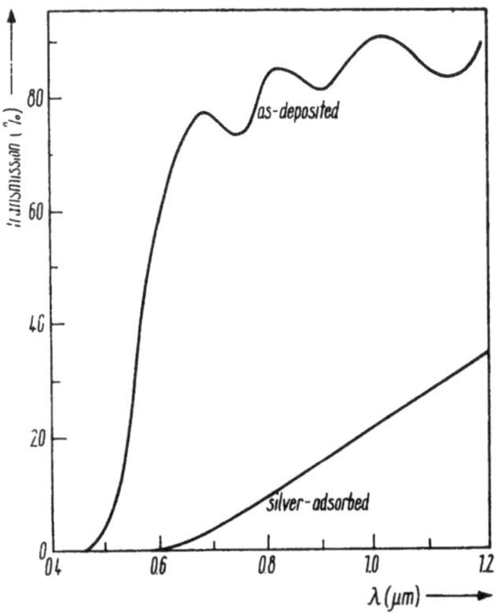

Fig. 1. Transmission spectra of 80°-incidence deposited a-$Se_{75}Ge_{25}$ film before and after the silver adsorption from aqueous solution of silver nitrate (0.025M)

position produces a large change in optical transmission of the film. This paper describes the application of this effect to store high contrast, submicron optical images in amorphous $Se_{1-x}Ge_x$ films using a bilayer process.

Amorphous $Se_{1-x}Ge_x$ films were prepared with different deposition angles by thermal evaporation of melt-quenched bulk alloys as reported earlier /13/. The films (3000 Å thick) were deposited on thin glass substrates at room temperature. The deposition of silver was obtained by immersing the films in a 0.025M solution of silver nitrate ($AgNO_3$). The transmission of amorphous $Se_{1-x}Ge_x$ films with and without silver deposition was measured using a Pye Unicam SP700A double-beam spectrophotometer.

Fig. 1 shows a typical example of the transmission versus wavelength plot for an 80°-incidence amorphous $Se_{75}Ge_{25}$ film after silver deposition from aqueous solution of $AgNO_3$ at room temperature. The transmission edge shifts toward longer wavelengths due to the formation of low band gap silver selenide domains.

The change in optical transmission between virgin and silver adsorbed regions has been utilized for optical imaging. The transmission contrast (transmission difference between virgin and silver adsorbed $Se_{1-x}Ge_x$ film) can be achieved to the required level by simply changing the following parameters; (a) composition of amorphous $Se_{1-x}Ge_x$ film, (b) angle of deposition, and (c) time and concentration of aqueous solution containing silver.

The process involves the following steps as shown in the schematic diagram of Fig. 2. After taking the films from the vacuum system a thin layer of organic photoresist was deposited on the amorphous $Se_{1-x}Ge_x$ film by spinning. The photoresist layer was exposed and developed using a test target

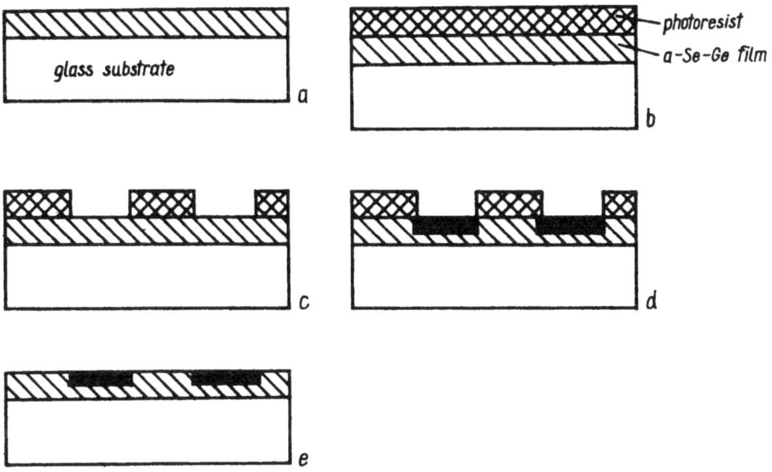

Fig. 2. Schematic diagram of optical imaging process in a-Se-Ge film by using a conventional photolithography technique. a) Vacuum deposition of a-Se-Ge film on glass substrate; b) spin coating of photoresist; c) pattern transfer by photolithography process; d) silver sensitization in aqueous solution of $AgNO_3$ (0.025 M), e) photoresist striping

mask by means of conventional photolithography. Subsequently, the deposition of silver effected by immersing the patterned bi-layer system into an aqueous solution of silver nitrate (0.025M) for 10 s. The adsorption of silver took place only in the exposed regions of Se-Ge films through the photoresist mask. Then photoresist was stripped off with the help of stripper. In this way an optical image was transferred to the amorphous Se-Ge film. Micropatterns down to 1.5 μm linewidth were easily resolved with well-defined edges. Fig. 3 (see on the photo pages before the Short Notes part) shows the optical microphotograph obtained by this process.

It is known that oblique deposition /14/ of thermally evaporated films induces a columnar growth in the film and this leads to a less densely packed material with an increasing number of voids and effective surface area. The film density was found to decrease with increasing deposition angle with a reduction as high as 25% for the films deposited at 80°. The primary cause of this columnar growth has been speculated to be the shadowing effect of the deposited material on the substrate. Our SEM and TEM investigations have indeed shown the presence of a columnar microstructure in a Se-Ge films, which increases with obliqueness. The enhanced silver adsorption can be attributed to the presence of columnar interfacial regions which act as diffusion pipes. A deep penetration of the silver is confirmed by depth profile observations using a SAM/ESCA instrument /8/.

Following are the advantages of the proposed bi-layer optical imaging system:

(1) It is a simple and non-conventional photographic process. Both positive and negative optical images can be registered by using the same system.

(2) The use of wide composition range and different angle of deposition of amorphous Se-Ge films offer tailored material properties for optical images as well as masks.

(3) Submicron resolution may be achieved by using e-beam or X-ray lithography.

The authors would like to acknowledge the helpful discussion with Dr. C.D.W. Wilkinson and thank Dr. R. Walker for his assistance in preparing the manuscript.

References

/1/ K. L. TAI, L. F. JOHNSON, D.W. MURPHY, and M.S.C. CHUNG, Electrochem. Soc. Meeting, Boston, May 6 to 11, 1979 (Abstract 94).

/2/ A. YOSHIKAWA, O. OCHI, N. NAGAI, and Y. MIZUSHIMA, Appl. Phys. Letters 31, 161 (1977).

/3/ K.D. KOLWICZ and M.S. CHANG, J. Electrochem. Soc. 127, 137 (1980).

/4/ A. WAGNAR, T. VENKATESAN, D. BARR, W. S. CRANE, V.E. LAMBERTI, K.L. TAI, and R.G. VADIMSKII, 16th Symp. Electron, Ion, and Photon Beam Technology, Dallas (Texas), May 26 to 29, 1981 (Abstract X-3).

/5/ A. YOSHIKAWA, D. OCHI, H. NAGAI, and Y. MIZUSHIMA, Appl. Phys. Letters 29, 677 (1976).

/6/ BHANWAR SINGH and K.L. CHOPRA, J. appl. Phys. 52, 428 (1981).

/7/ C.H. CHEN and K.L. TAI, Appl. Phys. Letters 37, 605 (1980); 37, 1075 (1980).

/8/ S. RAJAGOPALAN, K.S. HARSHAVARDHAN, BHANWAR SINGH, and K.L. CHOPRA, Presented at 9th Internat. Conf. Amorphous and Liquid Semiconductors, Grenoble, July 2 to 8, 1981.

/9/ BHANWAR SINGH, S. RAJAGOPALAN, P.K. BHAT, D.K. PANDYA, and K.L. CHOPRA, J. non-crystall. Solids 35/36, 1053 (1980).

/10/ F. JANSEN, Thin Solid Films 78, 15 (1981).

/11/ T. VENKATESAN, 16th Symp. Electron, Ion, and Photon Beam Technology, Dallas (Texas), May 26 to 29, 1981 (Abstract X-4).

/12/ K. BALASUBRAMANYAM and A.L. RUOFF, ibid. (Abstract X-5).

/13/ S. RAJAGOPALAN, BHANWAR SINGH, P.K. BHAT, D.K. PANDYA, and K.L. CHOPRA, J. appl. Phys. 50, U89 (1979).

/14/ A.G. DIRKS and H.J. LEAMY, Thin Solid Films 47, 219 (1977).

(Received August 21, 1981)

phys. stat. sol. (a) 68, K11 (1981)
Subject classification: 1.5 and 20.1; 21.6

Coventry (Lanchester) Polytechnic[1)]

The Optical Properties of Discontinuous Silver Films

By
W. T. MC KENNA and L. WARD

Introduction Jarrett and Ward /1/ reporting the results of an investigation into the optical properties of discontinuous gold films, observed an anomalous absorption band in the near infrared which moved to longer wavelengths as the film become more dense. It was possible to correlate the shapes of the film particles with the wavelengths of the absorption peaks by treating the islands as ellipsoids of revolution and using the shape factors defined by David /2/ modified by a term arising from the interaction between the islands.

Experimental The experiment has been repeated using thin silver films. These were deposited on glass substrates by evaporation from a tungsten wire inside a cylindrical stainless steel chamber maintained at a pressure of about $5x10^{-7}$ Torr. Optical measurements on the films were made in situ at an angle of incidence of 45° via three windows each offset at 5° from normal to eliminate any back reflection effects. A full description of the equipment is given in the paper by Jarrett and Ward.

The measurements, over the wavelength range 400 to 1000 nm, were made by a silicon photovoltaic cell attached to an arm pivotted at the centre of the vacuum chamber; fine adjustments of the angular position and height of the cell enabled the intensity measurements to be optimised and a phase sensitive detector system was used to minimise background noise levels. The quantities measured were: I, the intensity of the incident beam, T, that of the transmitted beam, and R, that of the reflected beam. Abelès /3/ has given expressions for the reflectance and transmittances of very thin films ($d \ll \lambda$) in the form of polynomials of $2\pi d/\lambda$ up to $(2\pi d/\lambda)^2$ and using these it may be shown that $(I - R - T)\lambda/T = (2\pi \varepsilon_2'/n_s)d$, where ε_2' is the imaginary part of the effective dielectric constant, d is the effective thickness of the film, and n_s is the refractive index of the substrate. Thus, for a film of constant thickness, the

1) Priory Street, Coventry, Great Britain.

quantity (I - R - T)λ/T is directly proportional to ε_2'. As explained later, the average shapes of the islands of which the film is composed may be deduced from the peak in the curve of (I - R - T)λ/T against λ . One considerable advantage of this method is that it does not depend on the height of the peak but only on its wavelength and hence it is not necessary to know the value of the effective thickness of the film.

Fig. 1 (see on the photo pages before the Short Notes part) shows (I - R - T)λ/T plotted against λ for a range of films whose structure is illustrated in the corresponding electron micrographs. The distinctive feature is again the shift in the anomalous peak to longer wavelengths as the particle size and density increase.

Theory The average shape factor of the film particles may be calculated as follows: Jarrett and Ward showed that

$$\varepsilon_2' = \frac{Q\, \varepsilon_a \varepsilon_2}{\left\{\varepsilon_a + (\varepsilon_1 - \varepsilon_a)F\right\}^2 + F^2 \varepsilon_2^2} , \tag{1}$$

where Q is the volume fraction, ε_a is the dielectric constant of the space between the islands, and F is a numerical factor made up of two parts, f_1, David's shape factor which is related to the ratio of the semi-axes of the ellipsoid and f_2 a term arising from interactions between the islands; thus the shape factor $f_1 = F + f_2$. ε_1 and ε_2 are the real and imaginary parts of the dielectric constant of the bulk metal. Values of these latter two quantities for silver have been reported by Ehrenreich and Phillipp /4/ and, over the range of wavelengths covered by the peaks in Fig. 1, both ε_1 and ε_2 are linear with λ and may be represented by the equations

$$\varepsilon_1 = 20.4 - 57.3\lambda , \tag{2a}$$

$$\varepsilon_2 = -0.466 + 1.73\lambda , \tag{2b}$$

where λ is measured in μm. Using equations (1), (2a), and (2b) the following quadratic equation may be obtained for the peak wavelength, λ_m:

$$0 = 5685\, F^2 \lambda_m - 3063\, F^2 \lambda_m + 1.73(\varepsilon_a\{1 - F\} + 20.4F)(\varepsilon_a\{1 - F\} - 10.5F).$$

This may readily be solved to give values of λ_m against F.

As there is some debate as to the correct value to use for ε_a solutions were obtained for values of ε_a, of 1.00, 2.25, and 1.63 corresponding to vacuum,

Table 1

Shape factors deduced by various methods

film No.	λ_m (nm)	shape factor		
		electron micrographs	David's theory (F)	David's theory plus interaction (f_1)
1	525	0.21	0.16	0.23
2	520	0.24	0.16	0.23
3	510	0.21	0.17	0.22
4	505	0.22	0.18	0.24
5	490	0.25	0.18	0.23
6	480	0.23	0.19	0.23
7	450	0.24	0.22	0.24

substrate, and an equal mixture of vacuum and substrate; Fig. 2 shows the three curves of F plotted against λ_m.

The experimental values of λ_m were used to read off the corresponding values of F and hence to derive the shape factors of the particles. Assuming no interaction (David's theory) $F = f_1$ and leads to the prediction of flat ellipsoids which become increasingly prolate as ε_a decreases from 2.25 to 1.00. Electron micrographs of the particle profiles on the films were obtained by folding the electron microscope grids and viewing the edge of the fold so enabling estimates to be made of the ratios of the semi-axes of the particles. These profiles indicated that, in all cases, the shapes of the particles were less prolate than predicted by David's non-interaction theory. The above results are set out in the first four columns of Table 1. The factors calculated by David's theory used $\varepsilon_a = 2.25$ which gave the best agreement with the measured particle shapes and supported the contention held by many workers in this field that the effective inter-island dielectric constant is the same as that of the substrate.

A new shape factor may be calculated by taking account of the interactions between the islands. For ellipsoidal particles the interaction term $f_2 = N_v V f_1 = Qf_1$ and hence $f_1 = F + Qf_1$, giving $f_1 = F/(1 - Q)$. For spherical particles $f_1 = 1/3$ and the formula reduces to that of Doremus /5/, i.e. $f_1 = F + Q/3$. Q, the volume fraction of the film, was calculated by drawing an arbitrary

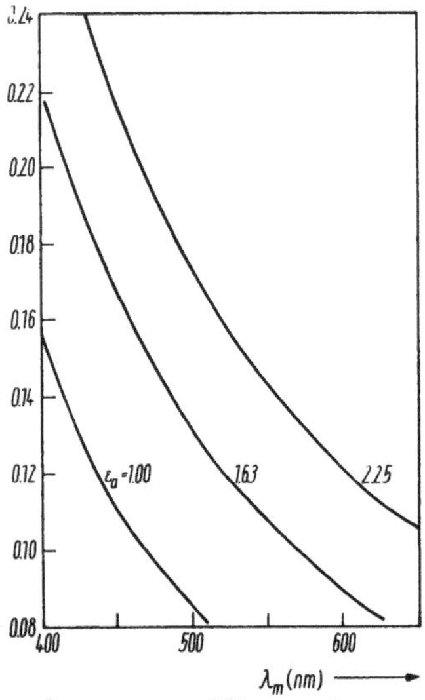

Fig. 2. F against λ_m for $\varepsilon_a = 2.25$, 1.63, and 1.00

straight line on the electron micrograph and determining the linear fraction L as the ratio of the length intersecting particles to the total length. Q = 2 L/3 (Underwood /6/).

The results of applying this modification are also given in Table 1, again using ε_a = = 2.25, and it is seen that good agreement is now obtained between the calculated and measured shape factors.

Accuracy The peak wavelengths could be determined to ±5 nm and, from Fig. 2, this would give an error in David's shape factor of less than 0.005. The main error was in Q which was calculated from six measurements of L taken at different parts of the electron micrograph, the standard deviation in each case being ±0.02. The shape factors determined from the electron micrographs also had standard deviations of ±0.02.

Conclusions The wavelengths of the anomalous absorption peaks of island silver films have been used to deduce the shapes of the film particles employing David's shape factors modified by an interaction term. In all the cases investigated the shapes predicted were in good agreement with those determined from profile electron micrographs. The shape factors did not show a significant variation with particle size and indicated roughly hemispherical particles (f_1 = 0.25) in all cases.

References

/1/ D.N. JARRETT and L. WARD, J. Phys. D 9, 1515 (1976).
/2/ E. DAVID, Z. Phys. 114, 389 (1939).
/3/ F. ABELÈS, Progr. Optics 2, 249 (1963).
/4/ H. EHRENREICH and H. R. PHILLIPP, Phys. Rev. 128, 1622 (1962).
/5/ R.H. DOREMUS, J. appl. Phys. 37, 2775 (1966).
/6/ E.E. UNDERWOOD, Quantitative Stereology, Addison Wesley, London 1970 (p. 27).

(Received July 29, 1981)

phys. stat. sol. (a) 68, K15 (1981)
Subject classification: 10

VNIINauchpribor, Leningrad[1)]

The Effect of Dislocations on the Pendellösung Fringes in a Bent Crystal

By
P.V. PETRASHEN and N.S. YAROSLAVSKAYA

The phenomenon of the contraction of Pendellösung fringes is well known /1 to 5/. The essence of it is that the number of fringes on a section topograph of a strained crystal is always greater than that of an unstrained one, whereas the distance between the fringes is correspondingly smaller. Physically it is explained by the curvature of the Bloch wave trajectories in a strained crystal which leads to an increase in the optical path distances along the trajectories and in consequence to an increase of the phase differences between the trajectories belonging to the two different branches of the dispersion surface /5/. This effect reveals itself in particular as the appearance of additional fringes on the section topographs near the sources of localized strain fields in otherwise perfect crystals - near dislocations /4/, near the edges of the oxide films on the surface /3/, etc.

The number of the additional fringes gives a quantitative information about the diffractional power of the strain field source - as, for example, the scalar product $(\vec{h} \cdot \vec{b})$ of the diffraction vector and the Burgers vector of a dislocation /4/, but does not give any data concerning the sign of the strain field since the strain field of any sign leads only to an increase of the number of fringes compared with a perfect crystal.

However, it is physically clear that in the macroscopically strained crystal, when the number of fringes is already greater than in a perfect one, the local lattice distortions are capable in principle of either increasing or decreasing the number of fringes compared with the remaining part of the crystal. Thus, the Pendellösung fringes in a strained crystal may be sensitive to the sign of the local strain field.

---

1) Stakhanovtsev str. 1, Leningrad 195112, USSR.

In the present paper this possibility is studied on the example of a screw dislocation in a bent crystal.

The number of Pendellösung fringes may be easily estimated in the Eikonal approximation /5, 6/ especially taking into account that for our purposes the dislocation strain field may be considered as a perturbation imposed on the elastic strain of the uniform bending. In order to obtain the Eikonal equation in a general (asymmetric) case of diffraction, we shall use the Takagi equations written in the quasi-Cartesian oblique-angled coordinate system /7, 8/ with the x axis directed along the crystal surface in the diffraction plane and the z axis along the median of the Borrmann triangle. The scales along the axes are chosen so that the Cartesian vector $\vec{R}$ may be expressed in terms of (x, z) in the form

$$\vec{R} = \frac{\Lambda}{2\pi}\left(\frac{z+x}{\gamma_o}\vec{s}_o + \frac{z-x}{\gamma_h}\vec{s}_h\right) , \tag{1}$$

where Λ is the extinction length, $\vec{s}_o$ and $\vec{s}_h$ are the unit vectors in the directions of propagation of the incident and the diffracted waves, $\gamma_{o,h} = (\vec{n}\cdot\vec{s}_{o,h})$, where $\vec{n}$ is the internal normal to the surface of the crystal. The Takagi equations in this coordinate system take the form

$$i\left(\frac{\partial}{\partial z} + \frac{\partial}{\partial x}\right)E_o + \sigma_{-h}\, e^{i\vec{h}\vec{u}}\, E_1 = 0 ,$$

$$i\left(\frac{\partial}{\partial z} - \frac{\partial}{\partial x}\right)E_1 + \sigma_h\, e^{-i\vec{h}\vec{u}}\, E_o = 0 . \tag{2}$$

The Eikonal substitution $E_o = e_o \exp(iS)$, $E_1 = e_1 \exp(iS - i\vec{h}\vec{u})$ leads to the Hamilton-Jacobi equation for S(x, z):

$$\left(\frac{\partial S}{\partial z} - \alpha(x, z)\right)^2 - \left(\frac{\partial S}{\partial x} - \alpha(x, z)\right)^2 = \sigma^2 , \tag{3}$$

where

$$\sigma^2 = \sigma_h \sigma_{-h}, \quad \alpha(x, z) = \frac{1}{2}\left(\frac{\partial}{\partial z} - \frac{\partial}{\partial x}\right)(\vec{h}\cdot\vec{u}) .$$

The common way of solving (3) includes the construction of the wave field propagation trajectories and the subsequent evaluation of the Eikonal S through the phase integrals along this trajectories /6/. In our case of the uniformly bent crystal the Eikonal depends just on the expression $(z^2 - x^2)$ quite so as in the case of a perfect crystal /5, 7, 8/, the distribution of the Pendellösung fringes is symmetrical around the median of the Borrmann triangle and $\partial S/\partial x\big|_{x=0} = 0$. Taking into account the dislocation strain field as a small perturbation one can assume that this symmetry property is weakly affected and the value of $\partial S/\partial x\big|_{x=0}$ remains small compared with σ^2 and α. In this case (3) may be easily integrated along the median of the Borrmann fan, the wave field phases evaluated on the central line of the section topograph, and the problem of counting of the fringe number solved.

One has from (3)

$$S(0, z) = \int_0^z dz'(\alpha(0, z') \pm \sqrt{\sigma^2 + (\alpha^2(0, z')}) \, . \tag{4}$$

We are interested only in the evaluation of the phase $\Gamma = \mathrm{Re}(S)$, so that in the case of small absorption it may be neglected, which gives $\sigma^2 = 1$. The two signs in (4) stand for the two wave fields belonging to different dispersion surface branches. The phase difference between the wave fields on the excit surface in the middle of the topograph is

$$\Delta\Gamma = 2 \int_0^z \sqrt{1 + \alpha^2(0, z')} \, dz' \, . \tag{5}$$

The expression for $\alpha(z)$ consists of two parts – the first one, α_b, is due to the crystal bending and the second, α_d, is due to the dislocation strain field. The only essential term caused by bending has the form /7, 8/

$$(\vec{h} \cdot \vec{u}) = B(z^2 - x^2) \, , \quad \alpha_b = B(x + z) \, . \tag{6}$$

For the sake of simplicity we shall neglect the asymmetry of the diffraction geometry in the evaluation of the second part, α_d. Qualitatively the result will be the same even when this asymmetry is large. In the case of a screw dislocation parallel to the surface one has

$$(\vec{h} \cdot \vec{u}) = \frac{(\vec{h} \cdot \vec{b})}{2\pi} \Theta \, , \quad \alpha_d = \frac{(\vec{h} \cdot \vec{b})}{4\pi} \, \frac{y'}{(y')^2 + (z - z_o)^2} \, , \tag{7}$$

where Θ is the polar angle in the dislocation coordinate system, y' is the shortest distance between the dislocation line and the median of the Borrmann fan, z_o is the depth of the dislocation under the entrance surface.

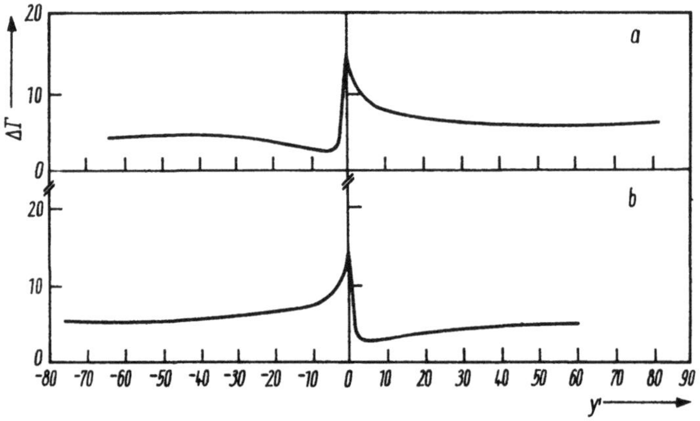

Fig. 1. The dependence of the additional phase difference on the distance y' from the dislocation line; a) $(\vec{h} \cdot \vec{b}) = 6\pi$, b) $(\vec{h} \cdot \vec{b}) = -6\pi$

The total distortion is given by the sum $\alpha_b + \alpha_d$.

It is easily noticeable that when the signs of B and $(\vec{h} \cdot \vec{b})\, y'$ are the same, the phase difference $\Delta\Gamma$ increases near the dislocation as well as the number of the fringes $\Delta\Gamma/2\pi$, whereas when $B\, y'(\vec{h} \cdot \vec{b}) < 0$ the phase difference decreases.

Far from the dislocation ($y' \rightarrow \pm\infty$) one has the well-known result for the bent crystal /5, 7, 8/

$$\Delta\Gamma_b = z\sqrt{1 + B^2 z^2} + \frac{1}{B} \ln\left(\sqrt{1 + B^2 z^2} + Bz\right) \tag{8}$$

for a perfect crystal (B = 0) one has $\Delta\Gamma_b = 2z$.

Fig. 1 shows the dependence of the additional phase difference $\Delta\Gamma - \Delta\Gamma_b$ caused by a dislocation and by bending on the distance y'. The parameters are the following: a) $(\vec{h} \cdot \vec{b}) = 6\pi$, b) $(\vec{h} \cdot \vec{b}) = -6\pi$, $z = 30$, $z_o = 20$, $B = 0.025$.

The asymptotic value of $\Delta\Gamma$ at $y' \rightarrow 0$ may be easily obtained from the integral (5):

$$\Delta\Gamma(0) = \Delta\Gamma_b + \frac{(\vec{h} \cdot \vec{b})}{2} \; . \tag{9}$$

A similar result was obtained in /4/ for a perfect crystal. In contrast to the case of a perfect crystal the dependence of $\Delta\Gamma$ on y' in the bent crystal is essentially asymmetrical: on the side $y'B(\vec{h}\cdot\vec{b}) > 0$ $\Delta\Gamma$ increases smoothly with the decrease of y' up to the maximum value (9), whereas on the opposite side has a minimum located approximately at

$$\left|y'_0\right| = \left|\frac{(\vec{h}\cdot\vec{b})}{4\pi B z_0}\right| \quad \text{(in our case } \left|y'_0\right| \approx 3\text{)},$$

which corresponds to the turning to zero of $\alpha(z_0)$.

The results of the computer simulations of the section topographic images /9/ for the parameter values listed above are shown on Fig. 2a, b (see on the photo pages before the Short Notes part). The change in the positions of the fringes on the opposite sides of the dislocation is well seen. Thus it is shown that the bending of the crystal enables one to determine the sign of the localized strain field from the locations of the Pendellösung fringes.

The authors are grateful to Dr. E.V. Suvorov for helpful discussions.

References

/1/ M. HART, Z. Phys. 189, 269 (1966).

/2/ N. KATO and Y. ANDO, J. Phys. Soc. Japan 21, 964 (1966).

/3/ J.R. PATEL and N. KATO, J. appl. Phys. 44, 971 (1973).

/4/ E.V. SUVOROV, V.I. POLOVINKINA, V.I. NIKITENKO, and V.L. INDENBOM, phys. stat. sol. (a) 26, 385 (1974).

/5/ N. KATO, J. Phys. Soc. Japan 19, 67 (1964); 19, 971 (1964).

/6/ F.N. CHUKHOVSKII and A. A. SHTOLBERG, phys. stat. sol. 41, 815 (1970).

/7/ P.V. PETRASHEN and F.N. CHUKHOVSKII, Zh. eksper. teor. Fiz. 69, 477 (1975).

/8/ F.N. CHUKHOVSKII and P.V. PETRASHEN, Acta cryst. A 33, 311 (1977).

/9/ P.V. PETRASHEN, Fiz. tverd. Tela 18, 3729 (1976).

(Received July 20, 1981)

phys. stat. sol. (a) 68, K21 (1981)
Subject classification: 12.1; 10.1; 21.1

Institut für Angewandte Physik der Universität Münster[1)]

X-Ray Topographic Investigations of Copper Whiskers during the Registration of Stress-Strain Curves

By

H.H. MENDE and H.-J. KAISER[2)]

Introduction Whiskers are filamentarily grown monocrystals with a nearly perfect lattice sometimes without any lattice defects. Therefore whiskers can be used as a model substance for an ideal solid to study the different properties, for example the influence of the dislocation arrangements on the stress-strain curve. The small thickness of the whiskers allows the application of the Lang X-ray transmission topography technique /1/ to image crystal defects and lattice distortions /2 to 7/.

Experimental The copper whiskers were grown using the method of Brenner /8/ by reducing copper iodide with hydrogen at about 650 °C. The growth axis was generally the ⟨111⟩ direction, the transverse dimensions extend from a few μm to about 50 μm, and the length from a few mm to about 10 mm.

A newly developed tensile apparatus has been constructed in such a way that the whole device could be attached to the goniometer head of an X-ray equipment. The load was measured by the deformation of an annular spring using strain gauges and the elongation of the specimen by an inductive displacement transducer. The stress-strain curves were directly registered by an X-Y recorder (for further details see /9/).

The deformation of the whiskers was performed in small steps without unloading the specimen during the whole process. A topogram using the MoK_α radiation was imaged at every adjusted stress-strain state. The exposure time for the topograms lasted for about 10 min.

Results and discussion The stress-strain curve of a ⟨111⟩ copper whisker received by the tensile apparatus, is shown in Fig. 1. The shape of the curve is characteristic for the stress-strain behaviour of copper whiskers, as also

1) Roxeler Str. 70/72, D-4400 Münster, FRG.

2) Present address: Max-Planck-Institut für Eisenforschung, Max-Planck-Str. 1, D-4000 Düsseldorf, FRG.

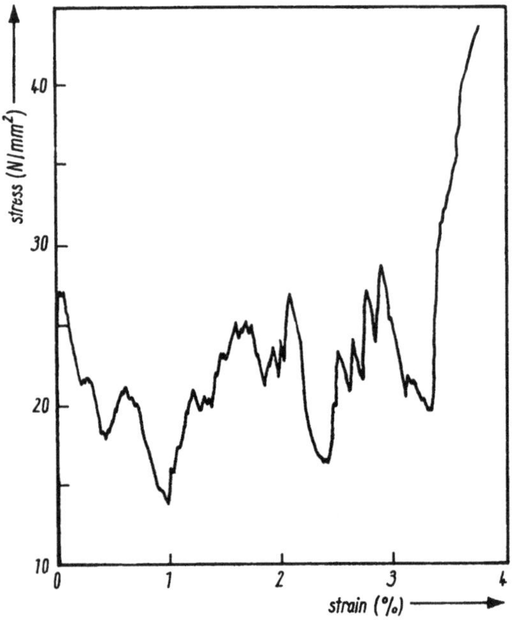

Fig. 1. Stress-strain curve of a ⟨111⟩ copper whisker, transverse dimension 27 μm

determined by other authors /14/. The elastic deformation of the whisker up to the yield point is followed by the Lüders strain region corresponding to a plastic deformation of the whisker. In this part of the stress-strain curve the elongation occurs with a stress oscillating about a mean value. In the following region the stress increases considerably, and the specimen is strain hardened until the whisker breaks, for instance the whisker of Fig. 1 at about 165 N/mm^2.

To study the relation between the different parts of the stress-strain curve and the generation as well as the movement of the lattice distortions, especially dislocations, the stress-strain curves and the X-ray topograms must be taken from the same specimen and during the same experiment. Fig. 2 (see on the photo pages before the Short Notes part) gives an example for many other investigations not mentioned here.

All investigated copper whiskers already show lattice deformations in the unloaded initial state (see No. a) in Fig. 2b). These contrasts were always observed near the glue-points indicating that the defects are generated by the specimen preparation. The deformations of the lattice were so high that these defects, and respective contrasts in the topograms, could not be resolved further more, they only appeared as dark contrast.

The topograms of No. b) to h) in Fig. 2b were taken after the whisker being fixed in the tensile testing apparatus and during a stepwise elastic deformation (see also the points No. b) to h) in Fig. 2a). The linear relationship between stress and strain could essentially be confirmed. Other results, not specified in this paper, sometimes showed local plastic deformations by small movement of dislocations in the linear elastic section of the stress-strain curve. Further indications to a plastic deformation below the yield point are given by the Young moduli determined from the linear region of the stress-strain curve. These values for monocrystalline copper are mostly lower than those mentioned in the corresponding literature (see /11/). Yoshida et al. /12/ obtained similar results.

The yield points of the stress-strain curves sometimes appear stronger than shown in Fig. 2a. The beginning of the Lüders strain region is characterized in the X-ray topograms by the movements of the lattice deformations - contrasts - of the initially unloaded state (see the topograms No. i) and j) in Fig. 2b). This is a sensitive method to indicate the movement of dislocations in the volume of a bulk specimen and therefore to determine the limit of the linear elastic region. At this part of the stress-strain curve no variation on the surface, for example Lüders bands, could be observed. The own values of the yield point stresses were much lower and less marked than those obtained by Yoshida et al. /12/. This is a result of the lattice distortions in the initial state of the whisker. According to Johnston /13/ the initial dislocation density is the main reason for observing no sharp stress decrease at the yield point as described in literature.

In the Lüders strain region (points No. i) to n) in Fig. 2a) the oscillating stress in the stress-strain curve occurs. The corresponding topograms (No. i) to n) in Fig. 2b) show an expansion of the plastically deformed section until the whole volume of the whisker is deformed. The strong stress oscillations in the Lüders strain region can be explained by the obstruction of dislocation movement by Lomer-Cottrell dislocations /10/.

Finally the whisker did not show any undisturbed regions - see for instance the topograms No. n) in Fig. 2b - i.e. the disturbed lattice sections from the two fixed points of the whiskers get into contact with one another, a further expansion of the crystal defects being obstructed. The strain hardening region of the stress-strain curve is reached. A higher stress is required for further deformation of the copper whiskers (see points No. o) to r) in Fig. 2a). The contrast in this strain hardening region of the stress-strain curve is so dark that no further resolution of the plastic deformation processes is possible.

The present results point out the possibility of studying the influence of the dislocation arrangements and movements on the different parts of the stress-strain curve using the same specimen and during the same experiment. Further experiments are planned according to this correlation. This method will be used for example to investigate the influence of the internal oxidation on the mechanical properties of copper whiskers. The internal oxidation reduces for instance the residual electrical resistivity of about a factor ten /15/.

References

/1/ A. R. LANG, in: Modern Diffraction and Imaging Techniques in Material Science, Ed. S. AMELINCKX, R. GEVERS, G. REMAUT, and J. VAN LANDUYT, North-Holland Publ. Co., Amsterdam 1979.

/2/ B. K. TANNER, X-Ray Diffraction Topography, Pergamon Press, Oxford 1976.

/3/ H. GALINSKI and H. H. MENDE, phys. stat. sol. (a) 27, 35 (1975).

/4/ H. H. MENDE and H. GALINSKI, Appl. Phys. 5, 211 (1974).

/5/ H. H. MENDE and W. HAGEN, phys. stat. sol. (a) 42, K95 (1977).

/6/ W. HAGEN and H. H. MENDE, Z. Metallk. 68, 550 (1977).

/7/ F. W. YOUNG, Jr. and F. A. SHERRILL, J. appl. Phys. 42, 230 (1971); 43, 2949 (1972).

/8/ S. S. BRENNER, Acta metall. 4, 62 (1956).

/9/ H.-J. KAISER and H. H. MENDE, Z. Feinwerkt. Meßtechn. 86, 228 (1978).

/10/ O. NITTONO, Japan. J. appl. Phys. 10, 188 (1971).

/11/ H. BIEHL and H. H. MENDE, J. Phys. Chem. Solids 35, 37 (1974).

/12/ K. YOSHIDA, Y. GOTOH, and M. YAMAMOTO, J. Phys. Soc. Japan 24, 1099 (1968).

/13/ W. G. JOHNSTON, J. appl. Phys. 33, 2716 (1962).

/14/ S. S. BRENNER, in: Growth and Perfection of Crystals, John Wiley and Sons, New York 1958.

/15/ G. THUMMES and H. H. MENDE, phys. stat. sol. (a) 26, 243 (1974).

(Received July 10, 1981)

phys. stat. sol. (a) 68, K25 (1981)
Subject classification: 19; 22.8.2

Departamento de Física, Universidade de Brasília[1)]

Microwave Resonance Absorption in Ferrites by the Photoacoustic Effect

By
O. A. C. NUNES, M. A. AMATO, and K. SKEFF NETO

The photoacoustic effect /1 to 4/ is simply the generation of an acoustic signal by a sample exposed to modulated light. The sample is heated to the extent that it absorbs the incident light and the energy gained is lost through non-radiative processes. Because the light is chopped, the temperature rise is periodic at the chopping frequency, and it is this periodic temperature rise at the surface of the sample that in turn causes a modulation of the gas pressure in the enclosed chamber. This is how the sound originates.

Even though the photoacoustic technique in the visible has been an extremely active field, to our knowledge, only recently an extension of the above described method to the microwave region has been reported by Nunes et al. /5/, for investigating the photoacoustic effect of a well known phenomenon in solids at this region i.e., the ferromagnetic resonance (FMR). In this note, we show experimental evidence that ferrimagnetic resonance can also be investigated using the photoacoustic effect as a probe.

Ferrimagnetic resonance can be pictured as the coupling of a microwave field with the Larmor precession of the net magnetization inside a ferrimagnetic sample. The equation of motion for a two-sublattice ferrimagnetic /6, 7/ of which the magnetizations are assumed to be antiparallel provides for the resonance condition

$$\hbar \omega_{res} = \gamma_{eff} \hbar H_{eff} , \qquad (1)$$

where γ_{eff} is the effective magneto-mechanical ratio and H_{eff} is the effective field "seen" by the spins in the specimen. We can expect to induce spin energy absorption as the precession frequency of the net magnetization about the applied dc magnetic field equals the frequency of the incident microwave field. This energy is subsequently transferred to the lattice as heat by the spin-lattice relaxation mechanism and this will be the source for the generation of the photo-

---

1) 70.910 Brasília, D.F., Brazil.

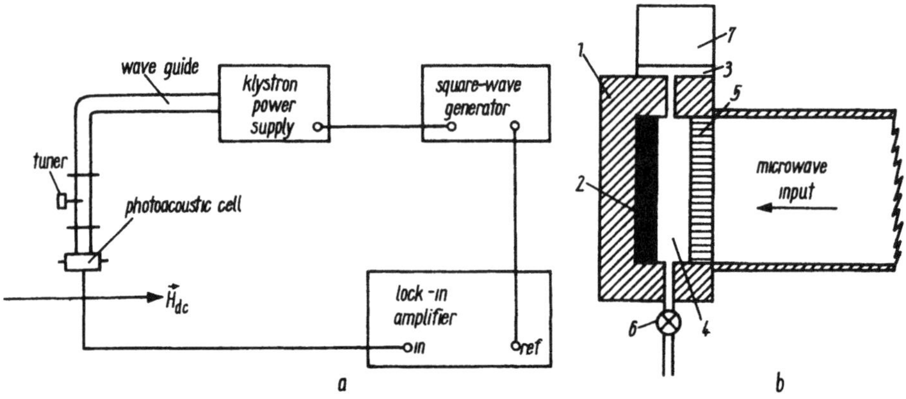

Fig. 1. a) Microwave-photoacoustic system for exciting the sample. b) The photoacoustic cell: 1 brass walls of the sample chamber, 2 ferrite sample, 3 microphone cavity, 4 sample chamber, 5 teflon window, 6 needle valve, 7 microphone

acoustic signal.

On a thin plane (200 μ inch thick) specimen of a magnetic oxide in form of magnetic recording tape (γ-Fe_2O_3), measurements were made at 11.4 GHz with the specimen inside the photoacoustic cell fitted across the wave guide. The field of an electromagnet was located parallel to the plane of the specimen and perpendicular to the microwave magnetic field. The photoacoustic signal generated inside the sample was measured as a function of the applied dc magnetic field by a lock-in amplifier tuned to a modulation frequency of 100 Hz.

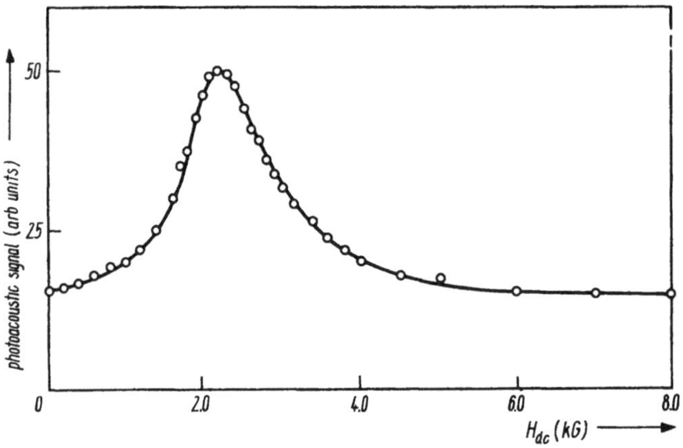

Fig. 2. Photoacoustic signal versus dc magnetic field H_{dc}

In Fig. 1a we show the microwave rig for exciting the sample. Fig. 1b shows the photoacoustic chamber. In Fig. 2 we present the photoacoustic signal as a function of the dc magnetic field and illustrate the resonance phenomenon.

The feasibility of the photoacoustic technique and the fact that the photoacoustic signal is a direct measurement of the amount of microwave power absorbed of the sample reveals this technique as a powerful tool for further investigations. Furthermore, because of the greater skin depth of ferrites caused by the high resistivity a relatively large volume of the specimen participates in the absorption process allowing a most detailed investigation of the inner layers of the specimen.

References

/1/ A.G. BELL, Amer. J. Sci. 20, 205 (1980).
/2/ J.G. PARKER, Appl. Optics 12, 2947 (1973).
/3/ A. ROSENCWAIG, Analyt. Chem. 47, 592 A (1975).
/4/ A. ROSENCWAIG and A. GERSHO, J. appl. Phys. 47, 64 (1976).
/5/ O.A.C. NUNES, A.M.M. MONTEIRO, and K. SKEFF NETO, Appl. Phys. Letters 35, 656 (1979).
/6/ R.K. WANGSNESS, Phys. Rev. 86, 146 (1952).
/7/ S. FONER, in: Magnetism, Ed. RADO and SUHL, Academic Press, New York 1963 (p. 415).

(Received April 30, 1981)

phys. stat. sol. (a) 68, K29 (1981)

Subject classification: 2 and 19; 22.6.1

Sektion Physik der Friedrich-Schiller-Universität Jena[1] (a) and
Sektion Chemie der Friedrich-Schiller-Universität Jena,
Otto-Schott-Institut (b)

^{11}B NMR Study on Glasses of the System TeO_2-B_2O_3

By

R. GÖRING (a), H. BÜRGER (b), H. NASS (b), and B. SCHNABEL (a)

Introduction The glasses of the binary system TeO_2-B_2O_3 studied in this work belong to a group of multicomponent glasses with extreme optical properties, which were formerly investigated with respect to decomposition behaviour and phase diagram /1 to 3/. Within the glass-forming region between 26.4 and 11.8 mol% B_2O_3 /4/ the composition range between 25 and 20 mol% B_2O_3 is characterized by crystal-free, homogeneous, clear, and colourless glasses, which are formed rather simply. Our ^{11}B NMR investigations were carried out to study the molecular structure of the glasses.

Experimental In the glass-forming region of the system B_2O_3-TeO_2 glasses were prepared with 80, 77.5, and 75 mol% TeO_2. The preparations were carried out with extremely pure substances (TeO_2 p.a. and H_3BO_3 p.a.) melted in gold crucibles (100 g batches) at 1173 K (cooling temperature 693 K, cooling rate about 1 K/min). The chemical analysis of the glasses shows nearly the same composition as the batches (experimental error ± 0.1 mol% TeO_2, ± 0.1 mol% B_2O_3).

All ^{11}B NMR measurements described here were performed at 21 MHz using the NMR wide-line spectrometer KRB 35/62 (ZWG, AdW of the DDR). From the glass samples the first derivatives of the ^{11}B NMR absorption spectra were recorded. The modulation amplitude was 0.1 mT, the modulation frequency 70 Hz. Cylindric samples ($\approx$ 8.7 g) were used to realize a filling factor of the NMR coil of about 1. For separating the central peak in the ^{11}B NMR spectrum, which is due to boron atoms in BO_4 tetrahedra, and the broad line belonging to boron atoms in BO_3 triangles we performed a visual comparison with spectra from standard samples (B_2O_3, BPO_4, $BaO\cdot 4B_2O_3$). Spectra were recorded in analogue form, then digitized by hand and, using a programmable desk calculator,

---

1) Max-Wien-Platz 1, DDR-69 Jena, DDR.

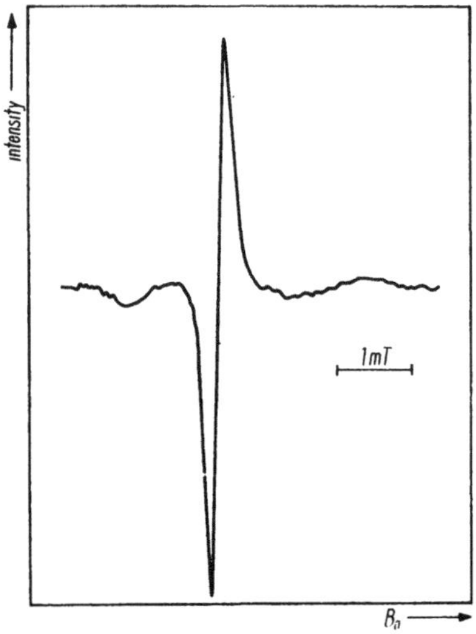

Fig. 1. ^{11}B NMR (derivative) spectrum for the sample containing 22.5 mol% B_2O_3 and 77.5 mol% TeO_2

the fraction (N_4) of boron atoms in BO_4 units was determined by the so-called "area method", where N_4 is calculated as the ratio of the area under the central peak and the total signal area. This procedure of separation and digitization gives only about twice the error as the method of spectrum fitting.

Results Fig. 1 shows the first derivative of the ^{11}B NMR absorption spectrum of the sample containing 22.5 mol% B_2O_3 and 77.5 mol% TeO_2. The two other samples investigated exhibit nearly the same spectra. The spectrum shape is typical for most of the binary and ternary borate glasses /5, 6/ and indicates the existence of boron atoms both in BO_4 and BO_3 units. The values of N_4, determined by the procedure given above, are summarized in Table 1.

From its shape the broad line due to the BO_3 units (Fig. 1) does not give any information for an appreciable existence of BO_3 units containing non-bridging oxygens. The quadrupole coupling constant is (2.70 ± 0.05)MHz. The width of the central peak is rather small compared with other borate glasses (inflexion point separation 2.58 kHz).

Table 1

glass composition synthesis (mol%)		analysis (mol%)		N_4
TeO_2	B_2O_3	TeO_2	B_2O_3	
80	20	80.3	19.7	0.56 ± 0.03
77.5	22.5	77.7	22.3	0.45 ± 0.03
75	25	75.1	24.9	0.43 ± 0.03

Discussion As seen above, our results give evidence for the presence of a considerable fraction of boron atoms in BO_4 units in the TeO_2-B_2O_3 system unlike other binary systems such as B_2O_3-GeO_2 /7/ or B_2O_3-SiO_2 /8/. Additionally, the spectrum parameters cited above indicate that both the BO_3 and BO_4 units obviously exhibit a symmetry typical for other borate glasses. The formation of BO_4 tetrahedra is probably accompanied by a change of coordination of at least some part of the tellurium atoms. This conclusion is suggested from the results of X-ray studies of crystalline tellurium(IV)-oxygen compounds /9, 10/, which indicate that the stereochemistry of the tellurium is changed rather easily by one of the four oxygen atoms. Starting from α-TeO_2 with a coordination number 4 and the coordination polyhedron of a trigonal bipyramid (one of the equatorial sites is occupied by lone-pair electrons) the coordination of tellurium changes from 4 over a mixed coordination 4 and 3 to a coordination number of mainly 3 in the tellurite (1:1) compounds.

The RDF curves calculated from neutron diffraction measurements in the TeO_2-P_2O_5 system /11, 12/ indicate also a change of the tellurium-oxygen coordination.

The present ^{11}B NMR results as well as the results of the above mentioned different methods of investigation lead to the conclusion that the Te-O structure may be modified rather easily even in glasses. Probably one of the axial oxygens in the TeO_4 polyhedron is consumed for building up BO_4 tetrahedra, and thus the tellurium coordination number is lowered from 4 to 3. Assuming the existence of BO_3 units only with all bridging oxygens, merely a rather small part of the tellurium-oxygen polyhedra ($\approx$ 1/4) exhibit such a coordination number lowering at the given TeO_2 to B_2O_3 ratio of about 4 and according to a $N_4 \approx 0.5$ of boron measured.

The small linewidth of the central peak indicates that the BO_4 tetrahedra are bonded mainly to TeO_3 or TeO_4 units. This is consistent with some chemical considerations, which also permit the existence of BO_4-TeO_3, BO_4-BO_3, and TeO_3-TeO_4 arrangements, but exclude the BO_3-TeO_4 arrangements because of the considerably different field strengths of boron(III) and tellurium(IV).

Some recent infra-red spectroscopic investigations, which are still continued, also give evidence for the existence of both BO_3 and BO_4 units in our samples.

Thanks are due to Mrs. D. Bechstein for analysis of the spectra.

References

/1/ Y. DIMITRIEV and E. KASHCHIEVA, J. Mater. Sci. 10, 1419 (1975).

/2/ H. BÜRGER, L. HORN, W. JAECKEL, and W. VOGEL, Silikattechnik 32, 177 (1981).

/3/ H. BÜRGER, K.-J. HOFFMANN, and W. VOGEL, Silikattechnik 32, (1981), in the press.

/4/ W. VOGEL et al., Silikattechnik 25, 209 (1974).

/5/ P.J. BRAY and J.G. O'KEEFE, Phys. Chem. Glasses 4, 37 (1963).

/6/ P.J. BRAY, Silikattechnik 19, 307 (1968).

/7/ J.F. BAUGHER and P.J. BRAY, Phys. Chem. Glasses 13, 63 (1972).

/8/ R. GÖRING and H. NASS, to be published.

/9/ J. ZEMAN, Monatsh. Chemie 102, 1209 (1971).

/10/ G. BAYER, Fortschr. Miner. 46, 42 (1969).

/11/ S. NEOV, V. KOZHUKHAROV, I. GERASIMOVA, K. KREZHOV, and B. SIDZHIMOV, J. Phys. C 12, 2475 (1979).

/12/ S. NEOV, I. GERASIMOVA, V. KOZHUKHAROV, and M. MARINOV, J. Mater. Sci. 15, 1153 (1980).

(Received June 26, 1981)

phys. stat. sol. (a) 68, K33 (1981)

Subject classification: 14.2; 21

Institute for Low Temperature and Structure Research, Polish Academy of Sciences, Wrocław[1)]

Potential Distribution on the Surface of a Thin Cylindrical Superconducting Film of In-Pb Alloy

By

A. BOHDZIEWICZ and M. KAZIMIERSKI

We investigated the distribution of the electric potential on the surface of a superconducting sample, through which current was passed, with the aid of several stationary voltage contacts. The sample was in form of a thin film cylinder, in contradistinction to bulk cylinders investigated by Irie, Ezaki, and Yamafuji (IEY) /1 to 3/. The latter workers presented a model of the resistive state in the form of a helical two-domain structure, for the purpose of explaining the observed "triangular wave" of the potential, obtained from six contacts positioned on the circumference of the bulk sample.

Our observations, with the use of some more contacts, showed the onset of a more complicated structure of the resistive state. The substrates of the samples investigated here were made of insulator (acrylic glass) or anodized aluminium. The dimensions of the substrates and the distribution of contacts are shown in Fig. 1 (inset). The potential contacts (18 to 40) were embedded equidistantly in three or four planes perpendicular to the substrate axis; there were also two additional contacts at the ends of the central, uniform part of the substrate. The contacts were in form of 0.1 mm diameter Cu wires embedded radially in the substrate wall and the ends passed through the hollow substrate core. After setting of epoxy the protruding ends were cut down; the substrate was then turned and polished. A thin film of the investigated material was evaporated finally with constant rate onto the uniformly resolving substrate up to a thickness varying in particular samples from 0.15 to 1.0 μm. Since the substrate makes several hundred revolutions during the constant rate evaporation, the uniformity of the film thickness is rather high.

The investigated alloy anneals at room temperatures. Thus any imperfections of our samples may be caused only by geometrical errors of the substrates. The

---

1) Próchnika 95, 53-529 Wrocław, Poland.

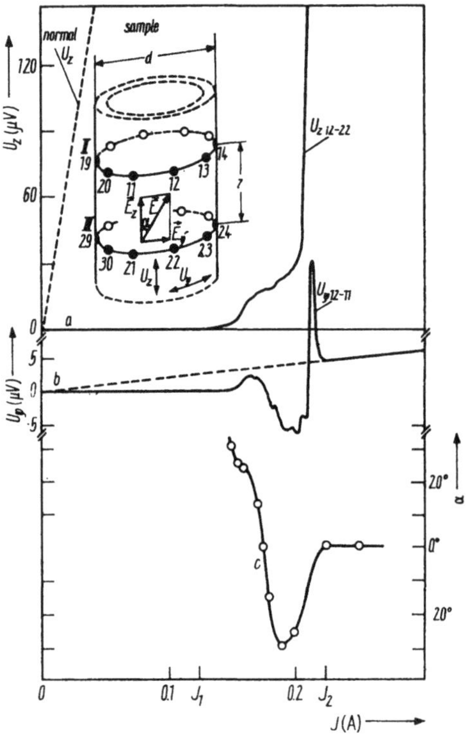

Fig. 1. Inset: Central part of the substrate of the 40-contacts sample, illustrating the distribution of the voltage probes. (a) The dependence of the longitudinal voltage U_z (between contacts 12 and 22) on the current J flowing along the sample. (b) The dependence of the azimuthal voltage U_φ (between contacts 12 and 11) on the current J. (c) The dependence of the angle of inclination of the mean electric field vector with respect to the sample axis in the region between the contacts 11-12-21-22. $T = 0.959\ T_c$, film thickness $t = 1\ \mu m$, $d = 4$ mm, $z = 3$ mm

material was an In + 5 at% Pb alloy (the alloy data were $T_c = 3.73$ K, $T_c = 0.02$ K, $\varkappa$ about 0.9 /4/). The critical temperature of the films varied from 3.75 to 3.80 K; the typical transition width was not greater than 0.02 K. Current leads were arranged coaxially. Measurements were conducted at temperatures close to T_c in the magnetic laboratory field, ordered and partially removed with the aid of permalloy screens. The current-voltage characteristics U_φ and U_z (Fig. 1, inset and curve a) were registered on an X-Y recorder, simultaneously for few appropriately chosen contacts (at small distances) as well as for the outer contacts (U_p).

From the form of the obtained characteristics follows that for the contacts, which in the normal state are positioned on the equipotential lines, we get the azimuthal voltages U_φ when the current increases (at onset of the resistive state). Exemplary results of U_φ measurements are presented in Fig. 1 (curve b) for the pair of contacts (11, 12). The absolute value of the voltage U_φ for that pair increased with increasing current, beginning from a value at J_1 to some maximum value and then decreased when the sample underwent transition to the normal state at current J_2. In the range of existence of azimuthal voltages U_φ (i.e. between J_1 and J_2) we observed a subrange, beginning at J_1 (about 1/3 part of the whole range J_1 - J_2), where U_φ was stable and practically reproducible. Fluctuations in the rest of the J_1 - J_2 range were caused predominantly by nonuniform transport of heat out of the sample surface. An inclination of the "normal" part of the characteristics was due to the errors in positioning of the potential contacts with respect to the perpendicular plane, nominally equipotential.

The mutual correlation between U_φ and longitudinal voltages U_z in the "stability range" of currents implies that the value of the electric field $\vec{E}$ increases weakly with current, while the angle between $\vec{E}$ and the sample axis may change pronouncedly. From simultaneous measurements of U_φ and U_z one can determine the mean values of E_φ and E_z in the region between respective contacts and obtain the approximate course of the dependence of α on current, as it is shown in Fig. 1 (curve c). It should be pointed out that not in all cases one observes a change of sign of the angle α. An exemplary series of voltages U_φ for one circumference (group of contacts) and for a specified value of current $J_1 < J < J_2$ is presented for different samples with different number's of contacts.

If one ascribes to consecutive intervals of voltage U_φ the corresponding values of the electric field E_φ, one notes that, depending on the current, one can obtain different numbers of regions with that field having different polarity. It seems plausible that an increasing number of contacts or employing sliding contacts will lead to the discovery of an even more complicated structure of E_φ. The analysis of the course of the voltages U_φ and U_z in our samples showed, as in the papers of IEY and also of Cave and Evetts /5/, the existence of regions of the sample surface where E_φ has opposite polarities. We found that there could exist several such regions distinguishing by the sign of E_φ, their number depending on the value of the current. The investigations by Cave and Evetts showed that the structure of the regions could be very complicated indeed, even in bulk samples.

We believe that our method of making contacts with the inner surface of the cylindrical thin film is appropriate since it changes neither the geometry nor the structure of the sample. The potential differences were determined with virtually no current passing along the wires. Thus we think that the presence of the contacts has no - or very little - influence on the current distribution in the resistive state, and the patterns observed have to be due to a structure in the current flow itself. Increasing the number of potential contacts (cf. Fig. 2 a and b) uncovers new details of the complicated structure, what makes a unique

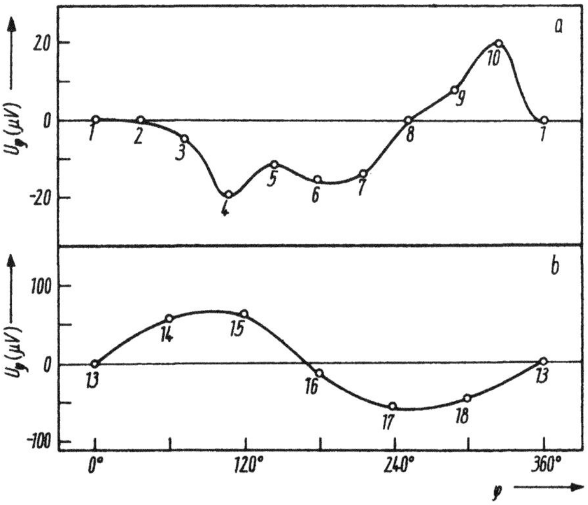

Fig. 2. The voltages between consecutive contacts positioned in one plane perpendicular to the sample axis for: a) sample with 10 contacts spaced equidistantly on the circumference; group I, d = 4mm, t = 1 μm, T = 0.995 T_c, z = 3 mm, J = 0.05 A; b) sample with 6 contacts on the circumference; group III, d = = 10 mm, t = 0.5 μm, T = 0.931 T_c, z = 5 mm, J = 0.01 A

interpretation of the results harder to reach. The helical two-domain structure in superconducting cylinders in resistive state postulated by IEY was generally not obtained in the thin cylindrical films investigated by us.

The authors are very much indebted to Prof. B. Makiej for numerous and thorough discussions.

References

/1/ F. IRIE, T. EZAKI, and K. YAMAFUJI, IEEE Trans. Magnetics 11, 332 (1975).

/2/ T. EZAKI and F. IRIE, J. Phys. Soc. Japan 40, 382 (1976).

/3/ T. EZAKI, K. YAMAFUJI, and F. IRIE, J. Phys. Soc. Japan 40, 1271 (1976).

/4/ E. TROJNAR, B. MAKIEJ, and A. SIKORA, Acta phys. Polon. 34, 311 (1968).

/5/ J.R. CAVE and J.E. EVETTS, Phil. Mag. B 37, 111 (1978).

(Received June 5, 1981)

phys. stat. sol. (a) 68, K37 (1981)
Subject classification: 9 and 10

A. F. Ioffe Physico-Technical Institute,
Academy of Sciences of the USSR, Leningrad[1]

On the Analysis of Diffusion with Association

By
R. SH. MALKOVICH

A number of problems of impurity diffusion with association in the case of local dynamic exchange equilibrium between the impurity particles and associates can be reduced to an equation of concentration dependent diffusion type. Among these are the problems of diffusion of a divalent impurity in NaCl-type crystals /1/, of impurity diffusion in crystals with uniformly distributed doping (binding) impurity /2, 3/, etc. In all these cases the one-dimensional diffusion can be described by the partial differential equation of the type

$$\frac{\partial C}{\partial t} = \left(D(C)\frac{\partial C}{\partial x}\right) . \quad (1)$$

In the considered case of a half-space with constant surface impurity concentration this equation can be reduced by means of the Boltzmann transformation $\lambda = x/\sqrt{t}$ to the ordinary differential equation

$$-\frac{\lambda}{2}\frac{dC}{d\lambda} = \frac{d}{d\lambda}\left(D(C)\frac{dC}{d\lambda}\right) . \quad (2)$$

The exact solution of this equation, $C(\lambda)$, can be generally found by finite difference methods, iteration methods, and variational methods /4 to 6/. Because of the known difficulties in the use of these methods it is desirable to find the approximate analytic expressions for the impurity diffusion profile.

In this paper approximate formulas for the impurity diffusion profile are given for the two cases of diffusion with association - diffusion of a divalent impurity in crystals of the NaCl type and impurity diffusion in a crystal with uniformly distributed doping impurity. These formulas are derived by means of the method of approximate solution of a non-linear diffusion (or heat conduction) equation /7/. They express the concentration profile in terms of a function $\lambda(C)$ rather than $C(\lambda)$. The comparison of the values determined from these formulas

1) Politekhnicheskaya 26, Leningrad 194021, USSR.

with the values found by the numerical solution of (2) /4/ shows that the accuracy of the given formulas does not exceed one or a few per cent.

In the case of diffusion of a divalent impurity it is supposed /1/ that the associates are formed between the impurity ions and the cation vacancies. It is assumed that the impurity concentration is equal to the vacancy concentration and that part of impurity ions is free and the rest forms pairs with vacancies. It is supposed also that the free impurity is immobile and only the associates can diffuse.

In the case of diffusion of the impurity in a crystal with uniformly distributed doping impurity it is assumed that the associates are formed between the diffusing and doping impurities and that the doping impurity and associates are immobile. It must be noted that in the earlier analytical expressions /8/ the diffusion of both impurities was taken into account but these expressions are valid only in the limiting cases of weak or strong association.

Diffusion of a divalent impurity in a crystal of NaCl type /1/ In this case the diffusion is described by equation (2) with

$$D(C) = D\left(1 - \frac{1}{\sqrt{1 + 4KC}}\right) , \tag{3}$$

where D is the diffusion coefficient of the associates, C is the total impurity concentration, K is the constant of the dynamic equilibrium: $K(C - S)^2 = S$ (S is the concentration of the associates).

From the relationships (A1), (A2) (see Appendix) we get

$$\lambda(c) = 2D\left[\frac{1}{Q_1 Q_2}(1 - c)\left(1 - \frac{2}{\sqrt{1+a} + \sqrt{1+ac}}\right) \ln \frac{2(\sqrt{1+a}+1)+a}{2(\sqrt{1+ac}+1)+ac}\right]^{1/2} , \tag{4}$$

where

$$Q_1 = D^{1/2}\left[1 - \frac{3}{4a^2}\left(2 + (a-2)\sqrt{1+a}\right)\right]^{1/2} , \quad Q_2 = D^{1/2}\left[2 + \frac{4}{a}\left(1 - \sqrt{1+a}\right)\right]^{1/2} ,$$

$$c = \frac{C}{C(0)} , \qquad a = 4KC(0) . \tag{5}$$

The error given by formula (4) does not exceed 3 to 4% for values of $a \lesssim 2$. When $a \gtrsim 2$ it is advisable to use the correction factor $\gamma(c)$ (see Appendix, (A3)); for $a \gtrsim 20$ the corrected values $\lambda_{corr}(c) = \gamma(c)\lambda(c)$ are accurate to better than 2%.

Impurity diffusion in a crystal with uniformly distributed doping (bounding) impurity /3/ In this case D(C) can be given by

$$D(C) = \frac{D}{2}\left\{1 + \frac{1 - K(S_o - C)}{\sqrt{[1 - K(S_o - C)]^2 + 4KS_o}}\right\}, \qquad (6)$$

where C is the total impurity concentration ($C = C_i + C_s$; C_i is the mobile impurity concentration, C_s is the bound impurity concentration); S_o is the doping impurity concentration, K is the constant of dynamic equilibrium: $KC_i(S_o - C_s) = C_s$; D is the diffusion coefficient of the mobile impurity.

Using the relationships from the Appendix we have

$$\lambda = (\lambda_1 \lambda_2)^{1/2}, \qquad (7)$$

where

$$\lambda_1 = \frac{D}{Q_1}\left[(1 + g) - (c + z)\right],$$

$$\lambda_2 = \frac{D}{Q_2}\left[\ln\left(\frac{1 + g + b}{c + z + b}\,\frac{1}{c}\right) + \frac{b}{h}\ln\left(\frac{z + h + \frac{b}{h}c}{g + h + \frac{b}{h}}\,\frac{1}{c}\right)\right],$$

$$Q_1 = \left[\frac{D}{2}\left(1 + bh + (1 - b)g + a\ln\frac{b + h}{1 + g + b}\right)\right]^{1/2}, \quad Q_2 = \left[D(1 + g - h)\right]^{1/2},$$

$$g = \left[a + (1 + b)^2\right]^{1/2}, \quad h = (a + b^2)^{1/2}, \quad z = \left[a + (c + b)^2\right]^{1/2}, \qquad (8)$$

$$c = \frac{C}{C(0)}, \quad C(0) = 1 + \frac{k}{k + 1}s_o, \quad a = \frac{4}{k}\,\frac{s_o}{C(0)^2}, \quad b = \frac{1}{C(0)}\left(\frac{1}{k} - s_o\right),$$

$k = KC_{io}$, $s_o = S_o/C_{io}$ (C_{io} is the surface concentration of the mobile, diffusing impurity).

In the case of strong association, $C(0) \geq 100$, the maximum error of (7) does not exceed $\approx 1\%$ ($C(0) = 100$) or a few per cent ($C(0) = 1000$). For values of $1 \leq C(0) \leq 10$ it is desirable to use the correction factor $\gamma(c)$ (see Appendix, (A3)); in this case the accuracy of the values $\lambda_{corr}(c)$ is better than 1%.

The author wishes to thank Dr. O.M. Chekmareva for discussion of the results and V.A. Deedeek and A.N. Tyureena for help in calculations.

Appendix Equation (2) under the boundary conditions $C(0) = C_o$, $C(\infty) = 0$ can be reduced to /7/

$$\lambda = 2 \int_c^1 \frac{D(c)}{q(c)} \,, \tag{A1}$$

where $c = C/C_0$, $q(c) = \int_0^c \lambda \, dc$. It can be shown that irrespective of the type of the function $D(c)$ the function $q(c)$ varies only between the limits $Q_0 c$ and Q_0, where $Q_0 = \int_0^1 \lambda \, dc$. Substituting in (A1) $q(c)$ by $Q_0 c$ and Q_0 gives two functions, $\lambda_1(c)$ and $\lambda_2(c)$, respectively. The approximate values of the function $\lambda(c)$ are derived by means of arithmetical, $\lambda = (1/2)(\lambda_1 + \lambda_2)$, or geometrical, $\lambda = (\lambda_1 \lambda_2)^{1/2}$, averaging of the values λ_1 and λ_2. Instead of the one value Q_0 needed for the evaluation of λ it is desirable to calculate two independent values, Q_1 and Q_2, according to the formulas $Q_{1(2)} = \int_0^1 \lambda_{1(2)} dc$. Then

$$Q_1 = \left(2 \int_0^1 c\, D(c)\, dc\right)^{1/2}, \quad Q_2 = \left(2 \int_0^1 D(c)\, dc\right)^{1/2} . \tag{A2}$$

In the case of functions $D(c)$ that vary between constant limits not too much differing one from another the accuracy can be improved by means of the correction factor $\gamma(c)$. This factor is the ratio of the exact solution $\lambda_{ex}\{D_0\}$ corresponding to the diffusion coefficient D_0 = const,

$$\lambda_{ex}\{D_0\} = 2\, D_0^{1/2} \arg \operatorname{erfc} c \,,$$

and the approximate solution $\lambda_{app}\{D_0\}$ found for $D(c) = D_0$ from the expressions $\lambda = (\lambda_1 \lambda_2)^{1/2}$ and (A1), (A2),

$$\lambda_{app}\{D_0\} = 2^{3/4} D_0^{1/2} \left[(c-1)\ln c\right]^{1/2} .$$

Therefore

$$\gamma(c) = \frac{\lambda_{ex}\{D_0\}}{\lambda_{app}\{D_0\}} = 2^{1/4} \frac{\arg \operatorname{erfc} c}{\left[(c-1)\ln c\right]^{1/2}} . \tag{A3}$$

The corrected values λ_{corr} are determined from the relationship

$$\lambda_{corr} = \gamma(c)(\lambda_1 \lambda_2)^{1/2} . \tag{A4}$$

References

/1/ A. LIDIARD, Hdb. Phys., Vol. 20, Part 2, Springer-Verlag, Berlin 1957 (p. 246).

/2/ H. REISS, C.S. FULLER, and F.J. MORIN, Bell Syst. tech. J. 35, 535 (1956).

/3/ R.SH. MALKOVICH and V.A. POKOEVA, phys. stat. sol. (b) 82, 421 (1977).

/4/ J.R. PHILIP, Trans. Faraday Soc. 51, 885 (1955).

/5/ M.A. LITOVSKII and R.SH. MALKOVICH, phys. stat. sol. (a) 36, K145 (1976).

/6/ L.A. KOZDOBA, Metody resheniya nelineinikh zadach teploprovodnosti, Nauka, Moscow 1975.

/7/ M.A. LITOVSKII and R.Sh. MALKOVICH, Zh. tekh. Fiz. 50, 2007 (1980).

/9/ V.V. VASSKIN and V.A. USSKOV, Fiz. tverd. Tela 11, 1763 (1969).

(Received September 29, 1981)

phys. stat. sol. (a) 68, K43 (1981)

Subject classification: 1.3

Materials Science Centre, Indian Institute of Technology, Kharagpur[1)]

Classification of Computer Generated Random Binary Alloys

By

J.K. MUKHERJEE and S.K. PRASAD[2)]

The computer generated random binary (cgrb) alloys have been discussed in an earlier paper /1/. A random binary alloy is a binary A-B alloy with the probabilities of occupancy of A and B atoms at each lattice site, P_A and P_B, given by the atomic fractions, m_A and m_B, respectively. For the classification of cgrb alloys, a cubic lattice of size 3a x 3a x 3a and several types of distributions on this lattice, namely ordered distribution, random distribution, and clustered distribution are considered.

The ordered distribution for a given A-B alloy is defined as one that has a maximum of AB bonds. These have been obtained for $m_A = 0.5$ and 0.879. The number of AB bonds, n_{AB}, has been plotted against the atomic fraction, m_A - the AB bond-atomic fraction plot with the help of the values for these alloys, Fig. 1. The plot has been completed by symmetry.

The random distribution of an alloy is defined by the Cowley (Warren) short-range order parameters /2, 3/, α_i, where i = 1, 2 ... is the number of spherical shell from a given lattice point. For a random distribution all

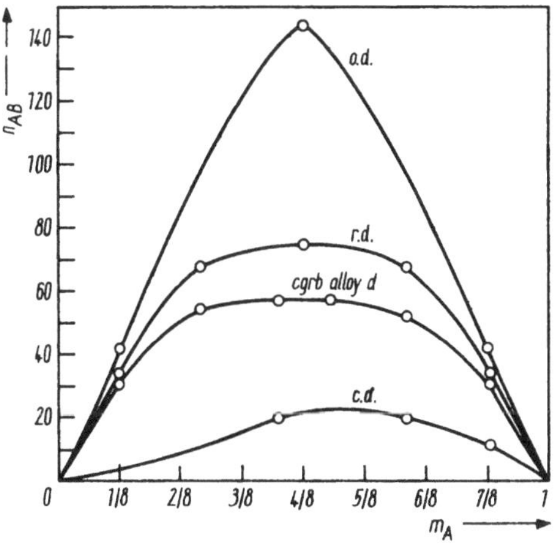

Fig. 1. Plots of the number of AB bonds, n_{AB}, against the atomic fraction, m_A, for o.d. (ordered distribution); r.d. (random distribution); cgrb alloy d (cgrb alloy distribution) and c.d. (cluster distribution)

---

1) Kharagpur 721302, India.

2) Presently at Physical Research Wing, Fertilizer (Planning and Development) India Limited, Sindri Dhanbad, Bihar, India.

$\alpha_1 = 0$ from all lattice points of the lattice. The distribution of atoms on the above mentioned lattice has been obtained by hand from several lattice points obtained from the lattice points (111), (112), (122), (222), (212), (221), and (121) with $\alpha_1 \approx 0$ for $m_A = 0.5$, 0.708, and 0.879. The AB bond-atomic fraction plot has been completed by the help of symmetry, Fig. 1.

The cluster distribution is defined here as one which has B clusters with a minimum of AB bonds. The distribution for the three values of $m_A = 0.447$, 0.708, and 0.879 has been obtained. The AB bond-atomic fraction plot is given in Fig. 1. The curve is not symmetrical because of preference of B clusters.

The AB bond-atomic fraction plot for cgrb alloys is also given in Fig. 1. The plot is obtained from values at $m_A = 0.447$, 0.708, and 0.879 and required symmetry. The plots show that the cgrb alloys are clustered. They are not random. The same inference is obtained from the AA and BB bond-atomic fraction plots.

The Cowley (Warren) short-range order parameter has been calculated as an average from the lattice points (111), (112), (122), (222), (212), (221), and (121) for cgrb alloys. It is found that $\alpha_1 = -0.20523$ for the alloy having $m_A =$ $= 0.879$, $\alpha_1 = 0.2865$ for $m_A = 0.708$, and $\alpha_1 = 0.10864$ for $m_A = 0.447$. Since they are essentially positive, clustering occurs in cgrb alloy. The inference is the same as above.

References

/1/ J.K. MUKHERJEE and S.K. PRASAD, J. Comp. Phys., submitted for publication.

/2/ C. BARRETT and T.B. MASSALSKI, Structure of Metals, McGraw Hill, Inc., New York 1966 (p. 289).

/3/ B.E. WARREN, X-Ray Diffraction, Addison-Wesley Publ. Co., Inc., 1969 (p. 229).

(Received September 10, 1981)

phys. stat. sol. (a) 68, K45 (1981)
Subject classification: 14.2; 21

Institute of Physics, Hanoi[1)]

Current-Induced Negative Resistance in Al-Al_2O_3-Al Superconducting Tunnel Junctions

By
H. K. PHAN

The phenomenon of negative resistance due to a linear dependence between the dimensions of the current-induced resistive region and voltage drop across long superconducting microbridges has attracted much attention, particularly in connection with the investigation of the transport properties of non-equilibrium superconducting films /1 to 4/. The I-U characteristics of such microbridges are usually S-shaped with discontinuities in the transition region /3/. Recently Eru et al. /4/ have observed new features of long superconducting microbridges with a normal bank: The absence of discontinuities and the continuous transition to the negative resistive section.

In the present work we report the results of a study of the I-U characteristics of Al-Al_2O_3-Al superconducting tunnel junctions of very low impedance. This study shows that a similar phenomenon of negative resistance can be observed in any symmetrical superconducting tunnel junction which has an impedance in the superconducting state lower than that in the normal state. Some other striking effects observed in such junctions are also presented.

The junction under consideration is prepared as follows: The lower and upper Al films with the same thickness of about 1000 Å are deposited in a vacuum of 2×10^{-6} Torr on aluminium single crystal substrate covered by a 5000 Å SiO layer. After the evaporation of the lower film, the Al oxide layer is made by thermal oxidation at room temperatures in an oxygen atmosphere of 250 Torr. The oxidation is continued for 1 min. The size of the junction is $0.1 \times 0.1\ mm^2$. The sample is then mounted in a He^3 cryostat which provides a temperature range from 0.58 to 4.2 K by means of a mechanical pumping system. A magnetic field produced by superconducting coils is applied parallel to the films and is used for control of the superconducting transition. The resistance

---

1) Hanoi, Vietnam.

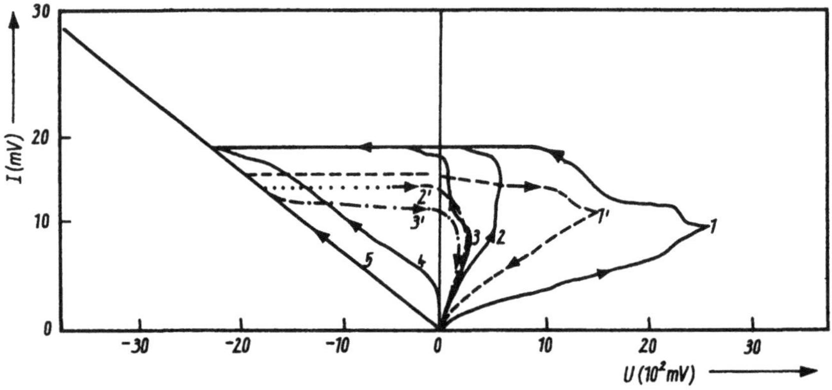

Fig. 1. Current-voltage characteristics of an Al-Al_2O_3-Al sandwich at different temperatures. (1) 0.57, (2) 0.9, (3) 1.15, (4) 1.25, (5) T ≥ 1.35 K

of the junction is about 1.5×10^{-2} Ω . It is defined as the ratio of the observed voltage drop at a given bias current which is fixed at 3 mA at 0.65 K (the transition temperature T_c is 1.25 K). The curves shown in Fig. 1 and 2 are dc I-U characteristics recorded for different reduced temperatures in the range 0.57 K ≤ T ≤ 1.35 K and for different magnetic fields in the range 0 ≤ B ≤ 500 G.

The main features exhibited in Fig. 1 and 2 are:

- the negative resistance in the superconducting transition region and the continuous transition from the positive to the negative section,
- the polarity reversal of the voltage drop across the junction when the junction is in the normal state,
- the considerable hysteresis of the I-U curves.

It should be noted that this anomalous behaviour of the I-U characteristics disappears if the impedance of the junction is high enough (above 10 Ω). The latter can be obtained by oxidation at higher oxygen pressure and for a longer time. This fact suggests that the effects mentioned above are related directly to the current-induced resistive phenomena in the superconducting films. This leads to a rise in the resistance of the film and therefore to a change in the distribution of the electric fields in the junction.

For a better understanding let us consider the electrical behaviour of a one-dimensional tunnel junction.

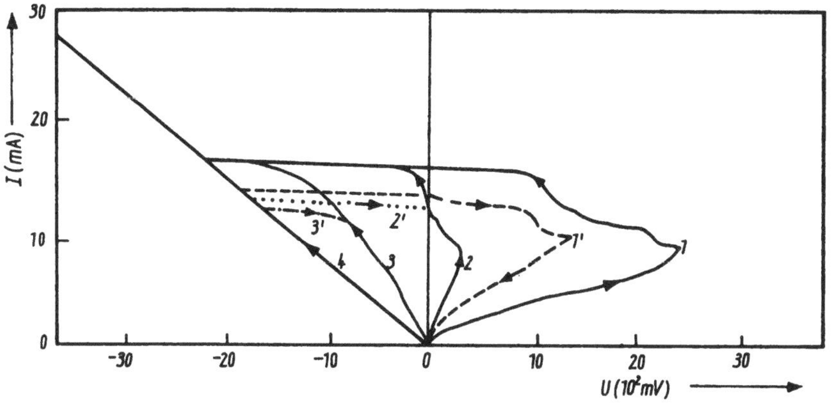

Fig. 2. Current-voltage characteristics of an Al-Al_2O_3-Al sandwich at different magnetic fields. T = 0.57 K; (1) B = 0, (2) 50, (3) 100, (4) ≧ 200 G

The equations governing the current transport and electric field distribution in the junction can be written as /5/

$$i_1 = -\frac{2L}{R_L}\frac{dU_1}{dx}, \qquad U_1 - U_2 = 2LR_s\frac{di_2}{dx},$$

$$i_2 = -\frac{2L}{R_L}\frac{dU_2}{dx}, \qquad i_m = i_1 + i_2, \tag{1}$$

where i_1 and i_2 are the currents through the lower and upper layers with the same resistance R_L; R_s is the differential resistance of the junction determined from the I-U curves in the superconducting state when R_L is assumed to be zero. The remaining parameters in (1) are defined in Fig. 3. With the Fourier-Laplace transformation, from (1) one obtains the resistance R_m of the junction:

$$R_m = \frac{U_m}{i_m} = R_s\alpha\,[\coth\alpha - \alpha], \qquad \alpha \equiv \left(\frac{R_L}{2R_s}\right)^{1/2}. \tag{2}$$

Expression (2) shows that a reversal of the voltage polarity in the superconducting tunnel junction may occur if its impedance R_s in the superconducting state is very small in comparison with the film resistance R_1 in the normal state. Since R_1 is a function of the bias current and magnetic field, it is possible to demonstrate the phenomenon of negative resistance and polarity reversal of the voltage drop by using the current-induced resistance in any symmetrical super-

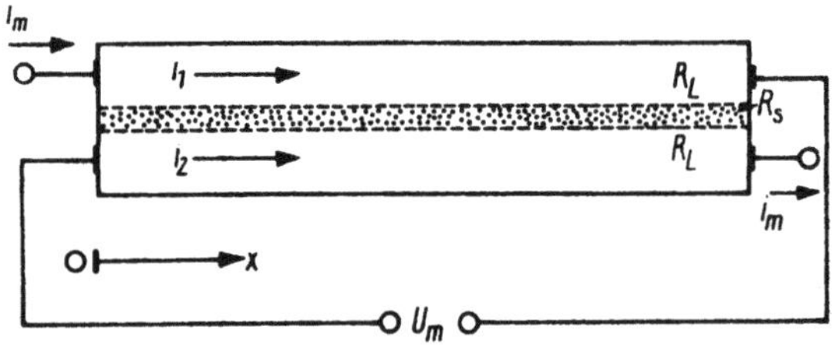

Fig. 3. Schematic arrangement of the one-dimensional tunnel junction

conducting tunnel junction which has a very low impedance. We believe that this is the case of our observation in Al-Al_2O_3-Al junctions.

Finally, we would like to point out here that an extreme hysteresis of the I-U characteristics at temperatures below the transition temperature T_c is observed. The arrows along the I-U curves in Fig. 1 and 2 indicate the development of hysteresis at different temperatures and magnetic fields. One of the possible interpretations of this hysteresis is that the effective critical current at a given T and B is reduced because of dissipation and heating effects /6, 7/.

The author wishes to thank the Groupe de Physique des Solides for their hospitality during the period in which this work was performed and Dr. J. P. Maneval and Dr. D. Huet for help with the instrumentation and for many useful discussions.

References

/1/ J. BINSLEV HANSEN, P. PERSEN, and P.E. LINDELEF, J. Physique c6/1, 500 (1978).

/2/ V.N. GUBANKOV, V.P. KOSHELETS, and G.A. OVSYANNIKOV, J. Physique c6/1, 535 (1978).

/3/ I.I. ERU, S.A. PESKOVATSKII, and A.V. POLADICH, Ukr. fiz. Zh. 20, 903 (1975).

/4/ I.E. ERU, S.A. PESKOVATSKII, and A.V. POLADICH, phys. stat. sol. (a) 53, K185 (1979).

/5/ D. HUET, Thesis, Paris 1972.

/6/ W.J. SKOSPOL, M.R. BEASLY, and M. TINKHAM, J. appl. Phys. 45, 4054 (1974).

/7/ W.H. PARKER, Phys. Rev. B 12, 3667 (1975).

(Received May 7, 1981)

phys. stat. sol. (a) 68, K49 (1981)

Subject classification: 14.3.3; 13.1; 22.1.2

Departement de Physique, Laboratoire de Physique Appliquée, Université de Perpignan[1])

Effective Electric Field in the n-Type Heavily Doped Emitter Region of Silicon p-n Junction Solar Cells

By

H. VAN CONG, S. BRUNET, and S. CHARAR

In our recent paper /1/, we calculated the effective intrinsic-carrier concentration n_{ie} of the n-type heavily doped emitter region of silicon p-n junction solar cells at thermal equilibrium, by using simplified models of parabolic bands,

$$n_{ie}^2 = n\,p = N_c N_v\, F_{1/2}\left(z = \frac{E_n}{kT}\right) \exp\left\{-\frac{E_g + E_n}{kT}\right\} = n_i^2 \exp\left(\frac{\Delta E_g + \Delta E_g'}{kT}\right). \quad (1)$$

n and p are the electron and hole concentrations, respectively, $F_k(z)$ is the Fermi integral function of order k, n_i is the intrinsic-carrier concentration of pure crystals which does not depend on the penetration of the Fermi level E_n into the parabolic conduction band and on the band-gap narrowing ΔE_g, and $\Delta E_g'$ is the "apparent" band-gap narrowing which represents the Fermi-Dirac statistics effect,

$$\Delta E_g' = kT \ln F_{1/2}(z) - E_n \; . \quad (2)$$

The aim of the present note is to investigate the built-in electric field E(x) (or the effective electric field) from (1) and (2).

In thermal equilibrium, E(x) can be defined as

$$E(x) \approx (D_p/\mu_p)\,(1/p_o)\,(dp_o/dx) \; , \quad (3)$$

where $D_p/\mu_p = kT/e$ for minority holes. From (1) and (3), one obtains

$$E(x) \approx (1/e)\, d(\Delta E_g + \Delta E_g')/dx - (kT/e)\,(1/n_o)(dn_o/dx) \approx$$
$$\approx -(kT/e)(1/n_o)(dn_o/dx) \left\{1 - (n/kT)\, d(\Delta E_g + \Delta E_g')/dn\right\}. \quad (4)$$

Furthermore, from (2) one has /2/

$$-(n/kT)\, d(\Delta E_g')/dn = \left(1 - \frac{F_{-1/2}}{F_{1/2}}\right)\frac{F_{1/2}}{F_{-1/2}} = \frac{F_{1/2}}{F_{-1/2}} - 1 \; . \quad (5)$$

---

1) Avenue de Villeneuve, F-66025 Perpignan Cédex, France.

Thus, (4) can be rewritten in following forms:

$$E(x) \approx -\frac{kT}{e}\frac{1}{n}\frac{dn_o}{dx}A(n) \quad , \tag{6.1}$$

$$E(x) \approx \frac{1}{e}\frac{d(\Delta E_g)}{dx} - \frac{F_{1/2}}{F_{-1/2}}\frac{kT}{e}\frac{1}{n_o}\frac{dn_o}{dx} \quad , \tag{6.2}$$

and

$$E(x) \approx -\frac{D_n}{\mu_n}\frac{1}{n_o}\frac{dn_o}{dx}\left\{1 - \frac{F_{-1/2}}{F_{1/2}}\frac{n}{kT}\frac{d(\Delta E_g)}{dn}\right\} , \tag{6.3}$$

where $A(n) = F_{1/2}/F_{-1/2} - (n/kT)d(\Delta E_g)/dn$ and $D_n/\mu_n = (kT/e)(F_{1/2}/F_{-1/2})$ for majority electrons /2/.

Some concluding remarks can be made here.

(i) In the non-degenerate case, ($E_n \ll 0$), one has /2/ $F_{1/2}/F_{-1/2} = eD_n/kT\,\mu_n = 1$, and thus $D_n/\mu_n = kT/e = D_p/\mu_p$. Equation (6.1) gives in this extreme case,

$$E(x) \approx -\frac{kT}{e}\frac{1}{n_o}\frac{dn_o}{dx}\left\{1 - \frac{n}{kT}\frac{d(\Delta E_g)}{dn}\right\} \equiv -\frac{kT}{e}\frac{1}{n_o}\frac{dn_o}{dx}A_S(n), \tag{7}$$

which is identical to the expression of E(x) recently given by Shibib et al. /3/, neglecting the Fermi-Dirac statistics effect in the n-type heavily doped emitter region of silicon p-n junction solar cells. This expression (7) is thus not satisfactory in n-type heavily doped semiconductors because in the non-degenerate case the Fermi level is located in the band gap.

(ii) In the highly degenerate case ($E_n > 4\,kT$), one has $F_k(z) \approx z^{k+1}/\Gamma(k)$. Thus, $F_{1/2}/F_{-1/2} \approx (E_n/kT)\Gamma(-1/2)/\Gamma(1/2) = 2E_n/3\,kT$ and therefore $D_n/\mu_n = (kT/e)(F_{1/2}/F_{-1/2}) \approx 2E_n/3e$.

Equation (6.1) is rewritten in the present case as

$$E(x) \approx -\frac{kT}{e}\frac{1}{n_o}\frac{dn_o}{dx}\left\{\frac{2E_n}{3kT} - \frac{n}{kT}\frac{d(\Delta E_g)}{dn}\right\} = -\frac{kT}{e}\frac{1}{n_o}\frac{dn_o}{dx}A(n) \,. \tag{8}$$

It is concluded that for a given temperature T, because of the extreme degeneracy condition ($F_{1/2}/F_{-1/2} \approx 2E_n/3\,kT \gg 1$ or $E_n > 4\,kT$), while the band-gap narrowing effect ΔE_g decreases the factor A(n), the Fermi-Dirac statistics effect represented by the term $2E_n/3\,kT$ increases it with increasing electron concentration n.

This conclusion is also valid for the factor A(n) given from (6.1) in the case of n-type heavily doped semiconductors at high temperature ($E_n/kT > 1$), because of the condition $F_{1/2}/F_{-1/2} > 1$ (see Table 1).

(iii) The above expression (6.2) of E(x) is in fact identical to that deduced from the transport equation (4.5) of a recent paper of Marshak and van Vliet /2/, if in their paper a change in the shape of the density-of-states function originally represented by $(\Gamma_n^o + \Gamma_p^o)$ can be neglected.

In other words, if taking into account this effect, (1) is rewritten as

$$n_{ie}^2 = n_i^2 \exp\left(\frac{\Delta E_g + \Delta E_g' + \Gamma_n^o + \Gamma_p^o}{kT}\right) . \tag{9}$$

(iv) Finally, from (6.3) one notes that the holes experience a retarding electric field represented by $-(D_n/\mu_n)(1/n_o)(dn_o/dx)$ only at sufficiently low values of n and sufficiently high temperatures T (i.e., $\Delta E_g = 0$). These physical conditions are identical to that given in the above non-degenerate case (i), in which $F_{-1/2}/F_{1/2} = 1$. It is shown that this retarding electric field effect cannot occur in our present model of heavily doped semiconductors; this conclusion is different from that of Shibib et al. /3/.

We now compare our results of A(n) with theoretical results of $A_S(n)$ given by Shibib et al. /3/, choosing n-type heavily doped silicon crystals at T = 300 K, as a typical example. Shibib et al. /3/ used a recent model of ΔE_g given by Lanyon and Tuft /4/,

$$\Delta E_g = \Delta E_g^{LT} \approx 2.6\text{x}10^{-20} (n/10^{26})^{1/6} \text{ (in J)} , \tag{10}$$

to calculate the factor $A_S(n)$. This expression (10) for the band-gap narrowing was recently discussed by Mahan /5/, and also in our recent paper /1/.

To calculate the factor A(n) given from (6.1) and (8), we use our recent result for ΔE_g /1/.

$$\Delta E_g = E_c^e + 2 E_o , \tag{11}$$

where the averaged exchange energy E_c^e is

$$E_c^e \approx 4.6\text{x}10^{-26} n^{1/3} \varepsilon^{-1} \text{ (in J)}$$

and the effective energy characteristic of band tails is

$$E_o = 6.1\text{x}10^{-34} (m_o/m_n')(n_t/g_c)^{2/3} \text{ (in J)} .$$

Table 1

n (10^{25} m^{-3})	$\frac{E_n}{kT}$	$\frac{F_{1/2}}{F_{-1/2}}$ /7/	$A_S(n)$	A(n)	
				from (6.1)	from (8)
5	1.1	1.567	0.068	-0.022	
7	1.6	1.757	0.014	-0.042	
10	2.2	2.020	-0.047	0.040	
50	7.1	4.975	-0.369	1.796	1.560

The meaning of other physical quantities is given in /1/. For simplicity, we also use the result of E_o which is approximately proportional to $n^{5/12}$ /6/. That gives

$$(n/kT)d(\Delta E_g)/dn = (1/3\ kT)\left(E_c^e + \frac{5E_o}{2}\right). \tag{12}$$

The numerical result of E_c^e, E_o, and E_n for various high values of n were already given in /1/. Thus, the values of $A_S(n)$ from (7) using the expression (10) and our results of A(n) from (6.1) and (8) for E_n > kT and E_n > 4 kT, respectively, are tabulated in Table 1 for different high values of n.

Some remarks are given here.

(i) The results of $A_S(n)$ illustrate well a reduction of E(x), including only the band-gap narrowing ΔE_g^{LT} and excluding the Fermi-Dirac statistics effect.

(ii) One notes from our results of A(n) that for 5×10^{25} $m^{-3} \leq n \leq 10^{26}$ m^{-3} the band-gap narrowing effect is more important than the Fermi-Dirac statistics effect, for $n \approx 10^{26}$ m^{-3} there is a compensation between these two effects (i.e., $A \approx 0$), and for $n > 10^{26}$ m^{-3} the Fermi-Dirac statistics effect is more important than the band-gap narrowing.

(iii) Finally, for $n = 5\times10^{26}$ m^{-3}, our results of A(n) from (6.1) and (8) are almost equal, because of a very high value of E_n/kT (= 7.1 > 4) - or of the highly degenerate case.

We would like to thank the reviewer of the original manuscript for his helpful suggestions which have greatly improved our results.

References

/1/ H. VAN CONG, phys. stat. sol. (a) 65, 683 (1981).
R. A. ABRAM, G. J. REES, and BL. H. WILSON, Adv. Phys. 27, 799 (1978).
/2/ A. H. MARSHAK and K. M. VAN VLIET, Solid State Electronics 21, 429 (1978).
/3/ M. A. SHIBIB, F. A. LINDHOLM, and F. THEREZ, IEEE Trans. Electron Devices 26, 959 (1979).
/4/ H. P. D. LANYON and R. A. TUFT, IEEE Trans. Electron Devices 26, 1014 (1979).
/5/ G. D. MAHAN, J. appl. Phys. 51, 2634 (1980).
/6/ H. VAN CONG, J. Phys. Chem. Solids 36, 1237 (1975); 41, 1231 (1980); 42, 95 (1981); 42, 193 (1981).
/7/ J. S. BLAKEMORE, Semiconductor Statistics, Pergamon Press, Oxford/ London/New York/Paris 1962 (p. 352 to 353).

(Received September 18, 1981)

phys. stat. sol. (a) 68, K55 (1981)
Subject classification: 1.5; 22.2.4

Institute of Radioengineering and Electronics, Czechoslovak Academy of Sciences, Prague[1]

RF Sputtering of $Al_xIn_{1-x}N$ Thin Films

By
K. STAROSTA

The preparation of binary compounds of $A^{III}B^{V}$ nitrides by sputtering is well known /1 to 4/. However, very few studies on the quasibinary systems of these compounds have been published /4 to 7/. Up to now sputtering has been used only for the BN-AlN system and only a small degree of solubility of BN in AlN has been found /4/. On the other hand, some of the III-V ternary semiconducting alloys have been prepared by sputtering /8/.

Here we report on the preparation of a new material - a ternary system of indium nitride and aluminium nitride. Thin films of these compounds were prepared by reactive multitarget sputtering. The metal targets (indium and aluminium) were simultaneously sputtered in a reactive atmosphere and the substrates rotated under the targets at a constant rate. The multitarget sputtering gives a possibility to obtain the alloys over the entire composition range with one set of the targets when deposition rates are controlled.

The sputtering apparatus used for the preparation of these films is the same as described previously /9/, with modification for sputtering of both targets and for substrate rotation /10/. The table carrying the substrate holders was able to rotate at rates ranging from 0.003 to 0.4 s^{-1}. The incident power was split to the targets through a power splitting network which made possible the control of the deposition rates at a desired ratio.

The targets were water-cooled discs with diameter of 100 mm, with a purity better than 5 N and were shielded to prevent enhanced sputtering and cross-contamination. As sputtering gas purified nitrogen was used, its operating pressure was 2.7 Pa. Films were deposited mostly at ambient temperature or at a temperature of 400 °C. As substrates microscope glass slides were taken, which were carefully cleaned before insertion into the vacuum system.

---

1) Lumumbova 1, 18251 Praha 8, ČSSR.

Prior to the deposition of $Al_xIn_{1-x}N$ films the targets were sputter-cleaned under the same conditions, but without substrate rotation. The substrate rotation was mostly 0.12 s^{-1}.

Both binary compounds have the same structure (crystallize in the wurtzite structure), but the formation of $Al_xIn_{1-x}N$ solid solutions is complicated due to the considerably different lattice constant (AlN: a = 0.3111 nm, c = 0.4980 nm and InN: a = 0.3544 nm, c = 0.5718 nm /1, 2, 7/). Osamura et al. /7/ found that in chemical equilibrium the $Ga_xIn_{1-x}N$ system did not form a complete solid solution and had a solid-phase miscibility gap (the lattice constants of GaN are about 3% greater than that of AlN). Despite this, $Al_xIn_{1-x}N$ solid solutions were prepared over the entire composition range. This is probably due to the fact that sputtering is a physical rather than a chemical deposition technique and so the deposition of the film needs not run through the conditions of thermodynamical equilibrium.

The formation of $Al_xIn_{1-x}N$ solid solutions was proved by X-ray diffractometry, RHEED experiments, and infrared spectroscopy /10/. A linear composition dependence of lattice constants with one structural phase was found for all layers deposited at sufficiently high substrate rotation rates. For these samples also one infrared transmittivity minimum was observed and its wavenumber has changed linearly with the composition. The film composition was determined by atomic absorption spectrofotometry.

All $Al_xIn_{1-x}N$ films deposited on amorphous substrates were polycrystalline. The texture was found to be the fiber texture with different preferred orientations parallel to the substrate surface. The orientation type depended on the sputtering conditions as well as on the composition of the film. The greatest degree of perfection showed the (0001) preferred orientation parallel to the surface.

The author is grateful to Dr. J. Kohout, who carried out the electron diffraction experiments and to Dr. V. Rosická for X-ray diffraction.

References

/1/ A.J. NOREIKA and M.H. FRANCOMBE, J. Vac. Sci. Technol. 6, 194 (1969).

/2/ H.J. HOVEL and J.J. CUOMO, Appl. Phys. Letters 20, 71 (1972).

/3/ N. PUYCHEVRIER and M. MENORET, Thin Solid Films 36, 141 (1976).

/4/ A. J. NOREIKA and M. H. FRANCOMBE, J. Vac. Sci. Technol. 6, 722 (1969).

/5/ B. BARANOV, L. DÄWERITZ, V. B. GUTAN, G. JUNGK, H. NEUMANN, and H. RAIDT, phys. stat. sol. (a) 49, 629 (1978).

/6/ J. HAGEN, R. D. METCALFE, D. WICKENDEN, and W. CLARK, J. Phys. C 11, L143 (1978).

/7/ K. OSAMURA, S. NAKA, and Y. MURAKAMI, J. appl. Phys. 46, 3432 (1975).

/8/ J. E. GREENE, C. E. WICKERHAM, and J. L. ZILKO, Thin Solid Films 32, 51 (1976).

/9/ K. STAROSTA, J. ZELINKA, D. BERKOVÁ, and J. KOHOUT, Thin Solid Films 61, 241 (1979).

/10/ K. STAROSTA, Thesis, Tech. Univ. Prague, 1980.

(Received July 17, 1981)

phys. stat. sol. (a) 68, K59 (1981)
Subject classification: 14.2; 21

College of General Education, Kyushu University[1)]

Quench-Condensed Thin Superconducting Pb Films Investigated by Electron Tunneling and Resistance Measurement

By
T. KAWAGUTI and Y. FUJIMORI

Amorphous or highly disordered metal films are formed by condensation of a vapour onto a substrate at a low temperature. This quench-condensation has influence on the superconducting properties. The Pb film is thought to be built up in a micro-crystalline structure by condensation onto the usual substrate even at liquid He temperature in contrast to Bi and Ga films in amorphous structure /1 to 3/. It was, however, shown that ultra thin Pb films quench-condensed onto nucleating monolayers of Mo or W had annealing curves of the electrical resistivity, characteristic of amorphous materials /4/. In this work, we have investigated by electron tunneling and resistance measurement whether quench-condensed thin Pb films make the transition from amorphous to a micro-crystalline structure with increasing thickness. The measurement of electron tunneling into the superconducting Pb film gives information on the phonon spectrum since Pb is a strong-coupling superconductor.

Tunnel junctions (Al-I-SiO-Pb) having fairly high resistance of 200 to 600 Ω/mm^2 were prepared. SiO of about 10 Å average thickness was deposited over the insulating oxide layer at the surface of the Al film electrode, prior to the quench-condensation of Pb (99.999%), after the substrate was cooled down to liquid He temperature in a combination cryostat-evaporator evacuated to $4x10^{-9}$ Torr. The Pb films on previously deposited SiO become continuous at very small thickness /5/. The film thickness was measured with a quartz-crystal thickness monitor. The deposition rate of Pb was 200 to 300 Å/min. The thickness dependence of the resistivity and the tunneling characteristic were measured by successive condensation of Pb on the previous film. Tunneling characteristics were measured by bridge techniques and harmonic detection. An ac voltage modulation was kept constant by the use of feedback /6/.

---

1) Ropponmatsu, Fukuoka 810, Japan.

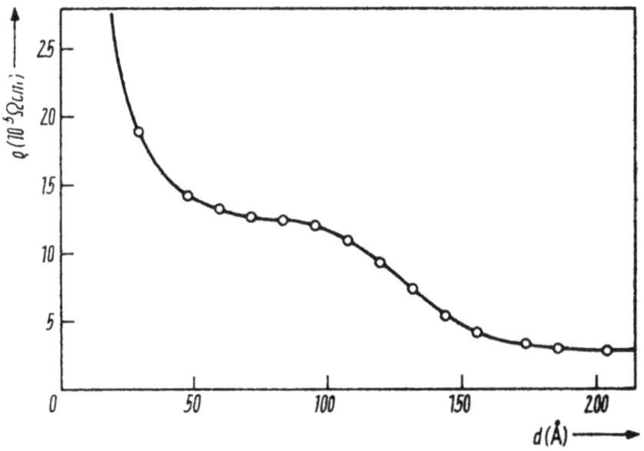

Fig. 1. Thickness dependence of the electrical resistivity of Pb film quench-condensed on previously deposited SiO

The broad resistive transition of Pb films to the superconducting state is in good agreement with the mean-field theory of thermodynamic fluctuations, by which the superconducting transition temperature T_c was determined. The residual resistivity of as-deposited Pb films onto previously deposited SiO is shown as a function of thickness in Fig. 1. Electrical continuity is obtained at the condensation of ≈ 12 Å Pb. The resistivity is scarcely dependent on thickness around 80 Å and decreases with increasing thickness. This thickness dependence of the resistivity suggests that the change takes place from amorphous or highly disordered structure to a crystalline one around 100 Å. A similar dependence was observed by Strongin et al. /5/. The T_c of as-deposited Pb films thinner than 100 Å decreases in inverse proportion to thickness. It becomes higher than the value extraporated from this linear dependence when increasing the thickness above 100 Å where the resistivity shows the kink in Fig. 1.

Fig. 2 shows the second derivative d^2I/dU^2 of the tunneling current versus dc bias voltage of an Al-I-SiO-Pb junction with different thicknesses of Pb. Here, a 90 Å thick Pb film shows a very smeared curves characteristic of amorphous superconducting films /2, 7/. The increase in the thickness to 210 Å causes a change in the second derivative curve around 4 and 8 mV. This indicates that the transverse and longitudinal peaks of the phonon density of states become sharp due to the change from amorphous to a micro-crystalline structure. This second derivative curve remains qualitatively unchanged when

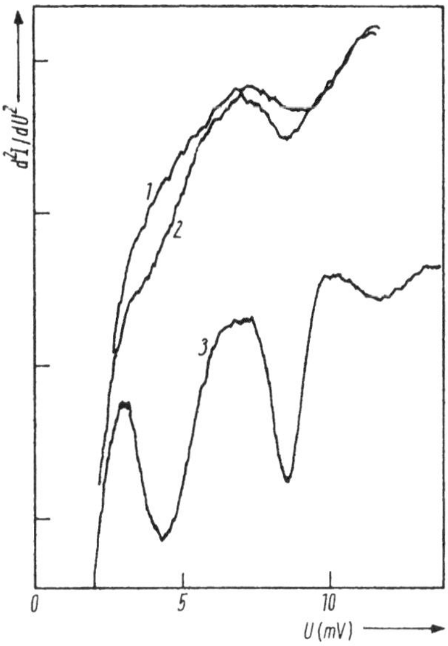

Fig. 2. Second derivative d^2I/dU^2 versus dc bias voltage of an Al-I-SiO-Pb tunnel junction with different thicknesses of Pb at 2.9 K. The bias voltage was counted from the superconducting energy gap of Pb film; (1) as-deposited Pb film of 90 Å, (2) as-deposited Pb film of 210 Å, (3) after annealing quench-condensed 1400 Å Pb film at 300 K. The zero of curve 3 is offset vertically

increasing the thickness to 1400 Å. After annealing this film up to 300 K, the d^2I/dU^2 versus U curve has sharp peaks peculiar to crystalline Pb. The condensation of 70 to 120 Å in average thickness was necessary for the electrical continuity of Pb films quench-condensed without the predeposition of SiO and these junctions showed the tunneling characteristics of micro-crystalline Pb as shown previously /1/. The anomaly due to the quantum size effect was observed in the tunneling characteristics of films thinner than 40 to 50 Å /8/, which was out of the thickness region where the d^2I/dU^2 versus U curve and the resistivity showed a noticeable thickness dependence in this work. The resistivity of very thin as-deposited Pb films shows a sharp decrease at a definite temperature during annealing, which indicates the transformation of the amorphous structure into a crystalline one. The temperatures become higher as the film thickness decreases /4/. The resistivity of the ultra thin film is affected by the patchy structure whose contribution should be distinguished from the increase due to lattice disorder. This study has shown by effective means of electron tunneling that Pb could be quench-condensed in "amorphous" structure below 100 Å in thickness.

A low substrate temperature reduces the atomic mobility during thin film growth, so that the number of nucleation centers increases. The predeposition of a very small amount of SiO may strengthen this tendency like Bi_2O_3 on glass substrates /9/, which increases the degree of disorder in the films and decreases the minimum thickness for the electrical continuity. Lazarev et al. showed that the amorphous state is stable only in films thinner than the critical thickness, for example, 600 Å for Bi and 50 Å for Fe /10/. The formation of new phases in thin films is due to considerable effects of the surface energy in the phase equilibrium conditions /11/. Therefore, Pb films could be formed in "amorphous" structure up to the critical thickness only by quench-condensation onto a specific surface.

We would like to thank Dr. R. Aoki for providing facilities for these experiments.

References

/1/ N.V. ZAVARITSKII, Soviet Phys. - J. exper. theor. Phys., Letters 6, 155 (1967).

/2/ K. KNORR and N. BARTH, J. low-Temp. Phys. 4, 469 (1971).

/3/ H. BÜLOW and W. BUCKEL, Z. Phys. 145, 141 (1956).

/4/ W. RÜHL and P. HILSCH, Z. Phys. B26, 161 (1977).

/5/ M. STRONGIN, R.S. THOMPSON, O.F. KAMMERER, and J.E. CROW, Phys. Rev. B 1, 1078 (1970).

/6/ A.F. HEBARD and P.W. SHUMATE, Rev. sci. Instrum. 45, 529 (1974).

/7/ C.G. GRANQVIST and T. CLAESON, Thin Solid Films 16, 65 (1973).

/8/ H. SIXL, J. GROMER, and H.C. WOLF, Z. Naturf. 29a, 319 (1974).

/9/ K.L. CHOPRA, Thin Film Phenomena, McGraw-Hill Publ. Co., New York 1969 (p. 172).

/10/ B.G. LAZAREV, V.M. KUZMENKO, A.I. SUDOVTSOV, and V.M. MELNIKOV, Soviet Phys. - Doklady 15, 846 (1971).

/11/ A.I. BUBLIK and B.YA. PINES, Dokl. Akad. Nauk SSSR 87, 215 (1952).

(Received June 30, 1981)

phys. stat. sol. (a) 68, K63 (1981)
Subject classification: 14.3.1 and 20.2; 22.2.1

Applied Physics Devision, AAEC Research Establishment, Sutherland[1)]

Laser Assisted Doping of n-GaAs from Sn Evaporated Films

By
S.J. PEARTON and E.M. LAWSON

Considerable effort has been directed towards the doping of GaAs by laser melting of thin evaporated layers /1/, usually to produce low resistance ohmic contacts. Difficulties have arisen in the control of various dopant layers owing to differing reflectivities and melting points. In addition, excess laser energy may result in the loss of As from the surface and produce problems associated with non-stoichiometry and Ga precipitation, although the evaporated layer may tend to reduce this effect. Previous work has described the formation of ohmic contacts by the laser irradiation of a double evaporated layer of Ag and Sn /2/. In the present work, a technique for laser-assisted doping from thin single layers of Sn is demonstrated. The contacts have remained stable to current densities of 300 mA/cm^2, with maximum diameters of 16 mm. It may be possible to produce larger area contacts by using multiple laser shots in an overlapping pattern.

The material was nominally undoped n-type GaAs, both bulk ($n = 2.2x \times 10^{16}$ cm^{-3}) and epitaxial (LP, $n = 1.8 \times 10^{16}$ cm^{-3}). Polished surfaces were obtained by lapping the samples on a slurry of 600 # SiC grit on glass, and etching for 3 min in a mixture of 3 HNO:2 H_2O:1 HF. Thin films of Sn, generally $\leq$ 50 nm thick, were evaporated onto the samples at $\leq 5 \times 10^{-6}$ Torr.

A Q-switched ruby laser ($\lambda = 0.6943$ μm, FWHM 25 ns) was used to irradiate the samples. The beam was homogenized by passing it through a quartz or Perspex light pipe of the type described by Cullis et al. /3/. Power densities after homogenization were in the range 80 to 240 MW cm^{-2} and uniform over diameters 7 to 16 mm. Evidence that the sample surface had melted during irradiation was sought by observing the typical /4/ melt pattern finish - an irregularly rippled 'orange-peel' appearance.

---

1) Private Mail Bag, New Illawarra Road, Sutherland 2232, Australia.

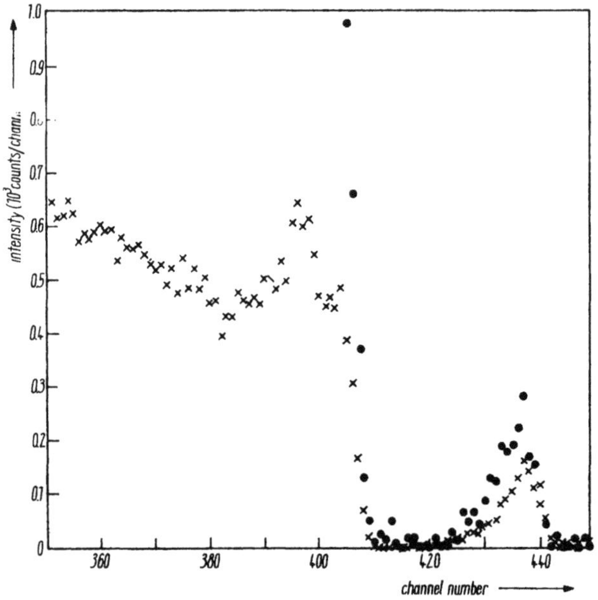

Fig. 1. RBS results from laser melted, Sn-doped sample showing random (●) and aligned (×) spectra, obtained at 170° scatter angle

Irradiations were performed both in air and in a gas cell with an inert (Ar) atmosphere. No effects due to the environment such as oxygen incorporation were discerned. After irradiation, excess Sn was removed by soaking for 15 min in HCl acid. The samples were then examined under the optical microscope for evidence of melting. A preliminary check for doping was performed with a thermal probe based on the Seebeck effect. The ten samples fabricated by the described procedure were all found to be heavily doped in the laser melted regions. Current-voltage (I-U) characteristics were recorded for some of these samples (measured between two laser doped spots of area 0.05 cm^2) and found to be ohmic for current densities $\leq$ 300 mA/cm^2. This current density is approximately an order of magnitude lower than that attained in actual discrete devices (e.g. junction field effect transistors) using diffused or ion-implanted contacts. The poorer performance of these laser melted contacts could be due to laser-induced lattice damage or the effect of excess dopant incorporated into the lattice, producing strain. Further study will clarify this. The degree of ohmicity can be determined by measuring the correlation coefficient from a least square analysis; coefficients of better than 0.997 were obtained. The contact sheet resistance, 3.8 Ω/□, was obtained from I-U characteristics recorded between evaporated gold layers placed a fixed distance apart on the laser doped region. It is not possible to compare this to previous experiments /2/. It should be noted that the contact areas in this work are several orders of magnitude larger.

The Sn laser doped contacts were further investigated by the Rutherford backscattering (RBS) of 2 MeV He^+ ions. Fig. 1 shows a set of RBS spectra obtained at 170°. A comparison can be made of aligned yields both off and on the Sn contact. Off the contact, χ_{min} is poorer than the values obtained by Kular et al. /5/, who laser annealed ion implanted GaAs, perhaps suggesting that our power densities may have been excessive.

The Sn penetration depth was estimated to be 53 nm from the FWHM of the impurity peak. This gives a lower limit for the melt depth assuming uniform incorporation of the Sn in the melt and no segregation. The melt depth is expected to be a function of both the laser output power density and the thickness of the evaporated Sn layer. Furthermore, the total Sn solubility (averaged over the penetration depth) is estimated to be $3.5x10^{20}$ cm^{-3} and the substitutional solubility to be $1.6x10^{20}$ cm^{-3}. It is not known what fraction of Sn is electrically active.

Using a scanning electron microscope (SEM) in the X-ray fluorescence mode we confirmed the Sn incorporation in the melt region. However, the sensitivity was such that the Sn X-ray peaks were only just visible above the background.

In an attempt to produce a p-type doped region on n-GaAs the procedure was repeated using Zn instead of Sn. The thermal probe indicated heavy doping, and more obvious Zn X-ray peaks were detected using the SEM. However, the Zn distribution was non-uniform, possibly because of a non-uniform evaporation. The junction I-U characteristics for these samples were markedly non-linear. These Zn-doping investigations will be continued.

In conclusion it has been demonstrated that thin, highly doped layers can be produced on n-GaAs by laser assisted doping from a thin film of Sn. There is some evidence that the upper limit oi the laser power density may have been excessive. It is probable that the best contacts will only be formed within narrow ranges of laser power density and film thickness.

The authors acknowledge Mr. A. Rose who performed the laser irradiations, and Mr. M. Scott who carried out the backscattering measurements. Mr. K.

Watson kindly provided the SEM services. The technical assistance of Mr. P.J. Lee and Mr. A.A. Williams is appreciated as is the continued interest and encouragement of Dr. A.J. Tavendale. Mr. S.J. Pearton acknowledges the support of an Australian Institute of Nuclear Science and Engineering (AINSE) postgraduate studentship.

References

/1/ G. ECKHARDT, in: Laser and Electron Beam Processing of Materials, Ed. WHITE and PEERCY, Academic Press, 1980 (p. 467 to 480).

/2/ R.S. POUNDS, M.A. SAIFI, and W.C. HAHN, Solid State Electronics 17, 245 (1974).

/3/ A.C. CULLIS, H.C. WEBBER, and P. BAILEY, J. Phys. E 12, 688 (1979).

/4/ G.BADERTSCHEV, R.P. SALATHE, and W. LATHY, Electronic Letters 16, 113 (1980).

/5/ S.S. KULAR, B.J. SEALY, M.H. BADAWI, K.S. STEPHEN, D. SADANA, and G.R. BOOKER, Electronics Letters 15, 413 (1979).

(Received July 16, 1981)

phys. stat. sol. (a) 68, K67 (1981)

Subject classification: 18.1; 18.2; 14.1; 21.1; 21.1.1

School of Science, Third Laboratory of Physics,
Aristotelean University of Thessaloniki[1)]

On the Atomic Ordering of Fe-Rich, $Fe_{1-x}Ni_x$ Binary Alloys

By

I. A. TSOUKALAS

Introduction It is a disputed matter whether at the iron-rich end of the $Fe_{1-x}Ni_x$ binary alloy system, the atomic ordering of the Fe_3Ni type does exist.

Several experimental methods have been so far applied in order to investigate this possibility, with contradictory results.

The methods used are neutron scattering /1 to 4/, X-ray diffraction /5 to 7/, high-energy particle irradiation /8/.

The present paper reports results of the electric resistivity and magnetic moment behaviour of several samples of the $Fe_{1-x}Ni_x$ binary alloy system.

Experimental procedures The samples used were prepared from 4N9 purity starting material, molten under high vacuum in an induction furnace. The initial ingots were remelted several times to ensure homogeneity.

Electrical resistivity measurements were conducted on rectangular shaped samples (6x0.5x0.1 cm^3), whereas the samples for magnetic measurements were small spheres of 3 mm diameter.

All samples used were initially annealed at 1323 K for 24 h in evacuated silica tubes and then quenched in cold water. After that, extended annealing was performed ranging from 1 to 1500 h at 823 K. The samples were quenched in cold water. Some of them were quenched in liquid nitrogen.

Heat treatments will be assigned to each sample in the text. X-ray and chemical analysis were used for the checking sample homogeneity.

A conventional four probe dc technique using a set-up described elsewhere /9/ was used to perform electrical resistivity measurements.

Magnetic measurements were conducted by means of a Foner-type vibrating sample magnetometer.

Experimental results and discussion The electrical resistivity behaviour of some of the samples examined is shown in Fig. 1a and b.

---

1) Thessaloniki, Greece.

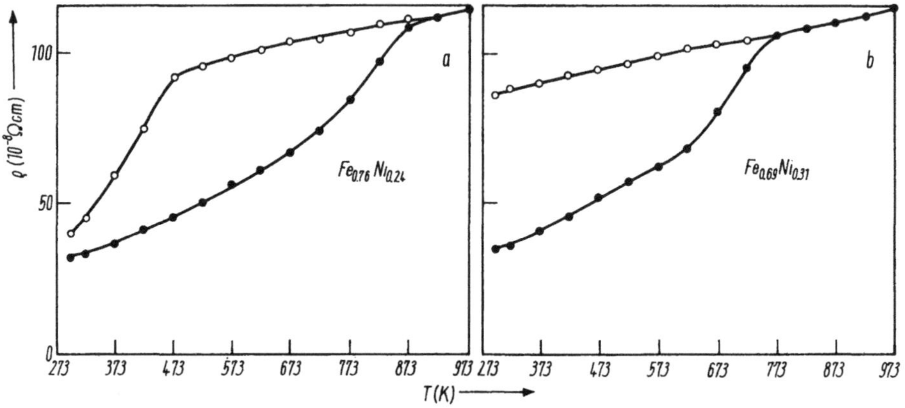

Fig. 1. Temperature dependence of the electrical resistivity of the a) $Fe_{0.76}Ni_{0.24}$ and b) $Fe_{0.69}Ni_{0.31}$ alloy, annealed at 823 K for 1500 h quenched in a) cold water and b) liquid nitrogen; ● heating, o cooling

Besides the martensite to austenite transformation which was produced for a 6 K/min cooling rate, no indication of Fe_3Ni-type ordering was found.

Should such an ordered state had been formed, a pronounced minimum in the resistivity versus composition curve should have appeared.

For the composition range (x = 0.21 to 0.31) we examined, no such variation resulted even for the most carefully heat treated samples.

The magnetic measurements were conducted with an external field of 1.2 T able to bring the samples into technical saturation. The spontaneous magnetization and Curie temperatures were estimated by means of Arrott plots /10/.

Samples quenched at a temperature inferior to their M_s temperature have shown irreversible changes in their magnetization upon cooling implying the presence of both ($\alpha + \gamma$) phases. Such a typical behaviour is shown in Fig. 2 for a $Fe_{0.69}Ni_{0.31}$ sample.

The variation of the reduced magnetization against reduced temperature of all the samples examined, shows that the mean field theory curve is not obeyed.

In Fig. 3 this behaviour is depicted for the $Fe_{0.69}Ni_{0.31}$ sample. The theoretical curve for J = 1/2 is also included.

Kouvel /11/ proposed an atomistic model in which only the first nearest-neighbor interactions are taken into account. Following Kouvel one can write for the individual atomic magnetic moments of Fe and Ni of the $Fe_{1-x}Ni_x$ alloy

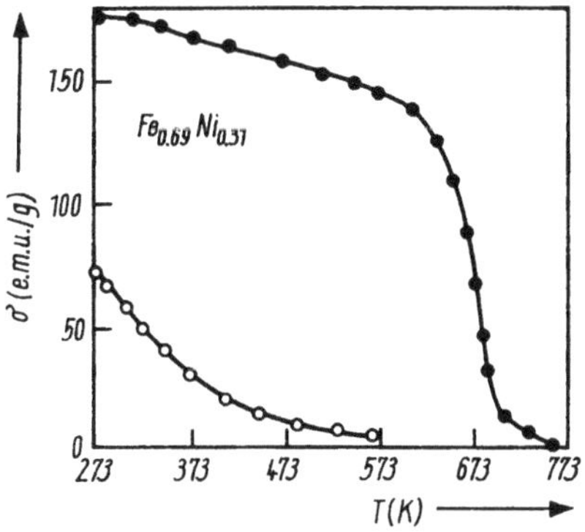

Fig. 2. Temperature dependence of the magnetic moment for the $Fe_{0.69}Ni_{0.31}$ alloy, annealed at 823 K for 1500 h quenched in liquid nitrogen; ● heating, o cooling

system

$$\mu_{Ni}(n_{Fe}) = a'_o + a'_1 \frac{n_{Fe}}{N} + a'_2 \frac{n_{Fe}(n_{Fe}-1)}{N(N-1)} ,$$

$$\mu_{Fe}(n_{Ni}) = b'_o + b'_1 \frac{n_{Ni}}{N} + b'_2 \frac{n_{Ni}(n_{Ni}-1)}{N(N-1)} ,$$

where $\mu_{Ni}(n_{Fe})$ is the atomic magnetic moment of a Ni atom which has n iron atoms as nearest neighbors and $N = n_{Ni} + n_{Fe}$ is the total number of iron and nickel nearest neighbors.

The coefficients a'_o to a'_2, b'_o to b'_2 were determined by our experiments and were found to be $a'_o = 0.55$, $a'_1 = 0.37$, $a'_2 = -0.36$, $b'_o = 2.20$, $b'_1 = 0.8$, $b'_2 = 0$.

In the case of Cu_3Au-type ordering for the Fe_3Ni sample (x = 0.24) each Fe atom has 4 Ni atoms and 8 Fe atoms as nearest neighbors, whereas each Ni atom has 12 Fe atoms and no Ni atom as nearest neighbor.

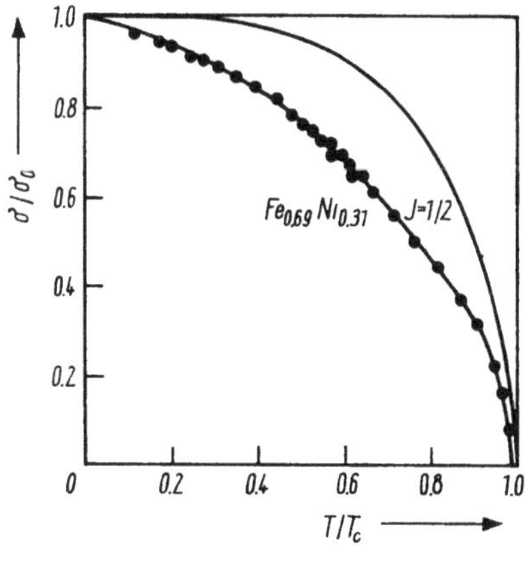

Using Kouvel's approximation the magnetic moment for the ordered state of Fe_3Ni comes out to be $\bar{\mu} = 2.01\ \mu_B$ against our experimental value of $\bar{\mu} = 2.15\ \mu_B$.

Fig. 3. Reduced magnetization against reduced temperature for the $Fe_{0.69}Ni_{0.31}$ alloy, annealed at 823 K for 1500 h quenched in liquid nitrogen

Thus in the ordered state of Fe_3Ni a decrease of 6% should be detected, which was not found in our experiments irrespective of the heat treatment given.

Such a decrease in the magnetic moment is thought to be due to an excess in Fe-Fe interactions which are antiferromagnetic in nature /12/. Preliminary transmission electron microscopy experiments give no evidence of superlattice formation.

We have therefore to conclude as a result of the experiments we have conducted on samples carefully heat treated, that no evidence of an ordered state in the Fe-rich end of the binary Ni-Fe alloy system is observed, as far as their electrical and magnetic properties are concerned.

References

/1/ A. Z. MENSHIKOV, S. K. SIDOROV, and V. E. ARKHIPOV, Zh. eksper. teor. Fiz. 61, 311 (1971); Soviet Phys. - J. exper. theor. Phys. 34, 163 (1972).

/2/ C.G. SHULL and M. K. WILKINSON, Phys. Rev. 97, 304 (1955).

/3/ M.F. COLLINS, R.V. JONES, and R.D. LOWDE, J. Phys. Soc. Japan 17, Suppl. B 111, 19 (1962).

/4/ B.G. LIASCHENKO, D.F. LITRIN, I.M. PUZEI, and YU.G. ABOV, Soviet Phys. - Cryst. 2, 59 (1957).

/5/ G. HAUSCH and H. WARLOMONT, Phys. Letters A 36, 415 (1971).

/6/ L.I. LYSAK and S.A. ARTEMYUK, Russ. Metall. 5, 116 (1973).

/7/ V.G. VEERARAGHAVEN, C.F. EAGEN, and P.G. WINCHELL, J. appl. Phys. 47, 4768 (1976).

/8/ P.L. GRUZIN, YU.L. RODIONOV, and V.A. PRYAKHIN, Dokl. Akad. Nauk SSSR 251, 1384 (1980); Soviet Phys. - Doklady 25, No. 4 (1980).

/9/ S. CHADJIVASILIOU, I.A. TSOUKALAS et al., Mater. Res. Bull. 15, 41 (1980).

/10/ A. ARROTT, Phys. Rev. 108, 1394 (1957).

/11/ J.S. KOUVEL, Magnetism and Metallurgy, Vol. 2, Academic Press, New York 1969 (Chap. 11, p. 523).

/12/ S.K. SIDOROV and A.V. DOROSHENKO, phys. stat. sol. 16, 737 (1966).

(Received September 18, 1981)

phys. stat. sol. (a) 68, K71 (1981)

Subject classification: 13.4 and 19; 22.8.1

Laboratoire de Physique des Solides, associé au C.N.R.S., Université Paul Sabatier, Toulouse[1] (a) and
Laboratoire de Chimie des Solides et des Hautes Pressions, E.R.A. C.N.R.S. 263, Tolouse[2] (b)

ESR of the F^+ Centre in B-Type Carbonated Hydroxyapatite

By

G. BACQUET (a), VO QUANG TRUONG (a), M. VIGNOLES (b), and G. BONEL (b)

B-type carbonated apatites, where the anion sites of the channels are occupied either by hydroxyl ions or by fluorine ions, have a structure isomorphous to that of the hydroxyapatite /1/. In these compounds, CO_3^{2-} ions substitute for PO_4^{3-} ions necessitating the existence of accompanying defects to insure the charge balance inside the lattice.

It was suggested in the case of B-type carbonated apatites prepared in a Na^+-free medium, that the substitution of a phosphate ion by a carbonate ion was associated with the appearance of vacancies in a Ca^{2+} and a OH^- (or F^-) adjacent site /2 to 4/.

In the case of B-type carbonated fluorapatites (B-CFAp) the evolution of the composition of these compounds was also accounted for by substituting groups (CO_3^{2-} + F^-) for PO_4^{3-} ions /5/.

Our previous X-band ESR studies have recently shown /6/ that these two substitution possibilities concurrently took place in B-CFAp. Effectively in powdered samples heated at 450 °C for 30 min, X-irradiated, and then annealed at 150 °C we have observed an "orthorhombic" spectrum which was attributed to the F^+ centre (electron trapped by the oxygen vacancy associated to the carbonate ion). Its intensity began to increase, reached a maximum, and then decreased with increasing amount of carbonate in the specimens. The spectral intensity was lower with B-CFAp prepared in a medium rich in fluorine ("direct" apatites) than in those obtained from a medium poor in fluorine ("reverse" apatites).

The same kind of ESR experiments were undertaken with B-type carbonated hydroxyapatites (B-COHAp) prepared following the method described by Labarthe et al. /2/. "Direct" and "reverse" B-COHAp were so obtained.

---

1) 118, Route de Narbonne, 31062 Toulouse Cédex, France.

2) 38, Rue des 36 Ponts, 31400 Toulouse, France.

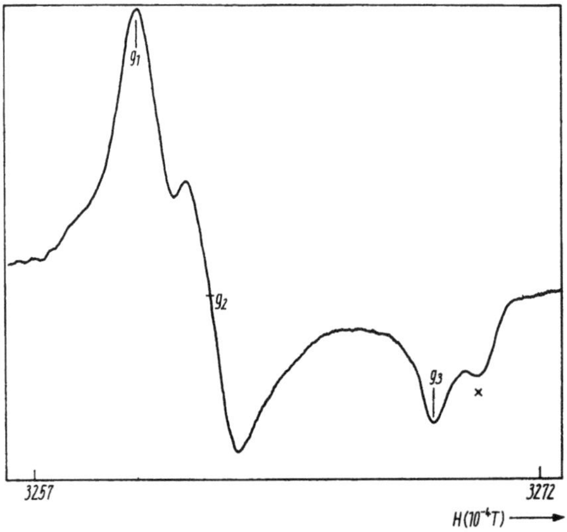

Fig. 1. X-band ESR spectrum of the F^+ centre in B-COHAp recorded at 223 K. The line marked by × is due to an unknown defect

The powdered samples were first heated at 450 °C for 30 min to eliminate adsorbed and zeolitic water. All the irradiations were made with an X-ray generator running at 50 kV, 10 mA, the time of exposure being 30 min. To destroy the paramagnetic defects having a poor thermal stability (specially the hole-trapped species) the powders were then annealed at 150 °C for 30 min.

With our ESR spectrometer, we record the first derivative of the absorption curve, and we call it spectrum.

As in B-CFAp we have observed the same "orthorhombic" spectrum (Fig. 1) due to the F^+ centre, with almost the same g-tensor components (Table 1 and Fig. 1) and with the same behaviour for the spectral intensity

Table 1

g tensor components for the F^+ centre in B-type carbonated apatites

compounds	g_1	g_2	g_3	⟨g⟩	T(K)
B-CFAp	2.0022 ± ±0.0001	2.0008 ± ±0.0001	1.9971 ± ±0.0001	2.0000 ± ±0.0001	80
B-COHAp	2.0028 ± ±0.0002	2.0013 ± ±0.0002	1.9975 ± ±0.0002	2.0005 ± ±0.0002	114

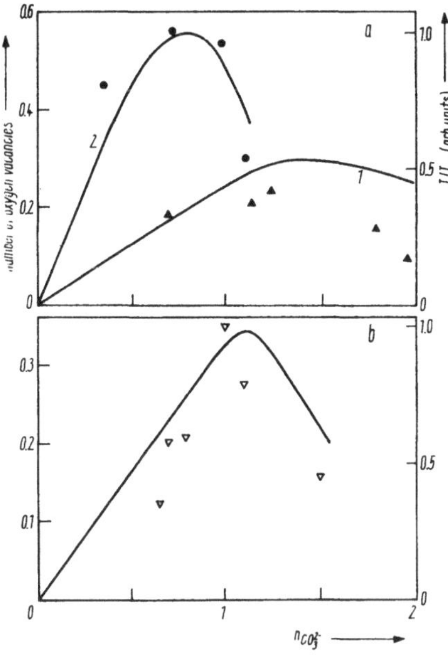

Fig. 2. Plot of the number of oxygen vacancies (curves) and the ratio I/I_{max} (points) for the intensity of the symmetrical line of the spectrum, as a function of the number of carbonates in the apatitic cell; a) curve 1 and black triangles correspond to "reverse" B-CFAp; curve 2 and black circles correspond to "direct" B-CFAp; b) "reverse" B-COHAp

(Fig. 2). In the case of "direct" B-COHAp the presence of a strong signal superimposed to that of the F^+ centre has not permitted to precisely measure the intensity of this last.

For each sample, the amount of each of the constituents was determined by chemical analysis.

In Fig. 2 are plotted, as a function of the carbonate amount, on the one hand, the spectral intensity and on the other hand, the number of oxygen vacancies calculated from the formula

$$Ca_{10-y+u}(PO_4)_{6-y}(CO_3)_{y-u}(CO_3X)_uX_{2-y+u} ,$$

with X = F^- or OH^- and assuming that the total number of PO_4^{3-} and CO_3^{2-} ions is always equal to 6 in the cell. In this formula y corresponds to the substitution mode involving one oxygen vacancy and u to that without such a vacancy. It can be seen that the greater the number of oxygen vacancies, the greater the spectral intensity.

Our ESR experiments thus show that in both B-CFAp and B-COHAp two mechanisms of charge compensation concurrently take place,i. e.

- substitution of a PO_4^{3-} ion by a CO_3^{2-} ion (existence of oxygen vacancy), and simultaneously apparition of a vacancy in a Ca^{2+} and a F^- (or OH^-) adjacent sites;
- CO_3^{2-} + F^- (or OH^-) substitute for PO_4^{3-} (no oxygen vacancy).

References

/1/ M.I. KAY, R.A. YOUNG, and A.S. POSNER, Nature 204, 1050 (1964).

/2/ J.C. LABARTHE, G. BONEL, and G. MONTEL, Ann. Chim. 8, 289 (1973).

/3/ G. BONEL, J.C. LABARTHE, and C. VIGNOLES, C.R. Coll. Internat. C.N.R.S. 230, 117 (1973).

/4/ C. VIGNOLES, Third Cycle Thesis, Toulouse 1973, unpublished.

/5/ M. VIGNOLES and G. BONEL, C.R. Acad. Sci. (France) C 287, 321 (1978).

/6/ G. BACQUET, VO QUANG TRUONG, G. BONEL, and M. VIGNOLES, J. Solid State Chem. 33, 189 (1980).

(Received August 7, 1981)

phys. stat. sol. (a) 68, K75 (1981)
Subject classification: 14.3; 13.1; 22

Institute of Radio Physics and Electronics,
University College of Science and Technology, Calcutta[1)]

Generalized Einstein Relation for the Diffusivity-Mobility Ratio in Multi-Band Degenerate Semiconductors

By
B. R. NAG, A. N. CHAKRAVARTI, and P. K. BASU

The Einstein relation for the diffusivity-mobility ratio (D/μ) of the charge carriers is modified /1, 2/ for degenerate semiconductors to

$$\frac{D}{\mu} = \frac{k_B T}{|e|} \left(n \bigg/ \frac{dn}{d\eta} \right) , \tag{1}$$

where e is the charge of the carrier, T is the lattice temperature, n is the carrier concentration, k_B is the Boltzmann constant, and η is the ratio of the Fermi energy E_F and k_BT. The relation has been used extensively in the literature /3 to 16/ to evaluate and study the characteristics of the D/μ ratio in degenerate semiconductors with single energy bands and also with multiple energy bands. The multiple energy bands may be the characteristics of the many-valley semiconductors with equivalent valleys as in Si and Ge or with non-equivalent valleys separated by small energy gaps as in GaSb. Multiple energy bands may also be formed in single-valley semiconductors by applying quantizing magnetic fields, or strong electric fields to produce quantized inversion layers or in size quantized thin films and superlattices. Relation (1) has been used to study the D/μ ratio for all the above kinds of materials.

The modified Einstein relation was first reported by Landsberg /17/ for a degenerate parabolic band with a single valley and has later been generalized /2/ from the Boltzmann transport equation for single-band materials considering the non-parabolicity of the bands and all kinds of scattering mechanisms. Nevertheless, the validity of the relation in multi-band semiconductors under situations where the edge of the conduction band is dependent on position (as in inversion layers) has not yet been examined. It is, therefore, necessary to derive the relation in its most generalized form for multi-band semiconductors. We give in

---

1) 92, Acharya Prafulla Chandra Road, Calcutta 700009, India.

this note the required derivation by assuming an isotropic band structure for the sake of simplicity. It may be pointed out here that the diffusion coefficient like the mobility may be a tensor under various physical conditions. For example, the diffusion coefficient is a tensor for electrons in the individual valleys of a multi-valley semiconductor in the bulk /18/ or under size and electrical quantization. It is a tensor also in high electric fields under hot-electron conditions /19/ and in quantizing magnetic fields. However, the D/μ ratio for low electric fields under various physical conditions of quantization which is the subject of this paper is not expected to be a tensor as it is determined only by the relation between the electron concentration and the Fermi energy. Our treatment here assuming an isotropic band would, therefore, be generally applicable.

The energy-wave vector relation for the carriers of the i-th band in multiband semiconductors may, in general, be written as

$$E_{ki} = \frac{\hbar^2 k^2}{2\, m_i^*(E)} + E_i = E_{ki} + E_i \ , \tag{2}$$

where E_i is the carrier energy in the i-th band for k tending to zero and $E_{ki} = \hbar^2 k^2/2m_i^*$, $m_i^*(E)$ being the effective mass. In many-valley semiconductors, E_i is the band-edge energy E_c or E_v and in multi-band cases it is the energy of the sub-band, i.e. $(i + \frac{1}{2})\hbar\omega_B$ (ω_B being the cyclotron resonance frequency) in the presence of a quantizing magnetic field and E_1, E_2, etc. in quantized inversion layers or size quantized thin films. We also note that k represents k_z, the component of the wave-vector in the direction of the magnetic field for magnetically quantized levels and it represents the transverse component of the wave-vector in two-dimensional systems as in inversion layers or thin films etc. The equilibrium distribution function of the carriers in the i-th band is

$$f_{oi} = \left\{\exp\left[(E_{ki} + E_i - E_F)/k_B T\right] + 1\right\}^{-1} . \tag{3}$$

The carrier concentration n_i in the i-th band is given by

$$n_i = \frac{1}{4\pi^3}\int f_{oi}\, d\vec{k} = n_i(E_{Fi}) \ , \tag{4}$$

where $E_{Fi} = (E_F - E_i)$. Thus, it is indicated that n_i is, in general, a function of E_{Fi}. We may show /20/, starting from the Boltzmann transport equation that the conduction current J_{ci} contributed by the carriers in the i-th band due to an electric field $\mathcal{E}$ in the z-direction is given by

$$J_{ci} = - \frac{1}{4\pi^3} \frac{e^2 \cdot \mathcal{E}}{\hbar^2} \int (\nabla_{k_z} E)^2 \tau_i(E) \frac{\partial f_{oi}}{\partial E_{ki}} d\vec{k}$$

$$= |e| \; n_i \mu_i \mathcal{E} , \qquad (5)$$

where μ_i is the mobility by definition and $\tau_i(E)$ is the relaxation time. For scattering mechanisms for which the relaxation time approximation is not applicable, the same expression remains valid /2/ but $\tau_i(E)$ is replaced by $\Phi_i(E)$, the perturbation in the distribution function defined by

$$f_i = f_{oi} - \frac{e \mathcal{E}}{\hbar} (\nabla_{k_z} E) \frac{\partial f_{oi}}{\partial E_{ki}} \Phi_i(E) . \qquad (6)$$

The total conduction current is then

$$J_c = |e| \sum_i n_i \mu_i \mathcal{E} = |e| \; \mu n \mathcal{E} , \qquad (7)$$

where μ is the average mobility of the carriers.

We also get a diffusion current when the carrier concentration varies with position. We assume that it varies in one particular direction, say the z-direction. We note further that, under these conditions, E_F as well as E_i will, in general, be functions of z. Following the same procedure as used for obtaining the expression for J_c from the Boltzmann transport equation, we can express the diffusion current contributed by the carriers in the i-th band as

$$J_{Di} = - \frac{1}{4\pi^3} \frac{e}{\hbar^2} \int (\nabla_{k_z} E)^2 \tau_i(E) \frac{\partial f_{oi}}{\partial z} d\vec{k} . \qquad (8)$$

We can also write /21/

$$\frac{\partial f_{oi}}{\partial z} = \frac{\partial f_{oi}}{\partial E_{Fi}} \frac{\partial E_{Fi}}{\partial z} = - \frac{\partial f_{oi}}{\partial E_{ki}} \frac{\partial E_{Fi}}{\partial z} \qquad (9)$$

and

$$\frac{\partial n}{\partial z} = \frac{\partial}{\partial z} \sum_i n_i(E_{Fi}) = \frac{\partial E_{Fi}}{\partial z} \Theta_i , \qquad (10)$$

where

$$\Theta_i = \sum_j \frac{\partial n_j(E_{Fj})}{\partial E_{Fj}} \frac{\partial E_{Fj}}{\partial E_{Fi}} .$$

Replacing $\partial f_{oi}/\partial z$ by using (9) and (10), we then get

$$J_{Di} = \frac{1}{4\pi^3} \frac{e}{\hbar^2} \frac{\partial n}{\partial z} \int (\nabla_{k_z} E)^2 \tau_i(E) \frac{\partial f_{oi}}{\partial E_{ki}} \Theta_i^{-1} d\vec{k}$$

$$= - \frac{e}{|e|} \frac{\partial n}{\partial z} n_i \mu_i \Theta_i^{-1} . \qquad (11)$$

Hence, the total diffusion current is given by

$$J_d = \sum_i J_{di} = - \frac{e}{|e|} \frac{\partial n}{\partial z} \sum_i n_i \mu_i \Theta_i^{-1} = -e D \frac{\partial n}{\partial z} . \qquad (12)$$

Thus, we finally get

$$D = \frac{1}{|e|} \sum_i n_i \mu_i \Theta_i^{-1} \qquad (13)$$

and

$$\frac{D}{\mu} = \frac{1}{|e|} n \frac{\sum_i n_i \mu_i \Theta_i^{-1}}{\sum_i n_i \mu_i} . \qquad (14)$$

Expression (14) is the most generalized form of the Einstein relation which will be valid for all kinds of scattering mechanisms and also for all kinds of band structures. It may also be noted that (14) reduces to (1) when the E_i's are independent of z. The E_i's are independent of z in uniform multivalley semiconductors with equivalent or non-equivalent valleys. These are independent of z also when the subbands are produced by quantizing magnetic fields. This is a valid approximation even for size quantization as in thin films /22/, provided the contribution of the charge density to the internal potential is negligible. The E_i's cannot, however, be considered as independent of z for quantized levels in inversion layers /23/ since these are functions of the carrier concentration which varies in the z-direction. For such layers, (14) will have to be used as such in the case of multi-subband occupation. However, in the electric quantum limit in these layers when i = j = 0, (14) will be simplified /8/ as

$$\frac{D}{\mu} = \frac{1}{|e|} n \Big/ \frac{dn}{d(E_F - E_0)} , \qquad (15)$$

E_0 being the quantized energy of the lowest electric subband.

We, therefore, conclude that (1) is valid for different kinds of multi-band semiconductors except for inversion layers and thin layers in heterostructures with very large charge densities /24/. For these exceptional cases the detailed expression (14) should be used to compute the D/μ ratio.

References

/1/ E. SPENKE, Electronic Semiconductors, McGraw-Hill Publ. Co., New York 1958.

/2/ B.R. NAG, Electron Transport in Compound Semiconductors, Springer Series, Solid State Sciences, Vol. 11, Springer-Verlag, Heidelberg 1980.

/3/ A.N. CHAKRAVARTI and D.P. PARUI, Canad. J. Phys. 51, 491 (1973).

/4/ D. MUKHERJI, A.N. CHAKRAVARTI, and B.R. NAG, phys. stat. sol. (a) 26, K27 (1974).

/5/ P.N. BUTCHER, A.N. CHAKRAVARTI, and S. SWAMINATHAN, phys. stat. sol. (a) 25, K47 (1974).

/6/ A.N. CHAKRAVARTI and B.R. NAG, Internat. J. Electronics 37, 281 (1974).

/7/ A.N. CHAKRAVARTI and B.R. NAG, phys. stat. sol. (a) 22, K153 (1974).

/8/ A.N. CHAKRAVARTI, A.K. CHOWDHURY, K.P. GHATAK, and D.R. CHOUDHURY, Acta phys. Polon. 58A, 251 (1980).

/9/ K.P. GHATAK, A.K. CHODHURY, S. GHOSH, and A.N. CHAKRAVARTI, Appl. Phys. 23, 241 (1980).

/10/ A.N. CHAKRAVARTI, K.P. GHATAK, A. DHAR, K.K. GHOSH, and S. GHOSH, Appl. Phys. (1981), to be published.

/11/ F.A. LINDHOLM and R.W. AYERS, Proc. IEEE 56, 371 (1968).

/12/ N.G. NILSSON, phys. stat. sol. (a) 19, K75 (1973).

/13/ B.A. ARONZON and E.Z. MEILIKHOV, phys. stat. sol. (a) 19, 313 (1973) and references cited therein.

/14/ B.K. KONONENKO, Zh. prikl. Spektroskopii 23, 354 (1975).

/15/ K.M. VAN VLIET and A.H. MARSHAK, phys. stat. sol. (b) 78, 501 (1976) and references cited therein.

/16/ P.T. LANDSBERG and S.A. HOPE, Solid State Electronics 20, 421 (1977) and references cited therein.

/17/ P.T. LANDSBERG, Proc. Roy. Soc. A213, 226 (1952).

/18/ E.M. CONWELL, Phys. Rev. 127, 1493 (1952).

/19/ R.E. ROBSON, J. Phys. B 9, L337 (1976).

/20/ B.R. NAG, Theory of Electrical Transport in Semiconductors, Pergamon Press, Oxford 1972.

/21/ R. STRATTON, IEEE Trans. Electron Devices 19, 1288 (1972).

/22/ R. DINGLE, Festkörper-Probleme, Vol. XV, Pergamon-Vieweg, 1975 and references cited therein.

/23/ F. STERN, Phys. Rev. B 5, 4891 (1972).

/24/ H.L. STÖRMER, Proc. Internat. Conf. Phys. Semicond., Phys. Soc. Japan, Kyoto 1980.

(Received August 12, 1981)

phys. stat. sol. (a) 68, K81 (1981)

Subject classification: 14.4.1; 18.4; 22.8.1

Department of Radiation Investigations,
Academy of Sciences of the Azerbaidzhan SSR, Baku[1)]

Investigation of the Magnetoelectric $(ME)_H$ Effect in Solid Solutions of the Systems $BiFeO_3$-$BaTiO_3$ and $BiFeO_3$-$PbTiO_3$

By

I.H. ISMAILZADE, R.M. ISMAILOV, A.I. ALEKBEROV, and F.M. SALAEV

In previous works /1 to 5/ it has been shown that the magnetoelectric $(ME)_H$ effect is a high-sensitive "detector" for the discovery of the mutual influence of electric and magnetic sub-systems in ferroelectric(antiferroelectric)-antiferromagnetic crystals, and for the determination of the electric ordering type in the electrically-magnetically ordered crystals. The electric polarization P_{in} induced by the external magnetic field H, and, particularly, the critical magnetic field strength H_c, at which the spin-flop phenomenon, i.e. the magnetic phase transition occurs, are highly sensitive parameters of the $(ME)_H$ effect. Thus, on the basis of the dependences P_{in}(mol%$BiFeO_3$) and H_c(mol%$BiFeO_3$) of the system $BiFeO_3$-$Pb(Fe_{1/2}Nb_{1/2})O_3$ it has been unambiguously established that in any case at positive temperatures $BiFeO_3$ as a whole is antiferroelectric possessing a polar space group /1/. Undoubtedly the magnetoelectric study of the other $BiFeO_3$-ABO_3 systems in which ABO_3 is also a perovskite-type ferroelectric material is of interest. In connection with this in the present work the $(ME)_H$ effect has been measured in solid solutions of the systems $BiFeO_3$-$BaTiO_3$ and $BiFeO_3$-$PbTiO_3$.

According to /6/ in the system $BiFeO_3$-$BaTiO_3$ (A) a continuous series of solid solutions is formed existing at room temperature in three modifications: in the interval of 100 to 67 mol% $BiFeO_3$ in the rhombohedral, from 67 to 7.5 mol% $BiFeO_3$ in the cubic, and below this concentration in the tetragonal phases.

The system $BiFeO_3$-$PbTiO_3$ (B) has been studied in /7/, and the formation of a continuous series of solid solutions was found possessing in the region from 100 to 70 wt% $BiFeO_3$ a rhombohedral and below 70 wt% $BiFeO_3$ a tetragonal symmetry. The authors have assumed that both modifications are ferroelectric ones.

---

1) Baku-143, Azerbaidzhan SSR, USSR.

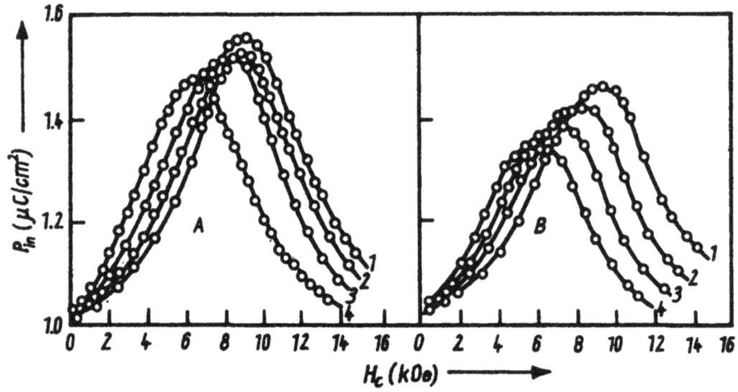

Fig. 1. The dependences P_{in}(H) for solid solutions of the systems $BiFeO_3$-$BaTiO_3$ (A) and $BiFeO_3$-$PbTiO_3$ (B) with 95 mol% $BiFeO_3$ at different temperatures; system (A): (1) 25, (2) 42, (3) 50, (4) 125 °C; system (B): (1) 25, (2) 60, (3) 100, (4) 140 °C

The solid solutions of systems (A) and (B) have been prepared by the usual ceramic technology from $BaCO_3$, Fe_2O_3, and PbO of "pure for analysis" grade, TiO_2 and Bi_2O_3 of "chemical purity" grade. Depending upon the $BiFeO_3$ contents the preliminary firing was carried out at temperatures from 650 °C to 800 to 850 °C (2 h), and the final one from 800 to 850 °C to 950 to 1000 °C (1 h).

The magnetoelectric $(ME)_H$ measurement technique has been described in /1/. The piezoelectric effect was examined using the reflecting galvanometer M1 7 with a sensitivity of 10^{-9} A.

The dependence of the induced electric polarization P_{in} upon the external magnetic field strength, P_{in}(H), has been measured at a few temperatures. Fig. 1 represents P_{in}(H) for solid solutions of the systems (A) and (B) with 95 mol% $BiFeO_3$ obtained at different temperatures. As can be seen, the shape of these curves is characteristic of the spin-flop phenomenon occurring at H_c. With increase of temperature the values of P_{in} and H_c regularly decrease. A remarkable property of the dependence H_c(T) is its strict linearity that allows on the basis of four to five experimental points of H_c(T) by extrapolation to H_c = = 0 to determine the Néel temperature T_N of a given antiferromagnet within 30 to 40 min with an accuracy of $\pm$2 to 3 °C /8/. The Néel temperatures of solid solutions of the systems (A) and (B) determined by this method are plotted in Fig. 2 (curves 1); the Néel temperatures taken from /6/ were included (curves

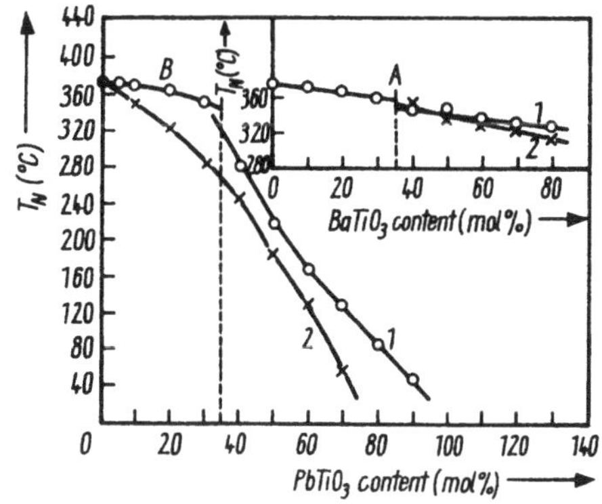

Fig. 2. The Néel temperatures of solid solutions of the systems $BiFeO_3$-$BaTiO_3$ (A) and $BiFeO_3$-$PbTiO_3$ (B); o from $(ME)_H$ measurements, × according to data of /6/

2). As is seen, the T_N values obtained by us for the $BiFeO_3$-$BaTiO_3$ system are in good agreement with those given in /6/, whereas for the system (B) a considerable divergence occurs which is pronouncedly manifested in solid solutions with 70, 80, and 90 mol% $PbTiO_3$. For example, extrapolation of the data /6/ gives for the solid solution with 80 mol% $PbTiO_3$ T_N = -25 °C, whereas according to the $(ME)_H$ data for this material T_N = +85 °C. On the other hand, observation of a remarkable magnetoelectric signal P_{in} at room temperature is clear evidence of the fact that the Néel temperatures of these solid solutions are higher than room temperature.

The dependences of P_{in} and H_c upon mol% $BiFeO_3$ at a constant temperature are of particular interest. The dependences P_{in} (mol% $BiFeO_3$) and H_c(mol% $BiFeO_3$) for systems (A) and (B) obtained at room temperature are given in Fig. 3. For the system $BiFeO_3$-$BaTiO_3$ (A) these dependences consist of two regions: (I + II + III), and IV which sharply differ from each other. Beginning with 100 mol% $BiFeO_3$, with decreasing $BiFeO_3$ content the value of H_c, which equals 5 kOe for pure $BiFeO_3$, strongly increases and for the solid solution with ≈ 93 mol% $BiFeO_3$ (region IV) reaches ≈ 11.7 kOe. A further decrease of the $BiFeO_3$ content leads to an opposite behaviour: the magnitude of H_c decreases at first moderately, and then beginning approximately with 65 mol% $BiFeO_3$ it diminishes more strongly, and at 9 mol% $BiFeO_3$ decreases abruptly. It is interesting to note that the boundaries of H_c(mol% $BiFeO_3$) (thick lines at the bottom of Fig. 3) are in a good agreement with those established by the X-ray method /6/ (thin lines). An analogous behaviour is exhibited by the $BiFeO_3$-$PbTiO_3$ (B) system as well (Fig. 3).

To explain the obtained magnetoelectric data let us consider the earlier known structure and dielectric data for these systems.

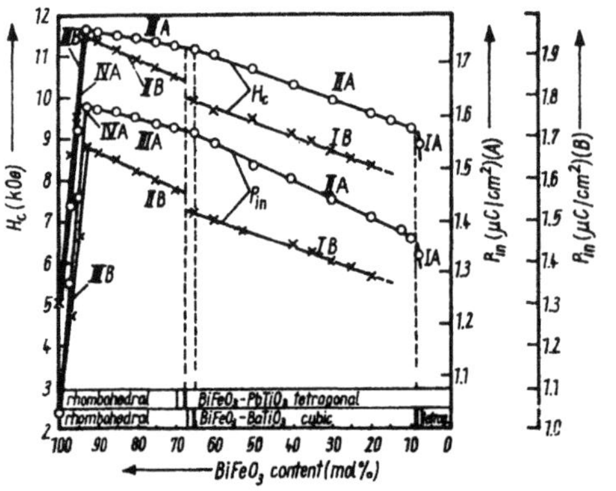

Fig. 3. The dependences H_c(mol% $BiFeO_3$) and P_{in} (mol%$BiFeO_3$) for the systems (A) and (B); at the bottom of the figure the phase boundaries established in /6, 7/ (thin lines), and determined by $(ME)_H$ measurements (thick lines) are given

In /6/ it was possible to measure the dependence $\varepsilon(T)$ of the system (A) only on several solid solutions from the $BaTiO_3$ side. Mainly on the basis of X-ray data the authors of /6/ consider that the rhombohedral solid solutions (from 100 to 67 mol% $BiFeO_3$) are ferroelectrics. However, in so far as in this region the magnetic ordering type remains invariable (antiferromagnetic), then, as in the $BiFeO_3$-$Pb(Fe_{1/2}Nb_{1/2})O_3$ system /1/, the observed opposite character of the H_c variation in regions IIIA and IVA can be explained only by the difference of the electric ordering types of these regions. The order of magnitude of H_c and the character of the H_c variation with $BiFeO_3$ content in regions IIIA + IIA and IV are identical with those of the regions 100 to 93, and 93 to 55 mol% $BiFeO_3$, respectively, of the system $BiFeO_3$-$Pb(Fe_{1/2}Nb_{1/2})O_3$ /1/, in which solid solutions between 93 and 55 mol% $BiFeO_3$ are ferroelectrics. Therefore, in the system $BiFeO_3$-$BaTiO_3$ an abrupt decrease of H_c with increase of the $BiFeO_3$ content (region IV) is also stipulated by the appearance of antiparallel (antiferroelectric) ordering in the unit cell at $\gtrsim$ 95 mol% $BiFeO_3$. Thus, the magnetoelectric study of solid solutions in the systems $BiFeO_3$-$BaTiO_3$ and $BiFeO_3$-$PbTiO_3$ (Fig. 3, regions IIB and IIB), confirms our earlier drawn conclusion /1/ that in any case at positive temperatures (T <

< 850 °C) $BiFeO_3$ as a whole is antiferroelectric possessing, like the antiferroelectric $PbZrO_3$, a polar space group.

The fact that in the system $BiFeO_3$-$BaTiO_3$ in the range between 35 and 92 mol% $BiFeO_3$ the values of H_c are equal to ≈ 9 to 11 kOe is of special interest, because these values are characteristic of the ferroelectric state. At the same time the polarized solid solutions of this concentration range show a piezoelectric effect indicating absence of a symmetry centre in these materials. Therefore, the above mentioned solid solutions are ferroelectrics, and therefore they cannot possess cubic symmetry, as was stated in /6/.

In conclusion it should be noted that the phase boundaries established by X-ray measurements /6/ and discovered by a magnetoelectric study in the systems $BiFeO_3$-$BaTiO_3$ and $BiFeO_3$-$PbTiO_3$ distinctly appear on the T_N(mol% $BaTiO_3$) and T_N(mol% $BiFeO_3$) curves obtained by us as well (Fig. 2).

The authors wish to express their sincere thanks to Mr. E.S. Safarov for his help in the preparation of the manuscript of the present paper.

References

/1/ I.H. ISMAILZADE, R.M. ISMAILOV, A.I. ALEKBEROV, and F.M. SALAEV, phys. stat. sol. (a) 57, 99 (1980).

/2/ I.H. ISMAILZADE, R.M. ISMAILOV, A.I. ALEKBEROV, and F.M. SALAYEV, Abstracts IV. European Meeting Ferroelectricity, Portoroz (Yugoslavia), September 1979 (p. 276).

/3/ I.H. ISMAILZADE and R.M.ISMAILOV, phys. stat. sol. (a) 59, K191 (1980).

/4/ I.H. ISMAILZADE, R.M. ISMAILOV, and T.M. STOLPAKOVA, phys. stat. sol. (a) 65, K25 (1981).

/5/ I.H. ISMAILZADE, R.M. ISMAILOV, A.I. ALEKBEROV, and F.M. SALAEV, phys. stat. sol. (a) 66, 119 (1981).

/6/ YU.N. VENEVTSEV, G.S. ZHDANOV, YU.E. ROGINSKAYA, S.A. FEDULOV, V.V. IVANOV, V.V. CHKALOVA, A.S. VISKOV, A.G. KAPYSHEV, V.S. KONDARENKO, and I.B. LADYZHINSKII, Izv. Akad. Nauk SSSR, Ser. fiz. 28, 683 (1964).

/7/ S.A. FEDULOV, YU.N. VENEVTSEV, G.S. ZHDANOV, E.G. SMAZHEVSKAYA, and I.S. REZ, Kristallografiya 7, 77 (1962).

/8/ I.H. ISMAILZADE, R.M. ISMAILOV, I.M. ALIEV, A.I. ALEKBEROV, F.M. SALAEV, and A.M. HABIBOV, J. Magnetism magnetic Mater. 22, 125 (1980).

(Received August 24, 1981)

phys. stat. sol. (a) 68, K87 (1981)

Subject classification: 1.5; 3; 20; 22.1.3

Institut für Experimentalphysik, Universität Innsbruck[1)]

Optical Loss in Selenium Thin Films as a Measure of Crystallinity

By

K. HARNISCH, S.M. NEUBACHER, and O. ENNEMOSER

Although the transmission spectrum of trigonal selenium single crystals shows only a slightly higher absorption in the MIR region (λ = 2.5 to 50 μm) compared with the amorphous samples /1/, the transmittivity of a-Se thin films is drastically reduced by the appearance of crystalline aggregates in the thin films. With an increasing content of these aggregates the characteristic absorption peaks of the amorphous phase (λ = 13.5 and 20.4 μm) disappear and the overall transmittance in the region λ = 2.5 to 10 μm is reduced regularly (Fig. 1). Even for highly crystallized samples (Fig. 1, curves d, e, f) the characteristic absorption peaks of trigonal Se at λ = 21.1 and 21.2 μm cannot be found. On the other hand, these samples show a clear increase of transmitted intensity at wavelengths λ > 15 μm. To our knowledge no investigations concerning the origin of this effect have been carried out up to now. It is our aim to show that the optical loss in the MIR region can be described by means of Debye-Bueche scattering and that this loss can be used to measure the crystalline content of the Se sample.

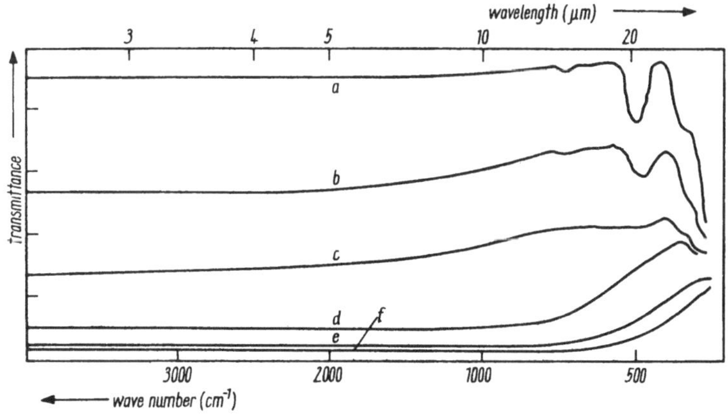

Fig. 1. Transmission spectra of selenium samples after annealing at 160 °C for (a) 0, (b) 1.25, (c) 1.5, (d) 2.0, (e) 2.5, and (f) 3.0 min

1) Schöpfstr. 41, A-6020 Innsbruck, Austria.

All films used were formed by vapor deposition of high purity Se onto carbon coated CsI tablets held at a constant temperature of 10 °C. The evaporation rate was 5 μm/min. The thickness of the films was 25 μm. The samples thus prepared were annealed at constant temperatures in a specially constructed furnace set-up in a Perkin-Elmer Model 547 spectrophotometer. This arrangement permitted continuous measurement of the transmittance of the unpolarized IR radiation during annealing.

As known from other polymer materials, crystalline aggregates within an isotropic amorphous material represent inhomogeneities due to density changes or orientation fluctuations and act as centers for light scattering within the total volume. This volume scattering has been investigated by several authors using statistical /2 to 4/ or model approaches /5, 6/.

Taking into account that the scattering objects in the Se thin films are of different size, shape, and orientation /7, 8/ the statistical model of Debye-Bueche appears to be the most suitable of the models proposed to describe the scattering phenomena. According to this model the scattered intensity i leaving a volume V under an angle Θ at a large distance R for a wavelength λ is given by

$$i \sim \bar{\eta}^2 V\omega/\lambda^4 R^2 , \tag{1}$$

where $\bar{\eta}^2$ means the mean-square value of the density fluctuations and ω is a correlation volume which depends on λ, Θ, and on the extension a of the inhomogeneities.

After correction of (1) for unpolarized light with $P = (1 + \cos^2\Theta)/2$ and integration over all directions in space except the area covered by the photometer one gets the total amount of optical loss which does not reach the detector. Dropping constant factors this loss is given by $\bar{\tau} = (a^3/\lambda^4)\int \omega P \, df$.

In Table 1 the values $\bar{\tau}$ thus calculated are listed together with the ratio a/λ. Up to $\lambda \approx a$ the optical loss is increasing moderately; for wavelengths $\lambda > a$ it shows a λ^{-4} Rayleigh-type dependence. Both is in good agreement with the transmission spectra of the highly crystallized samples (see Fig. 1, curves d, e, f).

Furthermore, assuming the crystalline fraction V_c of the sample as the actual scattering volume V in (1), an increase δV_c causes an increase δi of scattered intensity. Consequently, the intensity I transmitting the sample in forward direction is diminished, a fact which according to the above can be described by

Table 1

Calculated values of $\bar{\tau}$ as a function of a/λ

a/λ	$\bar{\tau}$
50	0.14
20	0.13
10	0.1
2	0.02
1	0.01
0.2	1.7×10^{-4}
0.1	2.7×10^{-5}
0.05	3.0×10^{-6}
0.02	9.9×10^{-8}
0.01	6.4×10^{-9}

the relation $\delta I = - \gamma I \delta V_c$ (γ being a proportionality constant). This finally leads to

$$I = I_a \exp(- \gamma V_c) , \tag{2}$$

where I_a means the intensity transmitted through the sample (substrate plus film) at the start of the annealing procedure.

Examining the transmittance at a fixed wavelength $\lambda = 2.5$ µm during annealing, a decrease of I can be observed (Fig. 2) - obviously as a consequence of the increasing crystalline fraction. Denoting the transmitted intensity for

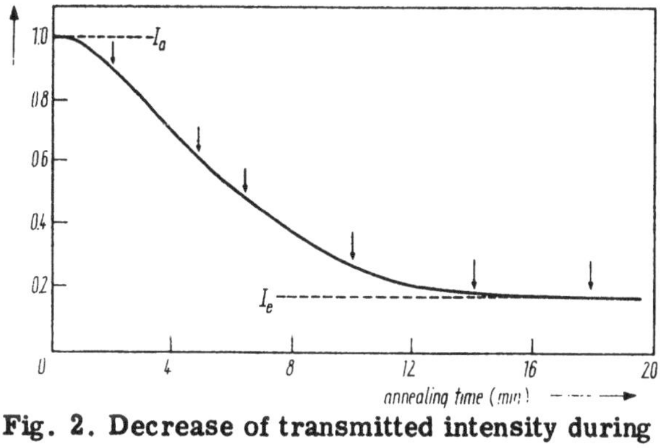

Fig. 2. Decrease of transmitted intensity during annealing at 120 °C

Table 2

Values α_s and α_c for samples annealed at 120 °C after different annealing times t

t (min)	α_c	α_s
2.0	0.08 ± 0.02	0.09 ± 0.01
5.0	0.26 ± 0.02	$0.24 + 0.02$
6.5	0.40 ± 0.03	0.39 ± 0.02
10.0	0.7 ± 0.2	0.83 ± 0.1
14.0	0.95 ± 0.05	1.0
18.0	1.0	1.0

the fully crystallized sample, i.e. $V_c = V_{total}$, by I_e, then substituting in (2) V_c by the crystallinity $\alpha = V_c/V_{total}$ and normalizing by $\alpha = 0$ for $I = I_a$ and $\alpha = 1$ for $I = I_e$, the course of α with annealing time can be evaluated from the corresponding intensities (Fig. 2) by

$$\alpha = \ln(I/I_a)/\ln(I_e/I_a) \ . \tag{3}$$

For experimental verification of this claim the isothermal annealing was interrupted after different annealing times (see arrows in Fig. 2). The samples thus obtained were broken up in different directions and α determined from the crystalline areas of the breaks (called α_s) by means of stereological analysis. In Table 2 the α_s-values thus obtained are compared with those calculated from the isotherms by (3), α_c, showing a good correspondence in all parts of the crystallization process. At other annealing temperatures a similarly high correspondence was found.

Summarizing these results, it can be seen that the optical loss in the MIR region can well be described by a Debye-Bueche mechanism and that this fact provides a simple and time-saving procedure for the determination of crystallization isotherms of Se thin films.

This work was supported by the "Fonds zur Förderung der wissenschaftlichen Forschung", Vienna, under the number 1503.

References

/1/ H. GOBRECHT and A. TAUSEND, Z. Phys. 161, 205 (1961).

/2/ P. DEBYE and A.M. BUECHE, J. appl. Phys. 20, 518 (1949).

/3/ M. GOLDSTEIN and E.R. MICHALIK, J. appl. Phys. 26, 1450 (1955).

/4/ G. ROSS and D.A. JARVIS, Optica Acta 27, 359 (1980).

/5/ R.S. STEIN and M.B. RHODES, J. appl. Phys. 31, 1873 (1960).

/6/ R.E. PRUD'HOMME and R.S. STEIN, J. Polymer Sci., Phys. Ed. 11, 1357 (1973).

/7/ K.S. KIM and D. TURNBULL, J. appl. Phys. 44, 5237 (1973).

/8/ M. KAWARADA and Y. NISHINA, Japan. J. appl. Phys. 16, 1531 (1977).

(Received September 17, 1981)

phys. stat. sol. (a) 68, K93 (1981)

Subject classification: 14.3 and 18.2; 22.8.2

Institute of Applied Physics, Academy of Sciences of the Moldavian SSR, Kishinev[1]

Influence of Anion Substitution on the Conductivity in Magnetic $Cu_yCr_2Se_{4-z}Br_x$ Semiconductors

By

V.V. TSURKAN, S.I. RADAUTSAN, and V.E. TEZLEVAN

The properties of the ferromagnetic semiconductor $Cu_yCr_2Se_{4-z}Br_x$ greatly differ from those of other chromium chalcogenide spinels /1, 2/. The main difference lies in the high Curie temperatures (T_C) and the high values of conductivity (σ) /3, 4/. The results of various authors on the conductivity of these materials disagree due to the lack of composition control and the polycrystalline character of samples /3, 5, 6/. Beside this, no data on crystals with pronounced semiconducting properties have been given so far.

In this note the results of an experimental investigation on the influence of anion substitution on the conductivity in ferromagnetic $Cu_yCr_2Se_{4-z}Br_x$ single crystals with controlled composition are presented.

The spinel single crystals were obtained by two methods: a) chemical transport reactions with iodine carrier for $x = 0$, and b) flux method for $x > 0$. The sample composition was determined by chemical and electron-probe microanalyses (with 1 to 2% relative errors for cations and 2 to 5% for anions). Molar concentrations were calculated assuming the Cr concentration to be equal to 2. The conductivity was measured by the standard probe dc technique on rectangular samples. High conductive materials were measured by the compensation method, low conductive ones - by an electrometer with 10^{11} Ω impedance.

The temperature dependence of the conductivity σ of a sample with $x = 0$ (curve 1) is represented in Fig. 1. The $\sigma(T)$ dependence (dashed line) of a $CuCr_2Se_4$ single crystal after Lotgering /7/ is also shown for comparison in the same figure. These data are in good agreement and $\sigma(T)$ has a metal-like dependence in the studied temperature range.

In spinels with copper deficit ($y < 1$) and Br substitution up to $x \lesssim 0.4$ a decrease of the conductivity has been noticed (curve 3, Fig. 1). There is no

---

1) ul. Akademicheskaya 5, 2 Kishinev, USSR.

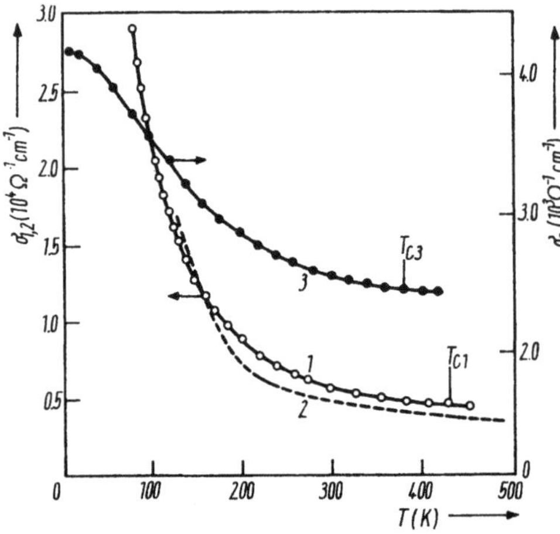

Fig. 1. Temperature dependences of the conductivity σ in:
(1) $Cu_{1.03}Cr_{2.00}Se_{4.0}$,
(2) $CuCr_2Se_4$ /7/,
(3) $Cu_{0.98}Cr_{2.00}Se_{3.85}Br_{0.28}$

anomaly near the Curie temperature. In all samples with 0.6 ≲ x ≲ 1.0 the semiconductor character of the σ(T) dependence appears above a definite temperature (Fig. 2, curve 1).

Spinels with Cu excess (y > 1) exhibit a semiconductor-type σ(T) dependence even for x ≲ 0.4 (Fig. 2, curve 2) /8/. We consider that it is the neglect of copper concentration deviations that could explain the discrepancies on the semiconductor-metal type transitions /3, 5, 6/.

We have observed a sharp increase of resistivity in samples with high Br concentrations x > 1.1. Fig. 3 shows the σ(T) dependence in one of these crystals, which has a well pronounced semiconductorbehaviour. The conductivity maximum near the Curie temperature is the main feature of this dependence.

The experimental investigations of the Hall effect and the thermoelectric power in these materials /9, 10/ have shown that crystals with x ≲ 1.0 are degenerate semiconductors. As far as samples of x > 1.1 are concerned the

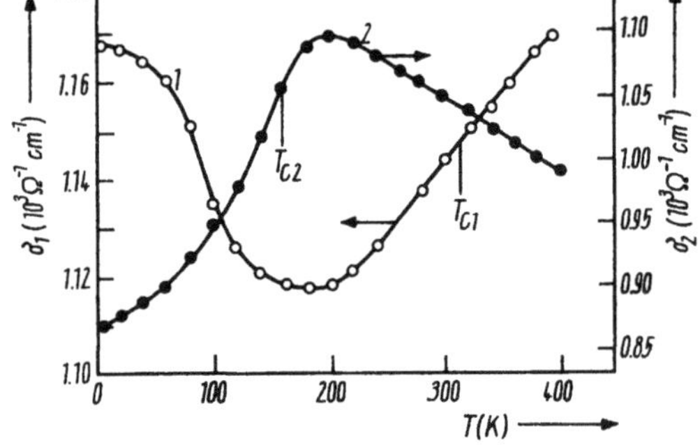

Fig. 2. Temperature dependences of the conductivity in
(1) $Cu_{0.89}Cr_{2.00}Se_{3.43}Br_{0.68}$,
(2) $Cu_{1.51}Cr_{2.00}Se_{3.89}Br_{0.42}$

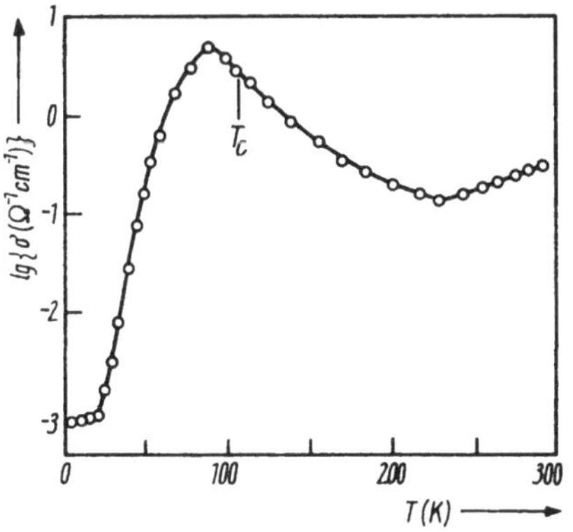

Fig. 3. Temperature dependence of the conductivity in $Cu_{0.73}Cr_{2.00}Se_{2.8}Br_{1.4}$

transition to the non-degenerate state seems to take place.

It should be noted that the large change of conductivity observed in crystals with x > 1.1 is not accompanied by the same change of magnetic characteristics. For example, the dependence of the Curie temperature on x is smooth for the Br substitution range $0.8 \lesssim x \lesssim 1.5$ /4, 11, 12/. It should be also mentioned for the sake of comparison, that samples with Cu excess (y > 1) exhibit a rather large decrease of T_C with the increase of y, but the conductivity changes slightly /8, 12/. These data could indicate evidently that charge carriers have little influence on the Curie temperatures of crystals and consequently on the exchange interactions between magnetic Cr ions. The small coupling of charge carriers and the magnetic system is also confirmed by the low values of the magneto-resistances in these substances /3, 10/.

Finally we want to note that conductivity maximum near the Curie temperature observed here (Fig. 3) is unusual in magnetic semiconductors. Further investigation of this phenomenon is now in progress.

References

/1/ S. METHFESSEL and D.C. MATTIS, Magnetic Semiconductors, Hdb. Phys., Vol. XYIII/1, Springer-Verlag, Berlin 1968 (p. 389).

/2/ S.I. RADAUTSAN (Ed.), Magnitnye poluprovodnikovye shpineli tipa $CdCr_2Se_4$, Shtiintsa, Kishinev 1978.

/3/ K. MIYATANI, K. MINEMATSU, Y. WADA, F. OKAMOTO, K. KATO, and P.K. BALTZER, J. Phys. Chem. Solids 32, 1429 (1971).

/4/ H. PINK. W.K. UNGER, H. SCHÄFER, and H. GÖBEL, Appl. Phys. 4, 147 (1974).

/5/ M. ROBBINS, P.K. BALTZER, and E. LOPATIN, J. appl. Phys. 39, 662 (1968).

/6/ A.W. SLEIGHT and H.S. JARRET, J. Phys. Chem. Solids 29, 868 (1968).

/7/ F.K. LOTGERING, Proc. Internat. Conf. Magnetism, Nottingham, Phys. Soc., London 1964 (p. 533).

/8/ V.V. TSURKAN and V.G. VESELAGO, Fiz. tverd. Tela 20, 3500 (1978).
/9/ V.V. TSURKAN, V.G. VESELAGO, V.E. TEZLEVAN, and S.I. RADAUTSAN, Fiz. tverd. Tela 21, 2710 (1979).
/10/ V.V. TSURKAN, V.G. VESELAGO, S.I. RADAUTSAN, and V.E. TEZLEVAN, Fiz. tverd. Tela 22, 1170 (1980).
/11/ H. PINK, H. SCHÄFER, H. GÖBEL, and W.K. UNGER, IEEE Trans. Magnetics 10, 623 (1974).
/12/ V.V. TSURKAN, V.G. VESELAGO, A.A. BABITSYNA, V.E. TEZLEVAN, S.I. RADAUTSAN, and V.T. KALINNIKOV, Fiz. tverd. Tela 20, 2863 (1978).

(Received July 24, 1981)

phys. stat. sol. (a) 68, K97 (1981)
Subject classification: 14.3.3 and 14.3.4; 22.4.1

Institute of Physics, Polish Academy of Sciences, Warsaw (a) and
Institute of Physics, Technical University of Poznań[1] (b)

The ZnS:Mn-Al Contact as a Part of the Multilayer Electroluminescent Element

By

J. RAUŁUSZKIEWICZ (a), B. SUSŁA (b), A. OKONIEWSKI (b), and W. POLESKA (b)

Introduction Metallic contacts on II-VI compound semiconductors have been widely investigated in recent years because of their potential possibility of applications in optoelectronic devices. The properties of the metal contacts on ZnS:Mn, especially the barrier region of ZnS:Mn, has become an important problem from the point of view of the electroluminescent devices based on ZnS. Recently electroluminescent multilayer light-emitting sandwiches of the composition SnO_2–ZnS-Cu_xS-ZnS:Mn-Al have been described /1/. It was assumed /2/ that the electroluminescence is connected essentially with the presence of the ZnS:Mn-Al contact which is a part of this multilayer device. Although the luminescent properties depend on the nature of this contact, no special efforts have been undertaken to obtain and investigate such a layer. The aim of this paper is to study the electrical properties of such a contact. This contact was investigated in the form of a SnO_2-ZnS:Mn-Al junction.

Experimental procedure Thin films of ZnS:Mn of the thickness range 100 to 400 Å, measured by the quartz monitor were evaporated in an oil diffusion pumped vacuum system (10^{-6} Torr) with control of the substrate temperature (300 to 450 °C), on a semitransparent and conducting layer of SnO_2 which has been obtained by the method of pyrolysis at high temperature. Successively evaporated thin films of Al of thickness 1000 to 2000 Å were the counter electrode of the junction.

Results and conclusions The current-voltage (I-U) characteristics and the dynamic conductivity dI/dU as a function of bias voltage were measured by means of four contact methods and a standard electronic modulation circuit at room, liquid nitrogen, and liquid helium temperature. The I-U characteristics presented in ln(I/U) = f(U) coordinates are shown in Fig. 1. It is seen

1) ul. Piotrowo 3, 60-965 Poznań, Poland.

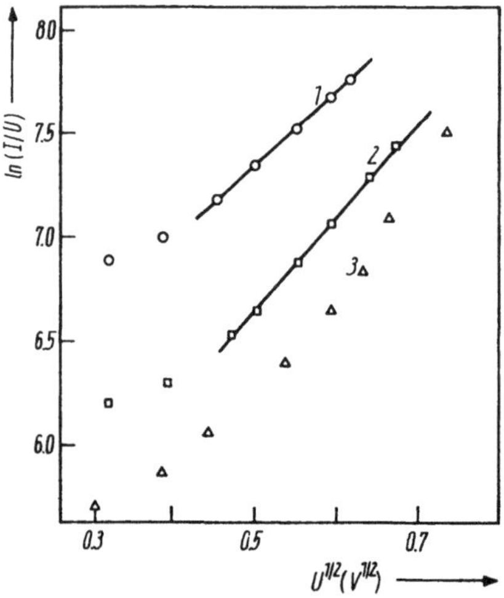

Fig. 1. Dependence of current-voltage characteristics on temperature; (1) 300, (2) 77, (3) 4.2 K

that the junction resistance increases with decrease of temperature, that is typical for a semiconductor barrier. It seems from the linear relation of these characteristics at 77 and 300 K that the dominant role in the current flow through the junction is played by the Frenkel-Poole mechanism /3/, which consists in the thermal ionisation of the impurity centers in the presence of a high electric field (of the order of 10^5 to 10^6 V/m), created in the barrier by the bias voltage applied to the junction. The conductivity related to the Frenkel-Poole mechanism is described by the following equations:

$$\sigma = \sigma_o \exp\left[-\frac{\Phi - \beta_{FP}\, U^{1/2}}{kT}\right] , \qquad (1)$$

$$\beta_{FP} = \frac{e^3}{(\pi\, \varepsilon_o\, \varepsilon_r\, d)^{1/2}} , \qquad (2)$$

where Φ is the impurity energy level, e the electron charge, $\varepsilon_o = 8.85\text{x}$ $\text{x}10^4$ F/cm, ε_r the ZnS permittivity, d the ZnS layer thickness.

The above equation well describes the results obtained at 300 and 77 K. The nonlinear shape of the ln I/U = f($\sqrt{U}$) relation (Fig. 1) at temperature 4.2 K shows that the current flow is due to another mechanism. The authors suggest that the current flow would be caused by the tunneling of electrons through ZnS, treated as a barrier between two conducting electrodes.

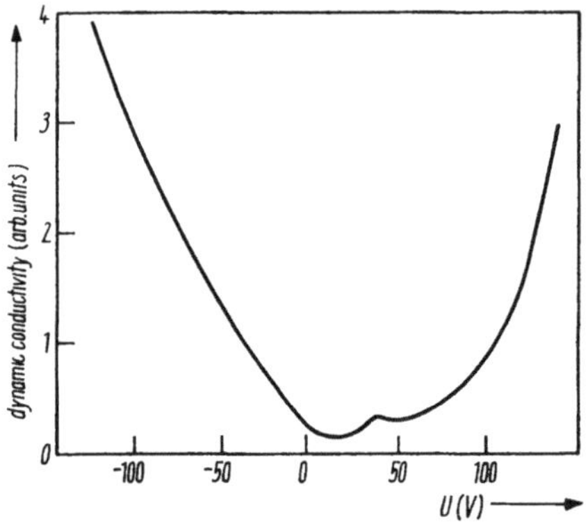

Fig. 2. Dynamic conductivity versus voltage at 4.2 K

Treating this barrier as rectangular /4/, the dynamic conductivity measurements as a function of voltage (Fig. 2) at 4.2 K were used to determine the SnO_2-ZnS:Mn-Al barrier parameters. The calculations were carried out according to the method of Brinkman et al. /5/

$$G(0) = \frac{3.16x10^{10}\ \bar{\varphi}^{1/2}}{d} \exp(-1.025\ d\ \bar{\varphi}^{1/2})\ , \tag{3}$$

$$eU_{min} = 0.649 \frac{\Delta\varphi}{d\ \bar{\varphi}^{1/2}}\ , \tag{4}$$

$$U_1 + U_2 = \frac{11\ \bar{\varphi}^{1/2}}{d}\ , \tag{5}$$

where G(0) is the conductivity at U = 0, U_{min} is the bias voltage related to the conductivity minimum, $\bar{\varphi}$ is the average height of the barrier, $\Delta\varphi$ is the barrier asymmetry, d is the barrier thickness and U_1, U_2 are the bias voltages for the conductivity minima. Thus we obtained the following barrier parameters: $\bar{\varphi}$ = 0.9 eV, d = 30 Å. Such a great discrepancy between the thickness obtained from the quartz monitor and method of Brinkman et al. suggests that the thickness of the evaporated ZnS:Mn the layer is nonuniform and the tunneling current flows through the points of minimum thickness, whereas the quartz monitors the average thickness of the incident mass. On the other hand, the logarithmic dependence of the junction resistance on the barrier thickness (Fig. 3), evaluated from the method of Brinkman et al. confirms its correlation with the value from the quartz monitor.

In the plot of the dependence of the differential conductivity on the bias voltage (Fig. 2) a small maximum at the energy (45 ± 5)meV is observed. This maximum would be connected with the inelastic tunneling of electrons through the ZnS barrier accompanied by the excitation of phonons of the same energy In Raman scattering measurements on ZnS monocrystals /6/ the energies of LO phonons are as follows: for the hexagonal structure: 43.1 meV and for the

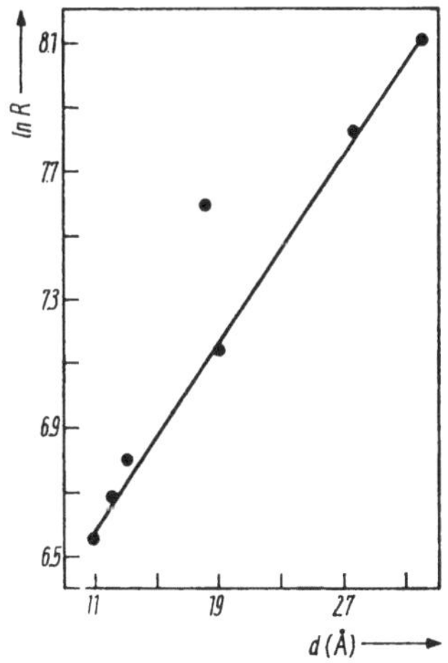

Fig. 3. Dependence of junction resistance on barrier thickness

cubic structure: 43.6 meV, what is close to the energy of the observed structure. The width of this maximum could be caused by the polycrystallinity of the ZnS layer.

This work is carried out under project number PR-3.7.04.2 coordinated by the Institute of Physics, Polish Academy of Sciences, Warsaw, Poland.

The authors express their gratitude to Dr. E. Chimczak and Dr. S. Szuba for valuable discussions.

References

/1/ E. CHIMCZAK, Ph.D. Thesis, Tech. Univ. Poznań.

/2/ E. WALENTYNOWICZ and S. SZUBA, phys. stat. sol. (a) 49, K201 (1978).

/3/ Hdb. Thin Film Technology, McGraw-Hill Book Co., 1970 (p.375).

/4/ P. TIRODKAR and R. PRATAP, phys. stat. sol. (a) 49, 637 (1978).

/5/ W. BRINKMAN, R. DYNES, and J. ROWELL, J. appl. Phys. 49, 1915 (1978).

/6/ G.B. WRIGHT, Light Scattering Spectra of Solids, Springer-Verlag, 1969 (p. 129).

(Received July 31, 1981)

phys. stat. sol. (a) 68, K101(1981)
Subject classification: 14.3.1; 22.6

Institut für Allgemeine Elektrotechnik und Elektronik der Technischen Universität Wien[1]

Time Periodic Dissipative Structures in VO_2 Films[2]

By

G. BADICS, I.A. SERBINOV[3], YU.D. KALAFATI[3], and W.F. FALLMANN

Recently growing interest was shown in investigations of nonlinear systems far from thermal equilibrium /1 to 3/. Forcing a system further and further away from equilibrium new structures may appear, so-called dissipative structures /2/ which are completely different from equilibrium structures. The realization of dissipative structures in VO_2 films was already discussed in an earlier paper /4/. In the present paper some results are presented concerning temporal dissipative structures (oscillations) in VO_2 films. Fig. 1 shows the arrangement for current-voltage measurements performed on VO_2 films. An In_2O_3 film as a heater was kept in thermal contact with the VO_2 film by a sapphire interlayer in order to produce the temperature variation for the semiconductor-metal phase transition (SMPT) /5/ in the VO_2 films. The nonlinearity of the system shown in Fig. 1 is due to the SMPT while the deviation from thermal equilibrium is produced by the current flow.

It is known that oscillations in VO_2 films can be generated by the connection of a PTC resistor with nonlinear current-voltage characteristic in parallel with VO_2 films /6/.

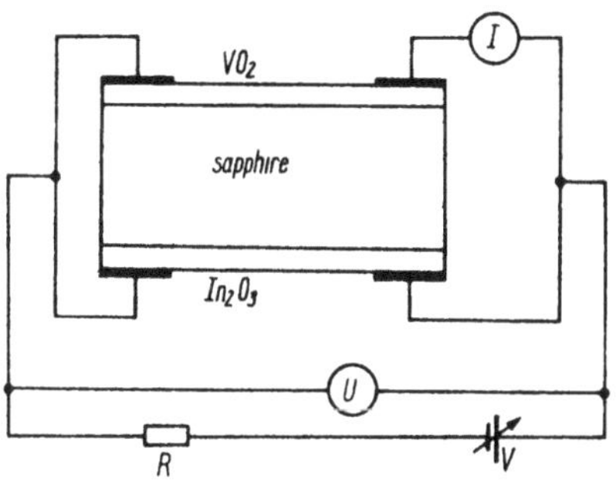

Fig. 1. Sample arrangement for generation of oscillations

1) Gußhausstr. 27-29, 1040 Wien, Austria.

2) This work was supported by the Fonds zur Förderung der wissenschaftlichen Forschung as part of project S-22/03.

3) Institute of Radio Engineering and Electronics, Academy of Sciences of the USSR.

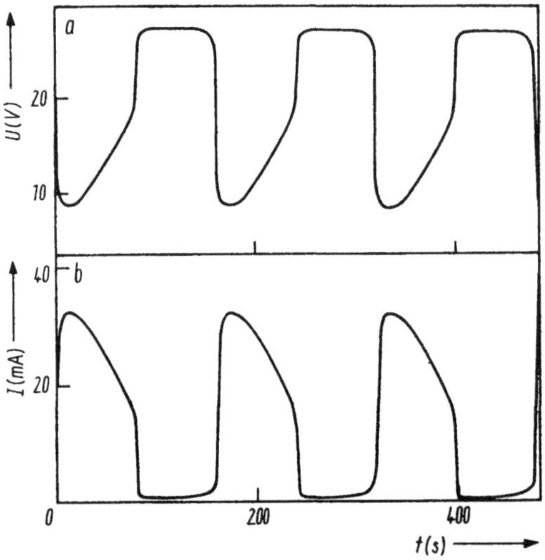

Fig. 2. Typical shape of a) voltage and b) current oscillations in the VO_2 films

However, in the present experiments the resistance of the In_2O_3 film was practically constant, independent of temperature. In VO_2 crystals oscillations were detected even without any external heat source /7/. In order to characterize the oscillations phenomenologically one can use a model for a two-position temperature regulator /8/. Fig. 2 shows the shape of the observed oscillations of voltage and current in our experiments. From these data the time dependence of the conductance was determined. The switching effect in VO_2 is known to be caused by the formation of a high temperature filament /9/. For the width h of the filament (c.f. insert in Fig. 3) it follows that

$$h = H_o G/G_o ,$$

where H_o is the width of the VO_2 film, and G and G_o are the conductances corresponding to width h or width H_o of the metallic phase region, respectively. Fig. 3 shows how the width of the metallic phase region varies with time for different bias voltages $U_1 < U_2 < U_3$. The phase

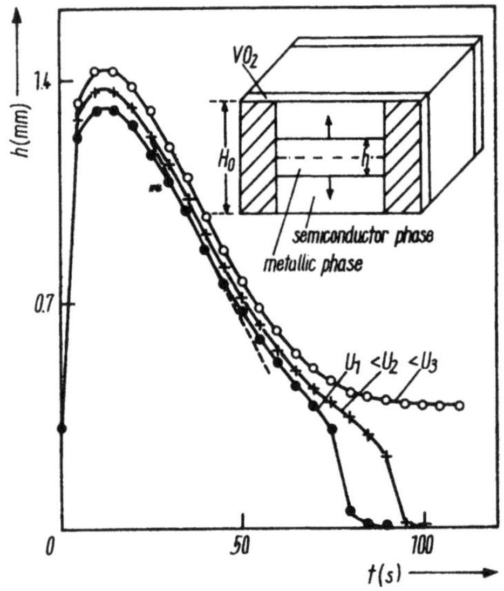

Fig. 3. Plot of the filament width versus time for different bias voltages $U_1 < U_2 < U_3$. The linear part of the slopes (e.g. the dashed line) indicates the period where the phase boundary moves at constant velocity

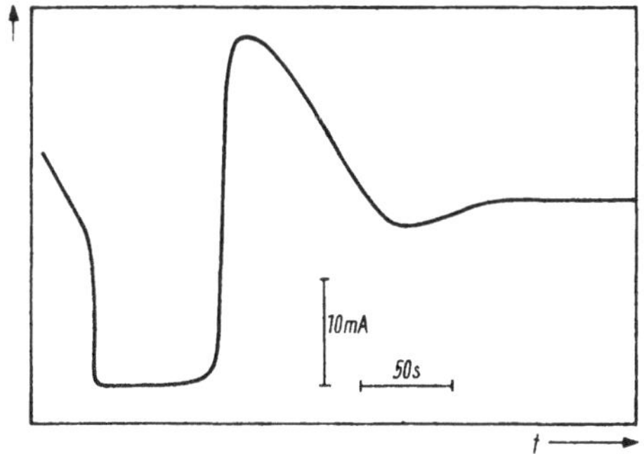

Fig. 4. Shape of the current trace due to the formation of a metallic phase of constant width occurring beyond a critical bias voltage

boundary moves with constant velocity at times when h(t) is linear as shown by the dashed line in Fig. 3 corresponding to a nonlinear temperature wave. The existence of such a wave at constant voltage and constant ambient temperature was discussed elsewhere /10/. In our experiments both voltage and temperature vary with time. For increasing voltage the temperature of the substrate decreases.

It should be pointed out that in the arrangement shown in Fig. 1 also transitions among different dissipative structures could be realized in an experimentally convenient way. Applying the voltage U first a transition from the state of quasi-equilibrium to a spatial-temporal structure (oscillations due to the movement of the phase boundary) occurs. Increasing the voltage further beyond a critical value a transition takes place from the spatial-temporal structure to a spatially inhomogeneous structure (supression of the oscillations by the formation of a stationary filament) as illustrated in Fig. 4. Varying the system parameters e.g. the resistance R also transitions between spatially inhomogeneous structures were realizable.

Thus VO_2 films with SMPT provide an excellent model for the investigation of different dissipative structures as well as transitions between them.

References

/1/ H. HAKEN, Synergetics, Springer-Verlag, Berlin 1978.

/2/ G. NICOLIS and I. PROGINE, Self-Organization in Nonequilibrium Systems, Wiley-Interscience, New York 1977.

/3/ W. EBELING, Strukturbildung bei Irreversiblen Prozessen, Teubner, Leipzig 1976.

/4/ I. A. SERBINOV, YU.D. KALAFATI, and L. A. RYABOVA, Zh. tekh. Fiz., Pisma 6, 196 (1980).

/5/ W. PAUL, Mater. Res. Bull. 5, 691 (1970).

/6/ K.H. MÜLLER, M. WOLF, F. MÜLLER, and K. JAENICKE-RÖSSLER, Nachrichtentechnik-Elektronik 27, 276 (1977).

/7/ Y. TAKETA, F. KATO, M. NITTA, and M. HARADOME, Appl. Phys. Letters 27, 212 (1975).

/8/ A.A. ANDRONOV, A.A. VITT, and S.E. KHAIKIN, Theory of Oscillators, Pergamon Press, Oxford 1966 (p. 235).

/9/ C.N. BERGLUND, IEEE Trans. Electron Devices 16, 432 (1969).

/10/ YU.D. KALAFATI, I.A. SERBINOV, and L.A. RYABOVA, Zh. eksper. teor. Fiz., Pisma 29, 637 (1979); Soviet Phys. - J. exper. theor. Phys., Letters 29, 583 (1979).

(Received July 9, 1981)

Pre-Printed Titles

Papers to be published
in "physica status solidi (a) - applied research"
Vol. 68, No. 2

Original Papers

Y. UMAKOSHI and M. YAMAGUCHI, Osaka
The Stability and Energies of Planar Faults in the DO_{19} Ordered Structure

G. DÖDING and R. LABUSCH, Clausthal-Zellerfeld
Anisotropic Conductivity of CdS after Plastic Deformation and Conduction along Dislocatons (II)

A. WITEK, Warszawa
The Effect of Illumination on Anomalous Ultrasonic Attenuation in Doped Single Crystals of $Bi_{12}GeO_{20}$

C.R. PICHARD, C.R. TELLIER, L. QUARBYA, and A.J. TOSSER, Nancy
Perturbations in Transport Properties of Thin Metals Films Due to Energy-Dependent Relaxation Time

A. BANDYOPADHYAY, S.B. PATIL, D.K. CHAKRABARTY, and C. RADHAKRISHNAMURTI, Bombay
Magnetic Clusters in Some Barium Lanthanide Cobaltates

H. ENDERT, E. FÖRSTER, K. GÖTZ, M.P. KALSHNIKOV, J.A. MIKHAILOV, A.V. RODE, G.V. SKLIZKOV, and W.-D. ZIMMER, Jena
Nanosecond X-Ray Topography of Crystalline Solids with a Laser-Produced Plasma Source

A.P. GALUSHKA, G.E. DAVIDYUK, and N.S. BOGDANYUK, Kiev
Interaction of 1.2 MeV Electron Irradiation Produced Defects with Cu Impurity Atoms in CdS Single Crystals

H. VAN CONG, S. CHARAR, and S. BRUNET, Perpignan
Minority-Hole Diffusion Coefficient in an n-Type Heavily Doped Semiconductor Region of Silicon Devices

M. NARAYANA and L. KEVAN, Houston
The Use of Electron Spin Echo Spectroscopy to Study the Ferroelectric Phase Transition in Ammonium Sulfate

O. BOSTANJOGLO and G. SCHLOTZHAUER, Berlin (West)
Impulse Stimulated Crystallization of Sb Films Investigated by Time Resolved TEM

H.-J. KAUFMANN and P.P. PAL-VAL, Berlin
Peculiarities of High-Amplitude Dislocation Internal Friction in Molybdenum Single Crystals of High Purity in the Temperature Range 5.9 to 300 K

N. SAITO and T. YAMAGUCHI, Hamamatsu
Optical and Compositional Properties of Amorphous $Si_xC_yF_z$ Films Prepared by Reactive Sputtering

A.I. KRIEGER, YU.N. MOSKVICH, A.A. SUKHOVSKII, and O.V. FALALEEV, Krasnoyarsk
The ^{77}Se High Resolution NMR in Antiferroelectric Single Crystals of Cesium Trihydrogenselenite

P. KRISPIN, Berlin
Thermochemical Model Applied to a Deep-Level Defect in GaP

B.S. SUNDERSHESHU and T. KENDELEWICZ, Warsaw
Temperature Dependence of the Absorption Edge of $Cd_{1-x}Mn_xTe$ Mixed Crystals

L. BROSSARD, L. THOMÉ, A. TRAVERSE, H. BERNAS, J. CHAUMONT, and F. LALU, Orsay
Stability of Non-Equilibrium Nickel Hydrides Prepared by Low Temperature Ion Implantation

D.M. VARDANYAN and H.M. MANOUKYAN, Erevan
The X-Ray Plane Wave Dynamic Reflection at a Laminar Crystalline Medium in the Bragg Case

M.P. SHEARER, S.K. SEN, and C.L. BAUER, Pittsburgh
Investigation of Interfacial Reactions in Bimetallic Thin-Film Couples of Al/Cu and Pb/Cu by Measurement of Contact Resistance

M. TOKUDA, J. KRATOCHVÍL, and Y. OHASHI, Prague and Nagoya
Mechanism of Induced Plastic Anisotropy on Polycrystalline Metals

D. MERGEL and R. LABUSCH, Clausthal-Zellerfeld
Optical Excitations of Dislocation State in Silicon

Papers to be published
in "physica status solidi (b) - basic research"
Vol. 108, No. 2

Original Papers

SATISH K. DAS, R.S. TRIPATHI, and S.K. JOSHI, Roorkee
Effect of Magnetic Field on the Susceptibility and the Specific Heat of Spin Glasses

A.C. SHARMA and S. AULUCK, Roorkee
Small Wave Vector $\vec{q}$ and Frequency ω Dependent Longitudinal Dielectric Function for a Model Semiconductor

D.D. ELEY, S. KINNEAR, and M.R. WILLIS, Nottingham
Lifetime of Photogenerated Carriers in TCNQ Single Crystals

A.M. BRODSKII, L.I. DAIKHIN, and M.I. URBAKH, Moscow
On the Polarization Dependence of Photoemission from Metals

S. SCHMITT-RINK and H. HAUG, Frankfurt/M.
Nonlinear Energy Transport of Radiation in Dielectrics Due to Virtual Biexciton Formation

S.K. MISRA and P. MIKOLAJCZAK, Montreal
EPR Linewidth Studies of Gd^{3+}-Doped $Yb_xY_{1-x}Cl_3 \cdot 6H_2O$ Single Crystals

V. NEKVASIL, Prague
Crystal-Field Analysis of Electronic Raman Scattering Data in YAG and YGG

E.A. KOTOMIN and A.L. SHLUGER, Riga
Semiempirical Calculation of Defect Properties in LiF Crystal (II)

S. NARA, A. FREUDENHAMMER, and G. HEBER, Duisburg
Dynamics of Two-Spin Cluster in Spin Glass (I)

J.A. TUSZYŃSKI and H. COFTA, Calgary and Poznań
Refined Model of Exchange Inversion in Spontaneous Metamagnetism

E.G. SIEVERTS, S.H. MULLER, C.A.J. AMMERLAAN, R.L. KLEINHENZ, and J.W. CORBETT, Amsterdam and Albany, New York
EPR Observation of an Au-Fe Complex in Silicon (II)

V. DRCHAL and J. KUDRNOVSKÝ, Prague

Theory of the Auger Spectra of Narrow-Band Metals with Impurities

A.M. SANDAUER and P. BYSZEWSKI, Warsaw

Evidence of the Conduction-Valence Band Overlap Induced by a Magnetic Field in a Zero-Gap $Hg_{1-x}Mn_xTe$

H. BŁASZAK, Poznań

Spin-Orbit Interaction in Superexchange with Orbital Degeneracy (II)

A. BALDERESCHI, K. MASCHKE, A. MILCHEV, R. PICKENHAIN, and K. UNGER, Lausanne and Leipzig

Thomas-Fermi Approximation for the Valence Electron Densities in Cubic Semiconductors and Insulators

A. A. ALYBAKOV, N. TOICHIEV, and SH. AKCHALOV, Frunze

Investigation of EPR and Thermoluminescence Glow Curves of X-Irradiated $KCl:CrO_4^{2-}$, Ca^{2+} Crystals

ZH. EGEMBERDIEV, S. ZAZUBOVICH, and V. NAGIRNYI, Tartu

Polarized Luminescence of Lead Ions Associated with Interstitial Radiation Defects in $KCl:PbCl_2$ Crystals

V. ČÁPEK and R.W. MUNN, Manchester

From Band to Hopping Diffusion of Excitons and Charge Carriers in Molecular Crystals (I)

K.S. SRIVASTAVA, M. HUSAIN, P. GUPTA, A.K. SRIVASTAVA, V. KUMAR, K. SINHA, and S. SINGH, Lucknow and Dhanbad

Shift in the Binding Energy of the Inner Electrons Due to Chemical Combination

W. NOLTING and A.M. OLÉS, Würzburg and Krakow

Simple Spectral Density Approach to the Problem of Mixed Valency

19. Magnetic Resonance

20. Optical Properties
    20.1 Spectra. Optical Constants (X-Ray Spectra See 20)
    20.2 Lasers
    20.3 Luminescence (X-Ray Spectra See 20)

21. Metals. Alloys
    21.1 Transition Metals (Group Sc to Zn) and Their Alloys
        21.1.1 Fe and Fe Alloys
    21.2 Alkali Metals
    21.3 Alkali-Earth Metals
    21.4 Rare-Earth Metals
    21.5 Actinides
    21.6 Noble Metals
    21.7 Semi-Metals (Semiconducting Alloys 22.7)

22. Semiconductors and Ionic Crystals
    22.1 Elements
        22.1.1 Germanium
        22.1.2 Silicon
        22.1.3 Group VI Elements
    22.2 III–V Compounds
        22.2.1 Arsenides
        22.2.2 Phosphides
        22.2.3 Antimonides
        22.2.4 Mixed Crystals
    22.3 IV–IV, II–V, and III–VI Compounds
        22.3.1 Mixed Crystals
    22.4 II–VI Compounds
        22.4.1 Sulphides
        22.4.2 Selenides
        22.4.3 Tellurides
        22.4.4 Mixed Crystals
    22.5 Halides
        22.5.1 Silver Halides
        22.5.2 Alkali Halides
        22.5.3 Alkali-Earth Halides
        22.5.4 Mixed Crystals
    22.6 Simple Oxides (CdO, ZnO See 22.4)
        22.6.1 Mixed Crystals
    22.7 Semiconducting Intermetallic Compounds
        22.7.1 Mixed Crystals
    22.8 Tertiary and Higher Compounds
        22.8.1 Salts of Oxy-Acids (Carbonates, Phosphates, Silicates Including Mica etc.)
        22.8.2 Spinels, Garnets, Ferrites
    22.9 Organic Semiconductors

23. Solidified Gases

Attention

The preceding classification scheme is closely akin to the scheme used by Physics Abstracts to classify the various fields of solid state physics. A paper will either be classified under one or more of the subsections or, if none of these are relevant, it will just appear under the general section heading (i. e. sections and subsections are equivalent in this sense).

The *subject classification* given at the head of each paper begins with the main subject matter of the paper, and this is followed, if necessary, by figures referring to subsidiary subject matter and the type of solids investigated.

The papers published in this issue are listed in the order of their subject classification at the end of the table of contents.

32914

Manuscripts and letters for physica status solidi (b) — basic research and physica status solidi (a) — applied research

should be sent to

DDR-1020 Berlin, Neue Schönhauser Str. 20

or

Internationaler Buchversand GmbH, 1000 Berlin 19, Eschenallee 21.

The object of "physica status solidi (b)" and "physica status solidi (a)" is to publish rapidly important and topical results concerning fundamental research in the field of solid state physics and aspects of application and experimental techniques of fundamental solid state research, respectively. The issues of "physica status solidi (b)" and "physica status solidi (a)" will be published on the 1st and the 16th day of each month, respectively. Either series will contain the following categories:

1. Review articles written by arrangement with the editors. They will be remunerated.
2. Original papers selected from those which have not been submitted for publication elsewhere, after being reviewed. (The latest date for acceptance of contributions is the 8th and the 23rd day of each month for "physica status solidi (b)" and "physica status solidi (a)", respectively, 50 days before the delivery of the journal.)
3. Short notes published in the off-set part of the journal. They are, if necessary after being reviewed, selected by the editorial staff. (The latest date for acceptance of manuscripts for this part is the 12th and the 27th day of each month for "physica status solidi (b)" and "physica status solidi(a)", respectively, 16 days before the delivery of the journal.)
4. Pre-printed titles of papers to be published in the next issue of both series.

Editorial Information

Manuscripts should be submitted preferentially in English. Besides, papers in German, French, and Russian are admitted. The final decision on the incorporation of the papers in the series (b) or (a) is taken by the editorial staff after consultation of the referee. Authors are asked to observe the following rules:

1. Manuscripts have to be written as concisely as possible. As most, 20 and 4 pages (28 lines per page; abstracts, tables, references, figures, and captions of figures included, where 3 figures are counted as about one page) are admitted for original papers and short notes, respectively. All manuscripts should be type-written on one side only. Two copies are required. Badly arranged manuscripts are returned.
2. The author's name, initials, and institution should be given.
3. One abstract of about ten lines in English and one in German, French, or Russian are required for original papers.
4. Because of lack of space conclusions (if any) must not have the character of a summary (i. e. substantially a repetition of the contents of the abstract).
5. Formulae should not be written too small and not with pencil. Separate lines for formulae are desirable. Indistinct writing will lead to a return of the manuscript.
6. Drawings should be enclosed on separate sheets of paper, not exceeding the maximum and minimum sizes of 20×30 and 10×15 cm^2, respectively. Photographs should be submitted, on white paper, only for review articles and original papers. Duplicates are required.
7. Captions for the figures should be submitted on separate sheets of paper, and special instructions with regard to the size of illustrations should be given on these.
8. References should be numbered in order of appearance and should be given, on a separate sheet, in the form used in physica status solidi.
9. SI-units should be used.
10. Each paper should contain a proposal for the subject classification according to the classification scheme appearing on the second and third cover of each issue.

The authors will be sent galley proofs of review articles and original papers. Proof reading of short notes and pre-printed titles will be done by the editorial staff.

For each original paper and short note a total of 75 reprints is available free of charge. The number of reprints free of charge per review article is 25. Requests for additional reprints (against payment) should be given together with the submitted manuscript. **The deadline for orders of additional reprints is the 15th and the 30th of each month** for "physica status solidi (b)" and "physica status solidi (a)", respectively, two weeks before the delivery of the journal.

Bei Fragen zur Produktsicherheit wenden Sie sich bitte an:
If you have any questions regarding product safety,
please contact:

Walter de Gruyter GmbH
Genthiner Straße 13
10785 Berlin
productsafety@degruyterbrill.com